Conceptual Issues in Modern Human Origins Research

EVOLUTIONARY FOUNDATIONS OF HUMAN BEHAVIOR
An Aldine de Gruyter Series of Texts and Monographs

SERIES EDITORS
Monique Borgerhoff Mulder, *University of California, Davis*
Marc Hauser, *Harvard University*

Conceptual Issues in Modern Human Origins Research

G. A. Clark and C. M. Willermet
Editors

ALDINE DE GRUYTER
New York

About the Editors

Geoffrey A. Clark is Distinguished Research Professor of Anthropology at Arizona State University, Tempe. He is the author or coauthor of over 125 articles and seven books on human biological and cultural evolution in the Pleistocene Old World. Professor Clark is also founding editor of the *Arizona State University Anthropological Research Papers* and the *Archeological Papers of the American Anthropological Association (1989)*.

Catherine M. Willermet is a doctoral candidate in anthropology at Arizona State University. A physical anthropologist interested in Middle Paleolithic hominids, speciation models, and fuzzy set theory, her current interests involve classification of Upper Pleistocene hominids in the Levant.

Copyright © 1997 Walter de Gruyter, Inc., New York

ALDINE DE GRUYTER
A division of Walter de Gruyter, Inc.
200 Saw Mill River Road
Hawthorne, New York 10532

This publication is printed on acid free paper ∞

Library of Congress Cataloging-in-Publication Data
Conceptual issues in modern human origins research / G. A. Clark and
   C. M. Willermet, editors.
      p. cm. — (Foundations of human behavior)
   Includes bibliographical references and index.
   ISBN 0-202-02039-8 (alk. paper). — ISBN 0-202-02040-1 (pbk. :
alk. paper)
    1. Man—Origin. 2. Human evolution—Research. 3. Human
evolution—Philosophy. I. Clark, Geoffrey A. II. Willermet, C. M.
(Catherine M.), 1968– . III. Series.
GN281.C586 1997
573.2—dc20   96-33415
                                               CIP

Manufactured in the United States of America

10 9 8 7 6 5 4 3 2 1

The study of the basic philosophies or ideologies of scientists is very difficult because they are rarely articulated. They largely consist of silent assumptions that are taken so completely for granted that they are never mentioned . . . [but] anyone who attempts to question these "eternal truths" encounters formidable resistance.

Ernst Mayr (1982:835)

Contents

III.   WESTERN PERSPECTIVES
Latin Europe and the Levant

IV.   WESTERN PERSPECTIVES
The Anglo-German Research Traditions

List of Contributors

Federico Bernaldo de Quirós Guidotti
Departamento de Prehistoria
Universidad de León
24071 León, Spain

Amilcare Bietti
Dipartimento di Biologia Animale
  e dell'Uomo
Università di Roma 'La Sapienza'
Piazzale Aldo Moro, 5
00185 Roma, Italy

C. Loring Brace
Museum of Anthropology
University of Michigan
Ann Arbor MI 48109

Günter Bräuer
Institut für Humanbiologie
Universität Hamburg
Allende-Platz 2
20146 Hamburg, Germany

Victoria Cabrera Valdés
Departamento de Prehistoria
  e Historia Antigua
Universidad Nacional de Educación
  a Distancia
Senda del Rey, s/n
28040 Madrid, Spain

Rachel Caspari
Department of Anthropology
University of Michigan
Ann Arbor MI 48109

Steven E. Churchill
Department of Biological
  Anthropology and Anatomy
Duke University Medical Center
Durham NC 27710

Geoffrey A. Clark
Department of Anthropology
Arizona State University
Tempe AZ 85287

David W. Frayer
Department of Anthropology
University of Kansas
Lawrence KS 66045

Dominique Gambier
Laboratoire d'Anthropologie
Université de Bordeaux I
Avenue des Facultés
33405 Talence, France

María Dolores Garralda
Sección de Antropología
Facultad de Biología
Universidad Complutense de Madrid
28040 Madrid, Spain

Joaquín González Echegaray
Instituto para Investigaciones
  Prehistóricas
Avenida de Pontejos, 9
39005 Santander, Spain

Colin P. Groves
Department of Archaeology and
    Anthropology
Australian National University
Canberra, A.C.T. 0200, Australia

Henry Harpending
Department of Anthropology
University of Utah
Salt Lake City, UT 84112

J. Brett Hill
Department of Anthropology
Arizona State University
Tempe AZ 85287

Manuel Hoyos Gómez
Departamento de Geología
Consejo Superior de Investigaciones
    Científicas
Serrano, 13
28001 Madrid, Spain

Susan G. Keates
School of Anthropology
University of Oxford
60 Banbury Road
Oxford OX2 6PN, United Kingdom

Marta Mirazón Lahr
Departamento de Biologia
Instituto de Biociências
Universidade de São Paulo
Caixa Postal 11.461, CEP 05422-970
01000 São Paulo, Brazil

Jane Maienschein
Department of Philosophy
Arizona State University
Tempe AZ 85287-2004

Jonathan Marks
Department of Anthropology
University of California
Berkeley, CA 94720

Charles E. Oxnard
Department of Anatomy and Human
    Biology
University of Western Australia
Nedlands, Perth, Western Australia
6907 Australia

Geoffrey G. Pope
Department of Anthropology
William Paterson College
Wayne NJ 07470

John H. Relethford
Department of Anthropology
State University of New York
    College
Oneonta NY 13820

Jean-Philippe Rigaud
Direction des Antiquités
    Préhistoriques d'Aquitaine
Université de Bordeaux I
33405 Talence, France

Michael Ruse
Departments of Philosophy
    and Zoology
University of Guelph
Guelph, Ontario N1G 2W1
Canada

Vincent M. Sarich
Department of Anthropology
University of California
Berkeley CA 94720

Betsy Schumann
Department of Anthropology
Washington University
St. Louis, MO 63130

Henry P. Schwarcz
Department of Geology
McMaster University
Hamilton, Ontario L8S 4M1
Canada

Lawrence G. Straus
Department of Anthropology
University of New Mexico
Albuquerque NM 87131

Christopher B. Stringer
Natural History Museum
Cromwell Road
London SW7 5BD, United Kingdom

Alan R. Templeton
Department of Biology
Washington University
St. Louis MO 63130

Bernard Vandermeersch
Laboratoire d'Anthropologie
Université de Bordeaux I
Avenue des Facultés
33405 Talence, France

Catherine M. Willermet
Department of Anthropology
Arizona State University
Tempe AZ 85287-2402

Milford H. Wolpoff
Department of Anthropology
University of Michigan
Ann Arbor MI 48109-1382

Wu Xinzhi
Institute of Vertebrate Paleontology
    and Paleoanthropology
Box 643, Academia Sinica
Beijing 100044
People's Republic of China

I

INTRODUCTION

> The data do not speak for themselves. I have been in rooms with data and listened very carefully. They never said a word.
>
> M. Wolpoff (1975:15)

Conceptual Issues in Modern Human Origins Research is about the effects of paradigmatic biases and regional research traditions on our capacity to make sense of that segment of the human past concerned with our immediate origins. The idea for the book originated in a series of international conferences attended by the editors over the past six or seven years. These were thematic, "big issue" meetings concerned with the biological and behavioral origins of modern humans, the archaeology of the Middle to Upper Paleolithic transition in Europe and west Asia, and the place of the neandertals in human evolution as seen from the perspectives of archaeology and physical anthropology. Some of these conferences resulted in edited volumes (Akazawa et al. 1992; Bräuer and Smith 1992; Cabrera 1993; Mellars 1990; Mellars and Stringer 1989, 1992; Nitecki and Nitecki 1994; Trinkaus, 1989), and in aggregate they provided the impetus for at least three semi-popular books (Lewin 1993; Stringer and Gamble 1993; Trinkaus and Shipman 1993b).

What impressed us most about these meetings were the enormous differences in the preconceptions, assumptions, and biases that different workers brought to bear on the resolution of problems that, on the surface at least, were thought to be held in common. At times, these differences were so great as to preclude any common basis for discussion. There also seemed to be little appreciation of the fact that such epistemological issues existed, and that they could have significant consequences for the research enterprise as a whole, affecting not only construals of pattern in the bones, stones, or genes which are the immediate objects of scrutiny, but also the meaning ascribed to pattern. It quickly became evident that people were "talking past one another," guided by different assumptions and preconceptions about what the human past was like.

Sparked by developments in molecular biology in the late 1980s (Cann et al. 1987), there was also a tremendous surge of popular interest in modern human origins, manifest in the print media (e.g., cover stories in *Time, Newsweek, Discover*) and on television (e.g., *Nova*), that tapped public fascination with neandertals, and with various construals of their relationship to ourselves. Be-

1

cause of their proximity in time, and because they have become part of western popular culture, it has never been possible, either in science or in the popular mind, to be quite as objective about neandertals as it is about more remote human ancestors. By the mid-1990s, molecular biology, genetics, population biology, and linguistics had entered the fray, alongside the historically important disciplines of archaeology and human paleontology. At stake in the argument is no less than what it means, biologically and culturally, to be human.

THE ASU EXHIBIT AND LECTURE SERIES

A more immediate catalyst for the book was a spectacularly successful museum exhibit and lecture series organized by Willermet at the Museum of Anthropology, Arizona State University, in the spring of 1993. "The Debate over Modern Human Origins: A Scientific Tug-of-War" discussed the place of neandertals in our ancestry and the influence that individuals' perceptions of the world of science had on their interpretation of data. The importance of the scientific worldview, or paradigm, has become increasingly prominent in discussions of modern human origins (MHO) research, which is the terminology used throughout, and is the central focus of this book. We would like to think that it provides a much-needed antidote to the strict empiricism that has dominated the debate since its inception more than 130 years ago.

One of the purposes of the museum exhibit was first to make visitors aware of some of the history of research on modern human origins, summarized in terms of "continuity" and "replacement" paradigms, and then to introduce them to data sets invoked in support of one or the other view. The visitor was given no answers. At the end of the exhibit was a large book in which visitors were invited to record their opinions about which they believed to be the more credible paradigm and, more generally, comment on the topic. A dialogue was thus established between exhibit and visitor, and among the visitors themselves. Visitors were often observed in the exhibit halls discussing various aspects of the exhibit; many also commented on what other people had written in the book.

The lecture series provided an exciting and spirited counterpoint to the exhibit. Seven speakers were invited to share their views on modern human origins (Stringer, Wolpoff, Clark, Turner, Stoneking, Sarich, Smith). These lectures, all videotaped, were extremely popular, each drawing near-capacity crowds from many different university departments as well as from the general public. The lecturers were chosen for their current research interests, stature, and reputation in the scientific community, as well as for their ability to speak to a lay audience. Both continuity and replacement advocates were represented, and archaeology, human paleontology, linguistic, and genetic perspectives were given. *Conceptual Issues* developed originally from a core of essays based on the lectures given by the participants in the lecture series. At the behest of reviewers, and of Aldine

Executive Editor Richard Koffler, we expanded the number of contributors four-fold, in order to achieve better geographical and intellectual balance.

THE NEANDERTALS

For historical reasons, neandertals are central to the MHO debate. The neandertals were a group of archaic *Homo sapiens* who lived in Europe and western Asia from roughly 100,000 to 35,000 years ago. They were very robust in their cranial and postcranial skeletons, had large and rather distinctive brow ridges, a high nasal angle, and a prognathic face. Some of them, at least, made and used Mousterian tools and buried their dead. Despite their rugged skeletons, they were clearly very "humanlike" in some of their behaviors. In Europe, neandertals seem to disappear at about the same time that modern humans such as Cro-Magnon *(Homo sapiens sapiens)* appear in the fossil record, ca. 40,000–30,000 years ago. People have questioned whether neandertals are ancestral to moderns primarily because *(a)* neandertals look somewhat different from modern humans in some craniofacial and dental features, and *(b)* the time interval between the latest dated neandertal (36,000 years ago at St. Césaire in France) and the earliest dated modern *H. sapiens* is thought by some to be too short for evolution to transform neandertals into moderns. One major point of disagreement is just how different, overall, the neandertals were? Disagreement about the taxonomic status of the neadertals is reflected implicitly throughout this book by whether or not the term 'neandertal' is capitalized. Despite strong editorial leanings towards standardization, we have left in tact the authors' original spelling, replete with symbolic loading as it is.

In the latter half of the nineteenth century, when only a few pre-modern human fossils were known, there was no consensus on the human status or age of the neandertals, let alone their place in then-emerging models of human evolution (Brace 1988b; Howells 1976; Spencer 1984). By World War I, and as new specimens were unearthed in Europe and in west Asia, two major conceptual frameworks began to develop to explain where neandertals supposedly "fit" in human evolution. These two frameworks, known historically as "neandertal" and "pre-sapiens" (and currently as "continuity" and "replacement"), have divided researchers on the issue of where to place neandertals on the human lineage.

ON PARADIGMS AND THEORIES

It is our contention that the conceptual frameworks of "continuity" and "re-placement" are paradigms—assertions about the way the world, or some relevant portion of it (like the domain of MHO research), is perceived to be (Kuhn 1962a). Paradigms are juxtaposed with theories—arguments invoked or constructed to explain *why* the world is as it appears to be. Paradigms exist at several conceptual levels, but the most common usage, after Kuhn (1962a), refers to paradigm at the level of the metaphysic. At least so far as explanation was concerned, Kuhn thought that metaphysical paradigms were closed logical systems, and that lower-order sociological and methodological paradigms were more or less directly derived from them—a view for which he was often criticized (e.g., Masterman 1970; Shapere 1971). The metaphysical paradigms of researchers are the intended objects of scrutiny here, although since paradigm boundaries are never wholly impermeable either "horizontally" (between metaphysical paradigms) or "vertically" (between the metaphysic and its constituent sociological and methodological paradigms), there is also considerable discussion of lower-order (especially sociological) paradigms (see Clark 1991, 1992a, 1993a, ed. 1991, and this volume, for further discussion).

THE CONTINUITY AND REPLACEMENT PARADIGMS

Assuming an African origin for the Hominidae in general, current advocates of the continuity (or "Multiregional Evolution") position argue that there was continual gene flow in peripheral areas of the world from the latter part of the *Homo erectus* stage right on up to that of anatomically modern *Homo sapiens* (Frayer et al. 1993; Smith, Falsetti, and Donnelly 1989; Thorne and Wolpoff 1992; Wolpoff et al. 1984). They postulate that *Homo erectus* radiated out of Africa between 1.8 and 0.7 million years ago and spread through the middle latitudes of the Old World. These colonizers developed regional morphological (racial) characteristics soon after their arrival in peripheral regions and evolved anagenically into archaic *Homo sapiens* (including neandertals) and then into anatomically modern humans *(H. sapiens sapiens),* all the while maintaining genetic continuity. The major implication of these models (for there are several variants) is that there were no cladogenic, or branching, speciation events subsequent to that which produced *Homo erectus.* From a continuity perspective, humans are seen as constituting a single polytypic species at any given point in their history from as early as the Lower Pleistocene up until the present.

The replacement ("Out of Africa," "Recent African Origin," or "Garden of Eden") position also holds that *Homo erectus* populations left Africa in the Lower Pleistocene and subsequently populated the middle latitudes of the Old World (Aiello 1993; Stringer and Andrews 1988b). However, the replacement view argues that early anatomically modern *Homo sapiens* evolved only in Africa late in the Middle Pleistocene (<200,000 years ago) from a *Homo erectus* population that had not migrated. By a process that has never been made clear, some of these modern populations also supposedly emigrated from Africa, displacing, extirpating, outcompeting, or, most generally, replacing neandertal and other archaic *Homo sapiens* indigenous populations that had evolved anagenically from the original *Homo erectus* colonizers. The more extreme advocates of this position hold that the origin of anatomically modern *Homo sapiens* was a cladogenic speciation event, and thus no admixture between the colonizing and indigenous populations would have been possible. Modern regional characteristics outside Africa were established after this replacement (e.g., Stringer 1992a). Others proceed from the assumption that neandertals differed from moderns at the subspecies level and that at least a minimal degree of admixture was possible, however unlikely it might have been (e.g., Bräuer 1992a).

ON THE INFERENTIAL BASIS FOR KNOWLEDGE CLAIMS

It is our opinion that the principal reason that the issue of modern human origins has not been resolved after more than a century of research is that the scholars involved—archaeologists, human paleontologists, molecular biologists—are mainly strict empiricists, relatively unconcerned with epistemology (*how* we know what we think we know about the human past). Many would disagree with this assertion, however, and would point to the empirical insufficiencies of the archaeological and paleontological records, claiming that "we do not have enough data to answer human origins questions," and that "with the eventual accumulation of more data, these issues will be resolved" (e.g., Klein 1992, 1995). However, from a certain philosophical point of view (that of hypothetical realism), "data" have no meaning (even existence) apart from the conceptual frameworks that define and contextualize them. Scientists have been trying to arrive at a consensus about our origins for 130 years, and the data have been accumulating, albeit slowly and sporadically, over that interval. Why haven't the scientists been successful?

We think they have not because scientists involved in MHO research pertain to different intellectual traditions, which in turn proceed from different assumptions about what the remote human past was like. These intellectual traditions are seldom subjected to critical scrutiny, and, while there is some overlap (especially in respect of methodology), they operate according to different paradigms, characterized by different biases and assumptions that determine what constitutes data, which questions are relevant to ask of data, and how data should be organized, measured, or, most generally, interpreted. Paradigmatic bias is *implicit* within research traditions, however, and since multiple research traditions are involved in MHO research, disagreements over what constitutes relevant data are inevitable and render MHO researchers unable to arrive at a consensus. At the operational level, and in a particular problem domain, they employ different basic concepts and terms or give to shared terms and concepts different meanings (Willermet and Clark 1995).

Once indoctrinated in the paradigm (and this occurs mainly through the for-

mal educational process), scientists typically dedicate themselves to the resolution of problems whose solutions tend to reinforce the credibility of the paradigm, rather than to question its validity. According to Kuhn, science in general
operates this way, without much explicit concern for the preconceptions and
biases that are the foundations of its knowledge claims. This all would appear to
work fairly well when a single intellectual tradition, based on a single metaphysical paradigm, is involved in the investigation of a single research domain, as is
the case with much experimentally based, highly axiomatized "big science"
research. It doesn't work nearly so well when multiple intellectual traditions,
based on multiple metaphysical paradigms, are involved in the investigation of a
single research domain (as is the case with astrophysics). It doesn't work at all,
we would submit, when research domains themselves are multiple but overlapping, and where findings in one field have implications for the research protocols
of another, as is the case with MHO research.

WHAT WE ASKED OUR CONTRIBUTORS TO DO

In light of these considerations, we asked our contributors to try to distance
themselves from the quotidian demands of "normal science," step back from the
data, and take a long, hard look at *why* and *how* they do what they do. In order to
constrain responses, and to keep the focus on the metaphysical paradigm itself,
we provided a series of questions related to how they conceptualize their research, and how they reach conclusions about pattern and the meaning of pattern
in the problem domains in which they work:

- Why do you use the kinds of data you do?
- What do you think these data tell you and why?
- What kinds of questions or hypotheses structure your approach to
  research?
- Why are you asking *these* questions instead of other questions?
- What are your operational definitions of major conceptual categories like
  "species," "cultures," etc.?
- How do you define a speciation event, and why do you define it the way
  you do?
- (and, most generally) What are the preconceptions, assumptions, and biases that govern your approach to research?

We submit that the answers to these questions outline the logic of inquiry that
researchers bring to the research domains in which they work. These nebulous
but no-less-real entities structure research in complex and subtle ways and provide loosely defined conventions by which we attempt to give meaning to pat-

tern. We are also of the opinion that paradigmatic biases exhibit a fuzzy but modal "regional" character—essentially the product of the intellectual traditions in which workers have received their training, combined with the compromises they must make in order to come to grips with the realities of evidence in actual, "real world" situations. It is interesting to note that there are differences of opinion as to whether or not it is possible to identify the parameters of national or regional research traditions, and even whether or not such things exist (Clark 1989a, 1989b, 1989c).

Readers will have their own opinions on these essentially philosophical matters, and they can judge for themselves the adequacy of the responses given. We were after "first person" accounts; we wanted contributors to try to articulate the basic, fundamental premises under which *they* conduct the research enterprise. What we sometimes got instead were personal reactions to what contributors perceived to be the ruling theoretical and methodological biases operative in their particular research domains, which is not the same thing. However, it is by no means a "bad" thing, since, by confronting the preconceptions of the dominant paradigm, they were also forced to confront their own.

ACKNOWLEDGMENTS

Clark wishes to thank the students in his modern human origins seminar (1994) for raising a number of the issues investigated here and for helping him to focus his thinking on various aspects of this latest incarnation of a century-old debate. An intellectual mentor has been the singular figure of Ernst Mayr, whose wide-ranging research over an astonishing 65 years is an inspiration to all of us concerned with the inferential basis for knowledge claims in evolutionary biology.

Willermet is indebted to the following ASU faculty and staff for support and guidance in creating the museum exhibit and lecture series: C. Michael Barton, Geoffrey A. Clark, Ann L. Hedlund, Peggy Lindauer, Mary W. Marzke, Charles L. Redman, and Christy G. Turner II. She also wishes to thank the lecture series participants, whose commitment to the book project made it a viable enterprise: Vince Sarich, Fred H. Smith, Mark Stoneking, Chris Stringer, and Milford H. Wolpoff.

Despite many competing demands on her time and energies, Marsha Schweitzer, the graduate coordinator for the Department of Anthropology at ASU, prepared the final manuscript and created the master bibliography from the most disparate array of word-processing languages imaginable. She did this in the course of her "regular job" with her usual competence, thoroughness, patience, and grace.

The chapters by Schwarcz, Vandermeersch, Gambier, Rigaud, González Echegaray, and Cabrera and colleagues are modified versions of papers published in 1993 in *El Origen del Hombre Moderno en el Suroeste de Europa* (OHMSE). We wish to acknowledge the Universidad Nacional de Educación a Distancia (UNED), Madrid, for allowing us to publish English translations of these essays which, with the exception of the paper by Schwarcz, originally appeared in Spanish and French. A special *agradecimiento, y un abrazo,* goes to Victoria Cabrera Valdés (UNED) who, as OHMSE editor, greatly facilitated this process. Inclusion of essays by these well-regarded Latin European scholars makes for a better-balanced volume and does much to counter the impression that the MHO debate is largely dominated by anglophone researchers.

We also wish to thank Aldine de Gruyter Executive Editor Richard Koffler for his unflagging and enthusiastic support of the project. Without Richard, the book would never have come into being. Aldine Copy Editor June-el Piper took on the forbidding task of editing the deathless prose, for which she deserves a long vacation in some tropical pleasuredome, all expenses paid! Aldine Managing Editor Arlene Perazzini coordinated and assembled the various components of the volume, and was an effective liaison between the press (in Hawthorne, New York), volume editors Clark and Willermet (in Tempe, Arizionia), and Piper (in Albuquerque, New Mexico).

Conceptual Issues is dedicated to the foundations for the promotion of science of the United States, Great Britain, Canada, France, Spain, Germany, Australia, and China. Without the support of these organizations, very little of the research which forms the basis for this book could have been accomplished.

G. A. Clark
C. M. Willermet

II
THE CONCEPTUAL FRAMEWORK

The papers in this section are conceptually the broadest in the collection—they identify the intellectual parameters of the modern human origins debate from the perspective of the 1990s. In the first essay, paleoanthropologist C. Loring Brace critiques the extreme replacement view, which assumes that "modern" form arose at a single geographical locus and spread from that locus—at the expense of "archaic" form—throughout the Old World. Reiterating Hrdlička's arguments of ca. 60 years ago, he concludes that the replacement view ultimately derives from the creationist assumptions and typological essentialism of medieval Christianity. He invokes archaeological evidence for vectored technological change to argue for reductions in selective force intensity over the course of the Upper Pleistocene. The consequences are reductions in morphological robusticity, mediated by the occurrence of mutations.

Milford Wolpoff and Rachel Caspari question what it means, morphologically and behaviorally, to be a "modern human." Proceeding from the population perspective of evolutionary biology, they dissect the major tenets of replacement, invoking genetics and archaeology to address the evolutionary processes that produced "modernity." Modernity, they argue, was not the appearance of a set of anatomical features, but rather a process and a pattern of change (they liken it to the intersecting ripple patterns created by throwing pebbles into a pond). Modern humans are not a "thing" because they cannot be both universally and uniquely defined.

Focusing on subspecific variation (race) from an historical perspective, molecular biologist Jonathan Marks challenges the notion that classification in human evolution is or ever can be the value-neutral process that it is in other zoological fields. By virtue of their close relationship to us, the classifications of our living and fossil relatives implicitly encode preconceptions and biases prevalent in the larger society at the time the classifications were made. Any scientific decision about our relatives, therefore, has political consequences that cannot be ignored.

In a similar vein, Geoff Clark's paper attacks the strict empiricism that underlies much paleoanthropological research. He argues that different intellectual traditions are involved in MHO research, which operate according to different metaphysical paradigms, characterized by different sets of biases and assumptions about what the human past was like. Since paradigmatic bias is implicit within research traditions, and since multiple research traditions are involved in this (and much other) paleoanthropological research, people tend to "talk past" one another. They literally do not understand what their colleagues are talking about.

Cathy Willermet and Brett Hill focus on the central zoological concept of the species and critique the conventional methodologies that paleontologists use to define fossil species which, they say, are basically Aristotelian in concept and bivalent in design. They argue that conventional systematics cannot acknowledge the possibility that a specimen of unknown taxonomic affinity can exhibit characteristics of multiple groups in varying degrees, and thus would have a high probability of being misclassified. An alternative method of classification uses fuzzy set theory, which better accommodates transitional specimens and thus more accurately models reality in many areas of paleoanthropological research.

The last chapter in this section, by geochronologist Henry Schwarcz, focuses on the radiometric methods all MHO researchers use to control for time. From the perspective of the anglophone research traditions, at least, time is a "reference variable" used to measure change attributed to other causes. The chronology for the appearance of modern humans in southwestern Europe is reviewed in terms of the different radiometric methods used to date this process or event. U-series dating emerges as the best overall discriminator amongst a series of models, but samples suited to U-series dating are not generally available, limiting its wider application.

1

Modern Human Origins
Narrow Focus or Broad Spectrum?

C. LORING BRACE

Years ago I suggested that an assessment of the selective forces that were involved in the transition from Neanderthal to modern form could be achieved by considering the evidence from archaeology (Brace 1964a). Many have assumed that this was a reapportion of the old view equating Neanderthal with Mousterian, and "anatomically modern" (to use the current and unjustified buzz words) with Upper Paleolithic. Recently, those who defend the typological assessment of the St. Césaire specimen as "classic Neanderthal" have noted with some satisfaction that its association with an Upper Paleolithic Châtelperronian assemblage discredits the view that there is any causal relationship between archaeology and human morphology (Lévêque and Vandermeersch 1980; Stevens 1991). The fact that Bordes and others have noted that the Châtelperronian is just a direct outgrowth of the local Mousterian has not been given fully considered recognition (Bordes 1968a, 1972; Harrold 1983).

Since I think there is good reason to believe that there is and should be some relationship between what is shown by archaeological data and what is visible in human morphology, it is of more than passing importance to explore these matters at greater length. First, it should be categorically denied that there was anything in their genome that compelled Neanderthals to produce Mousterian tools or that dictated that their early modern successors were genetically predisposed to produce Upper Paleolithic ones. Likewise, the production of Mousterian tools does not by that mere fact consign the makers to Neanderthal status, nor does the manufacture of Upper Paleolithic tools automatically confer modernity on their producers. The tools that an individual constructs, like the language he speaks, are aspects of learned behavior that are shaped by the sociocultural context within which that individual matures. That context in turn has its own dynamics, constrained but not created by environmental factors and the genetic limitations of those who sustain it.

But as those cultural dynamics have increased in complexity and extent through time—a matter that has been dramatically documented in the archaeological record—the cultural milieu has to an increasing extent become part of that very environment whose dimensions condition human chances for survival. This is what I meant when I suggested that human beings have become adapted to what can be called the "Cultural Ecological Niche" (Brace 1964a:14, 1967:56, 1991c:90, 1995b:chap. 9). It is only logical then that key cultural aspects should alter the forces of selection in specific ways. Note, however, that it is not culture as a whole that is the question, however important that actually is, but specific aspects of culture which have specific consequences.

Let us return again to the suggested mechanics for the transformation of Neanderthal into modern face form. Years ago I suggested that the key to this was to be found in the changing role played by the dentition, especially its anterior component (Brace 1962b, 1964a, 1967, 1979, 1995d). The development of special tools for special purposes, symbolized by the technological diversification shown in Upper Paleolithic as opposed to Mousterian assemblages, reduced the necessary role of the front teeth, which in turn led to the reduction of the teeth themselves and the supporting facial skeleton as well (Brace 1963, 1995d; Brace et al. 1987; Brace, Smith, and Hunt 1991).

However, the important thing is not whether the tools are typologically Mousterian or Upper Paleolithic but how they work. One archaeologist was quick to remind me that a Mousterian flake is just as good a cutting tool as an Upper Paleolithic blade (H. L. Movius, personal communication 1961). Trinkaus and Howells took this one step further and argued that the efficiency of Mousterian cutting tools effectively undermined my argument (Trinkaus and Howells 1979:127). They were absolutely correct about the effectiveness of Mousterian flakes as cutting tools, although they were quite wrong about the supposedly "modern" form and size of such Mousterian-affiliated specimens as Skhūl, whose teeth are perfectly intermediate in both size (Brace 1979:541) and form. In fact one would have to argue, as I have done (1979), that the selective force responsible for these changes had to have taken place in the Mousterian for the key morphological responses to have become apparent in the guise of the "modern" form that generally coincides with the appearance of the Upper Paleolithic. The keys to the appearance of modern form, then, have to be in the Mousterian itself. In many crucial respects, the Upper Paleolithic simply represents refinements of aspects of technology that had already been developed in the Mousterian. The stress on the typological distinctions, then, has been more misleading than useful in that it has turned our attention away from matters of real functional importance.

The first person to appreciate this in a general sense was none other than Aleš Hrdlička himself, but his discussion has been either ignored completely or dismissed in passing as the product of one who was archaeologically incompetent. It is part of the folklore of the field of anthropology as a whole (in the American sense), and especially of its archaeological component, that Hrdlička did not understand stratigraphy. This has led to the conclusion that Hrdlička was wrong in regard to his ideas on the antiquity of a human presence in the New World and, in addition, was to be disregarded on virtually anything else he said concerning archaeological interpretations in the Old World as well (Lorenzo 1978). I suggest, on the contrary, that this denigration of Hrdlička illustrates the intellectual myopia on the part of archaeologists—and, by extension, the rest of the anthropological world that is guided by their judgments—which is every bit as narrow as that which prevents paleoanthropology from appreciating the nature of a Darwinian perspective on the one hand or the real contributions to be gained from archaeology on the other.

Based on his assessment of the biological characteristics of living (and past) human populations, Hrdlička was convinced that Amerinds were derived from Asian ancestors, and that the time scale involved could not be more than that which separates modern Europeans from their Upper Paleolithic antecedents (Hrdlička 1913, 1915, 1918, 1925). Since the general (if erroneous) expectation is that no significant evolutionary change is involved in either case, obviously stratigraphic niceties can only result in finer divisions of what is essentially the same thing (i.e., a waste of time). Archaeologists, with their interests in far different kinds of problems, were predictably horrified, and their condemnation of Hrdlička echoes right on up to the present day (Alsoszatai-Petheo 1986:19). Evidently problems that can only be dealt with by looking at things in ten- or twenty-thousand-year chunks encourage a very different methodology from those that require a generational or even finer discrimination.

It should be obvious that the questions he raised concerning the circumstances surrounding the transition from the Neanderthal to the modern "phase" of human evolution have not been properly dealt with. In fact, the views against which he raised objections have recently been applied at an even earlier level of time, namely the emergence of *Homo sapiens* in the first place (e.g., Stringer 1990a; Stringer and Andrews 1988a). It seems appropriate, then, to resurrect Hrdlička's queries and give them the consideration they deserve.

WHAT'S IN A NAME?

To start with a point on which everyone is in agreement, Upper Paleolithic cultures in Europe are universally preceded by Mousterian cultures. And when human skeletal remains are discovered in association with the cultures in question, those found in Mousterian assemblages are noticeably less modern in appearance than those found in succeeding assemblages. Since these earlier specimens have brain cases of fully modern size and since this can be presumed to indicate fully modern levels of intellectual capability—the criterion by which we grant ourselves specific identity—these individuals are now generally accepted as belonging to the same genus and species as ourselves, namely *Homo sapiens*. In recognition of their less modern appearance, they can be given the subspecific designation *neanderthalensis* while more recent groups can rejoice in the title of *Homo sapiens sapiens*.

At this juncture we get to the point where opinions begin to differ—not seriously perhaps, but the seeds of disagreement are there. Some prefer to designate those earlier forms "archaic *Homo sapiens*" as opposed to the more recent "anatomically modern *Homo sapiens.*" Others prefer to call them "Neanderthals" after the place where the first such specimen to be recognized was discovered. Some have bowed to the revision of German orthography in 1901 and modernized the spelling to read "neandertal." I for one feel that this robs it of its nice archaic aura. Besides, the place was legitimately Neanderthal in 1856 and, in honor of the historic importance and status of the specimen, I see no good reason not to leave it that way (Brace 1991c:144–145, 1995b:197–198).

The use of the "anatomically modern/archaic *Homo sapiens*" dichotomy begins to introduce some more serious matters. If we use "archaic *Homo sapiens*" to describe Neanderthals, then how do we handle specimens that have been regarded as properly *Homo sapiens sapiens* but are noticeably archaic in appearance—like Skhūl at Mount Carmel, or Murray Basin Australian aborigines? In fact, the use of the descriptively typological "anatomically modern/archaic" dichotomy has as one of its consequences an implicit denial that a gradual transformation from the earlier to the later form was a possibility. For convenience of reference and useful flexibility, I myself prefer to depict the morphological spectrum involved by using the terms "Neanderthal" for the oldest examples of *Homo sapiens,* "Neanderthaloid" for those intermediate in time and/or appearance, while reserving "modern" for what is incontestably visible in post-Mesolithic populations throughout the world. Note that this would allow us to call the Australians, who were essentially Mesolithic at contact, either "archaic" or even "Neanderthaloid" as we might prefer.

By whatever name one might choose to give it, a recognizable morphological change accompanied the change in cultural manifestation. The basic question remains, as it did when Hrdlička first confronted it (Hrdlička 1927, 1929, 1930:328–349), whether those changes were the results of development in situ, or whether they were the consequences of invasion and replacement. With few exceptions, paleoanthropologists have preferred the invasion and replacement model (Boule 1911– 1913; Boule and Vallois 1957; Eldredge and Tattersall 1982; Graves 1991; Howells 1944, 1967, 1974, 1975; Stringer 1984a, 1985; Stringer et al. 1984), and it was towards this view that Hrdlička directed his queries.

HRDLIČKA'S QUESTIONS

1.   If moderns came to western Europe as invaders, where did they come from and what is the evidence to give credence to that place of origin (Hrdlička 1927:270)? Presumably a successful invasion could only be mounted from an

area where a substantial population base had already been established. When Hrdlička posed this question, there was no such evidence at all outside of Europe. Later the discovery of the Neanderthaloid, as opposed to fully Neanderthal, specimens of Skhūl, Mount Carmel, Israel—supposedly at a date prior to the onset of the last glaciation—gave a possible answer to this query. This however lost credibility with the redating of Skhūl to approximately 35,000 years ago (Jelinek 1982). The recent revision of the dating of the supposedly "modern" Qafzeh specimens once again was taken as evidence that Skhūl was more recent than fully "modern" form (Schwarcz et al. 1988; Valladas et al. 1988). Subsequently, however, the date for Skhūl has been pushed back by a comparable amount (Stringer et al. 1989).

Lately, one or another specimen from sub-Saharan Africa has been offered as indications of that supposed locus of ancient modernity—the two Omo Kibish skulls from southwest Ethiopia, for example (Leakey et al. 1969, 1984:91), Border Cave from Natal (Beaumont et al. 1978), and Klasies River Mouth in the Cape Province of South Africa (Singer and Wymer 1982:147). Even if there were no doubts concerning their dating and morphology—and there are many such—they are a very long way from Europe, and distance alone would give added strength to the objections implicit in Hrdlička's other queries.

2. If modern form did indeed develop separately from and parallel to Neanderthal form, what were the selective forces and circumstances that led to such differential rates of evolution (Hrdlička 1927:270)? In fact, one could go further and ask the holders of the traditional views to specify just what it is they feel produces modern form and what and where the evidence is to substantiate it.

3. If indeed the change was caused by invasion and replacement, why was it that the invaders took over the self-same living sites and pursued a life style that was essentially the same as that of their predecessors (Hrdlička 1927:271)? Of course, one has to recognize the fact that the Mousterian and the Upper Paleolithic do differ in many and important ways. Both, however, were hunting and gathering societies focusing on the same major prey animals. Upper Paleolithic technological innovations allowed the inclusion of quantities of smaller types of prey, a matter we shall turn to later when we consider the assessment of the difference in the forces of selection operating at the two cultural stages.

4. Hrdlička concluded his list of critical questions by asking whether history could produce any such example of wholesale replacement—implying that it could not (Hrdlička 1927:270). It was this last and in many respects least critical query that is the only one to have drawn any response from those who prefer the invasion and replacement model. North America and Australia, they note, both have a recent history that records the nearly complete replacement of an indigenous by an exogenous population (Howells 1944, 1974, 1975; Trinkaus and Howells 1979). Having noted this, presumably Hrdlička's position as a whole has been shown to be untenable, and the replacement of the European Neanderthals by invading moderns is left as the most likely interpretation for the available evidence.

ANSWERS TO HRDLIČKA

It is my contention, however, that this does less than full justice to the issues raised by Hrdlička and that, with some added emphasis, his objections are just as valid now as when he raised them more than half a century ago. I shall take the queries in reverse order because only the last one generated a response, and in discussing why that response failed to meet the issues one can show the basic strength of Hrdlička's position.

First of all, there was a cultural and technological discrepancy between the invading Europeans and the indigenous Australians and Amerinds that was orders of magnitude beyond the differences between the Upper Paleolithic and the Mousterian. Wrought iron and gunpowder alone gave the Europeans an edge in weaponry that has no parallel when the Mousterian and the Upper Paleolithic are compared.

Furthermore, post-Renaissance Europe, on the threshold of the Industrial Revolution, could sustain thousands of people in specialized roles that had nothing to do with basic subsistence behavior. Standing armies of professional soldiers were complemented by fleets of ocean-going ships manned by professional sailors. Backed by quantities of accumulated and storable staple foods, expeditions for exploration and conquest could be sustained for a matter of years. Neither Australia nor North America had anything comparable. Even though historians have noted that the reality was less dramatic than these words would indicate, it is still true that the contest was not one in which the protagonists were equipped with a comparable technology. If in all its unlikelihood there had been an Upper Paleolithic–Mousterian confrontation, the technological discrepancy would have been nothing like the iron-and-gunpowder versus bone-and-wood difference in the cases mentioned above.

Then too, the hunting and gathering economy of the Paleolithic could not generate the storable food surpluses that allow a percentage of the population to concentrate their activities on military matters. In fact, given the nature of their subsistence activities, population density itself was sustained at such a low level that anything remotely resembling major group confrontations that we might designate warfare was simply impossible. Neighboring kin-based bands of maybe a dozen or two individuals skirmishing over access to a desirable water source or hunting territory is a much more likely scenario for Paleolithic confrontations, but contention at this basically individual level is hardly going to result in a picture of invasion and replacement over an entire subcontinent, especially if there is no significant technological advantage on either side.

The matter of population density raises the final reason why the Australian and North American analogies are inappropriate answers to Hrdlička's question. The long history of agriculture, large animal domestication, and the consequent population increase in the Old World had been accompanied by the proliferation of endemic disease-producing microorganisms (Majno 1975:96). These in turn generated the mechanisms for their resistance in the affected populations. Australia and the New World were isolated from this history by ocean barriers and lower population densities. When Europeans appeared on their shores, they unwittingly brought samples of their diseases with them, and these had devastating effects on the indigenous populations. This as much as anything else contributed to the success of their efforts at conquest (Dobyns 1966, 1983).

But such circumstances certainly did not exist in the Paleolithic. Population density was never sufficient to sustain endemic diseases, and the cultural continuity from the fringes of western Europe all the way through Russia, the Balkans, and the Middle East indicates that isolation was never a factor. All of this should have been apparent even at the time Hrdlička asked his rhetorical questions.

DIFFUSION AND GENE FLOW

But if Hrdlička's stance is not contradicted by the evidence, does this in fact demonstrate that he was right? The answer is: "in itself, not entirely." If invasion and replacement at the Paleolithic level is the most unlikely of explanations, there is always the possibility of diffusion and gene flow operating to accomplish the transformation. This too, however, requires a major demonstrable source both for technological innovation and for the early appearance of "modern" form, and so far neither has been located. If such evidence becomes available, obviously that possibility will have to be entertained.

Australia provides a very good model for the operation of just such a process. The spread of the small tool tradition from the north some 3,000 to 6,000 years ago (Golson 1974:380) was followed by the spread of the spear-thrower (Davidson and McCarthy 1957), string technology (Davidson 1933), and nets and fishhooks (Bowdler 1976). All of this was accompanied by evidence both of gene flow (Macintosh and Larnach 1973) and the in situ effects of selective force changes instituted by the new technology. The morphological consequences were reductions in robustness pointing towards changes analogous to those that had taken place between Neanderthal and early modern form at the opposite end of the world some tens of thousands of years earlier (Brace 1980). How much of this change was due to gene flow and how much was due to selective force change in situ is still in doubt. Clearly both processes were involved, but a precise apportionment of their roles is still a matter of debate. In Australia, although there is a selective force time depth gradient running from earlier levels in New Guinea southwards with a parallel gradient in dental robustness, the nonadaptive aspects of craniofacial form show a close genetic relationship between the aboriginal inhabitants of Australia and New Guinea (Brace and Hunt 1990, Brace, Smith, and Hunt 1991).

A similar debate continues concerning the movement of rice agriculture into Japan during the first millennium B.C. (Akazawa 1982a, 1982b; Brace and Nagai 1982; Brace et al. 1989). But in both the Australian and Japanese cases, the prior existence of technological advance and population concentration in areas from which diffusion and/or migration could occur is an established fact. This is not the case for the issue of the Mousterian–Upper Paleolithic transition in spite of several generations of effort to discover such evidence, and, until the latter becomes available, it remains one of the less likely explanatory models.

THE MYTHOLOGY OF HUMAN EVOLUTION

If such is the case, then one might legitimately ask why so much effort has been repeatedly devoted to offering this as the scenario to be expected. Although no hard and fast answer can be given to this question, it would seem to be rooted in the history and traditions of the field. In the pre-Darwinian outlook of the focus that ultimately gave birth to paleoanthropology, the question of origins was considered an unanswerable and hence basically an unscientific matter (Broca 1862:314). It was enough that novelty arose somehow, perhaps by special creation, outside the area of immediate concern and investigation. The role of science, then, was in the documentation of the spread and establishment of the newly fixed entities. Among other things, the assignation of the proper name was one of the more important scientific activities (Broca 1860:435; Greene 1959:182–183).

There is something of an almost fundamentally biblical orientation in this approach. A localized Eden is assumed where living entities are created by divine fiat, a process that is essentially unexaminable. The scientist as Adam then gives them names and records their departure and subsequent spread over the face of the earth where they remain in stasis until Judgment Day—or the occurrence of another unexaminably instant "speciation event" which either exalts them to a new level of stasis or, as is more likely, consigns them to eternal oblivion. I suggest that it is the application of these views to the paleoanthropological evidence that has produced the prevailing "myth of human evolution" (cf. Eldredge and Tattersall 1982) and that it has more in common with that contradiction in terms, "creation science," than it does with post-Darwinian biology (Brace 1990, 1994). The assumption that modern human origins had a narrow focus that constituted an effective point of origin, then, is not rooted in scientific mechanism. Rather it is more the consequence of thought patterns that have had essentially sacred sanctions in the history of the western world. The result is something that I have referred to as the "Garden of Eden Hypothesis" (Brace 1979:539), but to give credit where credit is due, it was Vince Sarich who first noted that the interpretive traditions of paleoanthropology tended to conform to what he called "a 'Garden of Eden' school of thought" (Sarich 1971a:188).

IMMACULATE CONCEPTION IN THE GARDEN OF EDEN

This biblically based outlook has had a strong hold on the expectations of traditional biological anthropologists, and it has been largely responsible for the enthusiasm with which so many swallowed the premature claims put forth by the students of mitochondrial DNA that "modern" human form arose in Africa in the relatively recent past (Cann 1988; Cann et al. 1987; Stringer 1989a, 1990a; Stringer and Andrews 1988a).

The skeletal evidence for the existence of a population reservoir of modern form in sufficient numbers to generate an invasion has remained elusive. Nevertheless, the fact of the maternal transmission of mtDNA and the claim that it all stems from an African source has led to popular assumptions concerning a putative recent African "Eve" who served as the mother of us all within the past 200,000 years and possibly much less (Wilson and Cann 1992). Biological anthropologists have offered counter arguments based on appraisals of the skeletal remains (Thorne and Wolpoff 1992), and their vindication has recently been shown in spectacular fashion by the complete collapse of the supposed mtDNA evidence (S. B. Hedges et al. 1992; Templeton 1992). With the demonstration that there are 50,000 possible phylogenetic trees that are more parsimonious than the one offered by the proponents of a recent African Eve, it is clear that, at the present stage of our knowledge, the analysis of mtDNA can give us no answers concerning the approximate time or even the place of ultimate human origins.

A lesson should be learned from all of this. Anthropologists, by abandoning the indications provided by their own data in favor of evidence uncritically accepted from a field with which most have little familiarity, were in effect selling out on the validity of that which they are most competent to appraise. The biological anthropologists in particular made almost no use of the abundant archaeological record. Not only is there no archaeological trace of a major movement out of Africa in the past 200,000 years (Marks 1983, 1988, 1992), but there is no hint that the technology available to the inhabitants of Africa—or anywhere else—during the Pleistocene would have given its possessors an edge over their neighbors.

CULTURE, MECHANICS, AND HUMAN EVOLUTION:
SIGNIFICANT ASPECTS OF CHANGE

Since concern for the actual mechanics of change has been generally missing from discussions of the course of human evolution, especially that which has taken place in the past 50,000 years, it now seems only proper at least to make a beginning at an analysis of this sort. The first step in such a procedure is to identify the significant morphological changes that actually took place. By this I do not mean resurrecting the standard list of visible traits in which modern and Neanderthal groups can be observed to differ. Most of these are simply the by-products of differences in basic developmental processes which themselves may have important implications. Insofar as the peripheral indicators point to the existence of such basic matters, they can play a useful role in our understanding, but in and of themselves they are largely without significance.

Brow ridges, for example, have been the focus of an almost phrenological enthusiasm. In the view of what Gould has called the Panglossian hyperselection-ists, each nuance of form must have a specific selective reason for its existence (Gould and Lewontin 1979). Brow ridges then have been presumed to play some very specific role, but whether this is related to resisting chewing stress (Russell 1985), acting as a sunshade (Barton 1895:160), or serving to keep head hair from obscuring the vision (Krantz 1973) has yet to be demonstrated. There are no clinical or fossil instances of individuals who died because chewing stress caused the collapse of an unreinforced brow, and a scenario has yet to be suggested by which presence or absence of brow ridges would change an individual's life span or reproductive capacity. Consequently, rather than considering brow ridges to have any significance in and of themselves, it is more productive to regard them as peripheral indicators of basic developmental processes which are the real locus of adaptive significance (Moss and Young 1960; Ravosa 1988). In general, those populations where brow ridges are particularly prominent are characterized by other aspects of skeletal robustness and reinforcement which set them apart from groups in which brow ridges are less well developed. One can suggest that it is these basic differences in general levels of skeletal robustness that are more important from an adaptive point of view than are the peripheral manifestations that catch the eye.

This is not to say that some of the traits traditionally listed by which Nean-derthal and modern form can be distinguished do not have specific adaptive significance. The dorsal sulcus on the axillary border of the scapula has been associated with an emphasis on the musculature that attaches to the rotator cuff on the humerus (Churchill and Trinkaus 1990). Such a condition reinforces the shoulder against the possibility of dislocation, and, when one considers the evidence for the difference in Mousterian and Upper Paleolithic hunting tech-niques, the special importance of such a condition becomes apparent. Other details of the Neanderthal skeleton—joints, long bone shafts, digital flexor attachments—all point in the same direction.

Further, it is demonstrable that the dentition was larger, especially in its anterior component (Brace 1964a, 1979, 1995c). Dental change in fact is the most easily quantifiable and demonstrable aspect of the picture. Reduction pro-ceeded at the rate of one percent every two thousand years starting with the onset of the last glaciation. At the end of the Pleistocene, this rate doubles to approx-imately one percent per thousand years (Brace et al. 1987). The evidence for this is based on the mundane accumulation of measurements from the most abundant, well preserved, and measurable elements of the skeleton—the teeth themselves. The picture they present is one of gradual change through time with no significant gaps or jumps. On this basis alone it can be shown that the people of the early Upper Paleolithic, far from being indistinguishable from "moderns," were metrically closer to the Neanderthals (Brace 1979, 1995d). Subsequent change, although well-documented, has been so gradual and unremarkable that it has entirely escaped our notice (Brace et al. 1987). In general, it is apparent that the Neanderthals possessed a greater degree of skeletal and muscular robustness than more recent human populations, just as Hrdlička noted long ago (Hrdlička 1927:272–273), and "modern" form ensued simply as a result of the reduction of that robustness.

EVOLUTIONARY MECHANICS

The production of modern human form from Neanderthal antecedents, far from being the difficult and unlikely event assumed by the majority of analysts, is actually a very simple evolutionary process. All that is required is a slackening of the growth processes necessary to produce a Neanderthal level of skeleto-muscular and dental robustness. I shall argue that these two aspects have separate developmental and evolutionary histories, but that does not change the fact that modern form can be very easily achieved simply by reducing the amount of growth necessary to produce a Neanderthal.

This is easy enough to conceive in a general fashion, but there has been a conceptual stumbling block in the way of accepting this as what actually happened. The problem is in that aspect of neo-Darwinian dogma, stemming from the intransigence of the late R. A. Fisher, that all evolutionary change is the result of the action of natural selection—even if we cannot figure out how it works. As Gould and Lewontin (1979) have perceptively observed, there is a remarkable similarity in this to the outlook caricatured by Voltaire in the figure of Dr. Pangloss, who maintained that all is for the best in this the best of all possible worlds—even if we, in our less than infinite wisdom, are unable to understand how this can be so in specific instances.

Thus in the post-Fisherian view, "modern" human form had to be considered adaptively advantageous—an evolutionary "advance" positively produced by the forces of natural selection. Admittedly the spectacle of professional paleo-anthropologists defending in all seriousness the view that it is "better" to have more easily dislocatable shoulders, weaker hands, sprainable ankles, and friable teeth presents a picture that is ripe for treatment in the vein of Voltaire's Panglossian satire.

But rather than remaining constrained by the limits of Fisherian hyperselectionism, if we actually look at the conjunction of elements in the picture and consider them from a basically biological point of view, the answer to our question concerning process is almost transparently simple. In a word, the change from Neanderthal to modern form can be regarded as the product of entropy. In a manner similar to that represented in "Murphy's Law," left to themselves, things tend to run down (Brace, Smith, and Hunt 1991:43–44).

One can argue that the term *entropy* is only appropriate for usage in describing the realm of physics and not for biology. But if we really are trying to understand biological matters in terms of basic mechanisms, then we are forced to consider such things down at the level of the physical—or at least the biochemical. Molecular biology has produced the necessary background, and we now know what mutations are and what they are most likely to do—and indeed, most of them produce entropy. Years ago I noted that the most likely result of the most likely mutations is structural reduction, and where the forces of selection are reduced or absent, this is what one should expect to see as a matter of course (Brace 1963). I labeled the phenomenon "the Probable Mutation Effect" (Brace 1963), and I suggested that this was what was principally involved in transforming Neanderthal into modern human form (Brace 1964a, 1967, 1979, 1991c, 1995a, 1995b:chap. 13; Brace, Smith, and Hunt 1991:41 ff.).

For the evidence concerning the reduction in the forces of selection that had previously maintained full Neanderthal form, one has to turn to a selective reading of the archaeological record. Not all of the evidence is discernible from that record alone, however, since some is directly indicated by the skeletal morphology of the producers of that record. For example, the recent arguments over whether Middle Pleistocene hominids were hunters or merely scavengers have been restricted to a consideration of the archaeological and taphonomic evidence (Binford 1985, 1987; Klein 1987). The paleoanthropologist can add, however, that the extraordinary level of skeletal robustness evident in the hominids of that time span can only be accounted for by assuming that their survival depended on activities which required it. While one theory proposed to account for that robustness by suggesting that it was the consequence of long-term efforts to do each other in (Roper 1969), the most likely explanation was that they were in fact actively hunting prey animals of considerable size without much in the way of technological assistance.

ARCHAEOLOGY AND CHANGE IN SELECTIVE FORCE

If Middle Pleistocene levels of robustness were maintained as a result of engaging in major hunting activities aided by only the simplest of technologies, then we should expect that the subsequent decrease in that robustness should indicate that hunting techniques had improved to the point where raw brute force was less essential in the pursuit of basic subsistence than had previously been the case. The Mousterian provided a hint that something of the sort had indeed taken place. Compared with the preceding Lower Paleolithic, the typological differentiation and refinement of tool manufacture represents significant technological advance. How much this actually reduced the physical stress of hunting itself is still a matter of debate, but it can be suggested that one Mousterian innovation, the development of cooking, had important biological consequences (Brace 1979, 1995a; Brace, Smith, and Hunt 1991:46–47).

Archaeological treatment of this matter has been notably skimpy. While "hearths" have been recognized as features in Mousterian sites for more than a century now, it is apparent that they are fundamentally different from the camp-fires of the Kalahari hunters, the San, where the remains are hard to recognize a year or so later (Yellen and Harpending 1972; Yellen 1977a, 1977b). In the case of the Mousterian, however, even after fifty thousand years, "hearths" remain as identifiable concentrations of charcoal, bits of burned bone, and fire-cracked cobbles to a depth of half a meter or so. I have suggested that they in fact are the remains of "earth ovens," which allowed their makers to thaw and use the products of the chase even in winter during the last glaciation (Brace 1979, 1991c:155 ff., 1995b:226 ff.; Brace, Smith, and Hunt 1991:47, 51).

While this use of fire had important consequences for Neanderthal survival, it also had an unintended by-product. For the first time since hominids embarked upon a hunting way of life at the end of their australopithecine stage, the amount of chewing needed to process their food was significantly reduced. However much the wear patterns (Ryan 1980) and sheer size of the Neanderthal front teeth indicated an important manipulating role (Brace 1962b, 1979), the primary function of the dentition has always been in the necessary processing of food. When cooking technology, such as the earth oven, produced some reduction in the stress associated with that role, one could predict that entropy in the form of the Probable Mutation Effect would produce a subsequent reduction in the size and morphological robustness of the dentition in succeeding populations, and indeed this is the case (Brace 1991c:156–158, 1995b:277 ff., 1995c; Brace, Smith, and Hunt 1991).

So much for the teeth in a general sense. In addition, similar consequences for the postcranial skeleton follow from the separately demonstrable changes in hunting techniques. First, the use of projectiles to disable prey from a distance must have been a significant development. Chimpanzees will propel sticks and stones in their agonistic displays (Kortlandt 1972), but the act of throwing with accuracy, power, and lethal effect is a uniquely human accomplishment. When we find haftable projectile points in the archaeological record, we can legitimately infer that effective throwing had been previously learned. Of course, wooden-tipped spears can be equally lethal and must surely have been the first effective projectiles to be used, but they do not fossilize and are rarely preserved beyond the recent past. The abundance of bone projectile points at the beginning of the Upper Paleolithic tells us, however, that spear-throwing was one of the hunting techniques available (Brace 1988b:131, Trinkaus 1989:51).

Furthermore, the proliferation of small mammal, bird, and fish bones late in the Upper Paleolithic indicates that by this time they could be captured in quantity in a manner not previously possible (Clark 1987; Clark and Lindly 1989a:665; Straus 1983b). It seems most likely that this was the result of the development of what Hayden (1981) has called an "unobvious" piece of technology, in this instance, string. Snares, nets, fishlines, and carrying bags all depend on the existence of string, and once this is present, a whole world of food resources becomes available (Brace 1988b:131, 1991c:170, 1995b:272–273; White 1989a:92–93). Furthermore, the amount of raw brute force required in their procurement is very significantly reduced.

Big game continued to be hunted, but the use of spear-throwers and, later, bows and arrows reduced the frequency and necessity of direct contact with wounded and angry prey. As in the case of the dentition, entropy in the form of the Probable Mutation Effect should ensue and a reduction in robustness and muscularity should be a consequence. And so it is.

Archaeology, then, used in this way, can provide us with the evidence for those changes in selective force stress which are the immediate causes for the transformation of Middle Pleistocene levels of robustness into their modern manifestations. Furthermore, we should be looking to archaeology for the time when, and the area in which, such changes in stress first occurred because we should expect that the morphological changes by which modern form was produced first occurred in just those places.

South Africa can hardly be overlooked because the technology that enabled the utilization of quantities of small game was developed there perhaps earlier than anywhere else in the world (Singer and Wymer 1982). The morphological consequences to be expected also follow, and it is no surprise to find that levels of robustness and muscularity are as reduced among the South African aborigines as anywhere else. This, however, hardly means that at the onset of the last glaciation South Africans then massed together and resolutely marched towards the frozen north through the Middle East, invaded Russia, and then turned west to march to the Atlantic.

Initially unobvious aspects of technology quickly become very obvious once a demonstration is made—witness the proliferation of electronics in modern Japan. Diffusion to adjacent people at the hunter-gatherer level is far more likely than population movement (see Clark 1994b), but even in this case, archaeology provides no good evidence for the rapid northward spread of those early South African innovations. A better case can be made for independent invention in those parts of the world where the Upper Paleolithic succeeded the earlier Mousterian. If the seeds of that crucial technology were in the Mousterian, the flowering in the Upper Paleolithic in a broad band from Europe eastwards suggests that the changes in selective force that this represents should have produced the appearance of modern form in situ throughout this whole expanse. This, rather than the point of origin view, is what an appreciation of evolutionary mechanics and the functional significance of the archaeological record should lead us to expect.

The same expectations and logic should apply in other parts of the world as well. The implications of the Middle Stone Age in Africa (Klein 1983) and its counterparts in India (Allchin 1963) and farther east suggest that the in situ transformation from archaic to modern form should have occurred roughly simultaneously in each of those places for the same reasons. Although it has recently been stated that the "regional continuity model was first outlined in a broad theoretical context by Wolpoff et al. in 1984" (Smith, Falsetti, and Donnelly 1989:38), this would seem to be derived from the earlier view that the "broad spread of similar culture traits over wide areas of the Old World" (Brace 1964a:15) produced the "stage-to-stage evolution [that] occurred simultaneously in all parts of the inhabited world" (Brace 1967:95). The more recently proposed "multiregional" view has omitted mention of the mechanism offered to justify the model previous proposed (cf. Brace 1979:83).

SUMMARY AND CONCLUSIONS

It has recently been asserted that a simultaneous and more or less independent evolution towards modern human appearance from various regional Neanderthal populations "can hardly be considered likely" (Bräuer 1984a:332). This in fact represents a reassertion of the paradigm that has prevailed in paleoanthropology since its beginning and which has its roots in pre-Darwinian paleontology. It assumes that novelty arose by some unknown process in a manner not available for our examination, following which it spread in the form of an invasion which led to the replacement of the Neanderthal populations of the northern and western portions of the Old World. This view has been offered to account for the rise and spread of *Homo sapiens* in the first place, and of its post-Neanderthal representatives in the second. This stance, however, is simply a reassertion of the pre-Darwinian outlook that Marcellin Boule brought to the field from paleontology at the beginning of the century and hence deserves the label "hominid catastrophism" (Brace 1964a). In its failure to consider mechanism and its tacit assumption of a form of "special creation," this approach has much more in common with the view that refers to itself with such self-congratulation as "creation science" than it does with evolutionary biology.

The flaws in the logic of the invasion and replacement model were noted in telling fashion by Aleš Hrdlička well more than half a century ago. However, because he was perceived to have had a cavalier attitude towards stratigraphy in his field work, his credibility was impugned and his points of objection were generally ignored. Certainly they have not been given the consideration they deserved. It seems only right, then, belatedly but pointedly to provide answers to Hrdlička's questions.

1. There is no evidence for a major population source displaying modern human form contemporary with or prior to the Neanderthals which could have provided the reservoir for anything remotely like an invasion.

2. The reduction in selective force intensity produced by Mousterian techniques of food procurement and preparation should lead us to expect that it is in just those areas characterized by Mousterian cultural assemblages and their equivalents that we should expect to find the emergence of "modern" morphology. The added technological refinements of the Upper Paleolithic and its counterparts further contributed to the circumstances that allowed the reductions in robustness which converted Neanderthal form into the manifestation we term "modern."

3. The technological differences between the Mousterian and the Upper Paleolithic were not sufficient to have given the latter the kind of advantage that iron and gunpowder gave to Europeans invading Australia and North America in the recent historical past. Both the Upper Paleolithic and the Mousterian were hunting and gathering societies, preying on the same major game animals, and living sequentially at the same sites. Although the Upper Paleolithic eventually came to exploit a wider variety of food sources, initially it was not able to sustain a population density that would make it a threat, particularly if it were coming from outside and seeking to oust the entrenched and established inhabitants.

4. With no major population center as a source and a hunting and gathering subsistence base, there was no means of sustaining endemic infectious diseases that could play a role in any hypothetical confrontation.

5. For all of these reasons, the historical analogies used to answer Hrdlička's points can be considered to constitute inappropriate responses.

In sum, if one turns to the evolutionary dynamics by which modern form is produced, it is obvious that what is involved is simply a reduction in the degree of robustness both in faces and teeth and in the postcranial skeleton as Hrdlička perceived two generations ago. The change in life ways that allowed such reductions to occur is clearly indicated in the archaeological record. The use of earth ovens in the Mousterian, initially to help thaw frozen foods, reduced the amount of chewing required to process what was ingested. Entropy in the form of the Probable Mutation Effect then produced the transforming reductions in faces and teeth. The development of projectiles, symbolized by Aurignacian and possible Mousterian points, and the string technology used in snares, nets, and fishlines, indicated by the presence of quantities of small perforated objects and the great proliferation of fish, bird, and small mammal bones in the Upper Paleolithic, all significantly reduced the previously necessary levels of stress on the hunters' postcranial skeleto-muscular system. Again, entropy in the form of the Probable Mutation Effect did the rest.

Looked at in terms of the broader implications of life way rather than from the narrowly typological point of view, the archaeological record can give us the perspective we need to understand the dynamics of the transformation of human form from its Middle Pleistocene to its modern manifestations. If the Mousterian refinements in hunting and food preparation techniques represent changes in the selective forces operating on the portions of the physique previously necessary for the performance of these activities, then wherever we find Mousterian, or its functional counterparts elsewhere in the world, we should expect the beginnings of modern morphology to emerge in the course of time. And indeed, archaic modern form does succeed Neanderthal form in just those areas. Since the dynamics by which such a transformation can occur in situ are implicit in the Mousterian cultural remains themselves, then there is no compelling reason to look to other areas where the skeletal remains are skimpy and controversial and where little or no cultural information is available to demonstrate the existence of such transformational mechanics.

Given all of this, we should expect to find modern human form emerging in a broad spectrum across the entire northern expanse of the Old World in the Late Pleistocene. Southern Africa shows simultaneous and comparable developments. At the southeast end of the inhabited Old World, geographic factors played a role in delaying human occupation itself in the first place, and they also were important in the subsequent delay in the diffusion of those pieces of technology that reduced the forces of selection on particular aspects of the human physique. It is no surprise then to observe the survival of archaic morphology in Australia. What is interesting to note, however, is that following the introduction of earth oven cookery, projectile propelling devices, and string with all its implications, Australian physical form has been changing over the past 10,000 years in the same direction and at the same speed that was observable for the previous 50,000 years elsewhere in the Old World (Brace 1980).

The broad-spectrum view of the emergence of modern human form then is precisely what an appraisal of the mechanics would lead us to expect. In contrast, it is the traditional view of paleoanthropology which assumes that modern form suddenly sprang into being in one or another geographically restricted locale for reasons that are never specified—followed by spread, invasion, and replacement—that has little to recommend it. It is this and not the former that "can hardly be considered likely."

2

What Does It Mean To Be Modern?

MILFORD WOLPOFF and RACHEL CASPARI

The modern human origins debate rests on an assumption—that there is something called modernity, or an anatomically modern human, and that we can define it. All living people, of course, are modern humans. The question is about their ancestry. If modern humans all descend from a single population whose anatomical features spread because they were advantageous, then "modern human" must have a clear anatomical definition. There should be a unique set of features that all modern humans share, to the exclusion of other humans. This must be true regardless of whether the population spread and completely replaced indigenous aborigines or mixed with them.

Multiregional evolution addresses modernity quite differently (Thorne and Wolpoff 1992; Wolpoff et al. 1984). Multiregional evolution is a model of how evolution works in a widespread polytypic species. It stems from the basic observation that some of the combinations of features that show geographic variation today can be traced into the past, most often in the same geographic regions. We distinguish here between combinations of features and the distribution of individual features, because so many characteristics can be found in all populations. The model can be developed for our own species because more is known about its prehistory than any other. It is predictive (and thereby has the potential of being disproved) concerning the issue of patterning. For instance, if it could be shown that human evolution proceeded as a series of sweeping population replacements by successively better adapted forms, each with their own separate origin (something like what Teilhard de Chardin [1959] proposed), perhaps as the result of competition between parallel adapted human species as Tattersall (1994, 1995) and Stringer (1994) favor, multiregional evolution would be clearly disproved. But unlike theories such as the "Eve" theory or the "Out of Africa" theory (about modern human origins) or the single species hypothesis (about australopithecine species), it is not a focused theory about or interpretation of particular evolutionary events.

Multiregional evolution is based on the precept that the present is the best basis for modeling the past evolution of *Homo,* and posits that populations in polytypic species remain differentiated for the same reason that they evolve without speciation—because of the matrix of genic exchanges (and in the human case, information exchanges) they establish. Described as ethnogenesis in living populations (Moore 1994a), this process can be likened to the channels in a river that can separate and recombine numerous times. Human populations, like human languages and cultures, continue to change, divide, and merge in a matrix of changing patterns in which each population (or language or culture) can have several ancestors and several descendants. Extending this pattern into the past would explain how human evolution happened everywhere, because every area was always part of the whole (Thorne and Wolpoff 1992).

Most multiregionalists do not believe that the fossil record shows a single origin for modernity (although these ideas are not incompatible; see Smith, Falsetti, and Donnelly 1989). They liken the evolutionary process to throwing pebbles into a pond. Each pebble is an advantageous feature, appearing at different times and places. Ripples from the pebbles can spread across the entire pond surface; the interconnections of the populations provide the pathways for the traits to spread. The ripples interact to form complex waves whose patterns differ from place to place, and it is the spatially differing patterns generated by the same history of thrown pebbles that creates that shifting, varying pattern of anatomical variation we describe as modernity.

> . . . a polytypic species may still evolve as a single genetic system. Favorable mutants or gene combinations arrived at in one part (race) of such a species may, under the influence of natural selection, eventually spread to all other parts and thus eventually become a common property of the entire species. Thus, local autonomy of gene pools of racial populations does not preclude retention of the basic unity of the species as a whole. (Dobzhansky 1950:106–107)

This precept of modernity is that it is based on *features* that evolve independently, *not* because of a single ancestral population that links modern features together, and *certainly not* because of the independent evolution of modern populations in different regions (a view sometimes incorrectly attributed to multiregional evolution by those who disagree with it).

The modernity issue, as we see it, boils down to whether

1. the characteristics of modern humans have their origin in a single population that spread around the world with an integrated package of features that were not all advantageous (e.g., Stringer 1989a argues that the advantages of the population were enough to establish its features, whether or not each of these contributed to the population's advantage), or
2. modern human features have many independent origins and spread around the world as single, generally advantageous, items.

Therefore, the very existence of a unique series of characteristics that are modern, the ability to clearly distinguish when (and where) the set first appears, is a test of the single origin theory. If there is no unique definition, if modernity can neither be uniquely described nor clearly found, it would be a refutation of single origin.

Note, though, that the converse is not true. A definition and clearly distinguished place and time of origin is not a disproof of multiregional evolution *unless the mode of spread of modernity was complete replacement.* Any other variation of this model would be encompassed by multiregional evolution, a model after all that ascribes change in a polytypic species to clinal balances between selection and drift, and genic exchanges that include intermixtures and population movements.

SPECIES—WHAT'S IN A NAME?

Are modern humans a new species? To discuss this we must agree on what we mean by *species,* a term whose meaning has changed dramatically over the past quarter-century. The changes in species definition come from the ascent of the phylogenetic approach and the problems faced by paleontologists in applying the widely accepted biological species concept. If there is one change that is close to universally accepted, it is the realization that new species can only come from cladogenesis, the splitting of one lineage into two.

Living species are almost universally defined with the biological species concept. A biological species is a group of populations that can actually or potentially interbreed and produce fertile offspring, and that are reproductively isolated from populations in other species. The foundation of the biological species is in its breeding behavior, and this creates several problems when species are considered over time. One problem is that an interbreeding criterion cannot be applied to populations far removed in time; it cannot tell us whether 40,000-year-old Neandertals and modern humans living today are in the same species. A second problem is that fossil remains usually provide no information about interbreeding between contemporary populations. Although morphological differences between species are a result of gene exchange barriers, regional differences can also occur within a polytypic species that, like ours, is internally subdivided.

The currently accepted solution to these problems comes from a concept of species that is based on genealogy and considers species in their temporal dimension. Genealogical species have definite beginnings and ends, each with a unique ancestry. All members of a genealogical species share common descent; these species are monophyletic, consisting of an individual and all of its descendants. There are two species definitions that fit this description.

One of them is the phylogenetic species, a group of monophyletic individuals who share at least one unique feature. According to this definition the phylogenetic species is the minimum biological unit that can be found. Members of phylogenetic species are easy to recognize because each species is defined by the presence of one or more unique features. This makes the phylogenetic species concept particularly attractive to paleontologists. Many "splitters," those who argue that not enough past species have been named, apply the phylogenetic definition. Tattersall (1992) is a good example, as his precept of evolutionary mechanisms, species competition (1994), requires the expectation that there were numerous species to compete. Supporters such as Tattersall and Stringer (1992a, 1994) argue that since some biological species are separated only by reproductive habits, not by any anatomical difference, phylogenetic species must be valid because even they underestimate actual species diversity. If we use the phylogenetic definition we will err on the side of conservatism, they reason. Neandertals, for instance, are a phylogenetic species (Tattersall 1995), *Homo heidelbergensis,* because they are recognizably different from other humans. And this species concept requires that there are other past human species as well, all competing with each other and evolving in parallel (for instance, Tattersall's assertions require that *Homo heidelbergensis* attained a cultural capacity independently of *Homo sapiens;* Wolpoff 1994c).

One reason why all paleontologists have not embraced the phylogenetic species is that there is a serious problem of concordance with living biological species. Variation that is organized according to the phylogenetic species concept does not always conform to a taxonomy based on the biological species concept that forms the cornerstone of the modern synthesis. By erring on the side of conservatism, users of the phylogenetic species concept are unlikely to fail to recognize legitimate species. However, they may identify too many of them! The criteria that once led to human races being considered different species in so many of the polygeny theories (Gates 1944; Haeckel 1905; Keith 1936) would do so again. Of course the full acceptance of the biological species concept throughout all biology shows this isn't so. The discrepancy between the applications of the two species concepts shows us the phylogenetic species concept may be easy to use, but its value for understanding evolution is seriously limited.

There is a second genealogical species, the evolutionary species, defined as a single monophyletic lineage of ancestral-descendant populations which maintains its identity from other such lineages and which has its own evolutionary tendencies and historical fate. The evolutionary species has its beginning in cladogenesis and its end in another, later species split, or extinction, which is why it is also a genealogical concept. It retains the essence of the biological species—reproductive isolation, the reason why it can maintain its identity and have unique evolutionary tendencies. But at the same time it avoids many of the biological species' deficiencies: lack of time depth, absence of morphological criteria for diagnosis, and perhaps most important, emphasis on reproductive ties (genic exchanges). The evolutionary species differs from the phylogenetic species in that the species is defined by reproductive isolation rather than one or more shared unique characters; in fact, morphological distinctiveness is not a necessary requisite. Nevertheless, an evolutionary species can be morphologically diagnosed using criteria that are a consequence of its unique evolutionary pattern.

For instance, over the past million years, in all regions inhabited by human populations, no matter how isolated and otherwise distinct they were, the same changes in brain size (getting bigger) and tooth size (getting smaller) occurred. This is in spite of the fact that the changes were patterned differently from place to place, and both average tooth size and brain size differed between regions in the past, and do today.

So the question arises: to what extent are different species definitions related to the differences in theories? Stringer (1992a:19) describes the species definition he uses as "evolutionary," but elsewhere (1992c) as "morphological" and "phylogenetic," and "[it does] not depend on the presence or absence of hybridization" (1992c:201). Stringer is the only Eve theorist to be explicit about the species definition he uses. It is based on the recognition of "specific skeletal synapomorphies between recent and fossil human material" (1992a:19). Such a definition is a logical consequence of the single origin theory (and vise versa); synapomorphies would be expected if there was a single origin, and conversely support the single origin theory. The same data have different implications using an evolutionary species definition.

The difference in definitions is significant not only over how to determine whether modern humans are a new species, but also over the issue of what samples interpreted as intermediate, or hybridized, signify. Such samples would not disprove the Eve theory for Stringer, whereas they would for paleo-anthropologists who use the evolutionary species concept as described above. *But although the difference in definitions affects the issue, it is not the source of the problems surrounding the question of what it means to be modern.*

THE POLYTYPIC PATTERN

To understand the overall pattern of later human evolution and the appearance of modernity we need to consider the relevance of both population adaptations and histories, and resolve the paradox of global change and local continuity. Single source theories do not face many of these problems because they focus on the successful populations of modern humans, not the indigenous natives they replaced. In these theories the success of the modern humans stems from their superiority, and today's global pattern of variation reflects rapidly established adaptive divergence that followed their spread.

Multiregional evolution accounts for the characteristics of the colonizing human species during the earlier period, the spread out of Africa and establishment of interconnected networks through which ideas and genes were exchanged. The main assumption the multiregional explanation requires is that the evolutionary processes at work in relatively recent modern human populations were in operation in the past. Our species is unusual and difficult to model because it is polytypic, with extremely broad geographic and ecological ranges. Most polytypic animals are widely dispersed because they occupy an ecological niche that is broadly distributed, but the human pattern is of a widespread species with many different ecological niches.

The basis for this difference is human culture. The unique human capacity for information storage and transmission, the ability to communicate with other groups and obtain information from them, and the associated behavioral flexibility provide critical elements of what is necessary for human populations to occupy so many diverse niches. Culture also influences the demography of populations and plays an inestimably important role in shaping their evolutionary histories. While we do not think that culture appeared suddenly in its present form—its definition and evolution are obviously complex—the behavioral flexibility it allows was clearly significant during the colonizing phase in human evolution. Humans in this later phase should not be expected to follow the species proliferation or extinction patterns of other fauna that lack this degree of flexibility. This is why we assume the principle of uniformitarianism applies. We see no reason not to adopt the simplest hypothesis—that populational processes characteristic of our polytypic species, at least prior to sedentism, were operating similarly in the past. Modern humans themselves provide the best model to base the reconstruction of the multiregional pattern of human evolution, not just in the Late Pleistocene but during the entire colonizing phase.

Ethnologists such as John Moore (1994a) hold that populational, linguistic, and cultural relationships between populations today are multiple and complex, in part because each of them descend from, or are rooted in, several different antecedent groups. This pattern is not limited to a putative melting pot like the United States—it is a worldwide one. Roger Owen (1965) studied a series of native tribes across the world, from the American Great Basin to Australia to Tierra del Fuego, and everywhere he looked, there were no ethnically discrete groups. Instead, marriages across what seemed to be the most profound ethnic boundaries were common, and multilingualism prevailed. Susan Sharrock surveyed native American tribes from the Northern Plains and concluded that "the ethnic unit, the linguistic unit, the territorial coresidential unit, the cultural unit, and the societal unit did not correspond in membership composition" (1974:96). Moore calls the pattern of multiple rooting "ethnogenesis" and notes that the continued population contacts and exchanges it implies mean that there is no link between language, culture, and biology (Marks 1995; Wolf 1994).

The ethnogenetic perspective argues that humans have evolved as a species and not as unconnected regional populations, so that there always have been admixed populations. . . . The ethnogenetic perspective takes the multiregional position, arguing that human history has always been characterized by interaction across profound ethnic and cultural boundaries, by the amalgamation of linguistic traits, and by the recurrent hybridization of cultures. (Moore 1994b:937)

But single origin theories assume a link between language, culture, and biology, certainly for those moderns who, according to some theorists, were the first to use language of any sort. Supposing such a link is an assumption many make, for instance in the genetic studies which seem to assume that language or culture differences should mark biological boundaries. Linking culture, language, and biology is part of a different, opposite, model of population variation which is based on the idea that variation is created by continued treelike branching and divergences—a model "that must be assumed, explicitly or implicitly, whenever a branching analysis of population relationships is performed" (Morton and Lalouel 1973). But, as Moore (1995:531) puts it:

The idea that it is discrete ethnic populations that mix rather than individuals who mate or genes that flow is an outmoded one that does more to confuse than to clarify. . . . this distinction between admixed and unadmixed populations is illusory and a product of our ignorance of the ethnohistory of most of the world.

Now here is the point. If a single origin is not correct, no branching analysis of ancient and modern population relationships can be valid. To analyze data with a branching analysis in a test of single origin is to assume one's conclusions.

WAS THERE A POPULATION BOTTLENECK?

A population bottleneck for the human species would imply a specific time and place of origin. Although it would not prove modern humans are a new species, evidence for a single significant bottleneck would provide powerful support for this interpretation because of the expectation that new species arise as small, isolated, peripheral populations. In fact, the only aspect of the Eve theory that does not fit a peripatric speciation model is its place—Africa is hardly a peripheral region (Thorne, Wolpoff, and Eckhardt 1993).

Some geneticists use an Eve theory to explain both why human mtDNA and some nuclear DNA variation is so small, and why there is more genetic variation in Africa than there is elsewhere. The two theories, Eve and multiregional evolution, have to supply alternative explanations of the low level of human mtDNA variation and the pattern of variation in certain other genetic systems. The question is important because a bottleneck would provide a place and a time, and with a small population, the expectation of an anatomical complex, that could delineate what it means to be human.

When there is only a small amount of variation in a segment of DNA, it may be that variation was lost in the region because of a recent bottleneck, and there hasn't been enough time for mutations to cause other variations. If past population size was moderate or large, even rare variations would have a good chance of being passed on, some possibly even increasing in frequency. Were this the case for all of humanity, common recent origin would be the only explanation for adaptively neutral genetic systems with low levels of variation. But there are other explanations. It is also possible that variations were only rarely established because of small population size, or recurrent bottlenecks, and that the region has been variable for a long time, but the variations were not passed on to subsequent generations. Thus, if there had been small populations, the rare variations would almost certainly have been lost by chance. For instance, if a family name is rare in a small population, only a few men will have it, and as it can only be passed from father to son, it stands a good chance of getting lost. A name found in an equally small proportion of families in a larger population persists because, with more men holding it, the chances of at least a few sons are much better.

According to the multiregional model a recent common origin for all living people is not the explanation for the pattern of genetic variation we see today. The model is consistent with an expectation of small population sizes, and recurrent bottlenecking in the past. It predicts that different genetic systems will have different histories (if there were a common recent origin for all humanity, each genetic system would have the same history). From the perspective of multiregional evolution, the key issue is whether internal subdivisions of our species, with their genic exchanges and differing population size histories, can account for the pattern in which both differences and similarities persist for long periods. Several genetic studies, each addressing a different aspect of the model, suggest that the answer is positive.

GENETIC SUPPORT FOR MULTIREGIONAL EVOLUTION

Templeton (1993, 1994a), analyzing mtDNA data, showed that the worldwide distribution of genetic lineages *could only be attributed* to early genetic divergence between regions and subsequent genetic contacts between populations. The magnitudes of gene exchanges were controlled by the distances between the populations. This is a direct confirmation of the isolation by distance model. Templeton's analysis also showed that the genic exchanges have continued for the entire time period marked by the evolution of human mtDNA variations from their last common ancestor. This was a direct confirmation of the model of population divergences and relationships posited by multiregional evolution, and a clear demonstration that mitochondrial lineage histories are not population histories.

Also accepting the contention that mitochondrial and populational histories are not the same, Cann concludes:

> The Mitochondrial Coalescent date of 200,000 years BP for the most ancient maternal genotype now found in modern humans does not imply that modern human morphology evolved at that time, nor does it imply that a population bottleneck necessarily took place. (Cann 1993:311)

Xiong Weijun and colleagues (1991) examine the distribution of several very rare alleles in nuclear DNA found in two individuals, one of Japanese and the other of Euro-Venezuelan origin. Assuming these rare genetic variations were derived from a single mutant ancestor, their complexity indicates a time of origin of at least a half million years, but their persistence since then at a very low frequency shows that there was no population bottleneck over that period—in other words, no single, more recent origin for both populations.

Li Wenhisung and L. Sadler (1991) compare nucleotide and protein diversity in humans and *Drosophila*. The levels of protein diversity are quite similar, but nucleotide diversity is much lower in humans. They attribute this difference to a small but stable population size through most of human prehistory, rather than to a bottleneck. Their reasoning stems from the fact that because humans and chimpanzees share many common alleles for the major histocompatibility complex genes, if there had been a severe bottleneck long after the human-chimpanzee split, most of these shared polymorphisms would have been lost. If there was no bottleneck, a long period of small population sizes would account for the human data.

Excoffier (1990; Excoffier and Langaney 1989) has examined the distribution of mtDNA variations and showed that the more ancient variations were so widely spread there must have been restricted genic exchanges from an early time. This supports the notion of long-lasting population contacts, and it means that there could not have been a recent, severe bottleneck.

Ayala (1985, 1995) shows that some human genes have structures so similar to chimpanzees that the human and chimpanzee forms must be descendants of little-changed genes in the common ancestor. They diverged, when the species separated, some six million years ago. For instance, humans and chimpanzees share many common genetic variations for the major histocompatibility complex genes, an important part of the immune system. Ayala calculates that in the time since the chimpanzee-human divergence, certain of them such as the human leukocyte antigen could not have passed through bottlenecks of less than 100,000 copies—not much of a bottleneck, and certainly not the kind envisioned by the Eve theorists.

Finally, Relethford and Harpending (1994) examine the consequences of the possibility that greater African population size, not greater time depth for modern humans in Africa, may account for the greater variation found in Africa. With more people, rare genetic variations have a better chance to survive and their coalescence times are older. But this does not mean that African populations are older. They argue further that a unique African ancestry implies a bottleneck for all genetic systems, since modern humans and all their unique genes would trace their ancestry to only a small portion of humanity, as it existed then.

While this seems at first glance a reasonable notion, it soon becomes apparent that the actual effect of such a (bottleneck) event depends on both the magnitude and the duration of a shift in population size. Rogers and Jorde (1995a) show that given reasonable parameters for our species, the bottleneck would have to be more severe and long-lasting than is considered plausible. We would have to think of a population of 50 females for 6,000 years, for example (Relethford and Harpending 1994:251). What is much more credible is that recurrent bottlenecks of less severity kept the magnitude of genetic variation low. The difference can be important because if there are continued population size fluctuations, no single size reduction need be dramatic for the loss of genetic variation to be substantial.

If these interpretations are correct, it means that while some genetic systems went through bottlenecks and coalesced between 100,000 and 200,000 years ago, *others did not.* This shows that all genetic systems do not have the same history, and therefore the history of individual genetic systems is not population history—remember, if there was a common recent origin for all populations because a new species appeared, all of the genetic systems should reflect this and *there could be none that did not pass through a bottleneck.* But no population bottleneck means no single origin for modern human populations.

If not a single origin, what do genetic data show? We believe that two conclusions are now possible: (1) for a long period of time, geographically dispersed human populations have been exchanging genes with each other, and (2) there has been no recent population bottleneck in humans—modern humans do not uniquely descend from a single recent source.

The observations of independent gene systems histories tell us something important about origins—not when, not where, but something quite different. The genetic research implies that there were more people in Africa, possibly more than everywhere else combined, for most of human evolution. Perhaps this was roughly three times that of any other area, as Relethford and Harpending suggest, perhaps more, but the relation of Africa, the center, with the more peripheral populations on other continents has turned out to be much as the center and edge hypothesis proposed (Thorne 1981). In developing the multiregional hypothesis (Wolpoff, Wu, and Thorne 1984) we proposed that variation would be higher in the center than at the edges for most of human evolution, and that the predominant direction of genic exchanges would be from the center outward because of the higher population density there. These predictions have been proven quite correct, certainly well beyond our expectations in 1984. In fact, taken as a whole, the aggregate genetic data are quite in agreement with the fossil evidence used to support multiregional evolution. It is as Stringer had hoped:

The marriage of genetics and paleontology which has recently occurred in the study of *Homo sapiens* origins may seem to some to have been an unhappy one— even a shotgun affair—but after a difficult honeymoon, I think this will eventually prove to be a long and very fruitful union. (Stringer 1994:169)

THE EARLIEST MODERNS?

Single origins models often focus solely on the question of where the earliest modern humans are found. Determining whether the earliest moderns are found in sub-Saharan Africa is beset by two issues—the very problematic dating of the two allegedly early sites with more complete specimens, and the question of what it actually means to be modern. The three early modern sites with cranial material said to be diagnostic are Border Cave, Omo Kibish, and Klasies River Mouth. The first two are very questionable, to say the least.

At Border Cave, well-preserved remains have been found that are very modern in appearance (Corruccini 1992; Rightmire 1984). However, despite the best efforts of archaeologists, they are unprovenienced (Rightmire 1979), and chemical analysis this decade suggests they are probably quite recent. The situation at OmoKibish also leaves a lot to be desired. The three Omo specimens were found at different locations and may not even be penecontemporary. They cannot be clearly related to the single date that was determined for Site 1. This locality is reported as being well below what is said to be the Omo 1 find spot, and the dating technique used was recognized as being unacceptable, even by the scientists who applied it in 1967 (Leakey et al. 1969). These problems cannot be resolved because the findspot can no longer be located. Therefore, to regard specimens from these sites as "dated," in any sense, is to make a travesty of the numerous serious attempts underway to determine dates for the human fossil record.

The Klasies sample is the exception to the provenience and dating problems surrounding most Late Pleistocene African material, but unfortunately, the specimens from this site are far more fragmentary and their morphological interpretation is more problematic. Their anatomy has often been interpreted in the context of the more complete specimens from the other sites, as Smith (1992b) notes. The critical parts not represented in Klasies fragments, after all, are there at Omo or Border Cave, so traces that might reflect modernity at Klasies are interpreted as actually indicating it because they are found elsewhere. But the fossils at Omo Kibish and Border Cave are not sufficiently provenienced for this to be valid, and the Klasies specimens must be analyzed in their own right.

The fossils from Klasies River Mouth are the best dated of those said to be diagnostic, and are therefore a critical sample in the evaluation of the Out of Africa model (Stringer and Bräuer 1994). Many of those working on these remains have claimed that the Klasies people were anatomically modern, although most who have studied the sample were Out of Africa supporters before they came to Capetown. These assessments were made on the basis of features considered diagnostic of modern humans. These are said to include cranial shape—the possession of a high, rounded forehead—as well as small brow ridges, generally small facial size and reduced muscularity, small teeth, and the presence of a chin. Indeed, the frontal fragment (KRM 16425) lacks supraorbital development, and the mandible from Cave 1b (KRM 41815) has a well-defined chin. Even the interpretations of these observations are controversial; as Smith (1992b) suggests, the frontal may be juvenile, and Mann believes the prominence of the chin may come from recession of the upper symphysis, as the specimen had lost its front teeth during life. Other mandibles with symphyses lack chin development, and the zygomatic bone (KRM 16641) suggests a large midface with archaic features (Frayer et al. 1994a). While some features of the fragments are within the modern range, many of the others are nothing like normal variants in more recent Africans, or for that matter recent samples from anywhere else. Because of this mixture of features (Caspari and Wolpoff 1996) the interpretation of Klasies brings us face to face with the problem of how to recognize the earliest moderns (Caspari and Wolpoff 1996).

We think it important to make systematic comparisons with Holocene South African skeletal remains. Features regarded as diagnostic of modern humans are best considered regionally, since geographic variation can confound assessments of temporal trends. For example, many of the features considered diagnostic of modern humans reflect increases in gracilization (reduced size and muscularity). Specimens from areas of the world which exhibit elevated levels of robustness, for instance the cranial robusticity of aboriginal Australians, will appear to be archaic, and like Middle Pleistocene humans, lack modernity, while those with more gracile regional features will appear more modern than their contemporaries elsewhere. But small-sized ancient populations with gracile features are not necessarily more modern than their larger contemporaries, any more so than present-day gracile populations are more modern than present-day robust ones. It is all too easy to confuse regional and temporal characteristics. When comparisons ignore these regional signals, they can produce bizarre results. For instance, the cheek bones of Neandertals are more gracile than some Eskimos, and their brow ridges are thinner and less developed than some recent Australians. Do we conclude from this that the Neandertals are modern humans?

The interpretations of Klasies have fallen victim to just this problem. Some of the modern features in the Klasies specimens are mainly a consequence of their size. For instance, some of the gnathic remains are small and gracile—especially the AA43 and ZZ44 maxillae and the KRM 14695 and 16424 mandibles. Others are much larger. Postcranial remains are diminutive as well, among the earliest in *Homo sapiens* to be Khoisan-sized (Rightmire and Deacon 1991). This contributes to the impression of modernity. Some of the more archaic specimens are large, and the size range could contribute to the range of anatomical variation.

However, the situation is more complex. For instance, cross-cutting the size differences, both the largest and smallest mandibles completely lack chins. In fact, of the four symphyses preserved, these two (KRM 13400 and 14695) lack even a mental trigone and a third (KRM 21776) has only a weakly developed trigone (similar Neandertals are called "chinless").

In our preliminary work (Caspari and Wolpoff 1990; Wolpoff and Caspari 1990b), we compared the Klasies sample to a Holocene sample from the same region. After all, Deacon (1992) had suggested, from his archaeological analysis, that the recent Khoisan populations from southernmost Africa were their direct descendants. His evidence includes data that suggest the coastal and nearby populations of southern Africa have remained localized, with very little migration into and out of the region until quite recently. Therefore our assessment of modernity used prehistoric material from archaeological sites on the southern Cape (Hausman 1982). Fortunately, many of the collections at the archaeology department had been radiometrically dated as part of a recently completed Ph.D. dissertation by Judy Sealy, who very kindly gave us her unpublished data, and using her dates, we were able to isolate an appropriate mixed sex sample. Most of the specimens have been radiocarbon dated to between 1,000 and 9,000 years ago.

Our comparisons gave mixed results. For instance, while the supraorbital region of the sole frontal lacks brow ridges and fell well within the modern range, facial size of archaic proportions is indicated by the distance between its eye sockets and verified by the size and shape of the isolated zygomatic bone. In fact, some measurements of these facial remains are as large or larger than most Middle Pleistocene Africans, more than twice as old. An isolated proximal ulna was first described as modern, based on size and robusticity (Rightmire and Deacon 1991); however, its small size created a gracile appearance, and in another study, the morphology of the ulna was described as archaic, with only some elements that appear in some modern samples (Churchill et al. 1996). While individual fragments fell into the modern range for anywhere between a few and all of the features that we analyzed, when compared *as a sample* to the prehistoric modern Cape inhabitants, it was clear to us that the Klasies sample *as a whole* was not modern.

Of course, a mixture of archaic and modern features is exactly what one would expect in a transitional sample. And shouldn't early modern humans be more like their archaic predecessors? These considerations might make good sense, *but only within an evolutionary model in which modernity appears gradually, with its elements slowly increasing in frequency.* This could be a description of the multiregional explanation for modernity. But in the single origin theories, Klasies is a little late for a transitional African sample; for instance, it is the same age as Qafzeh, which is a modern sample according to the single origin interpretation. And the single origin theories account for the predominance of the early modern Africans, as they dispersed, by their modernity. Modernity, in this account, is a package of very successful, interacting anatomical features and behaviors that only evolved once and only in one way. Certainly, features may continue to change over time as populations adapt to local circumstances, but these are minor in comparison. Therefore, in the framework of the single origin explanation of modern humans, a few modern features mixed with archaic ones are just not enough to define a modern *people.*

If the validity of "Out of Africa" rests on the contention that the earliest humans are found there, the theory is in as serious trouble paleontologically as it is genetically (S. B. Hedges et al. 1992; Maddison 1991; Relethford and Harpending 1994; Templeton 1993, 1994a).

MODERNITY ELSEWHERE

The putative modern features in the Klasies remains are continuous ones that have variable expressions in both modern and fossil humans; at any given time in the past there was a lot of variation all over the world, just like there is today. What happens when we look beyond southern Africa? The world at about 100,000 years ago included other variable transitional samples. In the Levant, the Skhūl and Qafzeh samples are dated to about the same age as Klasies and are thought by many to represent modern humans. Some specimens have high, rounded foreheads; reduced brow ridges; and more linear builds postcranially. However, other specimens appear more archaic, specifically Neandertal-like, in these elements. Like Klasies, there is a mix of archaic and modern features in a single sample, and this mix is found at the Skhūl Cave alone, according to McCown and Keith (1939).

Other specimens with modern features are found outside Africa at the same age as or earlier than Klasies, and not just in western Asia (which is often seen as an extension of Africa). In China, the Xujiayao sample is extremely variable (Wu and Wu 1985), and Jinniushan, thought at least by some workers to be about this age (Pope 1992a), has modern features that make it clearly transitional between earlier samples such as Zhoukoudian and recent or modern Chinese (Chen et al. 1994). Jinniushan is no less modern human than penecontemporary Levantine samples (Pope 1992b). It is the relatively complete cranium and partial skeleton of a woman who demonstrates the earliest appearance of a very thin, large, and globular cranial vault; a vertically short, broad face; sculpted zygomatic bones; and the maxillary notch—all aspects of modern craniofacial morphology. Moreover, the Late Pleistocene Australian cranium WLH 50 shows so many marked and detailed similarities to the Indonesians from Ngandong that this sample may be transitional as well. The Australian is, of course, modern human in spite of features that appear "archaic," because of where it was found (Davidson and Noble 1992). Modernity in Australasia must be treated regionally, just as it is in southern Africa. The Australian version involves long sloping foreheads and large brow ridges, along with the significant brain size expansions and other changes that occur in all areas. The criteria for modernity are different in different samples. The earliest Australians, while they don't look much like Africans or modern Europeans, crossed the ocean in sea-going bamboo vessels, clear behavioral evidence of modernity not associated with any of the modern-looking Middle Paleolithic specimens discussed above. The focus on West Asians and the Middle Stone Age Africans as the earliest moderns reflects the fact that they have some modern European features, but it ignores the early appearance of modern features that characterize other regions.

However we define modernity, it must be a populational distinction. It is misleading and incorrect to use a single feature, a particular piece of modernity, to justify describing the entire sample as modern, or to argue that the first appearance of a modern feature marks the origin of modern humanity. This latter would assume that there was a single origin for modern humanity—the very issue that at the moment is most contentious. Anyway, where would we start? Perhaps with the chin of the Swartkrans specimen SK 74 or the flexed cranial base of ER 17400 from the region east of Lake Turkana, two much earlier African australopithecine specimens. Perhaps with the sharply curved maxillary notch of ZKT 11, a Middle Pleistocene Chinese cranium from Zhoukoudian, or the combination of a large and thin vault for Jinniushan?

MULTIREGIONAL ANSWERS TO THE MODERNITY QUESTION

The biggest surprise of the past decade is that there is actually no indication of modern behavior in any of the early populations generally considered to be "anatomically modern." The idea that a change in behavior causes a response in biological evolution, as populations adapt to meet their new circumstances, is fundamental to evolutionary thinking, but it wasn't working. There is no archaeological evidence that any of the populations called early modern enjoyed the kinds of advantages in terms of behavioral capacities that have been suggested would characterize anatomically modern humans and account for their subsequent dramatic success if they had a single recent origin (G. A. Clark 1992; Lindly and Clark 1990a; Wolpoff et al. 1988). How does multiregional evolution address this situation?

The multiregional explanation focuses on the fact that "modern human" has proven to be an elusive and slippery term to define. There is no consensus on definitions of modernity, a point that many think is because such definitions are tied to views of variation and of species. But this problem lies at the heart of the entire controversy over modern human origins. We don't know what a modern human is, and it is not just because we hold different definitions. When definitions of modernity have been proposed, the inadvertent consequence has been that definitions successfully excluding archaic groups also exclude some members of contemporary populations, and this of course will not do. We must begin with the precept that all living humans are modern!

Few workers have actually attempted to define modern humans morphologically, and those who have ran into difficulty. Day and Stringer (1982, and see 1991) proposed an anatomical definition that could be used on fossil remains—a combination of features that were given specific metric or morphological descriptions, like high foreheads, rounded crania, reduced or absent brow ridges. They ran into difficulty probably because the major purpose of their definition was exclusionary; that is, these paleoanthropologists were mainly focused on finding criteria that segregated fossil Neandertals from modern humans. They didn't focus on unique characters all humans share, and they did not take into account the gamut of modern human variation to arrive at some criteria for the species as a whole. Their attempt, in other words, did not begin with the precept of what modern *is,* but rather what it *is not.* Their definition showed that European neandertals were definitely not modern, but in doing so provided a definition that also excluded substantial numbers of recent (Wolpoff 1986) and living (Brown 1990) aboriginal Australians. The definition, of course, was withdrawn, but now two problems remain. Not only do we lack an operational definition for diagnosing modernity in the fossil record, but we must question whether such a definition is even possible.

If modern humans are a new species, with a recent common ancestry, reproductively isolated from their contemporaries, definition of this group should be possible. However, it may not be possible if the multiregional model is correct, because regional evolution makes the range of modern variation for all features deemed taxonomically relevant actually encompass the way these traits are expressed by many fossils. For example, take the Klasies frontal, which lacks a brow ridge. This makes it seem modern because the ancestral condition in Africa is to have a large brow ridge, and this would be an important, early change to a condition that is common in humanity today. But many contemporaries of Klasies have large brow ridges, just as some contemporaries of *living* Africans have large brow ridges. Modern human populations range from including people with large brow ridges to including people with none at all. If we apply the principle that the present is a valid guide for interpreting the past, the criteria we apply to interpreting human variation today would be used to interpret human variation 100,000 years ago, and brow ridge reduction at Klasies should not indicate that Klasies is more modern than its contemporaries.

In fact we believe it is probably impossible to arrive at a definition that simultaneously includes the variation of all living people and excludes all members of archaic groups. We can think of modernity as a cone, with its base in the present encompassing everyone, and the more ancient time slices producing smaller and smaller samples of features that are modern (Wobst 1990). A population approach to understanding the place of modern features in archaic populations suggests that even as they appeared, and as they increased in number and frequency, modern features were only part of the normal variation of populations. Were the people who possessed them more modern than their siblings who did not? Of course not. Any meaning of moderns, therefore, must encompass many ancients, and make it seem as though for long periods of time archaic and modern people were coexisting not just on the same continent, or in the same region, but in many cases within the same family.

More important than, and critical to, the issue of how to define modernity, we think, is understanding the evolutionary processes that produced it. The origin of modernity is like the origin of the English language. It has many sources, it is not well-defined, and it differs from place to place. It was not the appearance of a set of anatomical details, but a process and a pattern of change, that brought modernity to humankind. Above, we liken the process to throwing stones into a small pond. The ripples from each strike interact with each other. It is not the modern features and behavioral innovations, the ripples, that create modernity, but rather the interference patterns created as they overlap that define modern humanity, a definition based on variations in different regions. Each region is different, and yet it is a singular process that unites them, diffused through the interconnections between the populations. Each set of ripples spreads over the entire pond and guides the evolution of the species, but the angles and intensities of the ripples differ from place to place, and so their interactions create different patterns.

Modern humans are not a "thing," because they cannot be both *universally* and *uniquely* defined. People with a claim to early modernity do not appear much earlier in one place than in another; equally good claims for modernity are based on different sets of features; and when they are part of a sample, there are also individuals with distinctly archaic characteristics, forms that can never be found today. Modern people with a specific, unique set of anatomies did not spread from a single place because of an event. But what of modernity itself? Modernity does not grade slowly and insensibly away as we look into the past, as the anatomical features associated with it in different regions seem to. If discoveries of the past decade have taught paleoanthropologists anything, it must surely be that anatomy and behavior are independent of each other in the Late Pleistocene, just as other anthropologists contend that race and culture are independent. Modernity cannot be based solely on anatomical features. While it lacks a definition, we believe that modernity does have a specific meaning, one that extends beyond the description of all living people, but this may be hard or impossible to extricate from the record of the past. The archaeological record does not always reflect people's behavioral capacities. However, this is the only source of real information about modernity, for we contend that "human populations are modern when they behave in recognizably modern ways, no matter what they look like" (Wolpoff and Caspari 1990a:395).

3

Systematics in Anthropology

Where Science Confronts the Humanities (and Consistently Loses)

JONATHAN MARKS

Perhaps it would be better for the zoological taxonomist to set apart the family Hominidae and to exclude its nomenclature and classification from his studies.

G. Simpson (1945:188)

INTRODUCTION: SYSTEMATICS

Systematics is the most fundamental of the life sciences, defined by Simpson (1961a:7) as "the scientific study of the kinds and diversity of organisms and of any and all relationships among them." In other words, before one can legitimately lay claim to studying organisms scientifically, one must first impose some manner of order on the specimens, and establish both the natural groups into which they fall, and the relationships among those groups.

This is a cognitive task and, specifically, a cultural one. As anthropologists have long been aware, the organization of nature into manageable and meaningful units can be accomplished in many ways, with considerable local variation and significance (Berlin et al. 1966; Diamond 1966; Douglas 1966). The act of organizing nature *scientifically* is thus itself a cultural act, with the purpose of replicating evolution.

When it comes to applying these scientific principles to humans and their closest relatives, however, it has also long been noted that the system doesn't quite work properly (Campbell 1962; Evans 1945; Simpson 1945, 1963). Implicit in this observation is the assumption that the system should ideally work precisely in the same way upon humans as it does "on, say, angleworms or dung beetles" (Simpson 1963:7).

This assumes, of course, that the classification of humans is as value-neutral as the classification of angleworms and dung-beetles. I want to begin by establishing that this is not so, it has never been so, and it probably can never be so. That assumption reflects a classically "modernist" view of science—that we

45

scientists are out there collecting data objectively, and allowing our hopefully unbiased minds to make sense of the information. The only possibly unbiased mind, however, is that of the newborn, and the mere choice of which data to collect is itself subjective. In human systematics the classifier and the classified are alike; the classical distinction between subject and object becomes indistinct. As a consequence, living in a matrix of power, social relations, and cultural values, human systematists must invariably carry greater baggage about their subjects than angleworm systematists. To document this, we need only begin at the beginning.

TWO CENTURIES OF HUMAN SYSTEMATICS

Human systematics begins, as all systematics does, with Linnaeus. Bernier, in an anonymous article in 1684, had suggested a fundamental biological division of humans into Europeans, Africans, Asians, Lapps, and Americans. Linnaeus, however, formalized the distinction among peoples zoologically. Credited with the intercalation of humans into the natural order he discerned, Linnaeus formalized as well the subdivisions within the human species. In the tenth edition of *System of Nature* (1758), which modern systematics takes as its starting point, Linnaeus set forth five subspecies within *Homo sapiens*—four corresponding to continental landmasses *(americanus, europaeus, asiaticus,* and *afer)* and a fifth, grab-bag subspecies, *monstrosus.*

The contrast Linnaeus establishes among the geographic subspecies is very structured. Each subspecies begins with three words, corresponding to color, personality, and behavior (Table 3.1). Linnaeus proceeds to a terse description of the hair and face. Then a three-word description of intellectual properties, a contrast of body decoration, and a contrast of legal systems follows (Table 3.2).

Where is the objectivity, the dispassion, that a student of animal forms is expected to bring to the study of human forms? Linnaeus sees human variation as any well-educated European bourgeois liberal of the eighteenth century would, with one exception: he writes with zoological authority. Linnaeus succeeds in enacting the folk wisdom of the era as mainstream academic biology.

The contrast Linnaeus establishes among the four geographical subspecies is additionally noteworthy because it is the only use of that category within his

Table 3.1. Initial Subspecies Descriptions by Linnaeus in 1758

Trait	Americanus	Europaeus	Asiaticus	Afer
Color	red	white	yellow	black
Personality	ill-tempered	serious	melancholy	impassive
Behavior	subjugated	strong	greedy	lazy

Table 3.2.   Additional Contrasts by Linnaeus (1758)

Trait	Americanus	Europaeus	Asiaticus	Afer
Intellect	obstinate, contented, free	active, very smart, inventive	severe, haughty, desirous	crafty, slow, foolish
Body covering	paints self with fine red lines	wears tight clothing	wears loose garments	anoints self with grease
Governed	custom	law	opinion	caprice

Order Primates. In other words, humans are different from the very outset. The only other mammalian species to be distinguished by subspecies in this work were domestic dogs and sheep, and neither of these utilized criteria similar to those of *Homo sapiens.*

We may now skip ahead two centuries, from the work that initially formalized the science of human classification to the work that ended it: Carleton Coon's *The Origin of Races* (1962). In between, of course, were many diverse scholarly works on the subject, from Blumenbach's craniological (1775) through Boyd's sero-genetic (1950), but more significantly there was also a rising chorus of scholarly opposition to the subspecific classification of humans as reflecting any biological objectivity at all (Herskovits 1924; Livingstone 1962; Montagu 1941).

The Origin of Races was titled somewhat pretentiously to invoke Darwin and was intentionally relevant to issues of integration and civil rights which were hot at the time. The book made four major claims: first, that the human species was divisible into five fundamental constituent subspecies or races; second, that these units had a deep presence in prehistory, being discernible as equivalent subspecies of *Homo erectus;* third, that they evolved into *Homo sapiens* at different times; and fourth, that the economic and political dominance of contemporary Western Europeans and their descendants (and secondarily, East Asians and *their* descendants) was simply a consequence of their longevity as members of the species. The flip side, of course, was that the oppression of the dark-skinned peoples of the world was a consequence of their ephemerality as members of *Homo sapiens.* In other words, in Coon's scheme there was in essence a biological and naturalistic explanation for what appeared to be the facts of social and political history.

Interestingly, this was a very recent interpretation on Coon's part. Only eight years earlier, in *The Story of Man,* Coon had divided the world up into *six* basic races, did not make the order of their evolution significant, and in fact had them evolving in a *different* order. He casually suggested that sub-Saharan Africans were actually *older* than East Asians: "The Mongoloids are probably not as ancient as the Negroids, for they developed in extreme cold" (Coon 1954:198). This, of course, would contradict the interpretations in *The Origin of Races,* and the subsequent edition of *The Story of Man* was explicitly rewritten to bring it into harmony with the later book. That statement does not in fact appear in the second edition.

Whatever Coon's own social and political beliefs may have been, his scientific work gave succor to segregationists. In particular, businessman Carleton Putnam's *Race and Reality* (1967) not only invoked Coon's theories as a scientific basis for his own racist views, but reiterated as well his own thought that the opposing position of egalitarianism was the result of a conspiracy of communists and Jewish anthropologists. When challenged to repudiate Putnam's work by the geneticist Theodosius Dobzhansky (1968), Coon refused. Indeed, he steadfastly maintained the value-neutrality of his work (Coon 1968). But if he indeed saw Putnam's work as perverting his own, it is difficult to imagine why he would not have sought to distance himself from it, or denounce it. Alternatively, if he indeed appreciated Putnam's work as implicit in his own scientific writings, Coon's insistence on its value-neutrality must take on a disingenuously hollow tone. In any event, both Putnam on the political right and Dobzhansky on the political left recognized what Coon denied: that the scientific study of human systematics was not objective at all. It encoded cultural or personal values, which in turn could carry considerable social power as the validation of policy.

Where then is there evidence that human systematics could, even in principle, be carried out as one might carry out dung beetle systematics? With all due respect to the students of dung beetles, nothing of significance rests on the outcome of that work. *Of course* it can be done with dispassion. Where, on the other hand, complex social relations can be justified or refuted on the pronouncements of a scientist, the work itself is subject to deeper scrutiny, and to other pressures—and more important, the scientist takes on deeper responsibilities in presenting the work. It was true at the beginning of human systematics, it was true at the end of human systematics, and it is quite simply intrinsic to the system.

HUMAN RIGHTS FOR APES

It is not simply the systematics of humans themselves that is inextricably bound to social and political issues. *The Great Ape Project* (Cavalieri and Singer 1994) argues that because of the close genetic relationship between humans and the great apes, we should accord human rights to the great apes.

On the one hand there is a significant problem that requires attention: apes are threatened in the wild by the economic development of the human societies around them. They are often objectified in captivity by callous and cynical entrepreneurs, who neither regard them nor treat them as the sentient, emotionally complex creatures they are. They are generally disposed of when they lose their cuteness, usually less than 20% of the way through their life span. The lucky ones can live out their lives in the care of an enlightened or sympathetic zoo or primate research facility with sensitive caretakers and handlers. Most, of course, are not lucky. On the other hand, granting them human rights is a curious way to redress the situation. Two things seem to stand in its way: they are not human, and we can't even guarantee human rights to humans.

The authors overcome the primary obstacle with the argument that, genetically, the great apes are *nearly* human. This is the genetic basis for the argument of extending human rights outside the human species, and it is expressed most artfully in the essay by Jared Diamond (1994). Says Richard Dawkins (1994:82) baldly, "we *are* great apes."

But are we? After all, great apes have long arms, short legs, large canine teeth, copious body hair, prehensile feet, small brains, high degrees of sexual dimorphism, and a large suite of other anatomical features that are simply not found in humans. Dawkins's meaning lies in a significant word he has omitted: the word "genetically."

The observation that "we are apes" reflects what may be called The Great Overstatement of Molecular Anthropology, first articulated by the chemist Emile Zuckerkandl in 1963. Studying the blood molecules of humans and apes, Zuckerkandl noted that "from the point of view of hemoglobin structure, it appears that gorilla is just an abnormal human, or man an abnormal gorilla, and the two species form actually one continuous population" (Zuckerkandl 1963:247).

G. G. Simpson's response (1964:1536) was blunt and brutal:

> From any point of view other than that properly specified, that is of course nonsense. What the comparison seems really to indicate is that in this case, at least, hemoglobin is a bad choice and has nothing to tell us about affinities, or indeed tells us a lie. (It does show that men and gorillas are rather closely related, but that has long and more accurately been known from traditional morphological comparisons.)

In other words, one can say with some legitimacy that we are *genetically* apes, but the qualifier is crucial if we are decidedly distinguishable from apes anatomically, behaviorally, mentally, ecologically, and demographically—in fact, basically any other way in which specimens can be compared biologically.

Dawkins has taken the true statement "we are genetically apes" and dropped the significant qualifier. The implication is manifestly that the genetic comparison is all that matters, for it supersedes all other comparisons. That is an interesting philosophical position—reductionist and hereditarian—but fails to come to grips with the general biology of the organisms. In fact, we can express that biology, rephrasing The Great Overstatement as *The Central Paradox* of Molecular Anthropology: that *in spite of* the striking genetic similarity of humans and apes, we are actually quite different from apes. We are *bipedal, technology reliant, articulate, small-canined, slow-maturing, naked* apes—which renders

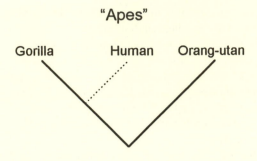

Figure 3.1. The category "ape" is paraphyletic.

the term "apes" largely meaningless in that context, and renders the genetic similarity exceptional and paradoxical.

Actually, Dawkins is exploiting the genetic knowledge to underscore a taxonomic point (and in turn using the taxonomic point to advocate a political stance). Genetics shows that regardless of how different we are from them, the apes are our closest relatives—so close indeed, we fall within the group that is constituted by the word "apes." Since human and gorilla are more closely related than gorilla and orangutan, the category "apes" is artificial if it includes gorilla and orangutan, but not human (Figure 3.1).

Technically the category is *paraphyletic,* for it subsumes descendants from a single common ancestor but omits one descendant from that common ancestor—humans. Why? Because humans have diverged radically in a short period of time and have therefore become different from the "other" apes.

Does this, then, mean that we are apes? Only if such paraphyletic categories are generally invalid and unique to this particular situation.[1] In fact, such categories are far from unique; they are ubiquitous, appearing every time a restricted group of organisms evolves a new fundamental adaptation and diverges from its close relatives. And far from being generally invalid, they are a central part of the conceptual apparatus of mainstream evolutionary systematics. Consider the turtle, the crocodile, and the pigeon: two reptiles and a bird (Figure 3.2). Although birds have diverged radically from their reptilian ancestry, they are more closely related to the crocodile than the crocodile is to the turtle. Thus the category "reptile" is paraphyletic and phylogenetically artificial to the extent that it fails to include the birds—perfectly analogous to the situation of apes vis-à-vis humans.

Nevertheless, to call pigeons "reptiles" is considerably less than a profound revelation about their nature and biology; it is simply a phylogenetic datum. Pigeons are birds, birds have divergent specializations, and "reptiles" is a paraphyletic category.

Or consider the relationships among a tuna, coelacanth, and dog (Figure 3.3). On account of its appendages, the lobe-finned coelacanth stands in a special

"Reptiles"

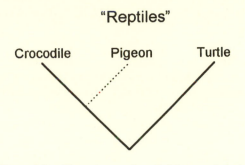

Figure 3.2. The category "reptile" is paraphyletic.

phylogenetic relationship to the tetrapods. Thus the coelacanth and dog are more closely related to each other than the coelacanth and tuna. The category "fish" is paraphyletic, because the tetrapods diverged radically from their ancestors (Forey 1991; Gardiner et al. 1979). Do we then proclaim that dogs are fish? Presumably not; there would be little value in ignoring the evolutionary changes that make tetrapods interesting as a biological group.

It is no more a profound revelation about human systematics to say that we are apes than it is to say that pigeons are reptiles and dogs are fish. It is, rather, a trivial deduction based on the nature of the "primitive" or "ancestral" paraphyletic categories we recognize their having evolved away from—the category of reptiles (lacking the specializations of birds), or fish (lacking those of tetrapods), or apes (lacking those of humans). Our apeness is real enough, but it requires a very narrow frame of reference to accept without qualification. Further, in the case of humans it is serving an overtly political purpose: the promotion of great ape rights. Thus, once again human systematics appears to be qualitatively different from general zoological systematics, for it is manipulated to validate social agendas.

"Fish"

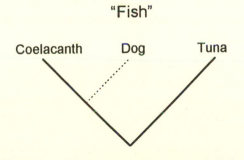

Figure 3.3. The category "fish" is paraphyletic.

PARAPHYLY AND HUMAN SUBSPECIES

Certainly the oldest research question in physical anthropology is the one that inquires about the number of fundamental subdivisions of the human species, and the relationships among them. This, we have noted, stretches in the modern scientific literature back to Linnaeus. In the past half-century, that question has generally been recognized to embody a series of false assumptions about the nature and pattern of human biological diversity. First, that human populations are divisible into a relatively small number of relatively discretely different types. Second, that there is greater homogeneity within each category than there is heterogeneity among them. Third, that such categories are roughly equivalent biological units. Fourth, that there is some significant aspect of human evolutionary history to be encapsulated by such division. And fifth, that such a division would be fundamentally biological in nature.

In fact all of these assumptions are wrong. The first is refuted by the observation of the fundamentally clinal nature of human variation (Buffon 1749; Hierneaux 1964; Livingstone 1962). The second was refuted by Lewontin's (1972) classic quantitative analysis of the genetic diversity within and between races. Hulse (1955) raised the issue of large demographic expansions of local populations clouding our notions of what the fundamental divisions of the species might be. Further, the inference of three races of the Old World certainly owes more to the biblical, pre-Darwinian conception of human biogeography involving Noah's three sons (Ham, Shem, and Japheth) colonizing the continents than it does to modern theory. And finally, the social and cultural nature of those divisions has long been noted by anthropologists: at their boundaries by the "one drop of blood" rule (Wright 1994) and more generally by the subjective choice of diagnostic traits and criteria of inclusion and distinction (Little 1961; Montagu 1964).

Nevertheless, in the cognate field of population genetics, one still occasionally encounters the question of the branching order among the "three races" (e.g., Cavalli-Sforza et al. 1988; Nei and Roychoudhury 1974). Cavalli-Sforza argues that the basic division among human populations in Old World is east-west, i.e., blacks and whites are more closely related to one another than either is to yellows. Nei argues the opposite, that the basic division is north-south, with whites and yellows being most closely related.

Again, however, a consideration of the nature of paraphyletic groups shows that the question as posed is a false one. The most popular model for the origins of modern humans, the Out of Africa model, directly implies that African populations are paraphyletic, and thus not strictly an evolutionary category strictly comparable to Europeans and Asians. Consider a simple historical model for human origins (Figure 3.4).

If the founders of the European population came from one part of Africa, then the remaining populations are no longer each other's closest relatives. They are

Figure 3.4. Left, a diverse set of populations in Africa. Center, one population divides and some members emigrate. Right, regardless of future expansions, the emigrants are phylogenetically most closely related to a specific African population; the others constitute a paraphyletic assemblage.

paraphyletic, for the founding African population would be more closely related to the descendant Europeans than to other African populations. Thus, to group *all* African populations together in studying their relationships to Europeans and Asians is to miss what might be the most salient feature of the biological history of the African populations. Indeed it represents quite literally the construction of a racial category that may not even exist in nature (Figure 3.5).

Further, this seems indeed to be the pattern we encounter genetically. The mitochondrial DNA analyses identify more extensive diversity among Africans, and find that diversity to subsume the genetic diversity of Europeans and Asians (Cann et al. 1987; Horai and Hayasaka 1990; Merriwether et al. 1991; for nuclear DNA, see Bowcock et al. 1994). If Africa is indeed the source of the founding gene pools of Europe and Asia, and its own human populations are consequently paraphyletic, it follows that attempting to calculate the phylogenetic relationships among those entities is impossible, for one of those categories is considerably

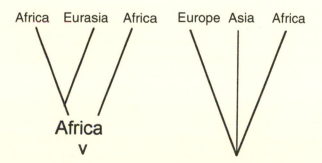

Figure 3.5. If all humans are a subset of Africans (left), then artifically dividing the world into continental groups constructs racial categories (right), rather than identifying them in nature.

differently constituted than are the other two—indeed, it subsumes them (cf. Gould 1995).

Once again, then, we see that lying just beneath the surface of a seemingly innocent zoological question about the human species are culturally loaded assumptions about the structure of human diversity. Is the study of human differences like the study of clam or fruitfly differences? No; it is much more difficult. It requires far deeper levels of introspection, for the implications and possible consequences for human lives and welfares are far more profound. Should human systematics take clam or fruitfly systematics as its model, as is often assumed? Again, I argue no—the problems faced in studying human systematics scientifically are different, and the responsibilities incurred are deeper. The basic principles are the same, but beyond that the similarities are illusory.

NAMING NEANDERTHALS

Neanderthals are diagnosably different from modern humans, and outside the range of variation detectable in living *Homo sapiens*. Should they be recognized at the species level? The standard interpretation prior to World War II was that neanderthals indeed represented a different species from ourselves, *Homo neanderthalensis* (Boule and Vallois 1957; Howells 1942; King 1864; Lull 1922; MacCurdy 1932). An alternative view was articulated by Weidenreich (1943b) and Dobzhansky (1944): that neanderthals were not a distinct species at all, but were properly recognized taxonomically as a subspecies of *Homo sapiens*.

The difficulties involved in assigning species status to fossil populations are well-known (Godfrey and Marks 1991; Kimbel and Rak 1993; Simpson 1943). In this case, the systematic decision was based upon McCown and Keith's (1939) interpretation of the Mount Carmel fossils as recent. Contemporaneous yet morphologically distinct, the fossils from Skhūl and Tabūn nevertheless also displayed a range of characteristics that appeared to be morphologically intermediate between neanderthals and modern humans. They consequently offered evidence of gene flow between neanderthal and human populations, which in turn implied the absence of a species boundary separating them.

This interpretation caught on widely in the postwar years, though the interpretation of the fossils is probably inaccurate. But it carried an added bonus as well. After all, there were still holdouts in the 1940s against the view that even living humans constituted a single species (Gates 1944, 1948; Osman Hill 1940). But if neanderthals are distinct at the subspecies level, and fall outside the range of modern human variation, and there are no valid taxonomic categories below the subspecies, then it follows that modern human "racial" variation can no longer be recognized taxonomically. In other words, acknowledging the neanderthals as a subspecies renders moot the question of how to recognize racial variation in a formal, scientific, taxonomic sense. The enterprise of racial taxonomy from Linnaeus on down becomes a technical artifact.

This turns out to be an extraordinarily beneficial consequence of hominid systematics. After all, the major issues of our lives are *not* wrapped up in what to call neanderthals; they are, rather, how to function in a biologically diverse democratic society, with uneven distributions of wealth and power. One of the generalities we encounter in a scientific society with such inequities is a strong tendency to explain the oppression of certain groups as being derived from nature, rather than from such human agencies as greed and malice. Where there is blame to be cast for injustice, invoking human agencies serves to blame the socioeconomically dominant classes, whereas invoking nature serves to blame the poor and oppressed for their own lot. This, of course, conjures the central arguments of such widely read and widely publicized tracts as Madison Grant's *The Passing of the Great Race* (1916), Carleton Coon's *The Origin of Races* (1962), and Richard Herrnstein and Charles Murray's *The Bell Curve* (1994).

Hominid taxonomy has something important to contribute to this social dialogue. If there are no formal taxonomic categories by which to recognize racial diversity, that helps to undermine any claim of profound constitutional differences separating these designated groups. Without available taxonomic categories, then, any attempt to subdivide the human population formally is not zoologically valid. Such racial classifications may still have cultural meaning, as classifications invariably do, but that meaning would not be derived from, or legitimized by, biological science.

In a society where there is strong and constant pressure to essentialize the differences among coexisting peoples, we are thus well served by *Homo sapiens neanderthalensis*—in ways that transcend the arcane science of paleoanthropology. Consequently, when we encounter the suggestion to re-elevate neanderthals to the species level (Rak 1993; Stringer 1993b; Tattersall 1986; Zollikofer et al. 1995), it is fair to ask what the benefits and costs of such a decision might be.

Clearly this would not be a value-neutral decision, for it would entail reopening the issue of the subspecific classification of *Homo sapiens.* So the salient question is: Does whatever *benefit* that might accrue to raising neanderthals to the species level outweigh the *cost* associated with re-essentializing the differences among living humans that would accompany it—indeed, that would be logically implied? Do we have any effective means of discouraging amateurs, or scientists in cognate fields, from pronouncing on the differences between taxa designated as *Homo sapiens europaeus* and *Homo sapiens afer?* Or debating whether the Polynesians, or the Khoisan, or the Indo-Dravidians, are entitled to their own taxonomic status?

This is by no means a remote scenario surrounding a dead issue. As the publicity surrounding *The Bell Curve* shows, Americans are exceedingly willing to acknowledge the existence of significant constitutional differences among their populations. And even the counterarguments are muddled: *Time Magazine* used a genetic compendium by Cavalli-Sforza et al. (1994) to bash *The Bell Curve* (Subramanian 1995). However, accompanying the story (subtitled "A landmark global study flattens *The Bell Curve* . . . ") was a computer-generated summary figure of genetic diversity in the human species, in which each of the four "ethnic regions" of the world was assigned a different color: "Africans"—yellow, "Caucasoids"—green, "Mongoloids"—blue, and "Australians"—red (cf. Cavalli-Sforza et al. 1994, color plate 1). As if the problem were semantic—calling them "ethnic regions" rather than "races" or "subspecies"—the technologically modern genetic study is a product of classic pre-modern anthropology, in which the human species is divisible into a small number of essentially different types of people.

Or from a recent paper by the geneticists Nei and Roychoudhury:

> Human populations can be subdivided into five major groups: (A) negroid (Africans), (B) caucasoid (Europeans and their related populations), (C) mongoloid (East Asians and Pacific Islanders), (D) Amerindian (including Eskimos), and (E) australoid (Australians and Papuans). (There are intermediate populations, which are apparently products of gene admixture of these major groups, but they are ignored here.) (1993:936–937)

Since the authors did not sample populations from geographically intermediate localities, such as eastern Europe, East Africa, or northwest Asia, it is not surprising that they found distinct geographical clusters. Racial clusters are loaded into the study by virtue of the sampling strategy (Bowcock et al. 1994). And the theory of genetic intermediacy coming simply as a result of admixture from pure races is archaic and false, a classically "folk" theory of human biohistory (Boas 1924; Hooton 1936; Shapiro 1961). The opportunity to reintroduce taxonomic formalism here could be so counterproductive to the progress anthropology has made in the past half-century as to merit long, sober reflection. To suggest that it is not an issue is to turn a blind eye to the history of the discipline, and to its ultimate significance.

But not only is the decision not a value-neutral one, it is also a largely arbitrary and nonobjective one. The boundary between paleo-species and paleo-subspecies is not nearly as clear as it can be neontologically. We certainly cannot distinguish the nature or extent of their interbreeding with modern humans, nor infer any differences between their mate recognition systems and ours. We cannot tell the extent to which their subtle, though diagnostic, anatomical differences from modern humans were due to life style and environment rather than to their gene pool. Thus the taxon *Homo neanderthalensis* must be based on judgment calls, the impressions of experts, and not on any objective test or experiment. Is it not fair to ask that the experts weigh as well the social implications of the judgments they render? Time was that a scientist could pretend to be thoroughly abstracted from society and social forces and prejudices, and could render scientific pronouncements as if they were indeed value-neutral and objective. In the wake of social Darwinism (Hofstadter 1944), eugenics (Kevles 1985), Rassenhygiene (Proctor 1988), and racial ranking (Tucker 1994), it is difficult to maintain this illusion into the twenty-first century.

Those times are gone, and we are now entering a postmodern era of historically and socially informed science, in which the authority of science is continually questioned, and in which scientists are held responsible for their words and for the implications of their words. Thus, we need to appreciate that *human* science is *humanistic* science, and its consequences far outreach the classroom. We need to be far more introspective, and far more diligent, about making and accepting scientific pronouncements about humans that are at best arbitrary decisions, and that have consequences stretching far beyond idle pedantry. I am of the opinion that the projectable costs of elevating neanderthals to the species level far outweigh its dubious benefits. It is not "just" an abstract scientific question: it is a fundamentally *unanswerable* scientific question, *any* answer to which has exceedingly non-scientific implications. Being a para-scientific issue in the first place, its resolution is largely if not entirely subjective. Therefore, all the possible implications of the decision should be weighed before rendering it.

CONCLUSIONS

Anthropology straddles the boundary between the humanities and the sciences, and intentionally partakes of both intellectual spheres. To the extent that there is a broadly perceived fragmentation occurring in anthropology, it is largely the result of the adoption of one or the other identity, instead of both. Physical anthropology, specifically, has traditionally imagined itself as strictly scientific, although every now and again we are reminded of the humanistic nature of the enterprise (Cartmill 1994; Landau 1991). Even so, this recognition is generally limited to paleoanthropology, and traditionally less introspective fields like anthropological genetics have only begun to appreciate the "anthropological" nature of their enterprise (Marks 1995).

The call to model anthropological systematics on zoological systematics, where the subjects and the objects of classification are very different from one another, is a trope, a literary device by which scientists can readily play upon popular insecurities to argue for ideological goals of various sorts.

> Let us imagine ourselves scientific Saturnians, if you will, fairly acquainted with such animals as now inhabit the earth, and employed in discussing the relations they bear to a new and singular "erect and featherless biped," which some enterprising traveler, overcoming the difficulties of space and gravitation, has brought from that distant planet for our inspection, well preserved, may be, in a cask of rum. . . . [W]e should undoubtedly place the newly discovered tellurian genus with [the apes]. (Huxley 1863[1959]:85–86)

> If an unbiased zoölogist were to descend upon the earth from Mars and study the races of man with the same impartiality as the races of fishes, birds, and mammals, he would undoubtedly divide the existing races of man into several genera and into a very large number of species and subspecies. (Osborn 1926:3)

> A zoologist from Outer Space would immediately classify us as just a third species of chimpanzee, along with the pygmy chimp of Zaire and the common chimp of the rest of tropical Africa. (Diamond 1992:2)

The merits of these political goals vary considerably, and the extra-terrestrial systematist calls in from different parts of the solar system each time, but the point is held in common: To see things clearly, we human systematists must ourselves become nonhuman (Landau 1991). But we are not nonhuman, we cannot be nonhuman, and we cannot know how a nonhuman would classify us or anything else. Further, the contention that a nonexistent classification by aliens is both knowable and intellectually superior is in no way validated by any manner of scientific knowledge, nor within any modern framework of scientific reasoning. What a grand paradox—the recruitment of science-fiction prejudices to make an ostensibly scientific point! If anything, however, it is refuted by the anthropological knowledge that another species of classifiers would probably utilize other criteria than those of modern Euro-American biology. Rather, science dictates that—as humans classifying humans—the best we can do is to acknowledge that classification is a fundamentally human procedure, to analyze *how* humans classify things, *why* humans classify things, to balance the merits of different classifications, and then to choose.

The science of human origins is the study of our intimate biological history. Yet history is political. History confers identity. And to the extent that physical anthropology represents the scientific study of human biological history, it must be appreciated to represent either a humanistic science or a scientific humanities. At the very heart of anthropology is the tension of a familiar symbolic dichotomy, "nature/culture." The mediation of this dichotomy not only helps to define an individual's place in a cultural universe of self-knowledge, but helps to define as well anthropology's place in a formal scholarly universe of self-knowledge. The acts of aggregating, dividing, and naming are cultural acts, even when they are performed by scientists on the products of biological evolution. As anthropologists, we are obliged to recognize the social power of these acts as we carry them out, and to insure that this power is applied in its most benign—if not beneficial—manner. Of course, we must do this without seriously compromising our scientific description of the natural world. To that extent, the tension between nature and culture must be negotiated within physical anthropology. One thing is clear: The only certain path to scientific error, as the lessons of history show, is to proceed as if no negotiation were necessary, as if nature were "out there" alone, to be recorded by the empty mind of the neutral observer, and then to speak for itself.

As anthropologists we know that is impossible. As physical anthropologists, we can and should use our knowledge of the cultural nature of science to enrich our scientific judgments. Seen in retrospect, the major problems in physical anthropology, indeed, have generally arisen by virtue of the self-delusion involved in denying the fundamental duality of the field. Culture is an organizing principle of nature, and it is intrinsic to the scientific anthropological endeavor— it can neither be ignored, nor peeled away like the skin of an onion. The basic difference between fruitfly systematics and human systematics is not that there is no culture in the former, but simply that it stands out less, and matters less. And ultimately it is, I think, a fully reflexive recognition and appreciation of the nature/culture duality within the *study* of human origins itself that will guide the next generation to a clearer comprehension of the scientific issues involved, and to their optimal resolution.

NOTE

1. A goal of one of the various schools of thought known as "phylogenetic systematics" or "cladistics" is the abolition of paraphyletic taxonomic terms, such as "fish," "reptiles," and "apes" (de Queiroz and Gauthier 1992; Hennig 1965; Wiley 1981). While attractive for its philosophical consistency, this would deliberately ignore the emergence of adaptive divergences in certain groups of organisms and is thus biologically of questionable value (Martin 1990; Mayr 1974).

4

Through a Glass Darkly

Conceptual Issues in Modern Human Origins Research

G. A. CLARK

In keeping with the intent of this book, and insofar as I can make them explicit, I provide here a personal account of the preconceptions, biases, and assumptions that underlie my approach to modern human origins (MHO) research. At stake in the MHO debate is an elaborate patchwork of ideas, theories, facts, and observations that make up the contemporary scientific view of the origins of our species. If anything is evident from the debate, however, it is that "the contemporary scientific view" of our origins is in no sense a unitary phenomenon, but instead a kaleidoscope of different colors and prisms that changes each time it is twisted. I take it as indisputable that human origins researchers pertain to different intellectual traditions that emphasize different sets of facts and theories differentially, and I argue here that it is this largely implicit selection process that shapes research protocols and conclusions for the various kinds of researchers involved in the debate.

Within a particular research tradition, current theory is perceived to do a more or less adequate job of explaining a multitude of fundamental research questions but, insofar as these theories are linked together in complicated ways that are seldom subjected to critical scrutiny, these explanatory models are little more than a house of cards—remove one card, such as the one on which the chronological framework is written, and the whole structure of inference is threatened with collapse. If these are acknowledged to be problematic issues within a single research tradition, imagine how complex the problem of inference becomes when it is recognized that MHO research comprises several research traditions and has, in addition, three major foci (archaeology, human paleontology, molecular biology) that crosscut these traditions and that are emphasized differentially in each of them. I am convinced that the broad parameters of the intellectual traditions to which MHO researchers pertain are identifiable, at least in general terms. To make them explicit (or more explicit) is the intent of this chapter. Others, who consider that MHO research comprises the work of a collectivity of individual scholars, each with a unique perspective, do not agree that it is possible to do this (e.g., Knüsel 1992).

Like all the contributors to this book, my approach to research is ultimately determined by my formal training in the intellectual tradition to which I pertain (that of anglophone, New World, anthropological archaeology), combined with the general kinds of problems and questions with which I have been concerned throughout my career (biological and cultural aspects of human adaptation in "deep time") and the necessity for modifying these intellectual components to satisfy the requirements of actual data sets obtained through field work. I think field work is important. No matter how compelling a theory might be, it must eventually be subjected to some kind of empirical test if it is to be accorded credibility in an explanatory framework. My approach to research has been accretional, partial at best, and marked by many false starts and wrong turns. In no sense a fully axiomatized, integrated philosophy, it is instead a cumulative, somewhat haphazard product of a twenty-year-long evolutionary process.

This chapter consists of two parts, and it proceeds from the general to the specific. Since I am of the opinion that MHO research is, or should be, a sciencelike endeavor, the first section presents my views of MHO research as I believe it fits into the context of western science. In light of a near-consensus among the contributors, it might seem unnecessary to do this, but the current assault on science as a paradigm for reality seems to require an affirmation of my conviction that a sciencelike view of the world is the only rational approach to understanding questions of long-term human biological and cultural evolution. The concept of the metaphysical paradigm is central here, and it is used to address what I take to be the crucial epistemological questions involved in MHO research—essentially *how* we know what we think we know about the past. In the second section, I tackle the three major MHO research foci—archaeology, human paleontology, and molecular biology—and give my construal of the status of the debate in each of them. Although I intend to make this as personal an account as I can, it was not always possible to avoid confronting the ruling theoretical and methodological biases which I take to be operative in the various research domains. By confronting the preconceptions of the dominant paradigm, however, I was also forced to confront my own.

THE PARADIGM CONCEPT: A TOOL FOR ANALYZING RESEARCH TRADITIONS

Scientists have been trying to arrive at a consensus about modern human origins for more than a century. Why haven't they been successful? In my opinion, it is because MHO researchers pertain to different intellectual traditions that proceed from different assumptions about what the remote human past was like. These traditions are based, in the final analysis, on the concept of the metaphysical paradigm—a collection of biases, preconceptions, and assumptions about the nature of our knowledge of the world of experience.

We owe the paradigm concept to the science historian and philosopher Thomas Kuhn (1962a, 1974, 1977). For him, the term means a prototypical mode of problem solving that implicitly defined for scientists how they must "see the world." He believes that scientists construct complex theoretical and meth-odological systems based always on a particular paradigm, but these inferential systems are seldom, if ever, made explicit. They are, in the final analysis, largely subjective. That does not mean, however, that they cannot be subjected to critical scrutiny. Paradigms act as a filter or a lens through which scientists determine which questions are important to ask, and how they should go about answering them. Once "indoctrinated" in the paradigm (and this occurs through the formal process of education), scientists typically dedicate themselves to the resolution of problems whose solutions tend to reinforce and amplify the credibility of the paradigm, rather than to question its validity. According to Kuhn, science in general proceeds without much explicit concern for the preconceptions and bi-ases that are the foundations of its knowledge claims. This all works fairly well when a single intellectual tradition, based upon a single metaphysical paradigm, is involved in the investigation of a single research domain. It doesn't work nearly so well when multiple intellectual traditions, based on multiple metaphysi-cal paradigms, are involved in the investigation of a single research domain. It doesn't work at all when research domains themselves are multiple, but overlap-ping, and where findings in one field have implications for the research protocols of another, as is the case with MHO research.

Despite their admittedly subjective nature, any metaphysical paradigm can be evaluated, or at least characterized, in terms of three criteria, which are perhaps best expressed as questions. Physicist and science philosopher Egon Guba (1990) argues that the first of these questions is concerned with *ontology:* according to the tenets of the paradigm, what is the nature of the "knowable"? What is the nature of the "reality" that is perceived to be the target of the research? The second criterion is its *epistemology:* what is the nature of the relationship between the "knower" and the "knowable"? The third is its *methodology:* how should the investigator go about the process of investigation? One could argue that the answers to these questions outline the basic belief systems according to which the logic of inquiry proceeds. It should be emphasized that these belief systems, while foundational, cannot be proven or disproven. They are simply rational assertions about how we might go about investigating the world of experience.

PARADIGMS IN WESTERN SCIENCE

In the past three centuries, essentially five metaphysical paradigms have emerged in western science, although the first, strict empiricism, is not usually considered part of modern conceptions of science. It is, however (and most unfortunately), very prominent in the history of two of the three branches of MHO research. These paradigms are, in a rough chronological order, strict empiricism, classic positivism, post-positivism, critical theory, and constructivism. I will briefly outline the basic tenets of the first three; the last two, while important in certain branches of social science after about 1930, are irrelevant for the purposes of this chapter (for extended treatments of these issues see Bernstein 1983; Casti 1989; Clark 1993a; Guba, ed. 1990).

Strict Empiricism

Strict empiricism proceeds from a realist ontology in the sense that it is held that an objective reality exists "out there," independent of our perceptions of it. Epistemologically, strict empiricism is characterized by an inductivist research protocol. Structure or pattern in data is considered to be intrinsic to them, and the significance of pattern is arrived at intuitively. The significance of pattern is considered to be more or less evident to the adequately prepared investigator, and the professional competence of the investigator is taken into account in the evaluation of the credibility of his or her conclusions, resulting in "appeals to authority" to settle contentious issues. Methodologically, strict empiricists have adopted an observational research protocol, devoid of any deductive component, lacking in any formal notion of a hypothesis, and of criteria for the evaluation of an hypothesis. While strict empiricism disappeared from the exact sciences more than two hundred years ago, it exhibits a remarkable and debilitating tenacity in archaeology and human paleontology. Probably something like 90% of all research ever conducted in these disciplines, in both the Old and New Worlds, has been done under a strict empiricist metaphysical paradigm. Even in 1995, strict empiricism is flourishing in *all* the intellectual traditions of archaeology and human paleontology, and it is in fact dominant in human paleontology and in the Old World research traditions generally (Clark 1993a, 1994a, 1994b). It is not so important (nor, therefore, problematic) in a highly experimental laboratory science like molecular biology, although, as the controversy over the meaning of mtDNA pattern searches amply demonstrates, molecular biologists have not been much concerned with epistemological issues either. In a manner of speaking, strict empiricism has no real epistemology, because the nature of the relationship between the investigator and the investigated is never subjected to critical scrutiny.

Classic Positivism

Classic positivism dominated western science from its beginnings in the Age of Enlightenment until the 1920s. It is also rooted in a realist ontology (i.e., it is held that a reality exists "out there," and that it is governed by time-and-context-free natural laws and mechanisms). It is the business of science to discover these laws, some (many) of which entail prediction and imply or require linear causality. Probably the best known example is the metaphysical paradigm of Newtonian physics. Epistemologically, classic positivism is both dualist and objectivist, which means that the investigator must adopt a noninteractive posture removed, supposedly, from the object of investigation. The duality is between the observer and the observed; the actions of the observer are not supposed to influence the results of observation in any way. Values, preconceptions, and other bias factors are, in theory, precluded from influencing the outcome of observation. From a modern perspective, this is ingenuous. Methodologically, positivism is experimental, which means that research questions or hypotheses are established beforehand in propositional form and then subjected to some kind of empirical test under carefully controlled laboratory conditions. The positivist paradigm characterized western science during the first three-quarters of the nineteenth century, when the "big science" disciplines (astronomy, physics, chemistry) began to assume recognizable form. It was never really important in archaeology or human paleontology, essentially because of the absence of an experimental research protocol.

Post-Positivism

Post-positivism, practiced today by the exact sciences, is a modified or evolved form of classic positivism. Ontologically, post-positivists move to a position of critical realism, which means that, although a real world governed by natural laws is held to exist, it is impossible to perceive it directly owing to imperfections in our sensory and intellectual capacities. Although "truth" in an absolute sense is held to be unattainable, the objective of post-positivist research is to arrive at better approximations of it. Realism (sometimes called "hypothetical realism") remains the central ontological bias. Epistemologically, post-positivism is usually described as "modified objectivist"; objectivity remains the regulatory ideal and is used to select among competing, alternative explanations, but it is acknowledged that it is impossible to arrive at total objectivity. There is also explicit recognition of the existence of a critical intellectual tradition, made manifest in a critical intellectual community (i.e., one or various "schools") within which the paradigm is more or less consistent and accepted by consensus, but outside of which it might be nonsensical. Methodologically, post-positivists have adopted a broadly defined experimental-manipulative research protocol, which aims to consider simultaneously multiple alternative hypotheses, which seeks to minimize the effects of bias in "natural" (nonobtrusive) settings, and which, in the exact sciences, depends rather heavily upon grounded theory (i.e., a body of theory already well-tested and confirmed). Processual archaeology in the anglophone world, and molecular biology in general, proceed from a post-positivist metaphysical paradigm and the preconceptions and biases that underlie it. With a few notable exceptions, and despite nominal acknowledgment of evolution, human paleontology remains mired in strict empiricism.

CONTINUITY VS. REPLACEMENT

Although archaeologists, human paleontologists, and molecular biologists have all contributed in important ways to the MHO debate, it is possible to take this literature and divide it into two opposing camps which seem to differ from one another in terms of the biases and preconceptions that underlie their knowledge claims. The major tenets of the "replacement" and "continuity" positions are summarized in Tables 4.1 and 4.2. It cannot be emphasized strongly enough that these positions represent central tendencies in the distribution of biases and assumptions. Probably no researcher in either camp subscribes to all of them, but were one asked to do it, it would be a simple matter to determine into which group most workers would fall (Willermet 1993, 1994).

Replacement advocates tend to make a distinction between archaic *Homo sapiens* and neandertals, and they postulate a series of adaptive radiations out of Africa, rather than the single, prolonged one favored by continuity theorists. They typically ignore grade-clade distinctions, emphasize cladogenic (or branching) over anagenic (or lineal) speciation, employ "splitter" taxonomies and dendritic (or branching) phylogenies. They seek support for a particular construal of the fossil evidence by invoking a mtDNA chronology that favors a rapid rate of base-pair substitution (e.g., the 2–4% per million years proposed in Cann et al. 1987). They argue that modern humans evolved only in Africa; that they spread out from there, displacing or extirpating archaic *Homo sapiens* and neandertals throughout the middle latitudes of the Old World, and that there was little or no genetic admixture between moderns and preexisting forms of humans, thus im-

Table 4.1. Major Tenets of the Replacement Paradigm

Replacement advocates
- make a distinction between archaic *Homo sapiens* (AHS) and neandertals;
- postulate a series of adaptive radiations out of Africa, rather than a single prolonged one;
- invoke a rapid mtDNA base pair substitution rate (2–4%/myr) to argue for morphologically modern humans (MMHs) evolving in Africa ca. 200 kyr B.P.;
- ignore or deemphasize grade/clade distinctions;
- emphasize cladogenic speciation over anagenic speciation (except in Africa);
- invoke "splitter" taxonomies and dendritic phylogenies;
- claim that archaic *Homo sapiens* and *Homo erectus* were largely or completed replaced throughout their ranges by MMHs between 300 and 100 kyr B.P.; and
- claim that there was little or no admixture between AHS and MMHs (except in Africa, where MMHs evolve from AHS through anagenic speciation).

Table 4.2.   Major Tenets of the Continuity Paradigm

Continuity advocates

- do not make a distinction between archaic *Homo sapiens* (AHS) and neandertals, the latter being one of several geographical clades within the former;
- postulate a single, prolonged hominid radiation out of Africa corresponding to the *Homo erectus* grade in human evolution;
- invoke a slower mtDNA base-pair substitution rate (<1%/myr) to argue for archaic *Homo sapiens* evolving in Africa ca. 750–500 kyr B.P.;
- emphasize grade/clade distinctions;
- emphasize anagenic speciation over cladogenic speciation;
- invoke "lumper" taxonomies and reticulate phylogenies;
- claim that MMHs evolved from AHS through anagenic speciation throughout the range originally colonized by *Homo erectus;* and
- claim that there was substantial genetic admixture between AHS and MMHs over time and space, and that local continuity, rather than replacement, marked the biological transition.

plying that the appearance of moderns was a speciation event. This is the replacement-without-gene-flow position favored by many Continental and Israeli scholars.

Continuity advocates consider neandertals to be the European and West Asian clades (geographical races) of archaic *Homo sapiens* and postulate a single, prolonged hominid radiation out of Africa corresponding to the *Homo erectus* grade in human evolution. They emphasize grade-clade distinctions, anagenic over cladogenic speciation, "lumper" taxonomies, and reticulate (or trellislike) phylogenies. They claim that moderns evolved from archaic forms of *Homo sapiens* throughout the range originally colonized by *Homo erectus.* They see many indications of morphological (hence genetic) continuity in the fossil evidence and invoke a mtDNA chronology that favors a relatively slow rate of base-pair substitution (e.g., <1% per million years), effectively pushing modern human origins back into the Middle Pleistocene (e.g., Wolpoff 1989b). They argue that admixture was not only likely but inevitable (assuming, for the sake of argument, the presence of "immigrants"), and that local continuity and gene flow, rather than replacement without admixture, characterized the evolution of modern humans everywhere. This is essentially the multiregional continuity position first advocated by Gustav Schwalbe in 1906 and, later, by Franz Weidenreich (1943b).

I suggest that the replacement and continuity positions are paradigmatic formulations, in the sense that they are grounded in different sets of preconceptions and biases about evolutionary process which result in different construals of what the human past was like. Each comprises a polythetic set of assumptions that privileges some suites of variables at the expense of others, and that weights variables held in common differently. By making explicit the tenets of the replacement and continuity paradigms, it becomes possible to generate test implications for each of them—patterns in the archaeological and paleontological records that should hold if in fact the paradigm is an accurate representation of reality.

ARCHAEOLOGY

The debate in archaeology turns on evidence for continuity in pattern, or the lack thereof, over the Middle to Upper Paleolithic transition, at ca. 30,000–40,000 years ago. Replacement advocates contend that it was during this interval that neandertals were supplanted by moderns. It is probably true to say that the consensus in the New World has always been in favor of clinal (although not necessarily regular) change over the archaeological transition, but there is a growing Old World literature that also supports the continuity position, and that represents a departure from previously established views. The argument is simply that if one looks at anything more comprehensive than the retouched stone tools traditionally emphasized by European typological systematics, there is continuity over the Middle to Upper Paleolithic transition on every single archaeological monitor of human adaptation. By continuity, I mean vectored, clinal change in *(a)* the major technological characteristics of lithic industries, *(b)* raw material procurement and use, *(c)* patterns of faunal exploitation (we can't say much about plants), *(d)* evidence for symbolic behaviour and ritual, and *(e)* settlement patterns (site numbers, types, settings, and distributions). Continuity advocates do not claim that all regions exhibit continuity on all variables in equal measure, nor that the rate of change is everywhere the same. It is clear and definite, however, that there are few if any instances of marked disjunction across a suite of variables that coincides with locally defined Middle to Upper Paleolithic transition boundaries. Disjunction would clearly be expected by all of the replacement scenarios published so far. *There is absolutely no evidence for it—anywhere.*

Empirical support for archaeological replacement rests on the notion, associated in recent years with Paul Mellars (e.g., 1989b), that Middle Paleolithic stone tools do not exhibit "imposed form" and "morphological standardization," whereas Upper Paleolithic stone tools do. It is an argument grounded in thinly veiled essentialism that goes back to the very beginnings of French paleolithic archaeology in the last quarter of the nineteenth century, and which shows up more recently in the Bordes-Binford debate of the early 1970s (e.g., Binford 1972; Bordes 1973). It treats pattern in retouched stone tools as if it were objectively real and intrinsically meaningful, and it uses pattern to identify the time-space distributions of identity-conscious social units analogous to those known from history. As I have tried to point out before, this is naive (Clark 1993a, 1994a, 1994b; Clark and Lindly 1991). The typological filter determines our perceptions of pattern and, since the Middle and Upper Paleolithic typologies are different, perceptions of differences on either side of the transition are inevitable. The notions of "imposed form" and "morphological standardization" are illusions. Exactly the same kinds of arguments about equifinality of form that Dibble (1987) has made so successfully in regard to Middle Paleolithic artifacts can and have been made about Upper (Barton et al. 1996; Coinman 1990) and even Epipaleolithic (Neeley and Barton 1994) artifacts.

If we don't get rid of Bordesian systematics altogether, we should at least uncouple them from the historicist biases that are invoked to explain them. Implicitly or explicitly, Bordesian systematics underlie the pattern searches that are the basis for all of the archaeological replacement scenarios, and historical process is invoked as the major kind of explanation for pattern. From an Americanist point of view, this is also naive. Bordesian systematics interpret techno-typological differences as evidence of tool-making traditions (i.e., modal ways of making stone tools transmitted by enculturation from one generation to the next in a social context). But it is immediately apparent that all paleolithic techno-typological modalities exist at a scale far beyond that of any conceivable social unit that might have transmitted them. So, whatever they mean, they have nothing to do with traditional ways of making stone tools. This, of course, implies that they have nothing to do with historical process, which in turn calls into question both the integrity and the utility of normative analytical concepts like Aurignacian, Mousterian, Perigordian, Châtelperronian, etc. In my opinion, the conventional archaeological units are typological artifacts (Clark 1994a, 1994b). They have no more objective reality than do "neandertals" and "moderns" (see below).

If not the products of "history," then what are the conventional paleolithic analytical units? What we think of as Middle and Upper Paleolithic technologies almost certainly constituted a range of options very broadly distributed in time and space, held in common by all contemporary hominids, and invoked differentially according to context. The challenge of future work is to determine what general contextual factors constrained choice amongst these options. Such factors certainly included (a) range and size of and distance to raw materials; (b) forager mobility strategies (essentially determined by the distribution of prey elements in the landscape and crucial resources like water); (c) anticipated tasks related to food procurement, which in turn would have affected (d) group size and composition, the structural pose of the occupants of a given site in an annual round, and more generally, (e) duration of site occupation (see, e.g., Kuhn 1990, 1991, 1995).

The assignment of meaning to pattern is, of course, an epistemological issue. There is a history of some concern with this in Anglo-American archaeology, and with efforts to identify paradigmatic bias at the level of the metaphysic and to distinguish paradigms (statements about the way the world is) from theories (statements about why the world, or some portion thereof, appears to be the way it seems) (e.g., Binford and Sabloff 1982; Clark 1993a; Trigger 1989; Watson et al. 1984). An explicit concern with epistemology is not particularly common

outside the anglophone research traditions, although there are occasional examples (e.g., Gallay 1989; Gardin et al. 1981).

HUMAN PALEONTOLOGY

So far as human paleontology is concerned, the chief modern advocates of
replacement tend to be Europeans and include British scholars, like Stringer and
Andrews (1988b), and Continentals, like Bräuer (1989) and Vandermeersch
(1993). Some Americans also subscribe to the replacement paradigm (e.g., Howell 1994; Klein 1992). Their adversaries in the continuity camp are mostly
Americans, led by Wolpoff (1992) and Brace (1964a, 1995b), and, interestingly,
anglos and Asians living and/or working far from the European "heartland" (e.g.,
Pope 1992b; Thorne 1981; Thorne and Wolpoff 1992; Wu 1992b). For the
purposes of this chapter, the controversy in human paleontology can be reduced
to the kinds of evolutionary models favored by the two camps—*dendritic,* in the
case of the replacement advocates, and *reticulate,* in the case of the continuity
advocates.

A typical dendritic phylogeny of the kind endorsed by replacement advocates
is shown in Figure 4.1. It postulates a series of relatively brief "adaptive radiations" (rather than a single, prolonged one), and a corresponding series of replacements. All but one of the branches (the one leading to ourselves) ends in
extinction. It is never clear what the adaptive mechanisms were, nor what biological or behavioral characteristics supposedly conferred an advantage, however
defined, on one form over another. The dendritic view of evolution has an explicit
theoretical justification in respect of modern humans in W. W. Howells's Noah's
Ark model, which held that all living populations had a single and recent origin
from a source that was already essentially modern (1976). Howells never indicated whether he believed that the "essentially modern" source was different at
the species level from preexisting hominids in any geographical region, nor did
he suggest a point of origin for the spread of moderns. He thought that *Homo
sapiens sapiens* colonized the globe, displacing or intermixing with other, more
primitive hominids who were already occupying the subtropical and temperate
latitudes of Eurasia. Racial differences developed only recently. The fact that he
did not take a position on whether differences were at the species or subspecies
level turns out to be crucial to understanding the evolutionary implications of
subsequent versions of the model. Obviously, if differences were subspecific,
admixture between indigenous and colonizing populations would have been possible (e.g., Bräuer 1992a). The problem is that we have no way of assessing the
magnitude of difference empirically.

The continuity position is based upon what I would call a "reticulate grade
model" of human evolution, where there is sufficient gene flow over time and

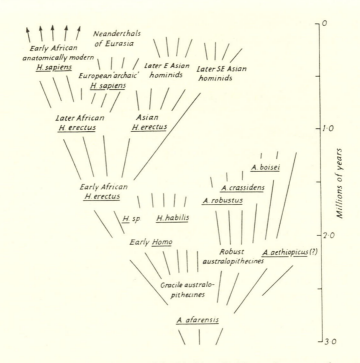

Figure 4.1. A dendritic phylogeny of the kind invoked by replacement advocates: hominid evolution portrayed as a series of radiations, rather than a single, prolonged one. Note that all lines but one (upper left) end in extinction (from Foley 1987b:384, used with the permission of the author and of *Antiquity*).

space to maintain genetic continuity over broad geographical areas. The reticulate view of human evolution is a modified version of another of Howells's evolutionary paradigms, the "candelabra model" (1976). It postulates an archaic *Homo sapiens* grade interposed between that of *Homo erectus* and that comprising modern populations, and it differs from the Noah's Ark model in that it assumes a single adaptive radiation out of Africa corresponding to the *Homo erectus* grade, with subsequent colonization of most of the uninhabited areas of the Old World during the Middle and Upper Pleistocene. A trellislike configuration is envisioned, with roughly contemporaneous development of parallel lineages through a succession of grades, but connected with one another across space by greater or lesser amounts of gene flow (Figure 4.2).

Whereas the Noah's Ark model is grounded in rather narrow characterizations of hominid taxa, and what might be called a "splitter" mentality, the reticulate view reflects an appreciation that *(a)* both *Homo erectus* and archaic *Homo sapiens* samples are temporally and spatially variable, *(b)* the nature and rate of change varied geographically and temporally, and *(c)* biological changes did not

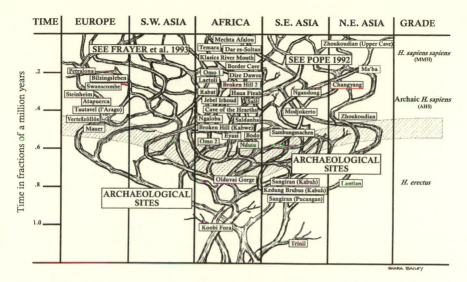

Figure 4.2. A reticulate phylogeny of the kind invoked by continuity advocates: hominid evolution characterized by broad and roughly contemporaneous similarities in organization (grades) and by characteristics indicating common descent within regions (clades) (from Clark 1992a:190). (Note: MMH = morphologically modern human)

take place in a behavioral vacuum and must have reflected adaptation in ways that are, at present, only partly understood (see Clark 1992a, 1992b, and Clark and Lindly 1989a, 1989b for an extended discussion).

Because adaptation is the focus of their research, and because biomechanical reasons for morphological changes are given by Trinkaus (1986), Oxnard (1987), Marzke (1983; Marzke et al. 1994), Tuttle (1988), and other functional morphologists, I consider their explanations of change to be more credible than those of cladists (and, in fact, those of *any* comparative study based exclusively on morphology). This is not to imply that functional morphological interpretations are not debated, or even sometimes completely wrong, but—despite their current popularity—there are so many conceptual and methodological problems with cladistic approaches that, in my view, they are worthless for analytical purposes (Clark 1988, 1989a, 1989b). Apart from very sticky problems with deciding what is in fact a derived character, it is necessary to show that derived characters not present in an ancestral group are shared by all members of a clade. This requirement alone introduces enormous circularity in any real-world paleontological context, given the absence of samples representative of biological populations.

This is more than a minor quibble. As I have argued for years, paradigms in human paleontology emphasize different suites of variables differentially. What actually happens is that practical considerations take over in the context of any real problem domain, influenced by a priori identification of taxonomic units (i.e., the groups to be compared, which differ from one worker to the next) and by biases that assign weights to particular suites of variables, even if the variables themselves are equally weighted. Ways to make these kinds of decisions are practically always implicit, and they are essentially subjective. In consequence, mechanical applications of cladistic analyses can sometimes produce absurd results (Stringer's contention that European neandertals are more like modern Europeans than they are like Upper Paleolithic Europeans comes to mind; Stringer et al. 1984). A common solution is to use outgroup comparisons of distantly related taxa to identify primitive characters, although this doesn't always work either. Very occasionally, quantified parameter estimates are generated for variables determined to be relevant to a particular problem. It should be noted that these observations are "generic" ones about research protocols in human paleontology in general. They are not exclusive to the MHO debate, nor are they directed at a particular MHO camp. Both sides count among their members card-carrying cladists; both contain individuals who would identify themselves as functional morphologists.

As is true of archaeology, essentialism is also a problem with human paleontology. Further, it seems to me that essentialist biases are more prevalent, and therefore more problematic, in the writings of replacement advocates than they are in the continuity camp. This is particularly true of Latin European workers who, it could be argued, have a stronger element of Lamarckian bias in their conceptions of evolutionary process than do the research traditions of the anglophone world (see, e.g., Brace 1988a, 1989). Although it is almost always implicit, the replacement position is based entirely upon the assumed credibility of the biological taxonomic units themselves—the assertion (for it has never been demonstrated) that neandertals and moderns were in some Aristotelian sense discrete natural types, that they were not only "different" but different enough to be different species. In a classic case of paradigm lag, the historical precedence of western European fossils and models has so dominated our conception of the neandertals that it is very difficult—even now—to form a global perspective on variability in archaic *Homo sapiens* that is not influenced to some extent by the skeletal hand of Marcellin Boule (1911–1913). These paradigmatic differences are thrown into sharp relief in the Levant, where fossils assigned to archaic and modern *Homo sapiens* are found in close proximity to one another and are, for all intents and purposes, contemporaneous. Although French and Israeli workers have argued that the archaeology and the human paleontology are separate research domains, and that the contemporaneity of neandertals and moderns on Mount Carmel is an indication of no direct relationship between the hominids and their cultural attributes, I think instead that it is much more likely that the biological taxonomic units themselves are suspect. To put it bluntly, there is absolutely no consensus on what constitutes "neandertals" and "moderns." This is more than a mere assertion. A recent study by Willermet (1993, 1994) inventoried cranial measurements used as criterion variables in 39 publications by continuity and replacement advocates, most of which postdated 1970, and found that only 11% of the variables were common to both paradigms. In short, replacement and continuity advocates are using two different sets of variables to assess differences and similarities, and they are defining differently variables held in common (Clark 1992a, 1992b, 1993a; Clark and Lindly 1989a, 1989b; Clark and Willermet 1995; Willermet and Clark 1995).

MOLECULAR BIOLOGY

Until the recent, devastating critique of Cann's use of the PAUP program (Templeton 1992, 1993), the sticking point with her scenario had been that no admixture could be accommodated (see also Vigilant et al. 1991). If true, this would imply that differences between moderns and earlier hominids had to be at the species level, and that the appearance of moderns corresponded to a speciation event. Replacement advocates picked up on the rapid base pair substitution rate published by Cann (2–4%/myr), which indicated a modern human origin somewhere in Africa at about 300,000 years ago. Initially, at least, there was little realization by archaeologists and human paleontologists that other base pair substitution rates were published and defended in the literature—notably that of Nei (0.71%/myr), which implied a hominid radiation out of Africa at ca. 850,000 years ago when, by a consensus rare in human paleontology, the only form of hominid present was *Homo erectus* (Wolpoff 1989b; cf. Wood 1993). The other, slower rates can be reconciled much more easily with the fossil and archaeological records than can that of Cann. Second, most workers didn't seem to realize that the Cann rate implied an absurdly late hominid-pongid split, at about 3.5 million years ago, for which there is no fossil evidence whatsoever. Finally, there seemed to be little appreciation of the fact that, just like any other field, molecular biology has an internal dynamic characterized by the same kinds of controversy and debate found in other disciplines. In short, replacement advocates jumped on the Cann bandwagon because it supported their biases about the course of human evolution, without having the foggiest notion just how incompatible her interpretation was with the fossil and archaeological records.

To judge from a number of recent exchanges in *Science,* the field of DNA clock models is undergoing some soul-searching in respect of its research protocols (Lewin 1988a, 1988b). Because they are relatively new, mtDNA clock models suffer from these conceptual and methodological anxieties to a greater extent than do nuclear DNA clock models, which have a much longer history. Although molecular clocks have been relatively successful in dating landmark divergence events like the initial appearance of the Hominidae (i.e., the hominid-pongid split), and although there has been a reconciliation with paleontological chronologies over the past thirty years, it is worth pondering whether they are equally effective in determining ages for much more recent, smaller-scale, "species-level" events or processes like the emergence of morphologically modern humans. In other words, is the assumption of a common rate of mtDNA base pair substitution valid over relatively "short" time spans, or could there be problems with the resolution of these techniques the closer one gets to the present? Nobody really knows.

Since I was curious about how these issues were perceived in molecular biology, and recognizing my own ignorance of the field, I asked an Arizona State University colleague, a molecular geneticist whose primary research area is the HLA (histocompatability antigen) system—about them. After noting a bias in favor of functional genes determined by recombinant DNA (mtDNA does not recombine), he pointed out that mtDNA researchers, in their efforts to develop clocks, tend to concentrate on the hypervariable region of the mtDNA genome, which is not a coding region for the mtDNA locus. As a result, it is difficult to know just what variation might mean in terms of a hypothetical speciation event. If they were working with a functional gene, he suggested, they might be able to infer a speciation event based on the magnitude of change, but if no function can be ascertained for the region analyzed, there is really no way to determine what variability might mean. To my "informant," the fundamental question is how much molecular variation is required in order to be able to infer that a speciation event had taken place. A genuine speciation event would presumably entail changes in protein functions, but there is no suggestion of that in any of the mtDNA literature (mtDNA mutations are assumed by most workers to be neutral; Stoneking and Cann 1989). It was his opinion that the whole debate about the meaning of pattern in mtDNA data is so subject to the particular scenarios set up for it, and the assumptions that implicitly underlie them, that the debate is almost completely circular. There is, in effect, no real debate because of the absence of a common basis for discussion.

This assessment of the state of mtDNA systematics by someone vastly more knowledgeable than I left me with the impression that, despite its experimental research protocols, mtDNA clock models were not in much better shape epistemologically than the other branches of MHO research. Predictably, I suppose, it did little to shake my confidence in the credibility of multiregional continuity. What, then, do I think is going on with the molecular biology? It seems to me that, while there is considerable disagreement about rates of molecular evolution, and stark differences in research protocols and in how data are interpreted, there is nevertheless a broad consensus that mtDNA lineages exist and go extinct at a rate that, in theory at least, is predictable. Could not "Eve," then, simply represent the deepest point in time from which there are surviving mtDNA lineages? In other words, the evolution of *H. sapiens sapiens* might have had nothing whatever to do with genetic "bottlenecks" (however defined), nor with replacement of one species by another, nor with genetic "swamping" of one subspecies by another, nor indeed with migration, radiation, or population movement of any kind. The greater diversity in the African samples seems to be real, however, and might simply reflect the fact that Africa is the presumed locus of evolution of the Hominidae, and greater DNA diversity (of all kinds) would be expected there than anywhere else (see also Ayala 1995).

CONCLUDING REMARKS

Historically, modern human origins research is entirely a product of western science. However, over the past century, and especially after World War II, it has evolved to the extent that there are now about four or five partly overlapping, partly discrete research traditions, some of which have taken root in nations formerly under the domination of the western imperial powers (e.g., China, southeast Asia, the Levant). These emergent research traditions can best be defined geographically and linguistically (the Latin European research tradition, that of the anglophone nations, etc.). This postcolonial amalgam of western science, coupled with the results of indigenous scholarship, has in turn produced eight to ten partly incommensurable theories that purport to explain different aspects of MHO research in the three main arenas of the debate—archaeology, human paleontology, and molecular biology. I have collapsed these theories into two polythetic sets labeled "replacement" and "continuity" (Tables 4.1, 4.2) and have argued here and elsewhere that they constitute two different metaphysical paradigms (Clark 1992a, 1992b, 1993a, 1994a, 1994b). Because they proceed from different paradigms, however, their constituent theories comprise different sets of biases and assumptions about the nature of the human past. No one evidential domain can be considered in isolation from the rest, and in the end, the theories that underlie them are wholly or partly incommensurable (e.g., one cannot argue simultaneously for and against regional continuity, nor for regional continuity in one context and against it in another). This is not an uncommon situation in many science contexts, but it is exacerbated in sciencelike endeavors where there is little or no explicit concern with the inferential basis for knowledge claims. I suggest that MHO research fits that description.

Science philosopher Richard Bernstein (1983) has argued that when researchers grapple with incommensurable theories, they cannot proceed by a linear movement from premises to conclusions, or—as here—from individual facts to generalizations, but must instead exploit multiple strands of diverse kinds of evidence to assess the overall credibility of a hypothesis or theory. This argument will doubtless provoke denunciations from the strict empiricists among us, and from those inclined to take issue with the paradigm concept itself. The important point, however, is that when there is no common ground for assessing the credibility of an argument, the cumulative weight of multiple strands of evidence can be rationally decisive (Bernstein 1983:69, 74). The cumulative weight of the evidence clearly supports multiregional continuity so far as modern human origins are concerned.

On the surface, the voluminous literature of the MHO debate paints a picture of informed and sophisticated interdisciplinary research in which data are absorbed and digested, arguments assimilated, and methodologies understood, compared, and evaluated. It is the way we all would like to think that paleo-anthropology "works." I suggest, however, that paleoanthropology has the form but not the substance of a science, and that the MHO debate is more of a caricature of science than a "portrait from life." We are, in effect, consumers of one another's research conclusions, but we select among alternative sets of research conclusions in accordance with our biases and preconceptions—a process that is, at once, both political and subjective, and to some extent unavoidable in any sciencelike endeavor. The problem with paleoanthropology in general, and MHO research in particular, is that those biases and preconceptions are seldom subjected to critical scrutiny. If there is no explicit concern with the logic of inference—*how* we know what we think we know about the past—there can be no consensus. The result is the kind of interminable discourse that has plagued MHO research for more than a century. All I am suggesting here is that paleo-anthropology should pay more attention to the inferential basis for its knowledge claims. Nature does not dictate the meanings we assign to it—that can only come from us. Human origins controversies stem from a failure to consider epistemological issues in the various disciplines involved in the research. The empirical and logical sufficiency of knowledge claims can only be established with reference to a paradigm. Merely acquiring more data won't resolve anything, for the simple reason that data have no meaning (some would say existence) apart from the conceptual frameworks that define them (Clark and Willermet 1995).

5

Fuzzy Set Theory and Its Implications for Speciation Models

CATHERINE M. WILLERMET and BRETT HILL

The aim of this chapter is to explore what we believe to be one of the most fundamental and deeply entrenched conceptual issues in science, including paleoanthropology. The logical system that is the foundation of our attempts to categorize natural phenomena and understand them is based on the presupposition that these phenomena can be defined only as either a member of a given category or not. In most scientific conceptual frameworks it is not possible for a phenomenon to be both a member and not a member of such a category. This presupposition is so deeply entrenched that most of us either scoff at potential alternatives or never consider them. We propose that by reconsidering the essential bivalence of our scientific tradition, namely by using a fuzzy set alternative, we might better understand issues of variation, transition, and boundary overlap that are such an essential part of the study of evolution and speciation.

The fossil evidence available to answer questions of modern human origins is fraught with sampling problems. It is fragmentary. It is unevenly distributed over vast geographic areas. It is difficult to date accurately. There is disagreement over both how many groups are represented, and how to assess and measure variability. Not surprisingly, interpretations of this evidence vary among paleoanthropologists.

A major issue confronting researchers of modern human origins questions is where to place the Neandertals in our human ancestry. Are they a separate species *(Homo neanderthalensis)* from modern humans *(Homo sapiens)?* Are they the same species, different only because of geographic and temporal variation? Should they be considered a different subspecies *(Homo sapiens neanderthalensis)?* Each of these species definitions has implications. They can bias the choice of research protocols or the kinds of research questions regarded as significant to ask. Whether one favors a replacement model (Bräuer 1984b; Stringer and Andrews 1988b) or a continuity model (Smith, Falsetti, and Donnelly 1989; Wolpoff et al. 1984), the definition of species used by the researcher biases interpretation of the data (Willermet 1993).

SPECIES CONCEPTS

One of the major problems facing human paleontologists is the difficulty in defining fossil species. Speciation in most modern populations is too slow to be directly observable, so researchers must attempt to reconstruct phylogenetic histories (Mayr 1988). Biologists use speciation models such as allopatric or sympatric speciation to help reconstruct these unobserved speciation events. For defining extant species, biologists utilize the reproductive (biological) species concept (Mayr 1963), the species recognition concept (Paterson 1985), the phylogenetic species concept (Tattersall 1992), or the morphological (phenetic) species concept (Ridley 1986). Using these methods, a biologist can make a determination of species for any individual subject based upon the known species' characteristics (and the range of variation of those characteristics).

The biological species concept (BSC) (Mayr 1942, 1963) states that groups of animals can be defined as species by their ability to breed with each other and produce viable offspring (if they cannot, then they are of different species). A variant of this definition is the species recognition concept (SRC), which argues that the "reproductive community" (a.k.a. species) is defined by a shared fertilization system. This differs from the biological species concept only by examining not whether the animals *can* interbreed, but rather whether they *do*. Otherwise they are similar in their approach to defining biospecies (Eldredge 1993; Paterson 1985). Application of these species definitions combines reproductive isolation with ecological, behavioral, and phenetic variables (Jolly 1993). Both the BSC and the SRC create discrete categories—either the animals interbreed (same species) or they do not (different species). Of course, there are situations where two distinctly defined species produce hybrids (Jolly 1993; see also Jolly et al. 1995 for an example of intergeneric hybrids). These populations are often considered to be in the process of speciating, although they are generally classified as two separate species. The BSC/SRC is of major conceptual importance to paleontologists, who attempt to apply it through the use of indirect evidence despite problems of applicability (Eldredge 1993; Szalay 1993).

Paleontologists work with bones, not living animals, and hence cannot observe their breeding habits. Another major problem facing the paleontologist is time depth, manifest in an interest in tracing evolutionary patterns. The researcher addresses this through the identification of character states for features which are held in common among many species and which are used to test relationships. These morphological features are often distinct from those used by zoologists to classify specimens from extant, observable populations.

Paleontologists often must construct categories of fossil species (or paleospecies) from extremely small samples of individuals or elements. This is done, out of necessity, by comparing the sample with those of other fossil species (defined the same way) using either the phylogenetic or the morphological species concept (Ridley 1986; Tattersall 1992). Species are defined on the basis of morphological characteristics (a feature's presence, the frequency of its presence, or its relative size and shape). Much debate is centered on how much variation one would expect in a living species that this sample (of only one or a few specimens) represents. There is also debate over how much difference between this specimen and others is required to justify the identification of a new species (Gingerich 1985; Tattersall 1986, 1992; Wood 1992).

Eldredge and Gould (1972) wonder if biospecies can be recognized from fossils, and even whether biospecies are "real" biological entities. They point out that Mayr's biological species are "nondimensional"—they are distinct at a particular instant of time, but the lineage can only be broken into segments arbitrarily. They also state that the taxonomic perspective, handed down to us from Linnaeus, cannot easily be reconciled with temporal change:

> . . . the hierarchical system of Linnaeus was established for his world: a world of discrete entities. It works for the living biota because most species are discrete at any moment in time. It has no objective application to evolving continua, only an arbitrary one based on subjective criteria for division. Linnaeus would not have set up the same system for our world. (Eldredge and Gould 1972:93)

Eldredge and Gould argue that instead of trying to retrofit the biological *taxonomy* to fossil species, we should classify fossil species using biological *concepts*. Mayr (1988) argues that this classification should be evolutionary in nature.

The complicating factor here is that species can evolve by both anagenesis and cladogenesis. If lineages are evolving anagenically, how does one define chronospecies—species at two points on a nonbranching lineage which are clearly different from each other? If lineages do show a break as a result of cladogenesis, then how much difference must develop before the two lineages can be distinguished using skeletal data?

From the standpoint of our conceptual framework or paradigm for studying modern human origins we must ask, How real are species? Do they have reality outside of our classification, or are they constructed for the purposes of classification, or both? Western science generally subscribes to realist ontologies and epistemologies, albeit with much discussion of what constitutes "reality." Many of us believe that there is some objective reality out there that exists independent of our perceptions of it (Clark 1993a, 1994a). This idea has a long and controversial history in philosophy of science contexts (Brace 1988a; also see Clark, this volume).

Hull (1970) argues that species represent a highly abstracted level of reality (more abstract than higher taxonomic units). To him, a species taxon is real in the same sense as an epidemic (an idea abstracted from the incidence of disease) or a cloud (a cluster of water vapor droplets) is real. Such entities, like species, are considered "individuals," have a finite life span, and can be given an identifying name (e.g., the 1919 flu) (Hull 1970).

Jolly (1993) continues this line of argument by suggesting that an epidemic can also be defined such that two independent observers arrive at the same conclusions about its parameters, but only if there is advance agreement on boundaries delineating the edges of the class. So too with species:

> Although we can treat species as real, we must recognize that each of the . . . current species concepts depends upon criteria (reproductive isolation, homologous morphological distinctiveness, and mate recognition . . .) that in nature are continuously distributed and multifactorially determined. Such data can easily be expressed as a list of entities called species, and all taxonomists should agree on the number and limits of these, as long as they also agree upon the diagnostic cut-off point on the appropriate continuum: how absolutely isolated, how distinctive, how similar in the mate-recognition system, do two populations have to be to be called different species? (Jolly 1993:69)

If we are dealing with populations that are "continuously distributed and multifactorially determined," then by drawing arbitrary and crisp boundaries between them we both lose information and run the risk of confusing a number of alternative classifications.

This problem is crucial in efforts to model the evolution and dispersal of modern humans. The organisms classified are, in this case, very close to one another genetically and behaviorally. The overlap amongst them is more than expected of most other groups of hominids. Because of this, researchers cannot agree upon the "diagnostic cut-off point on the appropriate continuum." Even if one does not agree that Neandertals and moderns formed a continuous, albeit highly variable population (the continuity view), one must still recognize that there is continuous evolution and that these populations shared a common ancestor fairly recently in evolutionary terms.

The problem of how to model continuous variability is a complex one. To model it mathematically, we need to utilize set theory to understand how we are classifying these organisms. "Neandertals" and "modern humans" are sets, and we want to place an unknown fossil into one set or the other. We need to understand how to construct these sets in order to place the unknown fossil. Our task is especially difficult because the boundary between these two sets is somewhat vague—many specimens overlap on some variables, making taxonomic distinctions difficult. There are several strategies for dealing with this problem (apart from ignoring it): *(a)* specimens that overlap two sets are called "transitional," *(b)* qualifiers are placed around segments of species designations (e.g., archaic *H. sapiens*), and *(c)* catchall categories are used (e.g., morphologically modern humans). How does one place an unknown specimen in a species with an unclear definition? The usual answer: first, measure every conceivable feature; next, recognize that the above-listed problems exist and that we should "proceed with caution"; and finally, interpret the results as if the problems were not present. What we argue here is that paleoanthropologists need to deal with this variability and overlap in a way that actively recognizes and addresses it, not as a statistical annoyance, but as a valuable and informative part of the record.

Regarding the data: what and how many variables should one use to classify an unknown fossil specimen? Some would take numerous measurements on the skull, teeth, and postcrania (e.g., Stringer 1989b; Wood 1992). Others would emphasize certain functional complexes (e.g., Rak 1986; Trinkaus 1993a). Once these data have been gathered, they are compared with all or some of the already classified specimens from other, established species. These comparisons are usually done statistically in order to determine the probability that the unknown specimen belongs to the set of A, B, or neither. In other words, the unknown specimen must either be or not be in one of these sets.

All of the commonly used statistical analyses (chi-square, *t*-test, ANOVA, and confidence intervals) operate using Aristotelian logic and either parametric or nonparametric probability theory (McNeill and Freiberger 1993). The tests are designed to answer the question "If the unknown specimen is a member of species A, how probable is it that it would differ as much as (or more than) it does from the specimens most typical of species A?" Either the unknown specimen is a member of species A or it is not (an example of this type of analysis can be found in Bräuer's (1992a) principal components analysis of the Dar-es-Soltane 5 cranium). These analyses address the probability of correctly assigning a specimen to one of several species. They do not accurately model the fact that the unknown specimen may share features partly with species A and partly with species B.

Why should this be a possibility? Several reasons immediately come to mind: *(a)* the specimen is in the "throes of evolution," meaning it is anagenically in the process of becoming a new species; *(b)* we have incomplete information about the number of species present at the time the specimen was alive and their range of variation; *(c)* the specimen exhibits the mosaic nature of evolution, and many traits seen as *characteristic* (not definitive) of species A may not be expressed in all its members, and traits *characteristic* of species B may be present in some members of species A (Wolpoff 1980), meaning they can be modeled as polythetic sets (Van Valen 1988); and *(d)* the specimen may be a hybrid—we already know from studies of extant species that many of them can interbreed and traits of these "species" can grade into each other. All of these are important issues, because fossil species are classified based on how they look, often by using indefinite descriptive criteria.

FUZZY LOGIC

In order to model membership of a specimen in more than one group, a new paradigm must be introduced that is not derived from Aristotelian logic. This paradigm is widely referred to as fuzzy logic. Fuzzy logic originates from the observation that the world is not composed of crisp sets (e.g., is not black and white, but shades of gray). Properties such as "size" form a continuum. How do we usually deal with these properties? We break up the continuum into discrete parts. Even this, however, is inaccurate. A pioneer in fuzzy logic, Max Black, described the problem this way:

> . . . imagine . . . an exhibit in a hypothetical Museum of Applied Logic. A line of countless "chairs" stretches toward the horizon. At one end is a Chippendale. Next to it is a near-Chippendale, virtually indistinguishable from the first. Succeeding items are less and less chairlike, until the row finally ends in a lump of wood. A normal person not only cannot draw a clean line between *chair* and *not-chair,* but performs this task uneasily. The concept *chair* does not permit this distinction. (in McNeill and Freiberger 1993:33)

This description could also apply to any classification system using variables with a continuous distribution (at what point does a flake become a blade? a cup become a bowl?). These distinctions are determined ultimately by the precision of the measurement instrument.

CLASSIC SET THEORY

Categories are modeled in mathematics by using set theory. Sets are collections of definite, distinguishable elements which can be conceptualized as forming a group. Any definition of a set also defines its complement, which is all elements that are not members of the set. If two sets, A and B, share some elements in common, we use terms such as *union* (the set of all elements which are in set A *or* set B) and *intersection* (the set of all elements which are in set A *and* in set B). The *empty set* is defined as the set which contains no elements. A *subset* is defined as a set whose elements are all elements of a second, larger set (Nanzetta and Strecker 1971; Zehna and Johnson 1962).

An important property of classic sets is that they are crisp. The idea is that, given any object, one should be able to determine whether or not it belongs to a given set (Zehna and Johnson 1962). Crisp sets are described using variables that are thought to have explicit operational definitions. Examples include "the set of red flowers," "the set of all true statements," and "the set of all animals in the species *Pan troglodytes.*" All of these sets are thought to have crisp membership; the complement would also be crisp (e.g., the set of all non-red flowers).

If the intent is to model a continuous phenomenon, then an arbitrarily line must be drawn between two points, defining two sets. An example is "the set of tall men." What does "tall" mean? Height is a variable with a continuous distribution, but this term must be defined crisply so that membership in this set is

Table 5.1. Truth Value Table for Crisp
Membership in the Set of "Tall Men"

Element	Truth Value
Jim (6'6") is tall.	1
Jon (6'2") is tall.	1
Tom (5'11") is tall.	1
Bob (5'10") is tall.	0
Bill (5'6") is tall.	0

The value 1 indicates that the element is a member of
the set, and conversely, the value 0 indicates that he
is not (adapted from McNeill and Freiberger 1993).

unambiguous. Let us look at five individuals who are candidates for membership
in this set (Table 5.1). If we choose the height of 5'11" or greater as "tall," then
we can determine who is and is not in this set. We ask the question "Is Tom tall?"
and the answer is "yes" (truth value 1) or "no" (truth value 0).

Bob (5'10") is almost the same height as Tom (5'11"), but this is not reflected
in the crisp membership of this set (Tom is tall, and Bob is not tall). A system
such as this is called bivalent; there are two values to the variable *height,* "tall"
and "not tall," or "present" and "absent," and every element is in one or the other
set. A continuum can also be split into an ordinal series, allowing a range of
values for each (e.g., not tall [<5'11"], tall [5'11" to 6'5"], very tall [>6'5"]) to
allow the categories to be more informative. In such a system there are more than
two categories, but each element can still only belong to one category.

Classic sets are founded upon Aristotelian logic, which is binary in nature
(Kosko 1993). Either something is A or it is not-A; there are no other alternatives. Aristotelian logic has become the foundation for western science. It is,
however, not the only way one can look at the world. Reality need not be either
black or white. Shades of gray must be accommodated, which is not easily done
using bivalent logic. The alternative is "A and not-A." This alternative is the
basis for fuzzy logic, of which fuzzy set theory is a part.

FUZZY SET THEORY

One problem with classic set theory is the sorites paradox: as one removes
sand grains from a heap, when does it cease to be a heap? Biology has Woodger's
paradox: if an animal can only be a member of one taxonomic unit (such as
species), then at certain points along an evolutionary trajectory, a child must
belong to a different species than its parents (McNeill and Freiberger 1993).
Classic set theory requires one to define a break point to delimit categories; fuzzy
set theory models the haze at the edges of the categories (Novák 1989).

A fuzzy set is a set of elements which are members to a degree (Kosko 1993; McNeill and Freiberger 1993; Novák 1989). A crisp set has elements which are members to a degree of 0 or 1. A fuzzy set has elements which are members to any degree from 0 to 1. A fuzzy empty set occurs when all its candidates for membership have zero membership (e.g., the set of living dodos).

The complement to the fuzzy set is the amount the memberships need to reach 1. For example, a glass that is one-third full requires two-thirds to be added to be considered full. A person who is satisfied with his job to a degree of 0.85 is not satisfied to a degree of 0.15 (Kosko 1993).

A fuzzy subset is a set whose elements belong less to the subset than to the larger set, or equally to both. For example, the fuzzy set of "tall men" has a subset of "very tall men." Ted, at 6'11" is a tall man and a very tall man, both to a degree of 1. Fred, however, is only 6'2". He is both a tall man (perhaps to a degree of 0.7) and a very tall man (perhaps to a degree of 0.2). Both of these men qualify for membership in the subset "very tall men," but they belong to the subset of "very tall men" to the same or a lesser degree than they do to the set of "tall men" (McNeill and Freiberger 1993).

Intersection and union of fuzzy sets involve the degrees of membership of their members. The intersection of the two sets "tall men" with "fat men" is the lower of each item's degree membership in either set. For example, if Ted is a member of "tall men" to a degree of 1.0 and "fat men" to a degree of 0.4, then the intersection value for Ted is 0.4 (he is a member of both sets to a degree of 0.4). In contrast, in a union of these two sets, Ted's membership would be 1.0 (he is a member of either set to a degree of 1.0) (McNeill and Freiberger 1993).

In fuzzy set theory, sets are described using language variables that are ac-knowledged to be imprecise. This point is important, because the grayness of the world is reflected in the language we use to describe it. The word *red* includes many shades, such as brick red, cherry red, and candy apple red. If you ask "Is this flower red or not-red?" and the flower is best described as pink (a mix of red and white), then whether you choose "red" or "not-red," you will be partially incorrect. However, each will also be correct to a degree corresponding to the amount of red in the pink flower. A more precise question might be, "To what degree is this flower red?" Therefore, "the set of red flowers" is more accurately modeled as a fuzzy set, although the same variable is used to describe it. The same can be said of "the set of all true statements," and "the set of all animals in the species *Pan troglodytes.*" This latter example is fuzzy because bonobos *(Pan paniscus)* are very similar to *Pan troglodytes* and overlap them in many features. All of these sets have elements which have fuzzy, not crisp, membership, and their complements would also be fuzzy.

Let us revisit our example of modeling the continuous variable of height, the set of tall men. In crisp set theory, the set "tall men" was defined as having members 5'11" or taller. In fuzzy set theory, the term "tall" can be defined so that membership in this set can be more realistically modeled. Let us look at the five

Table 5.2. Fuzzy Value Table for Fuzzy Membership in the Set of "Tall Men"

Element	Truth Value	Fuzzy Value
Jim (6'6") is tall.	1	0.95
Jon (6'2") is tall.	1	0.8
Tom (5'11") is tall.	1	0.6
Bob (5'10") is tall.	0	0.5
Bill (5'6") is tall.	0	0.2

Fuzzy values are the degrees to which each element is a member of the set (adapted from McNeill and Freiberger 1993). Fuzzy values can be calculated in a variety of ways, one of which is to break the range of possible heights observable into segments and to assign a fuzzy height value for each segment.

individuals who are candidates for membership in this set (Table 5.2). We ask not "Is Tom tall?" (the answer can only be yes or no) but "How tall is Tom?" (the answer can be any number between 1 and 0, Tom's *degree of tallness*).

The fuzzy values better reflect the closeness of Bob's and Tom's height. They differ by 0.1 rather than 1.0. This system is called multivalent; there are as many categories as there are points between 0 and 1, an infinite number.

Using classic set theory, researchers with an unknown fossil specimen would compare it with specimens of "known" species. They would ask the question, "Does specimen 1 belong to species A or B?" or, alternatively, "What is the probability that I would find the features of this specimen in known members of species A or B?" In reality, they are asking two separate questions: "Is specimen 1 a member of species A or species not-A?" and "Is specimen 1 a member of species B or species not-B?" Inclusion of more reference species adds more of these questions. The answers to these questions vary from researcher to researcher, based upon factors like the amount of variation tolerated within a species, what variables were chosen for analysis, and how these variables are defined. Ultimately, the answer to each of these questions will either be A or not-A, B or not-B. In the end, one species will be chosen as the set to which the unknown specimen belongs, or a new set will be defined, with the unknown specimen as its only member. This is the traditional way that most researchers have attempted to classify species.

EXAMPLES OF FUZZINESS IN PALEOANTHROPOLOGY

Examples of fuzzy data and the resultant problems associated with its analysis are widespread in paleoanthropology. Smith, Simek, and Harrill (1989) endeavor to model the continuously variable supraorbital torus size during the later Pleistocene. Many researchers struggle with the placement of the Skhūl and Qafzeh specimens in hominid phylogeny (e.g., McCown and Keith 1939; Trinkaus 1984; Vandermeersch 1989).

Wood's (1992) analysis of specimens from early *Homo* is a particularly illustrative example:

> any proposal claiming to have identified the earliest evidence for *Homo* needs to demonstrate a shift from the australopithecine to the hominine grade, but the closer to the cusp between the grades the greater the difficulty in distinguishing them. (1992:783)

In this extensive study on the origin and evolution of the genus *Homo,* Wood (1992) splits early *Homo* into two species, *H. habilis* sensu stricto and *H. rudolfensis.* He uses many variables to categorize specimens into crisp sets (e.g., *H. habilis* sensu stricto has complex suture patterns, foreshortened palate, etc.; *H. rudolfensis* does not). Many of his variables are continuous and descriptive, or fuzzy; examples include suture pattern (complex vs. simple), palate (foreshortened vs. large), malar surface (vertical or near vertical vs. anteriorly inclined), and mandibular fossa (relatively deep vs. shallow) (Wood 1992:786; see also Chamberlain and Wood 1987). Based in large part on this evidence, he splits the *Homo* clade into two, one containing the above two species, and the other containing *H. erectus, H. ergaster,* and *H. sapiens.* He defines the latter as containing the following character state changes (Wood 1992:789; emphasis added):

1. *increased* cranial vault thickness
2. *reduced* postorbital constriction
3. *increased* contribution of the occipital bone to cranial sagittal arc length
4. *increased* cranial vault height
5. *more* anteriorly situated foramen magnum
6. *reduced* lower facial prognathism
7. *narrower* tooth crowns, particularly mandibular premolars
8. *reduction* in length of the molar tooth row

Wood's criteria are all presented in the fuzzy language terms characteristic of ordinal variables (e.g., *increased* cranial vault height—note that there is no magic number, and in fact such variables often overlap taxa). All specimens have this trait to any degree between 0 and 1. The point is not that this interpretation might be wrong, only that it is trying to solve a fuzzy problem with a crisp solution. The result is conclusions that obscure important information about variability at the boundaries between groups. The genus *Homo,* we argue, is better modeled as a fuzzy set, the elements of which are members to a degree. Wood asks questions like "Does the hypodigm subsume more than one species?" and "Do (the species) belong to *Homo,* or is one, or more, an australopithecine?"

(Wood 1992). These are questions that are fuzzy in nature and have no single crisp answer.

DISCUSSION

Using a fuzzy model, we can argue that Neandertals are similar to modern humans "to a degree." By modeling these populations as fuzzy and not crisp, retaining information about group boundaries, we can *(a)* better describe the degree of similarity/difference observable among fossils attributed to Neandertal, archaic *Homo sapiens,* and modern *Homo sapiens; (b)* better model the relationships among fossils from these populations; *(c)* get at questions of degree of shared ancestry and potential for interbreeding; and *(d)* fine-tune models of anagenic and cladogenic speciation at the archaic to modern transition. This last point about variability is important: paleoanthropologists should be less interested in whether a specimen is a member of fossil group A or B and more interested in how crisp or fuzzy the boundary is between these fossil groups.

CONCLUSION

Most people feel that the world presents itself to us shades of gray and want to be able to model it accurately. Most researchers, however, have been choosing statistical packages designed according to the principles of Aristotelian logic. With the appearance and development of fuzzy logic and fuzzy modeling, the fuzzy paradigm can no longer be ignored. Its explanatory power has implications for numerous fields, including paleoanthropology. There are many questions that could be explored using the logic of fuzzy set theory. The role of the Neandertal (whatever his taxonomic status) in modern human origins is an obvious one. Wolpoff (1982) argues that "hominid traits" such as tool making/using, bipedalism, organized hunting, and communication predated hominid origins, but hominid origins involved (in part) a shift in importance of these characters. This "shift in importance" is, in fact, a fuzzy problem and can be so modeled.

Although fuzzy logic was formally introduced in the 1960s (Zadeh 1965), it has only recently begun to achieve wide acceptance in the West (although it has been heartily embraced in the East as it is in accord with aspects of several Eastern philosophies). One of the reasons for this is that fuzzy set theory directly opposes the Aristotelian principle of bivalence. Allegiance to the paradigm at the level of the metaphysic is very tenacious, however. Mayr observes that:

> The study of the basic philosophies or ideologies of scientists is very difficult because they are rarely articulated. They largely consist of silent assumptions that

are taken so completely for granted that they are never mentioned. . . . [But] anyone who attempts to question these "eternal truths" encounters formidable resistance. (quoted in McNeill and Freiberger 1993:45)

Fuzzy statistical packages are becoming available that will allow us to perform exploratory data analysis (Benson 1987), cluster analysis (Miyamoto 1990;
Zimmermann 1990), probability measures (Skala 1988), expert systems (Buckley
and Siler 1988), linear regression (Kandel and Heshmathy 1988; Tanaka et al.
1987), and other evaluations of fuzzy models (Gottwald 1993). Given the current
state of hominid systematics (e.g., Grine 1993 identifies six distinct hypotheses
of hominid phylogenetic relationships), the usefulness of fuzzy set theory must
be explored.

Despite the intuitively satisfying recognition of fuzziness, the bivalent paradigm has dominated science because it deals effectively with most aspects of
empirical reality. Categorization is an integral part of human behavior in general
and science in particular. The fewer the categories and the clearer the distinctions
among them, the more effectively we can communicate about them and place
them in an organizational framework. Although bivalent logic has greatly contributed to the advancement of science, we must not be misled by imagining that
a taxonomic system, imposed on reality by humans, necessarily equates with
reality. Even though a bivalent approach is generally efficient in dealing with
most of the phenomenological world, it is not as accurate as multivalent approaches, particularly when dealing with the boundary issues so important for
modeling the processes of speciation. The closer one gets to a boundary, the less
accurate bivalent approaches become. A bivalent approach obscures important
information about variability that is central to understanding evolutionary process. As we refine our observation of natural phenomena we will come to discover that classifications based on multivalent logic model them more accurately,
especially near the boundaries that are so problematic for Aristotelian approaches. Fuzzy logic is a conceptual framework that has important potential for
the advancement of scientific inquiry. We contend that paleoanthropologists will
be well served to consider this potential in our classification endeavors.

ACKNOWLEDGMENTS

We would like to thank C. Michael Barton, Mary W. Marzke, Geoffrey A.
Clark, George L. Cowgill, Milford H. Wolpoff, and C. Loring Brace for partici-
pating in stimulating discussions over the nature of species and fuzzy sets, and
for providing insightful comments on this manuscript.

6

Problems and Limitations of Absolute Dating in Regard to the Appearance of Modern Humans in Southwestern Europe

H. P. SCHWARCZ

Modern humans appeared in southwest Europe by at least 40,000 years ago. In this chapter we shall equate this event with the arrival of a particular hominid subspecies, *Homo sapiens sapiens.* We shall assume that the *sapiens* subspecies replaced (in some sense) another subspecies of *Homo sapiens,* namely, *H. sapiens neanderthalensis.* Other names have been used for these taxa (Stringer and Andrews 1988b) but, for the purposes of this paper, it will suffice to consider the event as the replacement of one by the other sometime around, or prior to, forty thousand years ago in southwest Europe. More or less synchronously with this biological transition there also occurred a major change in lithic industries in the same region, a transition from flake-based industries of the Middle Paleolithic to predominantly blade-based industries of the Upper Paleolithic. The earliest blade industry, the Châtelperronian, appears to be a distinct cultural manifestation (Mellars 1989b) and probably not a direct precursor of the earliest Upper Paleolithic industry, the Aurignacian, but I will not attempt to delineate the relation between these cultures here. The transition from Middle to Upper Paleolithic is broadly coeval with the transition from *H. s. neanderthalensis* to *H. s. sapiens* in southwest Europe, and it is commonly assumed that there is a linkage between these two phenomena. Again, I do not intend in this chapter to discuss the evidence for or against such a linkage, or the precise nature of the cultural transition. Rather, I shall take as my starting point the assumption that both of these transitions are well-defined, in the sense that a particular hominid skeletal component or a particular lithic assemblage (from one level of a given site) can be decisively assigned to one side or the other of the respective transitions. In the case of the skeletal remains this assumption cannot always be made because some specific hominid remains, especially fragmentary ones, present serious taxonomic problems.

The purpose of this chapter is to discuss some aspects of the chronostratigraphy of deposits in which these transitions have been observed. In particular, I

shall consider the limitations of absolute dating (to be defined below) in identifying some potentially interesting aspect of the transitions, and whether relative
dating can be used to complement absolute dating in resolving these transitions.

REGIONAL VS. SITE STRATIGRAPHY

These transitions have been recognized on the basis of data obtained at individual sites in southwest Europe. At each site the cultural stages occur in a
stratigraphic sequence which must also correspond to some chronological sequence. Likewise, hominid fossils have been obtained at particular stratigraphic
levels at some of these sites, and ages can, in principle, be assigned to these
remains. A few hominid remains have unfortunately been collected without
records of a well-defined stratigraphic context, but in some cases the context can
be inferred on the basis of the nature of the site, and the associated fauna and
artifacts.

It is useful to compare this situation with that normally encountered in geological stratigraphy. Conventionally, we assume that any particular stratigraphic
sequence (column) is a vertical section through a regionally continuous sequence
of laminar (sheetlike) strata. The stratigraphic columns represent fortuitous exposures in which these laminar bodies are exposed in cross section, permitting us to
determine their vertical sequence. Clearly this model is not strictly applicable to
archaeological sequences in most regions. The sections (columns) exposed at
particular sites are generally *not* remnants of once-continuous, superimposed
sheetlike bodies. Nevertheless, they have some characteristics in common with
the geological stratigraphic model. In particular, they represent a preserved record of what must have been regionally distributed patterns of culture or preva-
lence of various biological species (including hominids) over the region in which
the sites occur. Thus each site constitutes a long-term "trap" which sampled,
through a certain time interval, the regional distribution of culture, fauna, and
flora prevalent in the vicinity. The records preserved in these traps are imperfect
because they are commonly discontinuous, both because of noncontinuous deposition and because of erosion. Also, the presence or absence of faunal and
cultural evidence is partly determined by the behavior of the hominids who
occupied the sites and is not a simple reflection of the prevalent fauna or lithic
industries. However, we can treat them conceptually as more or less continuous
samples of hypothetical *paleosurfaces*. We define a paleosurface to be a map of
the areal distribution of culture and biology at a given point in time, over a
specific region. The records at individual sites (traps) can therefore be used to
reconstruct the successive paleosurfaces over that region. Given the uneven
distribution of these site-records, it is more useful to think of them as test samples
to be used to evaluate possible hypothetical regional distributions; any particular

site either should be consistent with a hypothetical regional pattern for a specific period or should be seen to lack evidence for that period because of discontinuous deposition or erosion.

Clearly if we are to attempt to reconstruct regional distribution patterns from individual sites, the stratigraphic subdivisions cannot be based on lithology, cultural features (industries), or biology; they must be constructed in a temporal framework that is independent of the local variations in lithology and other perturbing factors. Thus, a stratigraphic section of a site should have associated with it some estimate of the age of each horizon, either as a result of direct dating of that horizon, or by interpolation between dated levels (in situations where deposition appears to have been continuous).

To clarify these concepts, I shall now specify some possible regional models for the biological and technological transitions and how they would be reflected in the stratigraphy of sites in the region. For this purpose it is not necessary to specify how the chronology is obtained. This will be discussed in the next section where we consider the feasibility of discriminating between various regional models.

CHRONOSTRATIGRAPHIC MODELS

In order to clarify the nature of the problem, let us consider three models for the biological and technological transitions:

a. *Isochronous:* both transitions occurred everywhere at the same time, within the degree of temporal resolution given by available "clocks"; the biological and technological transitions are linked.
b. *Poly-isochronous:* The transitions are each isochronous (as defined above) but occurred at different times (the transitions are "disjunct").
c. *Diachronous:* The time of the transitions varied from place to place; in this model, the biological and technological transitions may have been either linked or disjunct.

These models can be represented diagrammatically in both plan and section, as shown in Figures 6.1–6.3. The plans are maps of paleosurfaces at particular times during the transition period. The sections resemble to some extent the conventional "fence diagrams" used in stratigraphy to construct linkages between stratigraphic sections as observed at a number of outcrops in a given region (Shaw 1964). The important difference to bear in mind here is that the stratigraphic sequence exists *only* at the points where the sections are observed (i.e., the traps). One could refer to this as golf-hole stratigraphy; the events of the past are only recorded as a result of material that has fallen into golf-holes scattered

SITES

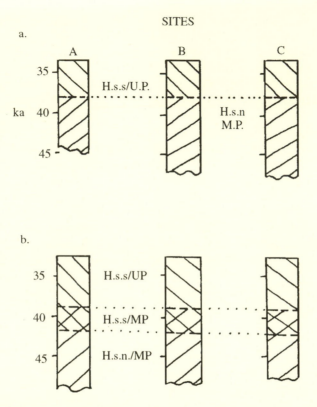

Figure 6.1. Schematic stratigraphic sections of three typical sites in southwest Europe illustrating possible isochronous models for the transition from *Homo sapiens neanderthalensis (Hsn)* to *Homo sapiens sapiens (Hss)* and from Middle Palaeolithic (MP) to Upper Palaeolithic (UP) industries: *(a)* simple isochronous transition (model a); *(b)* polyisochronous transition (model b). Time scale is for illustration only.

over the landscape. Any other surficial records of these events that may have existed have been eroded away. Again, remember that the filling of the traps is not a purely physical phenomenon. It is due in large part to the actions of hominids, especially as regards the components of the fill that interest us here: artifacts and hominid remains.

In stratigraphic section (Figure 6.1a) the isochronous model (model a) has a very simple appearance; in each section both transitions occur at the same stratigraphic position (i.e., at the same date). Figure 6.1a exemplifies a characteristic of all such sections: the vertical scale must be defined in terms of an absolute time scale (in years). Sequential variation in lithology at the site cannot be used to establish this scale, and it may even confuse attempts at correlation. Obviously major regional climatic changes could result in synchronized changes in the nature of the accumulating sediment (Laville et al. 1980), but the primary cor-

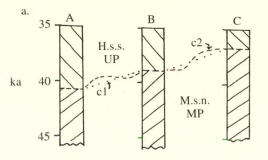

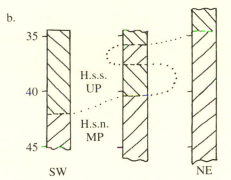

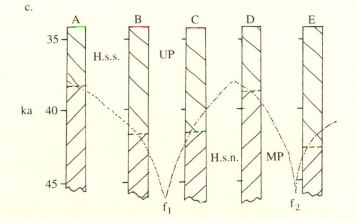

Figure 6.2. Schematic stratigraphic sections for three typical sites in southwest Europe illustrating possible diachronous models for the biological and technological transitions: (*a*) linear (planar) and nonlinear (nonplanar) diachronous invasion from the east (models c1 and c2, respectively); (*b*) model c3: reversed motion of boundary seen in part of region (near site B).

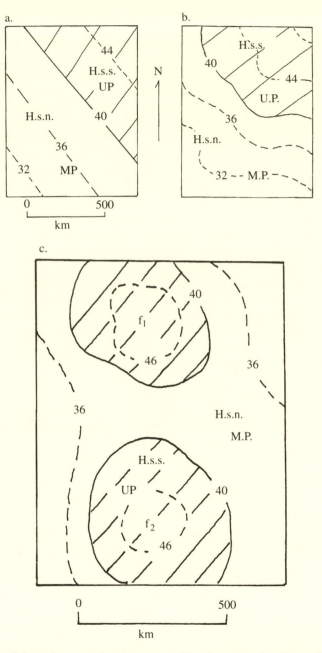

Figure 6.3. Schematic maps showing expected spatio-temporal patterns of the transitions in southwest Europe: *(a)* model c1: linear (planar) diachronous invasion from northeast; *(b)* model c2: nonlinear, diachronous invasion from northeast; *(c)* model c4: multifocal dispersion (foci f_1, f_2). Each map shows distribution of biological and technological types at 40,000 B.P., and positions of boundaries before and after this date.

relation cannot be based on these transitions since they may be diachronous or locally unrepresented. Only absolute (e.g., radiometric) chronology should be used when we are testing which (if any) of these models is applicable.

Model b differs from model a only in that the level at which the biological transition occurs is higher or lower than the level of the technological transition. As in model a, both transitions are always at the same level (i.e., date) at each site. If we consider how we might test these models, it is quite simple in principle to eliminate either "a" or "b" or both. If we found that absolute dates for the Middle Paleolithic at one site were younger than dates for the Upper Paleolithic at another site in southwest Europe, then the simple isochronous model for that transition could not be valid. This disregards the question of accuracy of the dates, however, to which we shall return. Maps of models a and b are rather uninteresting: at dates prior to the time of transition, the region is entirely covered by a Middle Paleolithic/*H. s. neanderthalensis* complex; after the transition, *H. s. sapiens* and Upper Paleolithic culture cover the region.

Model c (diachronous transitions) is really a family of models according to which, at any given time, part of the region was dominated by *H. s. neanderthalensis* + Middle Paleolithic while, in another part of the region, *H. s. sapiens* + Upper Paleolithic dominated. Therefore, some boundary must have existed which separated these two regions (Figure 6.2). The simplest form of this model (Figure 6.2a), assumes uniform (planar) diachrony, that is, that the boundaries migrated at a uniform rate across the landscape. If we sample a series of stratigraphic sections in a transect across the region, the date of the transition gets uniformly younger or older along the transect. Successive paleosurfaces representing this transition would show the boundaries migrating gradually and uniformly across southwest Europe. Each successive boundary is more or less parallel to its predecessor, and we can summarize a series of paleosurfaces on a single map showing the position of the boundary at successive times (Figure 6.3a).

In this example the earliest paleosurface shows the boundary to be located at one edge of the region. That is, the region is pictured to have been *invaded* by a successor species *H. s. sapiens* and its culture (Upper Paleolithic). This model is the most likely one to describe the transition in southwest Europe for the following reasons. First, we know that the subspecies *Homo sapiens sapiens* already existed long before the biological transition occurred in southwestern Europe (Schwarcz et al. 1988; Valladas et al. 1988). Therefore, there must have been an invasion of the new subspecies into this region from an external region of Europe in which it was already established. Secondly, it is observed that the transition from *neanderthalensis* to *sapiens* is closely coupled with the transition between Middle and Upper Paleolithic industries. It is therefore likely that the appearance of this cultural transition was also in the form of an invasion. However, unlike the successor hominid species, we do not have evidence of Upper Paleolithic traditions well-established outside southwest Europe long before their arrival here.

Therefore we cannot rule out the possibility that the Middle to Upper Paleolithic transition was the result of indigenous invention of a new lithic technology, even though the same industry is widespread in eastern Europe and western Asia.

The second version of model c is topologically identical to the first but slightly more complex. In this model, we assume that the *sapiens* and *neanderthalensis* regions are separated by irregular (nonlinear) boundaries on each paleosurface (Figures 6.2a, 6.3b). Individual stratigraphic sections throughout the region would hardly be distinguishable from those of model c1, and the irregularity of the boundaries is only made apparent when the sections are used to construct paleosurfaces.

A still more complex model (c3) assumes that a boundary can retreat to its former position, at least for a short time. As a result, for example, we could find sites at which the sequence MP/UP/MP/UP is observed (Figure 6.2b). Such reversals would not be seen in every site, of course, only at those fortuitously located near a region of reversed motion of the boundary.

Finally, according to a multifocal model, c4, the transition occurs not along a single boundary line (straight or curved) but spreading out from a number of discrete points or foci. This could have occurred if the populations or cultures were initially able to gain a foothold only at specific sites, from which they then expanded (Figures 6.2c, 6.3c). The climax of such a model would probably be indistinguishable from model c3, in which an irregular boundary separated the two regions.

All of these models (a, b, c1–4) assume that the transition is sharp. That is, there is no continuum between the Middle and Upper Paleolithic or between *neanderthalensis* and *sapiens*. For the transition between hominid species, however, this assumption is questioned by some scholars (Trinkaus 1986; Wolpoff 1989b). If either of these transitions were continuous rather than discrete, then defining the location of boundaries at particular times would be extremely difficult, given the other uncertainties that confront us.

These models have been erected mainly in order to assist in interpreting the chronological data obtained at individual sites. If we insist that the only relevant framework on which to reconstruct paleosurfaces is a set of absolute chronological dates for each site, then it is obvious that the choice of the correct model (if there is one) depends strongly on the accuracy and precision of these dates.

In models c1–3 we assume that the successor subspecies *(sapiens)* and the Upper Paleolithic culture gradually invaded southwest Europe. This invasion is most likely to have begun along the region's eastern margin, and the boundary would most likely have swept across the region in a westward or southwestward direction. Therefore, we should expect to find dates of the two transitions getting younger toward the Pyrenees and ultimately toward Gibraltar.

The choice between models a or b and the various models c1–3 is partly dependent on the time resolution of the clocks at our disposal. In effect, even model a (isochronous replacement) is a variant of model c, but with a rate of advance of the boundary too rapid to be resolved. That such an advance of a hominid population and culture across a region could be that rapid is attested to by the record of the arrival of modern humans into North America (Martin 1973). The spread of *H. s. sapiens* through North America essentially follows model a. From the time of arrival in Beringia to the earliest confirmed occurrences in South America, less than one thousand years elapses. This interval is fairly easily resolvable because it is a large fraction of the total age of the event, about twelve thousand years. At an age of 40,000 B.P. , however, an event with a duration of only one thousand years would be only slightly larger (2×) than the conventionally quoted error of ^{14}C dates, that is, ±1%. In fact, interlaboratory comparisons have shown that such quoted errors, based on counting statistics alone, tend to underestimate the uncertainty in the age. Therefore, model a could fit the *apparent* spatiotemporal patterning of the arrival of *H. s. sapiens* into southwest Europe (a much smaller region than the Americas) even though better chronological resolution might show that models b or c are more accurate descriptions of the events.

ABSOLUTE AND RELATIVE DATING

The remainder of this chapter will discuss the dating of the transitions in the light of these models. We must distinguish two types of dates which are both relevant here:

Absolute dates: the age of a stratigraphic level at a site given in years before the present (or before any other temporally defined event, e.g., the birth of a prophet). The term *absolute* in this context is not meant to connote any degree of precision or accuracy, but merely to distinguish such dates from relative dates. Most absolute dates are based on a radioactive decay process (e.g., ^{14}C) or on the cumulative effects of radiation on matter (e.g., TL, ESR).

Relative dates: the position of a stratigraphic level in some temporally defined (but not necessarily absolutely dated) sequence. Examples of such sequences include faunal or floral successions, regional lithostratigraphic sequences, amino acid (D/L) "chronologies," and paleomagnetic sequences (either secular variation or reversals). Such sequences may be valid only in a local context—for example, the climatically determined stratigraphic sequence of detrital fill in caves and rockshelters defined for southern France by Laville (Laville et al. 1980). They may, however, be of global extent, such as the oxygen isotope stratigraphy of deep-sea cores (Martinson et al. 1987).

In principle, we would of course prefer to define the chronology of the transitions in terms of absolute dates, and thus to distinguish between the proposed models. These models have been presented as discrete and mutually exclusive descriptions of the transitions. We may not be able to match chronological data to any of them, however, owing to the inaccuracy of the available dates. Thus, chronological inversions may appear to rule out models a or b (Figure 6.3), whereas the scatter of dates may not give a satisfactory fit to any of models c 1–4. Such ambiguities should be more common as we go further back in time, because the error attached to a date is approximately a fixed fraction of the age (e.g., 1–5%), a quantity which therefore must increase with age (Blackwell and Schwarcz 1993). Also, as we reach the natural limit of any dating method, the error in the age increases exponentially, and the method becomes more susceptible to effects of contamination, diagenesis, etc. In the present context, these problems are especially serious for the ^{14}C method.

For these reasons it may be preferable to complement absolute dates with evidence from relative dating which, if properly calibrated against absolute chronologies, could better define the timing of the transitions. I shall later show how this could be done for the case of the *H. s. sapiens*/Upper Paleolithic transitions, after first discussing the evidence available from absolute chronology.

METHODS OF ABSOLUTE DATING

^{14}C: The dating limit of this method lies between 40,000 and 50,000 years ago, depending on the background attainable in a particular laboratory. In practice, ^{14}C dates become highly susceptible to contamination for samples older than about 35,000 B.P., at which point less than 1% of the initial ^{14}C activity remains. An amount of modern carbon comparable to this could easily be introduced as a contaminant and would decrease the apparent age by one half-life (about 5,700 years). This may be of the same order of magnitude as the time scale of human migration. It is also much greater than the error usually ascribed to ^{14}C dates, based on counting statistics alone. Contamination with old ("dead") carbon is also possible and would raise the apparent age, but relatively large amounts of contaminant are required to have a measurable effect, and the older the samples are, the more resistant they become to this aging effect.

There is no particular advantage in using AMS dating for most samples in this time range, since the contamination effects may be as great or greater in using small samples. AMS analyses of biochemically unique substances (e.g., collagen) result in increased accuracy, since we can be assured that they date to the time of occupation of the site.

Both transitions occur, as it happens, very near to the point in time at which ^{14}C dates become highly susceptible to contamination, at or near 40,000 B.P. Relatively few high-precision ^{14}C dates have been obtained for this transition so far. Recent studies have shown the potential of AMS to define the timing of the Mousterian/Aurignacian boundary, which was shown to occur at 40,000 B.P. at two localities in Spain (Bischoff et al. 1989; Cabrera Valdés and Bischoff 1989).

Other older studies using conventional beta-counting for this boundary are highly suspect because of the high risk of contamination at such low activity levels. Paul Mellars (personal communication, 1992) is attempting at present to use AMS to date the transition across western Europe.

In addition to the problems of obtaining highly precise dates by ^{14}C we have the problem of a lack of a calibration for dates in this time range. This makes it difficult to establish precise correlations between ^{14}C dates and those obtained by other methods. Variations in apparent atmospheric ^{14}C activity of up to 50% are inferred from the U-series-based calibration of Bard et al. (1990), which could lead to uncertainties in dates of up to two to three thousand years, if uncorrected.

TL: Thermoluminescence dating of burnt flint (Valladas et al. 1988) is applicable for the time range of interest here. However, in that interval, this method has an intrinsic error much larger than that of ^{14}C dating. This error (or, rather, uncertainty) arises principally from errors in the determination of the environmental dose rate of radioactivity, and to a lesser extent from uncertainty regarding (and possible past variation in) the water content of the sediment surrounding the dated materials. Dose rates measured at the site today are only approximate estimates of the average dose rate that existed at the site during its total burial history. These uncertainties are not improved by higher precision of TL analyses.

ESR: This method is applicable to both tooth enamel and burnt flint, but the errors are comparable to, or worse than, those of TL. Added to the problem of uncertainty in the environmental dose rate is that of the uptake of uranium by enamel and dentine (Grün et al. 1987). This can result in an uncertainty of up to 50% in the age, although for younger samples such as those we are discussing here, there will generally only be a few percentage point differences between the EU (early uptake) and LU (linear uptake) dates.

U-series: This is potentially the most precise method applicable to the problem of dating the transitions, and least susceptible to errors. The method is ideally applicable to flowstones or stalagmitic layers which are interstratified in Upper Paleolithic or Middle Paleolithic sequences. The newly developed methods of mass spectrometric U-series (MS-US) dating can attain a precision of 1-2% (Edwards et al. 1987; Li et al. 1989). The more conventional method of alpha particle spectrometry can provide a precision of no better than 5% and commonly up to 10% of the date, which would not be adequate to define the transition.

While potentially capable of giving high-precision dates, the scarcity of suitable samples for dating is a major limitation. Only clean stalagmitic deposits appear to behave as closed systems for U and Th isotopes (Schwarcz and Blackwell 1991). Other potentially datable materials found in sites, such as bones and teeth, appear to be much less suitable, although some promising results have been obtained with tooth enamel. Stalagmitic deposits are especially rare in the time interval of the technological transition because it falls within a glacial stage. Speleothem deposition tends to slow or cease entirely in temperate regions during glacial stages as a result of the decrease of organic activity in the soils over the caves (Gascoyne et al. 1983). Nevertheless, some flowstones would be expected in caves of southern France and Spain, which could provide precise dates for these sites.

K/Ar (^{40}Ar/^{39}Ar) and fission track: These methods are applicable to volcanic rocks and have proven very important in calibrating the early stages of hominid evolution in East Africa. They are less widely applicable in southwest Europe, however, because of the rarity of volcanic rocks in archaeological contexts. Also, the precision of these methods is lower than required here when used to date such young rocks, although single-crystal laser fusion of feldspar grains in volcanic tephra has some promise (Lo Bello et al. 1987).

RELATIVE DATING METHODS

The previous discussion shows that the lack of precision and accuracy of absolute dating methods makes it difficult to establish the absolute timing of the biological transition with sufficient precision to test the models that we have discussed. We can now inquire whether any of the methods of relative dating could improve this situation at all. In particular, we can ask whether any non-radiometric methods provide time markers that can be used to correlate events between sites. To give an example of such a marker, consider reversals of the earth's magnetic field. Although no such reversal occurred in the time range of interest here, it typifies an event that happened on a global scale and that we can be reasonably certain occurred at the same time everywhere it is observed: it was an *isochronal* event. Are similar isochronal events preserved in the "golf-hole" record around the time of the biological transition?

Most of the relative chronometers consist of records whose character depends to some extent on climatic change. We can distinguish between two types, those that record a climate signal directly and those that are based on processes whose rates are dependent on climate and especially temperature. An example of the latter is amino acid racemization (AAR) as applied to the dating of shells (Miller and Mangerud 1985). This is based on the temperature-dependent rates of conversion of L-amino acids to their D-forms. The accuracy of such chronometers would therefore be dependent on knowledge of past changes in temperature. For both types of methods, if we were only interested in establishing the relative timing of events in our separate stratigraphic sequences, then it would suffice to know that the temperature history over the region had been uniform.

Paleoclimatologists believe that continental climate is coupled to the global climate as recorded in the oxygen isotope record of deep-sea cores (Martinson et al. 1987:Fig. 4). This is obviously a first approximation, since this record is believed to be principally controlled by changes in global, continental ice volume, and it is not clear that this should be tightly coupled to continental temperature variations. Various continuous records of continental climatic variation have been shown to correlate with the deep-sea record, notably the variation in magnetic susceptibility of Chinese loess deposits (Kukla et al. 1990). Taking this record as an example of such a land-sea correlation, we see a general agreement between the shape of the isotopic and susceptibility curves, but in any given interval of a few thousands of years there are significant intersite differences at supposedly correlative dates. Since there is no precise absolute chronology attached to such records, it is impossible to test regional isochronicity of comparable climatic stages (and in any case, the relation between specific climatic parameters [temperature, humidity, etc.] is unclear).

Indeed, it is likely that, over any period of changing global climate, the value of any specific parameter will differ across a region in much the same way that it does today. Therefore, any chronometer of the second type, which is based on the temperature-dependent rate of a chemical reaction, would run at different (but unknown) rates across the region and could not be used to establish isochronicity of cultural or biological stages. Even if the climate had remained relatively constant during the transitional interval (as we shall argue below), later temperature changes would also affect today's D/L ratio. Therefore the chronometric record would depend on the stability and uniformity over the *entire* record from then until the present, which cannot be safely assumed.

A relative chronometer based on direct records of climatic change might provide a more precise test of synchronicity. Changes in climate at a given site are likely to be coupled to changes in global climate with a response time shorter than one thousand years, even though the amplitude (i.e., the severity) of climate change may differ across a region. Thus, as climate worsens (becomes more glacial) on a global scale over a given thousand-year time span, it is assumed that isotherms will migrate across a region approximately parallel to their present-day positions, and likewise for other climate indicators. Therefore, if the paleotemperature record in the interval in question has varied in some characteristic form, these variations could be correlated from to place to place even though their amplitudes would differ.

Do archaeological sequences contain suitable paleothermometers capable of recording such temperature variations? The answer is a qualified yes. For example, we can use the oxygen isotopic composition of fossil bones to make inferences about climate where the animal lived (Luz et al. 1990). All such isotopic records are affected by more than one climatic variable, however. The $^{18}O/^{16}O$ ratio of bones is, for example, dependent on both temperature and humidity; if humidity changes were relatively minor in the region, then variations in $^{18}O/^{16}O$ of bones would be a good proxy record of temperature change. Fossil bones are found at many archaeological sites, and we could obtain isotopic records of climate at each site and use them to correlate between sites. There is, however, one more problem which effectively prevents us from using this or any other paleoclimate-based method to correlate between sites over the interval from

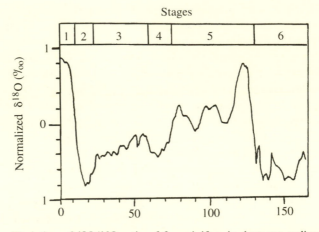

Figure 6.4. Variation of $^{18}O/^{16}O$ ratio of foraminifera in deep sea sediments over the past 250,000 years (data from Martinson et al. 1987).

approximately 50,000 to 30,000 years ago. This can be seen from closer inspection of the deep-sea record for that interval (Figure 6.4). From about 60,000 to 30,000 B.P., that is, during isotope stage 3, the deep-sea isotopic record is relatively flat. This is, indeed, the longest segment of relatively constant isotopic values over the past 250,000 years. Therefore, if local paleoclimate records resemble this one, then it would be exceptionally difficult to use such records in order to correlate between sites. The paleoclimate records at each site are likely to be comparably flat and featureless and lacking in characteristic signatures of climatic substages.

CONCLUSIONS

I have argued that the biological and technological transitions in southwest Europe should be looked upon as regional currents of culture and genes that moved across this region over some time interval in the past. According to this model, each individual site represents an "outcrop" of the total chronostratigraphic sequence produced by these currents. Unlike conventional geological stratigraphy, a continuous record of these stages never existed on the terrain that lies between these individual sites. Each site, rather, acted as a local trap (a "golf hole"), keeping a partial or complete record of these stages within its sedimentary layers. By connecting chronometrically correlative strata from site to site, it is possible to reconstruct the distribution of the biological and cultural domains through time, what we have called the paleosurfaces, as they were progressively altered by the arrival of new forms.

This model tends to emphasize the arrival of new forms (i.e., new genetic and cultural information) from *outside* the region of southwest Europe. The same analysis, however, will provide a test of the extent to which some or all of this information was generated locally within the region by in situ evolution. Certainly it is not necessary for each new technological feature to have arrived from outside; it is equally likely and, indeed, commonly assumed that populations within a region are constantly adapting to environmental change (including changes in other human populations and their cultures) by changing their life styles.

When we try to test these models, however, it appears that the cards are stacked against us. First, we recognize that only chronometric (absolute) dating of strata can be used to reconstruct the paleosurfaces. Correlation of such features as the cultural or biological transitions themselves clearly cannot be used to define the configuration of boundaries on the paleosurfaces, nor can stratigraphic features, which are too strongly controlled by local conditions of sedimentation at the site. When we turn, however, to the available set of absolute chronometers, each in turn is found to be too imprecise and/or inaccurate to provide the needed correlation. If we use the rate of advance of hunters through the Americas as a model, we would need a resolution of better than one thousand years to recognize a wavelike pattern of cultural or biological advance across successive paleosurfaces. Poorer resolution might give the appearance of an instantaneous transition over the entire region (model a, isochronal takeover).

While the resolution of radiometric chronometers is too poor to provide the needed correlation, it appears that some relative chronometers have the potential to provide this time scale—assuming there had been significant climate change during the time period over which the transition took place. If we use the deep-sea isotope record as a proxy for the continental paleoclimate record, however, it appears that this proposal is also doomed. This cannot be known with certainty until some exploratory studies have been made. I suggest that we explore the possibility of using records of $^{18}O/^{16}O$ in fossil bones at archaeological sites as a correlation tool between the sites. This work should begin by studying a more recent portion of the record, such as the transition between isotope stages 2 and 1, at around 12,000 years ago, where ^{14}C chronology is sufficiently precise and accurate to permit us to test whether the isotopic record is capable of providing time markers of equivalent precision.

ACKNOWLEDGMENTS

This research was partly funded by grants from the National Science Foundation (USA) to the University of California (BNS 8801699, to F. C. Howell) and the Natural Sciences and Engineering Research Council, Canada. This chapter is based on an essay first published in 1993 while I was at the Subdepartment for Quaternary Research, Cambridge University. My visit to England was made possible by a grant from the Royal Society of London. I am grateful to V. Cabrera for her encouragement, and to Paul Mellars for his comments on various versions of the manuscript.

III

WESTERN PERSPECTIVES
Latin Europe and the Levant

The chapters in this section are by archaeologists and human paleontologists who pertain to the intellectual traditions of Latin Europe—French, Spanish, and Italian perspectives are represented. Although a range of views is evident, the dominant paradigm is that of replacement, perhaps reflecting the historical importance and precedence of francophone research—replacement views have been strongly entrenched in France since World War I. A prominent replacement advocate, Bernard Vandermeersch rejects the possibility of neandertal-modern continuity and suggests the Near East as a possible point of geographical origin for moderns. He also takes up the question of the unity or lack thereof of the Aurignacian, associated with early moderns in Europe but not in the Levant. Whatever may be concluded from the archaeology, however, the biological origin of moderns is a separate issue.

Dominique Gambier reviews the human fossils in France associated with Aurignacian archaeological assemblages and identifies the characteristics of the sample known to date. Her analysis shows that the "substitution" model is best supported empirically and can thus best account for the appearance of moderns in Europe. However, the scarcity of fossils from pre- and early Aurignacian contexts cannot be ignored, casting doubt on the authorship of these transitional industries.

Amilcare Bietti looks at the archaeology of the Middle to Upper Paleolithic transition in Italy from the perspective of three hypotheses generated from human paleontology. Although it is clear that they were not synchronous, the fact that the Italian record is sparse, particularly in regard to human fossils, precludes a clear resolution of either the cultural or the biological transition. He makes the important point that perceptions of pattern are largely determined by the conceptual frameworks that contextualize data. Italian scenarios for the Middle to Upper Paleolithic transition are filtered through a conceptual framework that differs in important ways from that of the rest of Europe.

Spain's most prominent human paleontologist, María Dolores Garralda, discusses human fossil remains pertaining to the Middle to Upper Paleolithic transition interval in Iberia (ca. 30,000–50,000 B.P.). The circumstances of discovery, relative and absolute chronologies, and the morphological data themselves underscore both the presence and variability of *H. sapiens neanderthalensis,* in all cases associated with Mousterian archaeological assemblages. However, some few individuals with archaic morphology show up in contexts that could be considered to pertain to the earliest Aurignacian, calling into question the authorship of the early Upper Paleolithic.

The French archaeologist Jean-Philippe Rigaud analyzes the Middle to Upper Paleolithic transition according to diagnostic criteria invoked historically by European prehistorians on the basis of their more or less direct knowledge of the data base. Six scenarios for the archaeological transition are presented and discussed. Rigaud concludes that the origin of the Aurignacian lies outside both Europe and the Levant, and that its appearance in Europe took place while diverse local cultures with laminar technologies were already in the process of evolving into an Upper Paleolithic stage.

The following chapter, by Spanish prehistorian Joaquín González Echegaray, first critiques the historicist biases implicit in "invasion" hypotheses for early Upper Paleolithic origins in Europe and then provides a historical overview of the European conception of the early Upper Paleolithic from the 1920s up until the present. Two episodes of accelerated interest in the Aurignacian correspond to the mid-1920s, when a north African source was entertained, and the 1950s–1960s, when a Near Eastern origin was favored. He then turns to post-1990 research in southwestern Europe, with particular reference to Spain, and concludes that the notion of an east-west temporal gradient implied by Aurignacian invasion scenarios can no longer be sustained. The transition evidently was more of a "process" than an "event," and it lasted more than ten millennia.

In contrast to the rest, the chapter by Victoria Cabrera and colleagues is focused on the geological and cultural data from early Aurignacian level 18 at El Castillo cave, in Cantabria. El Castillo preserves the longest record of human occupation on the Iberian Peninsula (Acheulean to Bronze Age) and is one of the longest cave sequences in the world. Exceptionally well dated by ten AMS radiocarbon determinations to ca. 39,000 B.P., level 18 records the Middle to Upper Paleolithic transition in the north of Spain. The very early dates support a local transition between the Quina Mousterian and the early Aurignacian, without any external influence, and call into question the east-west temporal progression implied by most invasion scenarios.

7

The Near East and Europe
Continuity or Discontinuity?

BERNARD VANDERMEERSCH

Since the early 1980s, the origins of modern humans have become the object of intensive research and debate among paleoanthropologists and geneticists. Discoveries in the Near East and in sub-Saharan Africa have profoundly modified the paleontological and, especially, the chronological foundations upon which we depend in order to frame hypotheses. In Europe, the discovery of the St. Césaire neandertal has shown that at least in Charente, and perhaps for a short time, the neandertals survived the technological and cultural transition from the Middle to the Upper Paleolithic, as that transition had been conceptualized until recently (Lévêque and Vandermeersch 1980).

In this chapter, I review certain anthropological and prehistoric data bearing on the problem of the emergence and spread of modern humans. These data allow me to outline a hypothesis that I wish, first, to lay before the reader before exposing its weak points and murky areas, while reminding him or her that a hypothesis is only a provisional statement put forth as a possible explanation for certain phenomena of interest. Its role is to suggest new questions whose solutions might allow us to develop a new explanation that is closer to reality than its predecessors. The hypothesis, then, is a fleeting or transient thing, and the single problem it entails is to determine, at the moment it is framed, whether or not it can account for observed facts more adequately than its predecessors.

THE HYPOTHESIS

My hypothesis rests on two fundamental assertions:

1. The first is that there is discontinuity in the fossil record of Europe between moderns and neandertals. There is no evolutionary continuity between them; the former did not arise from the latter.

2. The second is that the biological and cultural record of Upper Pleistocene Europe can only be understood by taking its relationship with the Near East into account. The same is true for the Near East. The two regions have, during this period and probably before, exchanged both populations and cultures.

We will see below whether or not these two assertions can be accepted as stated, or whether there is room for nuance and subtleties in their formulation. The hypothesis, then, can be stated as follows:

> At the beginning of the Upper Paleolithic, morphologically modern humans arrived in Europe from the Near East (Vandermeersch 1992). They were the bearers of the earliest known Upper Paleolithic culture, the Aurignacian. Confronted with these new arrivals, the local populations of neandertals reacted by "inventing" the Châtelperronian, but this was insufficient to assure their survival, and they disappeared over an interval of several thousand years. This is the replacement hypothesis articulated by Stringer et al. (1984), among others.

IMPLICATIONS OF THE HYPOTHESIS FOR THE CONTINUITY POSITION IN CENTRAL EUROPE

We examine now some of the problems thrown into sharp relief by this hypothesis. The first and primary question, and that which affects in a certain manner all the rest, is this: Should one reject conclusively any possibility of the evolution of European neandertals in the direction of modern humans? There are two diametrically opposed answers to this question. Certain colleagues—mostly Americans—accept, following the work of Fred Smith (1984; Smith, Simek, and Harrill 1989), the possibility of neandertal-modern continuity in Europe, relying mainly on the fossils from central Europe in order to make this argument. Certain hominid skeletal fragments from a level considered early Aurignacian at the site of Vindija (Croatia) testify to a transitional stage between neandertal and modern morphology (Smith 1982, 1984). This is the case, for example, with the fragmentary frontal Vi 261/277/278, whose supraorbital torus was both reduced (especially laterally) and divided into supraciliary and supraorbital arcs. This would argue for a population in the course of losing its distinctive neandertal features (Malez et al. 1980; Smith and Ranyard 1980). In Czechoslovakia, the oldest modern fossils, also dated to the early Aurignacian, present traces of neandertal ancestry as well, essentially in the form of the persistence of robust morphology (Smith 1984).

We would have, then, in central Europe both evolved neandertals and moderns still in the course of transformation—that is, continuity between the two groups. What does this alleged continuity consist of, exactly? Excavations at

Vindija were not, unfortunately, conducted with the rigor characteristic of modern standards. Level G1, which contained the fragmentary fossils considered to mark the transition, has yielded only a single culturally diagnostic artifact—a fragment of a split-base bone spearpoint. This point has been used to link this level with the early Aurignacian, dated by comparison between 40,000 and 32,000 B.P. (Malez et al. 1980). How can we be certain that it is not an intrusive element introduced into level G1 from a higher, more recent level? The stratigraphy shows the presence of levels with many large blocks in them (roof fall); the interstices between them might have allowed for downward movement of isolated bones and artifacts, especially if the sediment were unconsolidated.

A CRITIQUE OF THE CONTINUITY POSITION IN CENTRAL EUROPE

With respect to the morphological characteristics of the fossils themselves, they are very difficult to interpret on fragmentary material, and what is generally invoked is a decrease in size vis-à-vis typical neandertals, not a change in shape. The supraorbital torus is considered to be smaller and less salient, the cranial vault more gracile, the anterior dentition smaller, and the chin more salient. It is true that the brow ridges preserved on the frontal fragments are not as developed as those of unquestioned neandertals, but a continuous torus is always present. In contrast, the main characteristic of supraorbital relief in moderns is not gracility—modern tori can be more pronounced—but rather the fact that the torus is divided into distinct supraciliary and supraorbital arcs. This latter (modern) configuration never shows up at Vindija. Only the thin cranial vault resembles the modern condition.

With respect to the vault fragments, it is difficult to determine whether they pertain to adult individuals. In addition, to cite only a single example, the variation among the vault fragments of neandertals from the Mousterian site of Marillac, dated to the early part of the last glaciation, is significant. With regard to chins, I have observed on the original specimens from Vindija that, while some mandibles do preserve a shallow linear depression below the alveolus, there is never a true chin, and the single fragment that preserves part of the ascending ramus also exhibits a large retromolar space.[1] Unfortunately, I have not had the opportunity to study the anterior dentition. They were, however, found as loose, isolated specimens; they are few in number; and they are not significantly smaller than their neandertal counterparts.

What is important to underscore here is that the arguments invoked by the partisans of continuity are arguments of size and scale: the last central European neandertals were more robust than their descendants. Although this is entirely possible, the important distinctions are not those of size, but of shape, so that—

were this argument to be substantiated—one would expect relatively gracile neandertal "morphology" to have preceded relatively robust modern "morphology." I submit that it does not. It is *morphology* that is diagnostically important, and, thus far, I do not see transitional morphologies (only differences in size) in the central European sample.

THE LEVANTINE DATA

It also appears to me that there are other, more theoretical arguments against the notion of a neandertal-modern transition. Possibly in sub-Saharan Africa, and certainly in the Levant, there are modern populations that long antedate the European transition interval (e.g., Border Cave and Klasies River Mouth in the Republic of South Africa, Qafzeh and Skhūl in Israel). These sites are dated (particularly well in the Near East) to ca. 100,000 B.P., some 60,000 years before the European transition (Valladas et al. 1988). It is also possible that modern populations very close to present-day mongoloids lived in Asia some 65,000 years ago, if the dating of Liujiang is correct. These populations were derived from local archaic *H. sapiens* stock. It would be necessary, then, to admit that modern humans, which comprise a single subspecies, developed in Europe from another population *(H. sapiens neanderthalensis)* than did moderns in Africa and Asia. In other words, the neandertal lineage would have progressively diverged from the rest of humankind over a period of some 300,000 years and would have had to rejoin the other modern populations abruptly, by a kind of evolutionary reversal regarded as exceedingly rare by most workers. Moreover, this phenomenon would have had to have been very rapid—perhaps only a few millennia—whereas some dozens of millennia would have been required in order for archaic *H. sapiens* to have evolved into modern *H. sapiens* elsewhere. However, the differences between neandertals and moderns are considerably greater than those that distinguish the latest representatives of archaic *H. sapiens.*

If the neandertal lineage had developed autapomorphies which resulted in a general skeletal architecture quite distinct from those of other groups of *H. sapiens,* it is highly improbable that, at the end of its evolution, it would have suddenly come to resemble those other groups. Morphological and metric comparisons show a strict resemblance between European Cro-Magnons and the Qafzeh and Skhūl fossils—a resemblance that would have been highly unlikely under the hypothesis of continuity, given their totally independent origins. To start from different morphologies in order to produce, at different rates, the same results seems to me to be an interpretation that is very difficult to accept. It would be necessary to put the neandertals in the same basket as the predecessors of modern Africans and Asians, which—if morphological analysis has any credibility at all—would be very difficult to do. Why derive modern Europeans from neandertals when these same modern humans existed beforehand at Europe's very gates? It seems to me that, for the moment, no data exist that would allow us to envision phylogenetic continuity between neandertals and moderns, and that numerous arguments can be raised against the idea.

THE CASE FOR A NEAR EASTERN ORIGIN FOR MODERN EUROPEANS

If we reject the idea of a neandertal origin for modern humans in Europe, the only alternative is to derive European moderns from the Near East. Two Levantine sites, Qafzeh and Skhūl, provide data on this subject. These two sites have yielded numerous human fossils, more than ten at Skhūl and more than twenty at Qafzeh. Even if most individuals are only represented by fragments, a number of relatively complete skeletons are well preserved, furnishing us with a certain appreciation of the internal variability within this population. The Qafzeh/Skhūl sample is fundamentally modern, and in fact very similar to Cro-Magnons, which led F. Clark Howell to suggest the designation "Proto-Cro-Magnon" in 1951, a term I have used since 1981 to describe the combined sample. I will not enumerate here the many morphological and metric similarities between the two groups (but see Vandermeersch 1981a, 1981b). When the term *Proto-Cro-Magnon* was proposed for the Near Eastern fossils, I thought (along with many other anthropologists and prehistorians) that they were chronologically near-contemporaries of the European Cro-Magnons. However, the recent TL and ESR determinations indicate an age of ca. 100,000 B.P. (Valladas et al. 1988), and it is evident that there is a 70,000-year gap between moderns associated with Mousterian industries in the Levant and those found with Aurignacian industries in Europe. This bioanthropological void must be filled, and it can only be filled in the Near East since we have good reasons to think that during this time interval nothing was going on in neandertal Europe. Unhappily, however, the only Near Eastern fossils that pertain to this period are those of the neandertals themselves (e.g., Kebara, Tabūn). What became of the "Proto-Cro-Magnons" there is an open question because we have, for the moment, neither chronological nor geographical continuity to back up the compelling arguments in favor of morphological continuity with Cro-Magnons. Between Qafzeh and Skhūl, and the earliest European sites with indisputable modern humans, there is a gap of 60,000 years and 2,500 km. Human fossils from the Upper Paleolithic of the Near East are recent; the discovery of Ohalo represents the oldest adult skeleton from this period, and it is dated to 25,000–20,000 B.P. It is still too soon to know whether it is possible to envision continuity between it and the Qafzeh/Skhūl fossils.

ARCHAEOLOGICAL IMPLICATIONS—THE LEVANT

It is perhaps possible, at least in theory, to supplement the scarcity of human paleontological data by taking the archaeology into account. If there is continuity in cultural evolution, one might suppose that this continuity rests, at least partially, on continuity between the populations that made the artifacts. It is indeed difficult to admit that cultural evolution set in motion by one population could be carried to completion by another, since it would be necessary to suppose that the former had abandoned its own technological and cultural traditions. However, cultural continuity without evidence of external contributions would seem to imply biological continuity. One Near Eastern site at least seems to document this kind of continuity—Boker Tachtit in the Negev Desert, excavated by Anthony Marks.

Four archaeological levels have been recognized at Boker Tachtit. Level 1, representing a final Middle Paleolithic, has yielded numerous blades and levallois points from non-levallois nuclei. Level 4 is early Upper Paleolithic; levels 2 and 3 represent the transition. Three dates have been published for level 1: 45,330 ± 9050 B.P., 44,980 ± 2420 B.P., and >43,360 B.P. However, these radiocarbon dates are at or near the limits of the method. The low terrace of the valley in which the site is located contains industries of the same type as those of level 1, and it has been dated to the same age by U/Th. The industries do not seem to be clearly connected with the Aurignacian.

The situation in the Near East is, then, as follows:

1. From the human paleontological point of view, modern humans are present at ca. 100,000 B.P., and then again at ca. 25,000 B.P. Neandertals, or "pre-neandertals," occur at Tabūn before 100,000 B.P. and up to about 50,000 B.P.
2. From the archaeological point of view, there is probable continuity between the Mousterian industries manufactured by both neandertals and moderns, and the Upper Paleolithic, which is always associated, when human fossils are present, with the remains of moderns.

It is interesting to note that an early Aurignacian at Qafzeh (Ronen and Vandermeersch 1972), unfortunately undated, has produced a frontal and some mandible fragments, all of which are completely modern in morphology. It seems reasonable to me to suppose that, from the beginning, the Levantine Upper Paleolithic was manufactured by modern humans.

ARCHAEOLOGICAL IMPLICATIONS—EUROPE

I turn now to the European situation, noting in passing that Turkey, which lies athwart the only possible migration route between the two regions, has both

Mousterian and Aurignacian industries that have so far not been dated and have produced only a few isolated teeth. The discovery by Lévêque in 1979 of a neandertal skeleton in the Charente cave of St. Césaire in a Châtelperronian level has overturned our ideas about the nature of the Middle to Upper Paleolithic transition, and the disappearance of the neandertals. But before discussing the consequences of this discovery, I wish to acknowledge the contribution of André Leroi-Gourhan who, in his study of the human remains from the Châtelperronian levels of Arcy-sur-Cure, first envisioned the possibility of a neandertal role in the manufacture of this assemblage (Leroi-Gourhan 1958).

It seems evident that, in Europe, the neandertals persisted into the beginning of the Upper Paleolithic; their remains have been recovered at St. Césaire and very likely at Arcy-sur-Cure. This is certainly true at least at St. Césaire, where the fossils are clearly those of a neandertal. This association is also supported circumstantially by arguments for cultural continuity between the Mousterian (more specifically the Mousterian of Acheulean Tradition, type B) and the Châtelperronian (Bordes 1968a). The cultural evolution is paired, then, with biological continuity, which seems eminently logical. The chronological break-point between the Middle and Upper Paleolithic is replaced by a cultural break-point between the Châtelperronian and the Aurignacian. What could be simpler? It is apparent that the two cultures are partially contemporaneous. However, the archaeological data are in fact somewhat equivocal and raise numerous questions. I will try to address some of these briefly here, without pretending to speak for my archaeological colleagues.

PROBLEMS RELATED TO THE CHÂTELPERRONIAN

The first problem that arises concerns the nature of the Châtelperronian as a cultural or an analytical unit. While it is clearly up to prehistorians to respond to these issues, perhaps I could, from my bioanthropological perspective, offer some observations that seem to me to be important ones. Does the Châtelperronian represent a simple transformation from a Mousterian base? I find the following fact striking: Mousterian industries, despite their uncontestable diversity, seem to have been relatively stable throughout the Middle Paleolithic. There are no assemblages within the Mousterian that mark chronological change in the same way that the Perigordian, for example, or the Magdalenian do. Yet suddenly, in the middle of the last glaciation, important changes occur—changes so technologically significant that they result in a different culture, much removed from the Mousterian, and unequivocally connected to the Upper Paleolithic. What could be responsible for these sudden changes? It immediately comes to mind that contact with the early Aurignacians might have forced the neandertals to change, to adapt in an effort to resist them. History, however, teaches us that when two cultures come in contact with one another, one always overwhelms the other. Could there have been Aurignacian influence on the Mousterian, with the result of this influence being the Châtelperronian? It seems to me that prehistorians are divided on this issue.

One could also inquire further into the question of who manufactured the Châtelperronian. At St. Césaire and at Arcy, it would seem to have been neandertals. But the radiocarbon determinations indicate coexistence with the Aurignacians that persisted for several millennia. Is it not possible for some Châtelperronian industries to have been manufactured by moderns? The best argument against this proposition is probably the distinct geographical distributions of Châtelperronian and Aurignacian sites. The Châtelperronian is geographically restricted and rare, and the same is true of the other Mousterian-derived industries (e.g., the Szeletian, the Uluzzian). The contrast between these "cultural isolates" and the very wide distribution of the Aurignacian can probably be explained by a dramatic reduction of neandertal populations over a comparatively short time. Conversely, it is possible that the Châtelperronian was never adopted by any *H. sapiens sapiens* group, which would raise the issue of just how the eventual technological superiority of the Aurignacian over the Châtelperronian came about.

With these questions, we confront the problem of contacts between the two populations and the two cultures. Do the radiocarbon chronologies really demonstrate contemporaneity? In nearly all the sites where both industries are found, the Châtelperronian occurs stratigraphically below the Aurignacian. Le Piage and Roc de Combe are the only two exceptions, and at both these sites, the two assemblages are interstratified. Unfortunately, neither site has been dated.

EX ORIENTE LUX

The Aurignacian also raises a number of questions. The first one (and it should not be me who addresses this) concerns the unity of this culture, or the lack thereof. Is what we call the Aurignacian in the Levant, in Turkey, in eastern Europe, in France, and in Spain the same culture? This question raises the issue of an eastern origin for modern Europeans. Moderns would have arrived in Europe from the east, bringing with them their new culture, which would have subsequently diffused throughout Europe in direct proportion to the geographical expansion of its bearers. Unhappily, however, the dates seem to argue against this proposition. While there is an early Upper Paleolithic in eastern Europe— Bacho Kiro in Bulgaria dates to ca. 43,000 B.P.—the Upper Paleolithic is also early at the other end of the continent, in Castillo and l'Arbreda caves in northern Spain, for example. If these dates are accurate, how do we explain the near simultaneous arrival in eastern and western Europe of the moderns who are supposed to have been the bearers of these industries? To recognize several

"Aurignacians" with different origins does not resolve the biological problem. Or one could admit the possibility that the European Aurignacian might have been the work of both moderns and neandertals—the phenomenon of an apparent convergence would thus be eliminated. But in this second case, it would be necessary to look for Mousterian-Aurignacian connections, and there do not seem to be any. Some prehistorians are now considering such a possibility, however.

In summary, then, one of the most acute problems has to do with the unity or plurality of the Aurignacian, and with its single or multiple origins. If the Aurignacian was indeed made only by modern humans, and is found at both ends of Europe a little earlier than 40,000 B.P., then it follows that it was nearly synchronous everywhere—that is to say, we lack the means to "track" the colonization of Europe by moderns, and the colonization process lies beyond the precision of our methods to date it. And, if it is acknowledged that the appearance of the Upper Paleolithic was a little earlier in the Levant, cultural diffusion must have been extremely rapid (given a single origin).

CONCLUSIONS

To return to some of the more strictly anthropological aspects of the debate, it is necessary to discern who might have made the earliest Aurignacian industries of western Europe. Only some bone fragments and isolated teeth exist to shed light on the matter, and these data are insufficient in and of themselves to allow us to recognize one or several populations. In aggregate these fragments and teeth are more robust than their more modern fossil counterparts, which is scarcely surprising. In central Europe, where human remains from the beginning of the Upper Paleolithic are better represented, they are also more robust than their successors. And, as Gambier (1993) has shown, there are no indisputable neandertal traits in early Upper Paleolithic fossils from western Europe. But to argue from these findings that all the industries called Aurignacian were, without exception, the work of moderns is an inferential leap that I would not be willing to risk.

Another anthropological problem sometimes figures in these discussions—the possibility of eventual admixture between moderns and neandertals. If they were two subspecies, there would, of course, be no biological obstacle to admixture, and if they were indeed contemporaries for several millennia in Europe, occasional admixture would seem all but inevitable. Could one imagine, in fact, that two populations would have lived in the same regions, depending upon the same game animals, without encountering one another from time to time and without interbreeding, even a little? But we know nothing concrete; no fossil discovered so far brings evidence to bear on the matter.

I close with some remarks that do not pretend to exhaust the issues that turn on the place of the Upper Paleolithic in Europe. I wish to underscore yet again that these issues seem to me to involve two central problems:

1. The phylogenetic origin of modern humans. If they have no evolutionary connection with the neandertals, they could only have come from the Orient.
2. The origin or origins of the Aurignacian. Is there a single Aurignacian without any link to the Mousterian, or are there several Aurignacians, some of which are derived from the European Middle Paleolithic?

ACKNOWLEDGMENTS

I wish to thank all those persons in the various European museums and research institutions that I have visited who have allowed me to examine the fossil specimens under their charge. I am also grateful to the organizers of the International Colloquium *"El Origen del Hombre Moderno en el Suroeste de Europa"* (Madrid 1991) for inviting me to participate. This essay is an updated version of a paper presented at that conference and published by the Universidad Nacional de Educación a Distancia (UNED) in 1993 (Vandermeersch 1993).

NOTE

1. The senior editor of this volume also has large retromolar spaces and is usually— although not always—considered to pertain to *H. sapiens sapiens.*

8

Modern Humans at the Beginning of the Upper Paleolithic in France

Anthropological Data and Perspectives

DOMINIQUE GAMBIER

According to data presently available, the expansion of modern humans into Europe coincided with, on the one hand, the development of Aurignacian culture, and on the other, with the "disappearance" of the neandertals—processes that extended over more than ten millennia (>43–30 kyr B.P.). From an anthropological point of view, two models will be discussed:

1. The Replacement Model: Modern humans replaced neandertal populations over the course of the biological transition (Stringer et al. 1984; Thomas 1972).
2. The Continuity Model: There was gradual, in situ evolution of neandertals in the direction of modern humans (Frayer 1986; Smith 1982, 1985; Wolpoff 1980; Wolpoff et al. 1981).

However, the discovery of a neandertal in a Châtelperronian context at St. Césaire in France (Vandermeersch and Lévêque 1989), and the existence of modern humans from 90 kyr B.P. in the Near East (Valladas et al. 1988), and perhaps from ca. 100 kyr B.P. in Africa (Bräuer 1984a), have a number of implications for the continuity hypothesis. This model has been expanded now to incorporate not only the possibility of neandertal evolution in the direction of moderns without external gene flow, but also the possibility of a significant genetic contribution to the neandertal gene pool from moderns who evolved elsewhere (Smith and Trinkaus 1991). Most immediately, this significant genetic contribution can be documented in central Europe, where some late neandertals (e.g., those from Vindija Cave, in Yugoslavia) show tendencies foreshadowing *H. sapiens sapiens,* and where the most ancient modern fossils show traces of neandertal morphology. In western Europe, however, the replacement hypothesis is, for these same workers, that best supported by both anthropological and cultural data.

This chapter shows that the anthropological data available for the Aurignacian in France lie at the very limits of our interpretive framework, and that these limits are linked in large part to the characteristics of the anthropological sample. Because of the inadequacies of the data base, the French Aurignacian anthropological sample is extremely sensitive to sample bias.

AURIGNACIAN SITES WITH HUMAN FOSSIL REMAINS

According to the various catalogues of fossil hominids published over the years (e.g., Hué 1937; Oakley et al. 1971; Vallois and Movius 1952), excavation reports, and miscellaneous old and recent publications, forty-two sites have produced human remains from Aurignacian contexts of variable credibility (Table 8.1).

An analysis of the circumstances of discovery and of the relevant archaeological data shows that eighteen of them should be discarded—definitely in some cases, provisionally in others. Generally these sites were excavated between 1860 and 1950, guided by research objectives and methodologies different from those of today. Reasons for disregarding the fossil remains from these sites include (1) the fact that the stratigraphic provenience of the human bone cannot be established, and the site in question contains a sequence that extends from the Aurignacian to the post-Paleolithic; (2) the associated archaeological industry is so impoverished that it cannot be accurately assigned to the Aurignacian, and (3) the stratigraphic provenience of the fossils is known but field observations suggest that the bones are more recent intrusions into Aurignacian levels. Consequently, only the remains from the twenty-four sites indicated in Table 8.1 will be considered in this chapter.

Geographical Distribution

Except for the Grotte du Renne at Arcy-sur-Cure (Yonne), the twenty-four caves and rockshelters are situated to the south of the Loire in west-central and southwest France. Most of the sites are concentrated in the Charente and in Dordogne.

Cultural Association and Chronological Age

It is not always possible to determine the stage in the Aurignacian sequence to which an associated industry should be assigned. Sometimes the collections from old excavations comprise mixed assemblages from different levels, or the industry is so impoverished or lacking in diagnostics that these determinations cannot

Table 8.1. Inventory of Aurignacian Sites with Human Remains

Site	Département	Credibility of Association with Aurignacian
Aurignac	Haute-Garonne	post-Paleolithic
Bayol	Gard	bad
Bellevaud	Charente	bad
Bize	Aude	good
Blanchard	Dordogne	good
Bouil Bleu (164)	Charente-Maritime	bad[†]
Brassempouy	Landes	good
Castanet	Dordogne	good
Chez Leix	Gironde	good
Combe Capelle	Dordogne	bad
Cro-Magnon	Dordogne	good
Duport/La Chaise	Charente	good
Font de Gaume	Dordogne	good
Fontéchevade	Charente	good
Four de la Baume	Saone et Loire	bad
Gourdan	Haute-Garonne	good
Isturitz	Pyrénées-Atlantiques	bad
La Balauzière	Gard	bad
La Combe	Dordogne	good
La Crouzade	Aude	good
La Ferrassie (grand abri)	Dordogne	good
La Ferrassie	Dordogne	good
La Gravette	Dordogne	good
La Quina	Charente	good
La Rochette	Dordogne	bad
La Souquette	Dordogne	bad
Laussel	Dordogne	bad
Le Flageolet	Dordogne	good
Les Battuts	Tarn	good (?)
Les Cottés	Vienne	bad
Les Morts	Corrèze	bad
Le Piage	Lot et Garonne	bad
Les Roches	Indre	good
Les Rois	Charente	good
Les Vachons	Charente	good
Pasquet	Dordogne	bad
Petit Puyrousseau	Dordogne	bad
Renne	Yonne	good
Solutré	Saone et Loire	post-Paleolithic
Tarté	Haute-Garonne	good (?)
Téoulé	Haute-Garonne	bad
Vallon des Roches	Dordogne	bad

[†] A particle accelerator date gives an age later than the Upper Paleolithic.

be made. In other cases, the exact provenience of specific human remains cannot be ascertained when there are several Aurignacian levels in a given site. With respect to the conditions of excavation, uncertainties arising from the stratigraphic position of levels and even the absence of environmental data do not permit the reliable placement of human fossils in a secure chronological framework. There is also a paucity of radiometric dates from these early Upper Paleolithic levels.

A single specimen—La Ferrassie b—would appear to be associated with a very early phase of the Aurignacian (Aurignacian 0). The others are probably associated with Aurignacian I/II. The most complete fossils (from Cro-Magnon) pertain to evolved Aurignacian contexts (Sonneville-Bordes 1960), and Bouchud (1966) does not rule out the possibility that the Cro-Magnon specimens actually pertain to the Upper Perigordian. The presence of Upper Perigordian deposits in the Cro-Magnon rockshelter is predicated mainly on the existence, in the Rivière collection, of gravette points and on the discovery of two engravings at the site during the post-1868 excavations. As noted by Movius (1969), these considerations suggest that the Upper Perigordian artifacts might have come from an area adjacent to the rockshelter, and not from the rockshelter itself. The only "modern standard" excavations were those of Denis Peyrony, who considered that the entire fill of the rockshelter pertained to the Aurignacian. Direct accelerator dating of the fossils themselves would be the only unequivocal way to determine their true age.

HUMAN FOSSILS

To date, an inventory of human remains from the more reliable French Aurignacian sites allows us to identify the remains of at least thirty-one adults and nineteen subadults.

State of Preservation

At least four adults (Cro-Magnon 1–4) are represented by more-or-less complete crania and mandibles, and parts of the postcranial skeleton. In this regard, it is worth recalling that the numbering of individuals at Cro-Magnon is in part arbitrary because, if anatomical connections existed, they were not recorded at the time of discovery (L. Lartet 1868a, 1868b). Commonly recognized osteological criteria for the determination of individuals from batches of loose bones include the recognition of symmetries, size, texture and patination of bones, and distinctive pathological lesions. In the case of the Cro-Magnon fossils, utilizing these criteria would not lead to more certain identifications essentially because the bones of the postcranial skeleton are, in all cases, fragmentary.

Table 8.2. Percentage Representation of Anatomical Elements

Anatomical Element	Theoretical Number	Adults (n = 31)	Percent	Theoretical Number	Infants (n = 18)	Percent
frontal	31	5	16.1	18	1	5.6
parietal	62	9	14.5	36	1	2.8
temporal	62	5	8.1	36	0	0.0
occipital	31	4	12.9	18	1	5.6
base	31	1	3.2	18	0	0.0
zygomatics	62	4	6.5	36	0	0.0
maxilla	31	6	19.4	18	0	0.0
mandible	31	7	22.6	18	5	27.8
permanent teeth	961	54	5.6	270	38	14.1
deciduous teeth	0			270	10	3.7
scapulae	62	2	3.2	36	0	0.0
clavicles	62	1	1.6	36	0	0.0
sternum	31	0	0.0	18	0	0.0
vertebrae	744	30	4.0	432	0	0.0
sacrum	31	1	3.2	18	0	0.0
coxae (innominates)	62	6	9.7	36	0	0.0
ribs	744	21	2.8	432	0	0.0
humerus	62	7	11.3	36	1	2.8
radius	62	7	11.3	36	0	0.0
ulna	62	7	11.3	36	0	0.0
carpals	512	0	0.0	0		
metacarpals	310	3	1.0	180	0	0.0
phalanges (hands)	868	4	0.5	504	0	0.0
femur	62	6	9.7	36	3	8.3
patella	62	1	1.6	0		
tibia	62	5	8.1	36	4	11.1
fibula	62	3	4.8	36	0	0.0
talus	62	2	3.2	36	0	0.0
calcaneus	62	1	1.6	36	0	0.0
tarsals	310	6	1.9	0		
metatarsals	310	7	2.3	180	0	0.0
phalanges (feet)	868	1	0.1	504	0	0.0

Theoretical Number = number of anatomical elements calculated from the minimum number of adults and infants.

Percent = (number of anatomical elements present / theoretical number of anatomical elements) × 100

The Flageolet I adult (excavations of J. Ph. Rigaud) consists of several portions of long bones. The twenty-six other adults are represented by cranial fragments or by very partial mandibles or isolated teeth.

The subadults (infants, for the most part) are even more poorly preserved. There is not a single complete skeleton. Crania are limited to several vault fragments, on the order of 1 cm across. Some pieces of mandibles, very rare long bone fragments, and isolated teeth constitute the remainder of the sample. Table

8.2 gives the percentage representation of anatomical elements by age class. It shows very clearly the fragmented character of the subadult sample.

Age at Death

Even in modern contexts, the methods for estimating the age at death using skeletal criteria are not very precise (Masset 1982). Applied to ancient, fragmentary skeletons, where we cannot guarantee that the processes and rates of growth and aging were comparable to those of modern populations, these methods give results whose reliability is still more doubtful.

For the adults, the degree of dental wear, the level of eruption of the third molar, or the stage of development of certain pathologies can sometimes allow us to distinguish young adults from older ones. Application of these criteria here enable us to determine that most of the Aurignacian adults were relatively young when they died.

For infants and adolescents, the degree of calcification of the milk and permanent dentition and the eruption of the permanent teeth are relatively feeble indicators of age. In the case of the Cro-Magnon and La Chaise infants, age estimates are based on the length of diaphyses of the long bones. Seven infants were less than five years old at the time of death, twelve were between six and fourteen years of age (Figure 8.1).

Sex

In my opinion, the best skeletal indicator of sex in adult humans is that based on the innominate. Two men (Cro-Magnon 1, 3) and one woman (Cro-Magnon 2) can be distinguished using this method. The robusticity of the bones grouped under the specimen number of the fourth individual (Cro-Magnon 4) indicates that it is a male. For the remaining adults (27 of 31 adult individuals), no determination of sex was possible. In the case of subadults and infants, no method exists for the determination of sex based on skeletal data.

Mode of Deposition

The intentional burial at the Cro-Magnon rockshelter is the only primary or secondary multiple burial known (further definition is not possible). Among the mixed remains of scattered teeth and bone, several show traces of intentional human action. One of the roots on two teeth—one from La Combe (MacCurdy 1914) and one from Brassempouy (excavations of Delporte, discovered July 1991)—has been perforated. Several mandible fragments show striae as a result of defleshing (Gourdan, Les Rois).

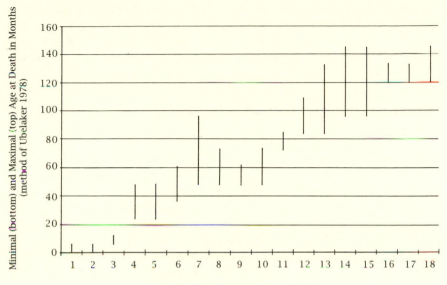

Figure 8.1. Estimation (in months) of the age at death of individual Aurignacian children.

1–3   Cro-Magnon
4   Font de Gaume
5, 7, 8, 10, 16, 17   Les Rois
6, 9   Fontechèvade
11   Renne
12   Chez Leix
13   La Quina
14   Brassempouy
15   Tarté
18   La Chaise

Dental Data: The Early Aurignacian Sample (Aurignacian 0)

The single human fossil associated with a very early phase of the Aurignacian is an upper left permanent incisor of an adult (Gambier et al. 1990). It comes from couche E', in the grande abri of La Ferrassie (D. Peyrony excavations). This level is considered to pertain to the Aurignacian 0 (de Sonneville-Bordes 1960). The lingual tubercle and the marginal crests are developed, but it cannot really be considered to be shovel-shaped. In its dimensions (buccal-lingual = 7.6 mm, mesio-distal = 9.9 mm) and the morphology of the crown, it more closely resembles modern human incisors from Qafzeh and Skhūl, and more recent Aurignacian specimens, than those of neandertals, where shoveling is often well developed. However, the very poor samples and the importance of individual variation, combined with the absence of traits specific to each of these groups, urge caution.

Dental Data: The Later Aurignacian Sample
(Aurignacian I and II)

Permanent Teeth.   Not all of the analytical categories of the permanent dentition are represented in the sample and some of them could not be recorded because of the degree of wear on the teeth. Compared with those of contemporary humans, Aurignacian dental crowns have values that are in the upper part of the range of present-day variation. In particular, the mesio-distal diameter of the lower teeth is generally greater than the corresponding measurement in modern dentitions. Comparison of crown dimensions with those of Mousterian groups— Qafzeh and Skhūl, and neandertals—does not reveal any sharp distinctions. The ranges of variability overlap extensively, as does intrasample variability, and the Aurignacian teeth are not clearly aligned with any of the other groups (Tables 8.3, 8.4).

From the point of view of occlusal surface morphology, the number and disposition of the cusps are fully comparable with the most frequently observed patterns in present-day dentitions. Some patterns described for neandertals and the early modern dentitions of Qafzeh and Skhūl are represented: (1) a single example of shovel-shaped incisors (Les Rois), (2) lingual tubercles and strongly developed marginal crests on the incisors, (3) a tricuspid configuration on the second lower premolars (Les Rois), and (4) enamel striae. These variations, also known from the teeth of earlier fossils, are probably archaic traits. Finally, the general degree of occlusal surface wear on the anterior dentition is comparable to that observed, for example, on the corresponding teeth from the late Upper Paleolithic.

Deciduous Teeth.   The deciduous teeth are poorly preserved, being confined for the most part to very worn, unmeasurable crowns. The deciduous lower second molar is the best represented. It is voluminous compared with its modern counterparts. However, on the basis of its crown dimensions, it cannot be distinguished from any of the comparison samples (Table 8.5).

Cranial Data

The cranial sample consists of seven adults and five infants from Cro-Magnon, Gourdan (1), La Crouzade, Fontéchevade, La Quina, Les Rois, and Les

Table 8.3. Lower Permanent Dentition

Teeth		French Aurignacian (*n* = 2–6)	Qafzeh-Skhūl[†] (*n* = 4–6)	Western European Neandertal[†] (*n* = 7)	Krapina[‡] (*n* = 3–14)
				Diameter (mm)	
I1	BL	6.1–7.0	6.4–7.7	7.2–7.5	6.7–8.1
	MD	7.0–7.2	4.5–6.7	5.0–6.9	5.0–6.3
I2	BL	7.0–7.5	6.7–8.0	7.0–8.0	7.3–8.2
	MD	7.0	5.7–7.5	5.7–6.6	6.4–7.5
C	BL	7.0–9.0	7.0–9.9	8.2–9.5	8.7–10.3
	MD	7.0–8.0	7.4–9.1	6.8–8.6	7.6–9.3
P1	BL	7.5–8.5	7.7–9.5	8.4–10.0	8.7–9.3
	MD	7.5–8.5	5.8–8.5	7.0–8.5	7.8–8.4
P2	BL	8.5–9.0	7.8–9.8	8.0–10.2	9.4–10.5
	MD	6.9–8.0	7.0–8.6	6.9–8.3	6.9–8.2
M1	BL	9.8–11.2	10.5–12.8	9.6–11.5	10.8–12.1
	MD	11.0–13.0	0.5–13.2	10.3–12.5	11.4–12.5
M2	BL	10.0–11.9	10.5–12.4	10.3–12.0	9.8–12.4
	MD	10.0–11.8	10.7–12.0	10.7–12.5	11.5–13.9
M3	BL	9.3–11.3	9.9–11.9	10.0–13.0	9.8–11.4
	MD		9.5–12.0	10.3–12.9	11.6–13.9

Ranges of variation in buccal-lingual (BL) and mesio-distal (MD) diameters (mm) of Aurignacian teeth compared with those of fossils from Mousterian contexts.

[†] Measurements from Vandermeersch 1981b

[‡] Measurements from Wolpoff 1979

Roches. Three types of modern and archaic character suites are recognizable in the crania of the Aurignacian adults. The modern characters, which dominate in the sample, have been adequately described elsewhere (Broca 1868; Gambier 1989; Vallois and Billy 1965; Vandermeersch 1981b). Both morphology and metrics are of interest.

The suite of modern craniometric and craniomorphological characteristics is well known (Billy 1970, 1972), namely the form and orientation of the orbits, the enlargement of the maxillary-malar region of the face, and the minimal increase in the height of the face compared with the maximal length of the cranial vault. These characters are not all equally expressed in each of the Aurignacian specimens (Gambier 1989).

The principal archaic features of the cranium and face are (1) general craniofacial robusticity, as attested by large dimensions and indices, rugged relief in various of the cranial landmarks, and thickness of the bone; (2) maxillary *de type à inflexion;* (3) a relatively low vault when compared with maximal cranial dimensions; (4) marked suborbital relief; (5) an occipital bun *(chignon);* and (6) an atypical occipital torus.

The mandible also is characterized by marked robusticity with strongly devel-

Table 8.4. Upper Permanent Dentition

Teeth		French Aurignacian (n = 3–4)	Qafzeh-Skhūl† (n = 4–7)	Western European Neandertal† (n = 7)	Krapina‡ (n = 8–16)
			Diameter (mm)		
I1	BL	6.8–7.3	7.3–8.7	8.0–8.5	8.1–9.7
	MD	9.4–10.2	9.0–11.1	9.0–12.0	9.4–11.1
P2	BL	8.5–9.5	8.4–11.0	9.8–11.0	9.5–11.7
	MD	6.0–6.7	6.5–7.6	6.6–8.0	6.8–8.8
M1	BL	11.0–12.8	11.2–13.2	11.5–12.5	11.3–14.2
	MD	10.1–11.0	9.9–12.4	10.4–12.5	11.5–13.6
M2	BL	11.8–12.5	11.5–12.3	10.1–14.5	11.6–14.2
	MD	9.0–11.0	8.6–12.2	9.5–11.5	9.3–13.1
M3	BL	11.5	11.0–12.0	11.6–13.0	12.0–13.5
	MD	8.8–9.5	8.6–10.9	9.5–11.5	9.9–11.3

Ranges of variation in buccal-lingual (BL) and mesio-distal (MD) diameters (mm) of Aurignacian teeth compared with those of fossils from Mousterian contexts.
† from Vandermeersch 1981b
‡ from Wolpoff 1979

oped surface relief. The mental foramen is sometimes located directly below the first permanent molar and the digastric fossae are sometimes oriented low on the horizontal corpus and not toward the rear, as tends to be the case with present-day mandibles. However, these archaic traits are not consistently represented. Thus La Crouzade, and Cro-Magnon 1 and 2, have weakly developed suborbital relief,

Table 8.5. Second Deciduous Molars

	French Aurignacian (n = 4)	Qafzeh-Skhūl† (n = 3)	Western European Neandertals†‡ (n = 11)	Krapina* (n = 7)	Modern Infants† (n = 16)
			Diameter (mm)		
BL minimum	9.1	8.1	8.4	8.6	
BL mean					8.7
BL maximum	10.0	9.4	10.5	10.0	
MD minimum	10.0	9.9	9.2	9.8	
MD mean					9.9
MD maximum	11.5	10.7	11.3	11.5	

Ranges of variation in buccal-lingual (BL) and mesio-distal (MD) diameters (mm) of Aurignacian teeth compared with those of fossils from Mousterian contexts.
† from Tillier 1979
‡ from Mallegni and Ronchitelli 1989
* from Wolpoff 1979

and Cro-Magnon 2 lacks both an occipital bun and an atypical occipital torus. Also, the Cro-Magnon 1 mandible does not exhibit the modern condition with respect to the placement of the mental foramen.

More "gracile" and thus more modern-appearing specimens (e.g., Cro-Magnon 1 and 2, La Crouzade) occur alongside more robust, "archaic" individuals (e.g., Cro-Magnon 3 and 4), and sexual dimorphism can only account for part of these differences since the first ("gracile") group comprises individuals of both sexes. The cranial bones of infants contribute little information because they are, in all cases, very incomplete.

A single specimen, Cro-Magnon 5, is worthy of further comment. It consists of a fragment of the vertical part of the frontal bone with the supraorbital region preserved. The absence of any torus on the specimen recalls the condition seen in present-day infants or neandertals of approximately one year of age. Tillier (1986) notes that the distinctive development of the supraorbital torus in neandertal subadults only becomes apparent after the age of six. This specimen, then, does not shed any light on the taxonomic affinities of its erstwhile owner.

While very fragmentary, the mandibles from La Quina, Fontéchevade, Les Rois 2, and Les Roches are more robust than those of present-day children of the same dental age. The Les Rois 1 mandible, whose age at death was between eight and eleven years, is characterized by a vertical external profile of the symphysis, a chin of modern type (albeit somewhat receding). The anterior portion of the dental arcade is rounded. Some archaic characters are also evident in the Les Rois 1 mandible, however. The robustness evident in the thickness of the mandibular corpus is greater than that of modern children of comparable age. There is a weak alveolar plane, the digastric fossae are found low down on the mandibular corpus, and the symphysis is very broad. Despite these archaic characters, however, the mandible is clearly assignable to *H. sapiens sapiens*.

Postcranial Data

Postcranial elements from five adults are known (Cro-Magnon 1–4, Le Flageolet 1); most are fragmentary. Various studies concur with respect to a general robusticity and muscle insertion marks on the bones that are more pronounced than those of moderns, resembling those of the Qafzeh-Skhūl hominids (Billy 1970, 1972; Gambier n.d.; Trinkaus 1983; Vandermeersch 1981b). Stature estimates based on the Cro-Magnon specimens appear to be high. Less robust than those of neandertals, in addition, the Aurignacian postcranial skeletons lack any characteristics specific to the neandertals (Trinkaus 1983). Again in contrast to neandertals, the Aurignacian postcranial skeleton exhibits elongated distal limb segments (Billy 1970, 1972).

CONCLUSIONS

Based on this somewhat cursory and incomplete review of the anthropological data, it is possible to underscore the following points:

1.   Chronological uncertainties combined with the poor state of preservation and the likely effects of sampling error with respect to the original biological populations make it impossible to reconstruct the demographic composition of Aurignacian populations, or to disentangle the separate effects on morphological variation of age, sex, and time and how they might have changed in relation to changes in the natural and cultural environment.

2.   Without rejecting the likelihood of postdepositional destruction and disturbance of the deposits containing the bones, the near-absence of evidence for intentional Aurignacian burial (at least in France) suggests that the typical Aurignacian practice regarding disposal of the dead did not emphasize primary inhumation in caves or rockshelters. The few bones with cutmarks suggesting defleshing, the degree of fragmentation of the bones, and their general state of preservation all suggest other funerary practices. The perforated teeth can probably be interpreted as evidence of symbolic behavior, at least in regard to certain kinds of artifacts, before 30 kyr B.P.

3.   The makers of Aurignacian 0 or "Pre-Aurignacian" assemblages are not known. Even if these assemblages were made by moderns, as perhaps suggested by the La Ferrassie upper incisor, the essentials of their morphology escape us.

4.   From the Aurignacian I, the rare fossils *actually known* are assignable to modern humans.

5.   There are no "derived neandertal characteristics" in the Aurignacian sample *as defined today based on the classic neandertals of western Europe.*

6.   Traits shared with the neandertals are plesiomorphies and *tell us nothing about phylogenetic relationships.*

7.   Except for occipital bunning, these traits can also be observed in the skeletons of early modern populations associated with Mousterian industries in the Near East. The Qafzeh-Skhūl assemblage can be divided into a gracile, more modern group and a robust, more archaic-looking group. Supraorbital tori can be pronounced (e.g., Qafzeh 6, Skhūl 9) or weakly developed (e.g., Qafzeh 9); Skhūl 5 and Qafzeh 3 are both characterized by atypical occipital tori (Hublin 1978; Gambier, personal observation). The teeth have voluminous crowns, and the occlusal surface presents certain archaic features found both in Aurignacian and in neandertal teeth. The chin (mental eminence) can be weakly developed, and there is a weak alveolar plane on the mandible of Qafzeh 4 (Tillier 1979) as well as on the subadult mandible Les Rois 1.

The same kind of variability characterizes the modern populations of Mousterian tool makers in the Near East and the modern Aurignacian populations from southwestern Europe. It is not necessary, therefore, to invoke strong continuity with the neandertals, as Wolpoff (1980), Frayer (1986), and Smith (1985) have done, in order to explain the persistence of archaic traits in the modern populations associated with the Aurignacian. As Vandermeersch (1981b) has shown, the narrow affinities of the Aurignacian sample with the Qafzeh-Skhūl group, along with other shared traits, are sufficient to explain the presence of archaic traits.

Given the present state of the anthropological data, the major component of the modern peopling of western Europe seems to be intrusive, and the Qafzeh-Skhūl group is perhaps a representation of the population of origin. Elsewhere in Europe, the examination of fossils linked to Aurignacian archaeological assemblages leads to similar conclusions. The central European sample, which I have also studied (Gambier n.d.), is comparable to the southwestern European sample.

In central Europe, one encounters the *same* lack of chronological resolution, the *same* state of conservation, and the *same* mode of deposition (absence of primary interment). The attribution of "Pre-Aurignacian" industries to modern humans is subject to *the same reservations.* The distinction between neandertals and moderns based on dental traces like those from Bacho Kiro and Istallosko in Hungary is not evident (Gambier et al. 1990; Tillier 1979). The most complete specimens (Mladeč, Zlatý Kůň, Hahnoffersand, Velika Pečina) present, on the same indubitably modern morphology, the same kinds of archaic variation as the western European sample. Central European Aurignacian cranial morphology also argues in favor of resemblances with the modern Mousterian toolmakers from Qafzeh and Skhūl, and the southwestern European Aurignacians (Gambier n.d.). However, on the basis of certain traits mainly recorded on the Mladeč series, the Aurignacians of central Europe appear to be more archaic than those from the southwest (Frayer 1986; Gambier 1992; Jelínek 1983). This would imply, if the greater antiquity of the central European fossils can be confirmed, an incursion of modern people toward western Europe.

Given the present status of the evidence, the "replacement" model would appear, then, to be the best able to explain the observed anthropological facts. However, this model does not presume to know the mechanisms of change in production technologies nor in sociocultural forms that characterize the transition in Europe because there is no necessary correspondence between anatomical types and archaeological cultures or industries. The stages of replacement and the causes of the "success" of the modern populations remain to be determined.

We cannot give a certain answer to the question: Was there admixture between the neandertals and the first modern humans? If the early or Archaic Aurignacian is in fact the work of moderns, the archaeological data plead in favor of a relatively long contemporaneity between moderns and neandertals. In the absence of biological or cultural barriers, the possibility of such exchanges cannot be excluded. *No decisive anthropological argument can be advanced,* since (operationally) we do not know how to identify such an admixture. Given the migration scenario presented above, however, it would behoove us to consider questions like:

- How can the occipital bun, traces of which persist in the Aurignacian sample, be found in Europe when it is absent in Qafzeh-Skhūl, the supposed source of the European Aurignacians?
- How can we explain the posterior position of the mental foramen on the Cro-Magnon 1 mandible?
- How can the weak development of the mastoid process on the Zlatý Kůň cranium (Gambier 1992) be explained, when the effects of admixture appear to be nonexistent or minimal on the skeleton, and when admixture has been so adventitious that its traces remain perceptible on a fragmentary, nonrepresentative, and already very late sample?

Did the transition to other production technologies have rapid and positive consequences on the living conditions of modern Aurignacian populations, conferring on them a biological dynamism superior to that of neandertal groups (e.g., reduction in the rate of infant mortality, increase in life span, improvements in health)? We cannot respond to this question for lack of the relevant skeletal series (cf. Tillier 1986). Even if we could, we still do not understand why the neandertals, who were also capable of technological innovation (whether accelerated or not by contact with moderns), did not also benefit from these innovations.

Discoveries of additional human fossils would certainly be welcome, especially for the earliest phases of the Aurignacian, but despite its limitations, the material presently available still has not been exploited to the maximum extent possible. Detailed comparisons between the Aurignacians of central and western Europe, on the one hand, and comparisons between the Aurignacian and the Gravettian, on the other, are necessary to understand the progression of these populations across Europe. The question of admixture, as well as that of the success of modern humans, should be taken up again in the context of the evolutionary meaning of the osteological variations among the neandertals, other archaic, and modern *H. sapiens* populations from the various regions involved.

Finally, little recent work on Aurignacian paleopathology or patterns of growth exists. These research domains should be developed as a means of addressing the impact of the technological and cultural "advances" that seem to coincide with the beginning of the Aurignacian on the life ways and behaviors of the transition interval populations.

ACKNOWLEDGMENTS

I wish to thank all those persons in the various European museums and research institutions that I have visited who have allowed me to examine the

fossil specimens under their charge. I am also and equally grateful to the organizers of the International Colloquium *"El Origen del Hombre Moderno en el
Suroeste de Europa"* (Madrid 1991) for inviting me to present the results of my
research. This essay is an updated version of a paper presented at that conference
and published by the Universidad Nacional de Educación a Distancia (UNED) in
1993 (Gambier 1993).

9

The Transition to Anatomically Modern Humans
The Case of Peninsular Italy

AMILCARE BIETTI

For more than a century, the modern human origins controversy has been a subject of debate in Europe and, after about 1940, in the Levant. However, in the latter area, the first appearance of anatomically modern humans seems to date to somewhere near the end of Isotope Stage 5 (i.e., to ca. 90,000 B.P.), thus indicating a long interval of contemporaneity between anatomical "moderns" and the neandertals, whereas in peninsular Europe the coexistence of the two species (or subspecies, if one accepts the definition of *Homo sapiens neanderthalensis*) seems to be confined to a time interval of only about 10,000 years. In this chapter, I will concern myself only with Europe, and in particular with the situation on the Italian Peninsula (it is well known that Sicily and Sardinia have never produced any evidence of a neandertal presence).

According to my reading of the modern human origins literature, there are three basic hypotheses that frame the anatomical transition: the Recent and Single Origin model (hereafter RSO), advocating a monocentric origin for anatomically modern humans, probably in Africa, with long-term "coexistence" between moderns and neandertals in the Levant and with a substantial and short-term "replacement" interval in Europe (e.g., Stringer and Andrews 1988b); the Multiregional Evolution model (MRE), an evolutionary polycentric approach wherein anatomically modern humans evolve from neandertals (and archaic *Homo sapiens*) wherever the latter are found (e.g., Frayer et al. 1993; Wolpoff et al. 1984); and a series of "intermediate" models which can be lumped under the rubric of Hybridization and Replacement (H and R), in which the African origin of anatomically modern humans is accepted but allowance is made for subsequent hybridization or gene flow in other areas, including Europe (e.g., Aiello 1993; Bräuer 1984a, 1992a; Smith 1992a).

I will first discuss the plausibility of these hypotheses from the perspective of fossil and archaeological evidence from Italy. As will rapidly become evident, relevant data are very scant, in particular for the human fossils. In the next section I will focus on the evidence for "late" neandertals, and the transitional archaeological complexes—the Uluzzian and the early Aurignacian. The last section will be devoted to a general discussion and some tentative conclusions. I am, of course, an archaeologist, and so my perspective must perforce be an archaeological one, based on a systemic and "contextual" (in the sense of the currently unfashionable "processual" archaeology!) view of human behavior as recovered from the archaeology of the various Italian sites that document the "transition" interval. It is my opinion, in regard to modern human origins and many other problems of general interest to the profession, that one must "think globally" but "act locally"—that is, try to squeeze as much information as possible out of local sequences before attempting more general comparisons at broader temporal and spatial scales. Needless to say, this paradigmatic view has a firm basis in the well-known evolutionary anthropology of an earlier generation, founded on the classic works of Leslie White (1959) and Sahlins and Service (1960). While I am not well-versed in the human paleontology of the human origins debate, I have the strong impression that some physical anthropologists, as well as—unfortunately—most European prehistorians, continue to subscribe to essentially historicist biases in regard to the debate and are willing to make global generalizations on the basis of single (or a few) supposedly diagnostic anatomical traits, and to invoke essentially historical, or quasi-historical, processes like migrations or acculturations to explain patterns far back into the Pleistocene (see Bietti 1991 in regard to Italy; also Clark 1993a,1994b, and Clark and Lindly 1991 for general critiques of these biases).

ITALIAN PALEONTOLOGICAL AND ARCHAEOLOGICAL DATA

In this section and in what follows, I will refer only to evidence thought to date to the 40,000– 30,000 B.P. interval (i.e., that temperate part of Isotope Stage 3 conventionally referred to as Hengelo/Denekamp). Figure 9.1 shows sites with well-established, relatively unambiguous stratigraphies, sometimes with absolute (radiocarbon, ESR) dates. In the caption, sites with human fossils are indicated by italics; those with absolute chronologies are underlined. Most are caves or rockshelters; only three are open-air sites: S. Francesco (2) in the town of San Remo, Monte Avena (3), and Serino (10). The archaeological sites are classified as late Mousterian, Uluzzian and related assemblages, and early Aurignacian. As regards the last, the modifier "early" does not necessarily refer to *early* absolute dates, but to the traditional typological designation employed by most Italian researchers (e.g., Palma di Cesnola 1993). According to the established typological systematics, the Aurignacian is divided into two major complexes: an allegedly earlier one where, in addition to the diagnostic carinated and nosed scrapers, there is a consistent presence of marginally backed bladelets (i.e.,

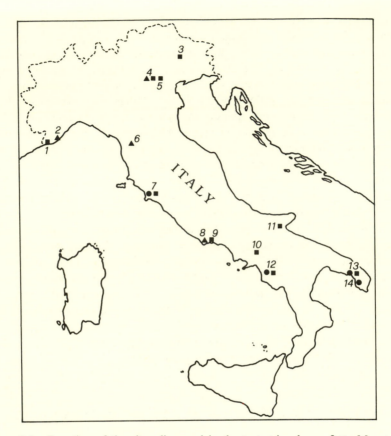

Figure 9.1. Location of the sites discussed in the text: triangles = Late Mousterian, circles = Uluzzian and related assemblages, squares = Early Aurignacian.

1   Balzi Rossi caves and *Riparo Mochi*
2   Stazione di S. Francesco
3   Monte Avena
4   Riparo Tagliente
5   *Riparo di Fumane*
6   *Buca della Iena*
7   Grotta della Fabbrica
8   *Grotta Breuil*
9   *Grotta del Fossellone*
10 *Serino*
11 *Grotta Paglicci*
12 *Grotta di Castelcivita*
13 Grotta di Serra Cicora and Grotta M. Bernardini
14 *Grotta del Cavallo,* Grotta di Uluzzo, and Riparo di Uluzzo C.

Dufour, Font-Yves, and Krems types); and a supposedly more recent one, also called typical (i.e., more or less in accord with French definitions), where the bladelets practically disappear and the diagnostic "type fossil" is the split-based bone point. In sites where both assemblages have been found (e.g., Riparo Mochi), the early Aurignacian always underlies the "typical" Aurignacian.

THE FOSSIL EVIDENCE

Fossil remains from the 40,000–30,000 B.P. interval are very rare, consisting of isolated and fragmentary finds clearly not representative of a biological population. In consequence, there is no way to discriminate among the three hypotheses for the biological transition solely on the basis of the human remains.

Late Mousterian

At present, Grotta Breuil (Figure 9.1, No. 8) is the only site where human remains with absolute dates have been found. This cave, discovered in 1936 by A. C. Blanc as part of a survey of the caves on Monte Circeo, was first tested by very small sondages because access to the deposit is possible only from the sea (Blanc 1938). Full-scale excavations were not undertaken until 1986, and by 1988 it was possible to identify a series of strata (Bietti et al. 1988, 1990–1991). The uppermost levels have been dated by ESR to ca. 36,000 B.P.; the determination comes from layer 3 (Schwarcz et al. 1990–1991). The human remains consist of a fragment of a left parietal (Breuil 1) and two molars (Breuil 2, 3). These remains have been attributed to neandertals by Manzi and Passarello (1990–1991, 1995). The parietal and one of the molars were found out of stratigraphic context, however, although they probably come from layers 4–6. The remaining molar was recovered from layer 6. Preliminary ESR assays from layer 6 give an average date of 33,000 ± 5,000 B.P. (Schwarcz, personal communication 1995), which, given the large standard deviation, agrees fairly well with the layer 3 determination.

In addition, a fragment of a mandibular corpus (Fossellone 3) was recovered from a Mousterian layer at the nearby site of Grotta del Fossellone (Figure 9.1, No. 9), in association with the remains of elephant and rhinoceros (Blanc 1954). There is no absolute date. On the basis of the industry, the so-called Denticulate Mousterian, the fossil is believed to be "late" (see Palma di Cesnola 1987; Vitagliano and Piperno 1990–1991). In my opinion, however, the presence of *Dicerorhinus mercki* and *Palaeoloxodon antiquus* more strongly indicates a chronological assignment to Isotope Stage 4.

The Uluzzian

The only human fossils associated with "transitional" Uluzzian industries come from Grotta del Cavallo (Figure 9.1, No. 14), the site where this cultural facies was first recognized (Palma di Cesnola 1967, 1993) The fossils consist of two deciduous molars from level EIII (possibly EII), with a difference in depth of about 15–20 cm (Palma di Cesnola 1993:115). According to Messeri, the level EIII tooth more closely resembles its modern human counterpart, while the stratigraphically later one (EII) is more neandertal-like (Palma di Cesnola and Messeri 1967). There is a Rome Radiocarbon Laboratory determination of >31,000 B.P. for levels EII–I.

Early Aurignacian

No definite early Aurignacian human fossils have been found in Italy. However, in the Balzi Rossi caves (Figure 9.1, No. 1), and in particular at Barma Grande, Basau da Ture, and Caviglione, a series of well-known multiple burials reported from excavations in the last quarter of the nineteenth century so far remain undated (Rivière 1887; Verneau 1892, 1899). Palma di Cesnola (1993) refers to these burials as Aurignacian sensu lato. The same holds true for a fragmentary maxilla and a piece of a scapula from Fossellone (Fossellone 1 and 2, respectively), in Latium. These remains purportedly show anatomically modern features, although the maxilla comes from an erosional feature within a Mousterian layer and the scapula from a more clearly Aurignacian deposit (Mallegni and Segre-Naldini 1992:Fig. 9.1). In any case, the Aurignacian of Grotta del Fossellone is not dated. Since it has split-based bone points, however, it pertains to "the more recent phase" of the Aurignacian on typological grounds (Blanc and Segre 1953).

THE ARCHAEOLOGICAL EVIDENCE

Although archaeological evidence is much more abundant than human fossil remains, a lot of the work was done very early on and, in consequence, is of very variable quality. And, except for a few recently excavated or restudied sites, most of what we know consists almost exclusively of the typological analysis of retouched stone tools, in accordance with the emphasis placed on typological systematics by several generations of Italian prehistorians (for an extended discussion of these epistemological issues, see Bietti 1991). Analyses of lithic assemblages oriented towards aspects of technology (reduction sequences, raw material procurement, etc.) are almost unknown, as are taphonomic and zooarchaeological analyses. In most cases not even the NISP for the large mammals has been reported, nor are data presented in sufficient detail to allow its calculation from published sources. The same holds for microwear analyses.

Late Mousterian

According to the chronotypological perspective adopted by most Italian prehistorians and well-illustrated by the review article of Palma di Cesnola (1987), in the various Italian sites the final Mousterian is characterized by a decrease in the incidence of levallois technique and by a substantial increase in the incidence of denticulates. These trends are supposedly more or less "interregional" in scope.

The only undated and stratigraphically ambiguous site presented in Figure 9.1 is S. Francesco (Figure 9.1, No. 2). I include it here because it is commonly attributed to the Würm II-III interstadial (de Lumley 1969; de Lumley and Isetti 1965) and because, although it is usually considered to pertain to a "Denticulate" facies of the Mousterian, it also shows a high degree of laminarity, and selection of laminar blanks for the rather large formal tools. Unfortunately, no work has been done thus far on the reconstruction of reduction sequences for this enigmatic site.

The upper Mousterian layers at Riparo Tagliente (Figure 9.1, No. 4), have also been attributed to the Hengelo interstadial (Palma di Cesnola 1987:148). The main interest of these layers is that they directly underlie early Aurignacian deposits (see below).

The site of Buca della Iena (Hole of the Hyena), in Tuscany (Figure 9.1, No. 6), has been dated by the U-Th method to <40,000 B.P. (Pitti and Tozzi 1971). According to Palma di Cesnola, it is a typical example of the late "Denticulate Mousterian" of Tuscany (1987:145). However, and as the name implies, this cave seems to have only occasionally been visited by humans, according to a series of taphonomic analyses of the faunal remains by Mary Stiner (e.g., Stiner 1990).

The last late Mousterian sequence considered here comprises the upper strata of Grotta Breuil (see above). A clear blade technique is well documented in layer 3 at this site, together with a persistence of levallois flaking, mostly unidirectional. In fact, as can be seen from a hypothetical reconstruction of the reduction chain (Figure 9.2), there is good evidence for what have been called "pseudoprismatic" cores (cores with plain striking platforms that resemble Upper Paleolithic prismatic blade cores). These cores, together with the "levallois" unidirectional cores, represent the majority of the core types. The centripetal cores—typical of "classic" Mousterian assemblages—constitute only 9.5% of the core fraction. The analysis of flake dimensions and, in particular, that of the predetermined flakes suggests that laminar debitage (i.e., that derived from pseudoprismatic and unidirectional cores) is the dominant kind of debitage present (Rossetti and Zanzi 1990–1991:358). As regards the formal tools, it is interesting

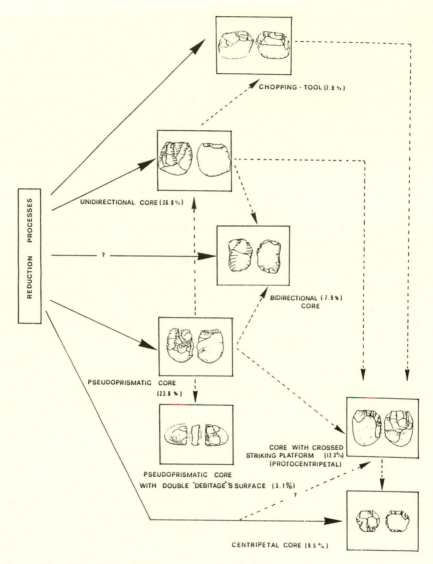

Figure 9.2. Tentative reduction sequence for the industry of layer 3 of Grotta Breuil, derived from the cores. The dotted arrows indicate alternative "continuous" reduction processes (after Rossetti and Zanzi 1990–1991).

to note that, in contrast with the traditional dominance of denticulates in late Mousterian Italian assemblages, at Grotta Breuil they are practically negligible. Moreover, the same technological and typological patterns continue in the industries of layers 4 and 5 (Bietti et al. n.d.).

Recent zooarchaeological analyses of the Grotta Breuil Mousterian faunas are also of interest. The mortality pattern of the ungulate remains suggests an "efficient" hunting strategy, targeting prime age adults (Stiner 1990–1991a, 1990–1991b). It resembles patterns observed in *late Upper Paleolithic* sites in the region and is very different from that observed in more ancient Mousterian sites, such as Grotta Guattari or Grotta dei Moscerini, where scavenging seems to be well documented. Moreover, evidence for carnivore activity at Grotta Breuil is scarce in the extreme when compared with that indicating a human presence, whereas the reverse is true for Grotta Guattari and Grotta dei Moscerini (Stiner 1990, 1994). Analysis of anatomical part frequencies and seasonality data at Grotta Breuil suggests that the site was a residential camp during the late Mousterian sequence, probably occupied from fall to spring (Stiner 1990–1991b:347).

The Uluzzian

The term *Uluzzian* comes from the Bay of Uluzzo, on the Ionian coast of Apulia, where the caves of Serra Cicora, M. Bernardini (Figure 9.1, No. 13), and Cavallo, Uluzzo, and Riparo di Uluzzo C (Figure 9.1, No. 14) are located. The Uluzzian is defined on the basis of the typology of the formal tools, as is customary in Italian prehistory (see discussion in Bietti 1991). Palma di Cesnola (1993) extends the geographical distribution of the Uluzzian well beyond the region of Uluzzo Bay and claims that it is present in Calabria, Campania, and Tuscany, where it is represented by cave sites (Figure 9.1). It is also supposedly found in various open-air sites, especially in Tuscany (Palma di Cesnola 1993:Fig. 9.10). These open sites are not considered here, however, because they have no stratigraphic context and have produced no absolute dates.

According to Palma di Cesnola, the main technological characteristics of the Uluzzian are *(a)* the presence of variable numbers of unidirectional or bidirectional cores (so far undescribed), along with many splintered pieces (possible wedges or exhausted cores, cf. *pièces esquillées* in French typologies) which sometimes cannot be distinguished from the bidirectional cores (1993:82); *(b)* a very low incidence of lamellar debitage, albeit commonly with faceted butts; and *(c)* local raw material procurement, particularly in the earliest levels (e.g., EIII at Grotta del Cavallo, where thin slabs of local calcareous chert were extensively used) (1993:82–88). As regards the formal tools (Palma di Cesnola 1993:Fig. 3-5), there are several endscraper types, sidescrapers, very few burins, and some backed pieces, mainly with curved backs, which resemble large crescents or lunates (well represented at Grotta del Cavallo). These last are traditionally considered the lithic "type fossils" of the Uluzzian. It is interesting to note that there are also some Upper Paleolithic–looking bone tools, which would usually be considered points or *sagaies* in French typologies.

Palma di Cesnola (1993) has proposed the following chronotypological sequence for the Uluzzian, mainly derived from the sequence at Grotta del Cavallo: Early Uluzzian, represented by layer EIII at Cavallo; Evolved or Middle Uluzzian (layers EI–II); and Upper Uluzzian (layers DII–Ib). The retouched tool component in the Early Uluzzian is dominated by sidescrapers and denticulates (70% on average of the retouched pieces), with a consistent presence of end-scrapers, and by the aforementioned crescents or lunates. The Middle Uluzzian supposedly differs both in the raw materials employed (artifacts on thin slabs of calcareous chert are rare; the dominant raw material consists of small pebbles of flint and jasper), and in the typology of the formal tools. Small backed pieces now dominate, and are increasingly of microlithic proportions. At the same time, the splintered pieces are much more abundant than they are in layer EIII at Cavallo. The Upper Uluzzian is characterized by a sudden decrease in the incidence of backed pieces, an increase in the frequency of denticulates, and the appearance of some "Aurignacian-like" retouched blades (Palma di Cesnola 1993:Fig. 5).

Layers EII–I at Cavallo have produced an absolute date of >31,000 B.P., as noted above. The fauna indicate a more temperate climate for layer EIII; a colder, steppelike situation from layers EII to DII, with a sharp increase in the frequency of equids; and a return to more forested conditions in layer DIb.

Only some of the phases defined at Grotta del Cavallo are found in the nearby caves of Uluzzo Bay. In the M. Bernardini cave (Figure 9.1, No. 13), the Early Uluzzian is present in layer AIV, the Evolved Uluzzian is practically absent, and the Upper Uluzzian is well represented in layers AII–I (Borzatti von Löwenstern 1970; Palma di Cesnola 1993). In the Uluzzo Cave (Figure 9.1, No. 14) only the Upper Uluzzian is present, in layer N (Borzatti von Löwenstern 1964), while in the Uluzzo C rockshelter, the Uluzzian layers (D–C) are very poor in lithic artifacts, so no chronotypological classification is possible (Borzatti von Löwenstern 1965; Palma di Cesnola 1993:98). There are no absolute dates for these sites; the fauna generally resemble those from the corresponding levels at Grotta del Cavallo (Palma di Cesnola 1993:90). An Uluzzian-like industry has also been found in the Serra Cicora cave (Figure 9.1, No. 13) in the D horizon of layer B (Spennato 1981). Palma di Cesnola (1993:98, 99) defines this industry as a "terminal" Uluzzian that supposedly postdates the Upper Uluzzian of Grotta del Cavallo. There is a further increase of "Aurignacian" types. The fauna seem to indicate a relatively temperate environment, perhaps indicating an Arcy interstadial date (Palma di Cesnola 1993:90).

A very important site in southern Italy (Campania) is Grotta di Castelcivita (Figure 9.1, No. 12) (Cioni et al. 1979; Gambassini 1982). As at Grotta del Cavallo, there is a gap, in layer rsi, between a Mousterian with a levallois industry (Cioni et al. 1979:Fig. 2) and the Uluzzian. Florence Radiocarbon Laboratory dates for the Uluzzian cluster around 33,000–32,000 B.P., with the exception of Sample F 106 (out of stratigraphic order and >34,000 B.P.). A recent date from layer rpi, still unpublished, agrees with a date of 33,220 ± 780 B.P. from the Sample F 107, also from rpi (Palma di Cesnola 1993:90). The faunal assemblage is rather different from that of the classical Uluzzian of Salento. Red deer, roe deer, and fallow deer are well represented, indicating more forested conditions, as would be expected for the Tyrrhenian coast of Italy, in contrast with its Adriatic coast, which typically has more open environments regardless of the paleoclimatic regimen.

The Castelcivita lithic industries are rather different from those of the Uluzzo Bay region. Backed tools are scarce (although a few crescents are noted, "atypical" when compared with those from the Apulian sites), while sidescrapers and denticulates are dominant. Pièces esquillées are also abundant, especially towards the top of the sequence (e.g., in layer rsa). Gambassini maintains that it is difficult to correlate the Castelcivita assemblages with those from Uluzzo Bay (1982). On the basis of preliminary data (Cioni et al. 1979; Gambassini 1982), it is my opinion that the domination of Uluzzian diagnostics in this assemblage is inadequate and that we are probably facing a Middle to Upper Paleolithic "transition" industry strictly related to local adaptation phenomena.

The same may be said of Grotta della Fabbrica (Figure 9.1, No. 7), in Tuscany (Pitti et al. 1976). The stratigraphy is rather complex, with many erosional episodes; five main stratigraphic units have been recognized (Pitti et al. 1976:Fig. 2). Layer 1 contains a scarce Mousterian industry, layer 2 the Uluzzian, layers 3–4 Aurignacian assemblages, and layer 5 (traces of which are preserved on the walls of the cave) an Epigravettian industry sensu lato. There are no absolute dates. Equids dominate the fauna through layers 1–4, with substantial numbers of red deer in layer 1, decreasing in layer 2. Equids decrease in frequency in layers 3–4, which are also marked by the appearance of chamois (Sorrentino in Pitti et al. 1976:184). The fauna have been taken to indicate paleoclimates for layer 2 that resemble those of the Middle Uluzzian of Grotta del Cavallo. Despite superficial qualitative similarities which led Tozzi (in Pitti et al. 1976:198) to affirm a close relationship with the Grotta del Cavallo industries, the lithic assemblages from Grotta della Fabbrica differ from those of Cavallo in many important respects. The backed tools are very rare, for example, and distinctive morphologically from the well-known crescents of the classic Uluzzian; there are also some Aurignacian-like carinated and nosed endscrapers. In fact, these differences led Palma di Cesnola to define a "Grotta della Fabbrica facies" of the Uluzzian that differs from that of the classic Uluzzian (1993:99).

Several surface sites, mainly in Tuscany (Indicatore, S. Romano, Maroccone), are also considered Uluzzian by Palma di Cesnola, but, like Fabbrica, are relegated to distinct Uluzzian "facies" (1993:99–110). My own impression, as for Castelcivita, is that these sites represent adaptive "transitional" situations peculiar to the region of Tuscany. The main difficulty with the notion of transitional industries, from the contextual and systemic approach to which I subscribe, is that the present lack of analyses of the lithic reduction processes, of "modern" quality zooarchaeological and taphonomic studies, and of absolute dates, precludes a better understanding of the subsistence and settlement patterns of the mobile foragers represented by these "transitional" assemblages.

Early Aurignacian

Aurignacian sites are not particularly common in Italy, and many were dug long ago. One of the most important is Fumane Cave in the Lessini Mountains, near Verona (Figure 9.1, No. 5)—one of the few recently excavated Aurignacian sites to provide us with an almost complete, modern-quality account of the stratigraphy, absolute dates, paleobotanic and faunal analyses, taphonomic studies, and the lithic assemblages (Bartolomei et al. 1992). As regards the archaeology, the most important part of the deposit is lower unit A and the lower part of unit D (Figure 9.3). Layer A4 II is the uppermost stratigraphic unit with Mousterian industry, while the early Aurignacian starts with layer A3. There is, however, a very thin intermediate layer (A4 I), not shown in Figure 9.3, with scarce archaeological remains of Mousterian type (Bartolomei et al. 1992:Fig. 14). Owing to the presence of two Dufour bladelets, Broglio and Peresani (in Bartolomei et al. 1992:156) suggest possible contamination of this layer from bracketing layers A3 and A4 II, although they also recognize an "Uluzzian" backed crescent (Bartolomei et al. 1992:Fig. 14, No. 1) in the assemblage. In my opinion, this artifact does not resemble classic Uluzzian crescents and is, instead, simply a classic *couteau à dos,* sensu Bordes.

As regards Aurignacian levels A3–A1, several AMS dates from the Utrecht laboratory range from 36,800 +1200/−1400 to 31,600 ± 400 B.P. in layer A2, and from 32,300 ± 500 to 31,700 +1200/−1100 B.P. in A1/A2. There is also a very old, probably unreliable, determination with a large standard error from layer A2: 40,000 +400/−300 B.P., Sample UtC 1774 (Bartolomei et al. 1992:156–161). These dates agree well with the paleoclimate indicated by the faunal remains and the palynology, taking into account the local environmental setting of the site (foothills of the Alps). Some site features, in the context of so-called living floors, have been identified (Bartolomei et al. 1992:Figs. 16–18, 20–21). The industry is a well-developed Upper Paleolithic one with an extensive bladelike technology and backed bladelets and points (Bartolomei et al. 1992:Fig. 26). There is also an industry on bone, as well as some ornaments (beads). It is clear that this is something very different from a transitional Late Mousterian or Uluzzian assemblage.

As noted above, an Early Aurignacian industry directly overlies a late Mousterian level in the nearby Tagliente rockshelter (Figure 9.1, No. 4). There is, however, some doubt about this stratigraphic continuity since, in other portions of the same shelter, erosional phenomena resulted in an almost direct contact between the Mousterian and the Epigravettian (Bartolomei et al. 1982; cf. Broglio 1994:43)! The Aurignacian levels at Tagliente are, again, characterized by carinated and nosed endscrapers and by Dufour-like bladelets.

According to Broglio (1994:43), a similar (but sparse) industry also occurs in layer 9 of the Paina cave (in the Grottina Azzurra), in the Berici Mountains, not far from Tagliente. It is dated by two AMS dates from the Utrecht laboratory:

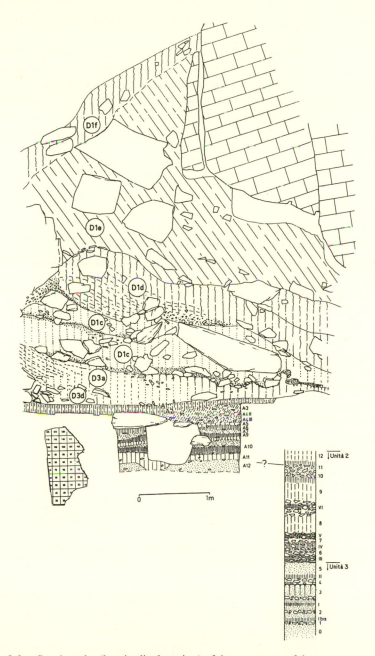

Figure 9.3. Stratigraphy (longitudinal section) of the upper part of the sequence at Grotta di Fumane (M. Cremaschi, from Bartolomei et al. 1992).

38,600 +1400/−1800 B.P. and 37,900 ± 800 B.P., which agree with the older dates from the Fumane cave.

An alpine Aurignacian site is Monte Avena, in Veneto, at 1430 m above sea level (Figure 9.1, No. 3) (Lanzinger 1984). The general features of the industry are characteristic of the Early Aurignacian, but the most important thing about the site is its functional interpretation, linked to raw material procurement (local flint). The excavated sample records the entire primary reduction process from extraction to the selection of nodules suitable for cores to core preparation itself (Lanzinger 1984).

On the Tyrrhenian coast is the very important site of Riparo Mochi, one of the Balzi Rossi group (Figure 9.1, No. 1). This site was excavated by the Istituto Italiano di Paleontologia Umana between 1938 and 1959, but only an illustration of the industries (Blanc 1953) and a typological classification of the Upper Paleolithic collections (Laplace 1977) have been published to date. A reexamination of the old collections from the Mousterian and Aurignacian levels is now in progress (Kuhn and Stiner 1992). In the general stratigraphy through the end of the 1949 excavation, layers A–D represent the Gravettian-Epigravettian sequence, layer E is almost sterile, while layers F and G are characterized respectively by the supposedly more recent "classic" Aurignacian split-based bone point (called Ancient Aurignacian in Laplace 1977), and by the Early Aurignacian industry with Dufour bladelets (called Proto-Aurignacian in Laplace 1977). Layer H is a nearly sterile "transition" to layer I, which contains a Mousterian industry. This part of the sequence resembles the situation found in layer A4 I of the Fumane cave: Kuhn and Stiner suggest that the archaeological remains in layer H are in fact derived from bracketing layers G and I (1992:81–82).

Although analysis is still underway, AMS dates on charcoal from layer G recently obtained from the Oxford laboratory range from 32,280 ± 580 to 35,700 ± 850 B.P. (Hedges et al. 1994). To the best of my knowledge, layers G and F in the Mochi shelter constitute the only Italian site sequence where there is stratigraphic continuity between Dufour bladelet and split-based bone point–bearing Aurignacian deposits. The later phase is also represented by the D layers in the Fumane cave (according to Palma di Cesnola 1993:143; but see Bartolomei et al. 1992).

There are also Aurignacian deposits in the Grotta della Fabbrica (Figure 9.1, No. 7), where layers 3–4 have yielded an Aurignacian assemblage with several splintered pieces, sidescrapers, and carinated endscrapers (Pitti et al. 1976). A few Dufour bladelets led the investigators and Palma di Cesnola (1993:131) to attribute this industry to the Early Aurignacian even though it is probably more recent than the assemblage from layer G at Riparo Mochi.

It is worth remarking that this chronologically "late" attribution of the Aurignacian of Grotta della Fabbrica seems to be confirmed at a few other sites in the southern part of the peninsula. Grotta di Castelcivita (Figure 9.1, No. 12) yielded a Florence radiocarbon laboratory date of 31,950 ± 650 B.P. from Aurignacian layer gic. The Aurignacian levels of this cave have been divided into two separate units: one with Dufour bladelets, in the upper part of layer rsa (and where the "Uluzzian" in the lower part has been dated to 32,900 ± 700 B.P.), and the other comprising layer gic, with a "Castelcivita" facies Aurignacian, characterized by small, "marginally-backed" points (Cioni et al. 1979; Gambassini 1982). The inland open site of Serino (Figure 9.1, No. 10) has also produced an industry with "marginally-backed" elements rather similar to those from Castelcivita. This industry has been dated to 31,200 B.P. and is associated with a forested environment with beech, pine, and birch, consistent with the Arcy interstadial and with the absolute date (Accorsi et al. 1979).

At Grotta della Cala, a site on the Tyrrhenian coast near Castelcivita, a portion of the deposit (spits 13, 14) between layers R (Mousterian) and Q (Gravettian) seems to record a transition towards a more steppelike configuration, with horse and *Microtus*. A date from between the spits of 29,800 ± 900 B.P. suggests a drier interval after the Arcy interstadial. The industry has some backed microliths of the Castelcivita type, but also a very typical "Uluzzian" crescent. Thus stratigraphic continuity with the underlying Mousterian layer R seems to me to be rather questionable.

In northern Apulia, in the famous site of Grotta Paglicci (Figure 9.1, No. 11), a series of Aurignacian deposits has been discovered recently just below the early Gravettian. Two AMS dates from the Utrecht laboratory are 34,000 +900/−800 B.P. for layer 24B1 and 29,300 ± 600 B.P. for layer 24A1 (although the faunal analyses by Boscato suggest a similar cold-temperate climate for both layers; Palma di Cesnola 1992). The industry is very peculiar, with some tools resembling those of other southern Italian sites with "marginally-backed" bladelets from layer 24A4, and some endscrapers and backed pieces of "Paglicci type" (Palma di Cesnola 1991) from layer 24A1 that resemble those from overlying Gravettian levels.

Finally, there is another Aurignacian facies which, in the opinion of Palma di Cesnola (1993), is contemporary with the "classic" Aurignacian with split-based bone points. Called the "Uluzzo-Aurignacian," it implies a kind of continuity between the Uluzzian and the Aurignacian (Palma di Cesnola 1993:150–156). It is only known from one stratified context—in horizons C–A in Serra Cicora cave (Figure 9.1, No. 13), where it overlies the "terminal" Uluzzian of the D horizon described above (Spennato 1981). Other sites attributed to this "Uluzzo-Aurignacian" facies consist of surface finds in southern Italy and Tuscany (Palma di Cesnola 1993). What this facies might mean in the context of the neandertal to anatomically modern human transition is discussed below.

DISCUSSION AND TENTATIVE CONCLUSIONS

From the preceding review it is evident that the transition interval in Italy is very poorly known. Besides the almost complete lack of human fossils, there are only two sites with relatively complete stratigraphic sequences: Grotta Breuil for the late Mousterian and Grotta di Fumane for the early Aurignacian. Nevertheless, we can establish some points and draw the following tentative conclusions.

1. There is no evidence of direct stratigraphic continuity between the Mousterian and the Early Aurignacian, with the possible exception of Riparo Tagliente and Grotta della Cala where the stratigraphic situation is ambiguous, according to the excavators. There are, however, some deposits with "intermediate" assemblages combining aspects of Mousterian and Aurignacian (e.g., A4I at Fumane, H at Riparo Mochi). These have been explained by invoking mixing between Mousterian and Aurignacian deposits (Bartolomei et al. 1992; Kuhn and Stiner 1992). Scarce as it is, the stratigraphic evidence appears to rule out the multiregional evolution hypothesis described in the introductory section of this chapter.

2. A "sudden replacement" hypothesis is also ruled out by the data, as indicated by the various transition assemblages (e.g., the late Mousterian of Grotta Breuil, the Uluzzian of Uluzzo Bay and the related industries from Grotta della Fabbrica and Castelcivita). Moreover, the scarce human remains associated with these industries do not allow us to conclude that *all* transition assemblages were necessarily produced by neandertals. This is most certainly true of Grotta Breuil and maybe also of Grotta del Cavallo.

3. The transition sites are all located (with the possible exception of S. Francesco in Liguria) in the middle and southern parts of the peninsula, where they are more or less contemporary with (or perhaps even later than) the Early Aurignacian of the north (e.g., Fumane, Mochi). The Early Aurignacian sites in the south are definitely more recent, which suggests a slow migration of anatomically modern Aurignacians from the north.

4. Italy is a geographic cul-de-sac, and it is not unique in that regard in the Mediterranean Basin. In Spain, for example, recent excavations at Zafarraya Cave (Malaga) produced fossil evidence of neandertals dated both by radiocarbon and U/Th methods to 32,000–29,000 B.P. (Hublin et al. 1995), suggesting coexistence of neandertals and moderns, and a late survival of neandertals, as is apparently also the case in Italy.

5. How can we explain all these different transition assemblages in central and southern Italy? Implausible as it sounds, the current favorite hypothesis is to postulate the existence of a series of local "acculturation" industries (e.g., Manzi et al. n.d.; Mellars 1989b, 1992). Contacts between local neandertals and Aurignacian "newcomers" supposedly produced more extensive use of blade technologies (e.g., at Grotta Breuil) and more "Upper Paleolithic" tools (e.g., endscrapers, backed pieces, and bone tools, such as those found in the Uluzzian). Contacts could also have existed in the form of gene flow, involving hybridization between the two

populations, in agreement with the hybridization and replacement model described above. However, in the Italian context, there is the serious difficulty that there is *no evidence at all of interstratification* between the transition assemblages and the Aurignacian, as apparently occurs in southwestern France at, for example, Roc de Combe (Bordes and Labrot 1967). This suggests to me that acculturation or hybridization occurred elsewhere, perhaps in northern Italy, and that the "acculturated" neandertals subsequently moved south! Another possibility is that the Uluzzian and related complexes are simply a series of local neandertal adaptations, without any particular contact with the Aurignacians. The "Uluzzo-Aurignacian" of Serra Cicora may be the only relatively plausible example of acculturation (of the Uluzzians, in this case) by the Aurignacians, as proposed by Palma di Cesnola (1993:500).

In my opinion, the data available at present are not sufficient to discriminate among these hypotheses. Much more comprehensive analyses of the archaeological record, aimed at a systemic understanding of local site contexts and formation processes, will be required. It is quite possible that all three processes (local evolution of neandertals, hybridization, and culture contact) took place simultaneously. However, I must acknowledge that the notion of Aurignacians "educating" neandertals seems implausible. It invokes a kind of quasi-historical explanation perhaps more appropriate to more recent transition scenarios (e.g., the Mesolithic to Neolithic transition) and not very well suited to the much more ancient time frames which are central to the modern human origins debate (see Clark 1994b).

In conclusion, I want to stress again the very great *diversity and variability* in the archaeological record of the transition, especially as regards technology and typology of the late Mousterian and Uluzzian lithic industries, compared with a certain uniformity evident in the Early Aurignacian. This variability underscores the need for a systemic perspective in the analysis of individual assemblages prior to any attempt at interassemblage comparison. It is, unfortunately, customary in the Italian research tradition to proceed more or less directly to interassemblage comparison, often on the basis of relatively cursory preliminary study of the retouched tool components of archaeological assemblages, too often based only on partial archaeological records. Needless to say, more absolute dates and, more important, a better sample of human fossils will be of substantial help in providing a more adequate understanding of the biological and cultural transition on the Italian Peninsula.

10

The Human Paleontology of the Middle to Upper Paleolithic Transition on the Iberian Peninsula

MARÍA DOLORES GARRALDA

I present here a detailed discussion and analysis of human fossils from the Iberian Peninsula pertaining to the late Würm II, the Würm II-III interstadial, and up to the mid-Würm III (ca. 50,000– 30,000 B.P.). This interval corresponds to the late Mousterian and the early Upper Paleolithic as conventionally reckoned in Spain, a crucial period from the standpoint of understanding modern human origins there. Although archaeological remains are relatively well documented for this period, human fossils are, unfortunately, very scarce and often problematic in terms of their chronostratigraphic placements.

BIOANTHROPOLOGICAL DATA—THE LATE MIDDLE PALEOLITHIC

The human remains attributed to the Würm II and associated with Mousterian industries are those of Axlor (Basabe 1973) and Lezetxiki (Basabe 1970) in the Basque country; some of the material from Cova Negra, in Valencia (Arsuaga et al. 1989); and probably Columbeira, in Portugal (Ferembach 1964–1965) (Figure 10.1). All these finds are extremely fragmentary—isolated teeth in the case of the Basque sites, an adult left parietal and a permanent incisor at Cova Negra, and one molar germ from Columbeira. By a broad, albeit not empirically well founded, consensus, they are assigned to *H. sapiens neanderthalensis,* but it is almost impossible to estimate the range of variability at either the population or individual level, nor—owing to chronological insufficiencies—to look at variability across time. Differences noted by Basabe among the teeth from the two Basque sites, Lezetxiki and Axlor, suggest larger size and more taurodontism at the former, while the Axlor teeth are smaller and tend to have smaller pulp cavities.

Fossils assigned with varying degrees of confidence to the Würm II-III interstadial and/or to Würm III include Banyoles (Catalunya), Abric Agut (Barcelona), Devil's Tower (Gibraltar), Carihuela (Granada), and Zafarraya (Málaga).

Figure 10.1.   The Iberian Peninsula—location of sites discussed in the text. Closed circles
= Mousterian, open circles = Aurignacian, dashes = unknown.

1   Agut
2   Axlor
3   Banyoles
4   Camargo
5   Carigüela
6   Castillo
7   Columbeira
8   Cova Negra
9   Devil's Tower
10 Lezetxiki
11 Morín
12 Zafarraya

Banyoles (Gerona, Catalunya)

The Banyoles (or Bañolas) mandible was recovered from a travertine deposit
in a quarry in 1887 and lacks an accurate provenience; consequently, its age
remains highly controversial. An initial ^{14}C date taken from a sample of matrix
adhering to the inside of the jaw is generally regarded as too recent (17,600 ±
1000 B.P.) (Berger and Libby 1966). A U-series assay resulted in two very
different determinations, both obtained from the same travertine matrix as the
^{14}C date: 73,000 ± 4000 B.P. (Yokoyama et al. 1987) and 45,000 ± 4000 B.P.
(Juliá and Bischoff 1991). Yokoyama et al. (1987) consider their determination
probably to be too recent because of the possibility of a radioisotopic contamination of the travertine with younger carbonates. The Yokoyama interpretation is

also contested by Juliá and Bischoff (1991), who insist on the validity of their dates based on having conducted numerous analyses of the travertines of the region. While the limits of our knowledge of, and some of the pitfalls inherent in, these dating methods are recognized (see Schwarcz, this volume), how to interpret the morphological and anatomical characteristics of the Banyoles mandible remains the primary anthropological issue. Should we consider it in the context of the range of variability of the so-called pre-neandertals *(antenéandertaliens)* or that of the more recent neandertal populations? The numerous controversies surrounding the fossil were reviewed recently in Maroto (1993), where its atypical neandertal characteristics are well documented morphometrically.

Abric Agut or Abric Romaní (Capellades, Barcelona)

Four isolated teeth recovered from this rockshelter by Vidal and Romaní in 1912 were assigned by Ripoll and de Lumley (1965) to the Würm II-III interstadial. De Lumley (1973) points to taurodontism in these incomplete and fragmentary specimens, considered to pertain to a single adult individual.

Devil's Tower (Gibraltar)

The earliest report of human fossils from this famous Mousterian cave site is that of D. Garrod in 1926 (Garrod et al. 1928). Unfortunately, the remains have an uncertain stratigraphic provenience, variously considered to date to the Würm II or to the Würm II-III interstadial (Garralda 1978; Tillier 1982). Tillier (1982) identified two young children: Gibraltar 2, comprising a frontal, left parietal, right maxilla, mandible, and isolated teeth; and Gibraltar 3, represented by an isolated right temporal. However, Zollikofer et al. (1995) have argued that all these fragments probably pertain to a single individual. According to Tillier (1982), G2 exhibits several "typical" neandertal characteristics (e.g., torus supraorbitalis, round vault profile) but also has a curved (rather than an extended) upper maxilla. She considers the maxillary configuration to be a plesiomorphic character because of its presence in the supposedly earlier Gibraltar 1 skull.

Carihuela (or Carigüela) Cave (Píñar, Granada)

The excavations of Spahni (1954–1955) in the Mousterian levels of La Carihuela produced hominid skeletal material that he initially assigned to Würm I (García Sanchez 1960). Subsequent work on the archaeological collections (de Lumley 1969) and new excavations in the late 1960s (Almagro et al. 1970) resulted in a reassignment of the fossils to late Würm II. Recent work at the site by Vega Toscano (1990; Vega Toscano et al. 1988) produced a third chronostratigraphic assessment which assorted the fossils as follows: (1) Píñar 3, C, found in chamber

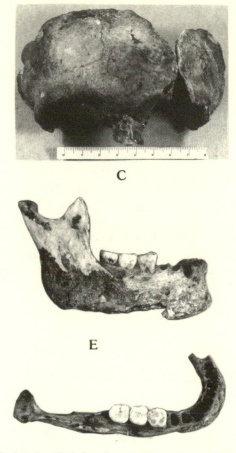

C

E

Figure 10.2. Carigüela: (C) infant frontal, (E) adult mandible (both reduced).

II—Würm I-II interstadial; (2) Píñar 1, A, found in chamber III—Würm II-III interstadial; and (3) Píñar 2, B, also from chamber III—Würm III.

The human remains consist of two small fragments of adult parietals (1, A and 2, B), and the incomplete frontal of a child (3, C). They were completely described by García Sanchez (1960); the endocranial surface of the frontal was also studied by Fusté (1957). These workers, and de Lumley and García Sanchez (1971), classified them as *H. sapiens neanderthalensis.* In the case of the child (3, C—Figure 10.2), de Lumley and García Sanchez (1971) identified what they regard as typical "neandertal" characters (apomorphies), like the accentuation of the torus supraorbitalis visible on an individual 6–7 years old at the time of death; the low frontal; and postorbital constriction, together with some traits they considered "advanced" or "modern" (e.g., small frontal sinuses, large orbits, short nasal bones, complex brain convolution patterns).

Subsequent excavations by Almagro and Irwin (Almagro et al. 1970) found six isolated teeth (Píñar 7, G) in the Mousterian levels at the entrance of Car-ihuela chamber I. The deposits containing the teeth were recently assigned either to the Würm II stadial or to the Würm II-III interstadial (García Sanchez et al. 1994). They pertain to a child 8–9 years old at the time of death; three of them exhibit enamel hypoplasia. The teeth are large and their overall morphology is not a priori in contradiction with a neandertal affiliation.

In the old excavations of Spahni, several human remains were recovered from levels 1–3. Level 1 was (and is) considered to be Neolithic; levels 2 and 3 were thought at the time to be Mousterian "with a weak Aurignacian influence" (Gar-cía Sanchez 1960). However, reinvestigations by Almagro (Almagro et al. 1970) and Vega Toscano (Vega Toscano et al. 1988) fail to document any clear Upper Paleolithic levels at the site, and it seems that, after the long Mousterian se-quence, Carihuela was rarely used or occupied until the Neolithic. The fossils were considered by García Sanchez (1960) to be *Homo sapiens fossilis,* but they remain without accurate archaeological or stratigraphic provenience.

The detailed anthropological study of García Sanchez (1960) shows com-pletely modern morphology and metrics for these isolated remains, which consist of a small parietal fragment (Píñar 4, D), one incomplete mandible (Píñar 5, E— Figure 10.2), and a partial tibia (Píñar 6, F). It is noteworthy that they are very similar to undisputed Neolithic remains from level 1, also studied by the same author. The following comments are added to those of García Sanchez (1960).

1.   The parietal fragment (4, D) shows some parallel incisions produced artificially postmortem, and attributed by García Sanchez (1960) to intentional human manipulation of the skull. Similar marks were identified on several (Up-per) Paleolithic fossils (see, e.g., Le Mort 1987), but we should bear in mind that García Sanchez and Carrasco (1981) also published an account of a very com-plete vault with indubitable (and very different) cutmarks clearly associated with the Neolithic levels at the site.

2.   Study of the mandibular dentition (5, E) revealed the presence of caries in two of the five preserved molars (García Sanchez 1960). Although caries are not unknown on Paleolithic teeth, they are not very common (Frayer 1989b). On the other hand, the Neolithic levels at Carihuela do show a high incidence of carious teeth.

3.   The molar crown dimensions are within the range of variation of modern European teeth, being smaller than comparable (though scarce) teeth from the European Upper Paleolithic (Frayer 1978, 1989b; Gambier et al. 1990; Garralda et al. 1992).

4.   The mandible (5, E), considered by García Sanchez (1960) to be that of an adult male, is nevertheless very gracile (Figure 10.2), lacking a torus mandi-bularis, and with dimensions similar to those of five isolated jaws found with some other human remains in the Neolithic levels of the same cave, Carihuela III. Table 10.1 compares Carihuela dental metrics with those of Zafarraya 2, associ-

Table 10.1. Dimensions of Several Fossil and Subfossil Mandibles from the Iberian Peninsula

| | Carihuela (Neolithic) | | | | | | Zafarraya ♂ | Azules ♂ | Levant ♂ | | | Levant ♀ | | | Central Plateau ♂ | | | Central Plateau ♀ | | |
	5, E ♂	1 ♂ ?	2 ♂	3 ♀	5 ♂	6 ♂			n	$\bar{x}$	σ	n	$\bar{x}$	σ	n	$\bar{x}$	σ	n	$\bar{x}$	σ
Mandible length	105?	—	—	99	—	115?	106.5	—	29	101.21	7.41	15	98.27	2.81	5	103.4	6.27	5	98.2	3.42
Bicondylar breadth	110?	—	120*	109*	—	122*	137	—	16	116	8.31	9	111.1	5.40	4	122		3	111	6.44
Bigonial breadth	96?	—	—	87	102*	98*	92	—	29	95.93	6.68	13	88.08	6.79	5	98	7.84	7	88.14	
Biment. breadth	42	—	—	—	—	—	53	—												
Symphyseal height	31	—	33	27	34?	28?	35	33	53	33.53	2.86	21	30.71	2.67	15	34.13	3.29	8	31.25	2.31
Body height†	32.5	—	32	28	32	—	34.5	32												
Thickness†	11	12	12.5	9.5	10	14	16	14												
Ramus height	67	54	—	59	—	67	72	—	32	60.38	5.68	17	55.41	4.35	8	63	4.87	8	54	4.66
Ramus breadth	32	35	—	27	—	30	44	—	38	32.71	2.67	20	30.55	2.80	10	32.40	2.46	9	29.44	0.88
Mandibular angle	118?	—	—	120	—	125	106	—	32	118.6	5.64	17	124.1	5.82	7	121.4	8.58	6	121.3	8.26
Symphyseal angle	81?	—	93	92	—	—	77	—												
Mandibular index	95.4?	—	—	90.8	—	—	77.74	—	17	88.3	8.11	9	89.60	5.22	4	84.01	5.22	3	92.72	
Robusticity index	33.8	—	39.06	33.9	31.2	—	46.38	43.75	31	54.08	6.27	17	55.01	5.98	8	52.25	5.62	8	54.99	5.12
Ramus index	47.7	64.8	—	45.7	—	44.7	61.11	—												

† Foramen mentale

* Estimated by symmetry

Sources: Carihuela, García Sanchez 1960; Zafarraya: García Sanchez 1984; Azules: Garralda 1986; Levant and Central Plateau: Garralda 1974.

Table 10.2. Comparison of Some Mandibular Measurements Using the *t*-test: Neolithic
Males from the Spanish Levant

Variable	df	Levant/Carihuela 5, E		Levant/Zafarraya	
		t	p%	t	p%
Mandible length	21	−0.503	>60	−0.702	<50
Bicondylar breadth	16	+0.700	50	−2.450*	2 < p < 5
Bigonial breadth	29	−0.010	>90	+0.578	<60
Symphyseal height	53	+0.876	<40	−0.509	<50
Ramus height	32	−1.148	<20	−2.014*	>5
Ramus breadth	38	+0.262	<80	−4.174**	<0.1
Angle	32	+0.115	>90	+2.210*	<5
Mandible index	17	−0.848	>40	+1.268	20 < p < 30
Ramus index	31	+1.001	>30	−1.103	20 < p < 30

* significant ** highly significant

ated with Mousterian artifacts (García Sanchez 1984); the Azilian male burial at
Los Azules (Garralda 1986); and two Neolithic series from the Iberian Peninsula
(Garralda 1974). Differences between Los Azules and especially Zafarraya, vis-
à-vis Carihuela and the Neolithic series, are clear. Both Los Azules and Zafarraya
present much higher values, while the other samples are statistically very similar
to one another. The *t*-test indicates that the small differences between the Carihuela E teeth and the male Neolithic series from the Spanish Levant (the Mediterranean region between Tarragona and Alicante) can be explained by chance
alone (Table 10.2).

   5.   The isolated tibia fragment (6, F), assigned to an adult female, is also very
gracile. The platycnemia observed cannot be interpreted as an exclusively Upper
Paleolithic character. Comparable tibias from various times and places in Iberia
have the same or similar morphology, as indicated by the cnemic indices for the
following individuals and series.

- Carihuela F (García Sanchez 1960)   58.10%
- Los Azules male, Azilian, left tibia (Garralda 1986)   57.44%
- Moita do Sebastião, Epipaleolithic (Ferembach 1974)
    male: $n = 35$, $\bar{x} = 64.0$ $\sigma = 4.25$
    female: $n = 23$ $\bar{x} = 63.3$ $\sigma = 4.96$
    male + female platycnemics: 43.10%
- Meseta Central, Neolithic (Garralda 1974)
    male: $n = 19$ $\bar{x} = 63.58$ $\sigma = 6.33$
    female: $n = 17$ $\bar{x} = 67.30$ $\sigma = 6.22$
    male platcynemics: 57.89%
    female platcynemics: 35.29%

   It is possible, of course, that the Carihuela remains represent modern humans
making Mousterian tools, as occurred in Israel and north Africa, but their

morphologies and the many uncertainties surrounding the circumstances of discovery argue against this interpretation. It seems more likely that they are simply Neolithic fossils associated (because of mixing, chronostratigraphic errors, etc.) with Mousterian stone artifacts. Absolute dates would go a long way to resolve this dilemma (as they did with respect to the Urtiaga crania—Altuna and Rua 1989), but the bottom line is that the stratigraphic provenience of many of the Carihuela fossils simply cannot be established with certainty.

Zafarraya Cave (Alcaucín, Málaga)

Zafarraya is a cave located in the hamlet of Alcaucín (Málaga), where Barroso and his colleagues recovered a fragment of a left femoral diaphysis and a nearly complete adult mandible in the early 1980s in level D, associated with Mousterian artifacts attributed to Würm III (Barroso et al. 1983, 1984). Both fossils were studied by García Sanchez (1984), who assigned them to *H. sapiens neanderthalensis*. The femur, which shows some traces of human intervention on the diaphysis, is that of a robust adult male, around 162.2 cm in height (estimate based on the Steele & McKern method for incomplete long bones).

The mandible, also thought to be that of a male, is robust and has well-developed muscle markings (Figure 10.3). Its metrical and morphological characteristics (e.g., the mandibular index, the verticality of the symphysis [with marked tubercula mentalia] and the rami, the position of the foramina mentalia, the absence of taurodontism) could be attributed either to variability within an evolving late neandertal population or to admixture with modern humans.

A *t*-test was used to compare the Zafarraya mandible with the more gracile (and much more recent) male Neolithic series from the Spanish Levant (Table 10.2) (Garralda 1974). A detailed biometrical study by Sánchez (1990) also confirms the unquestionable "neandertal + modern" characteristics of this interesting fossil. Unfortunately, neither García Sanchez (1984) nor Sánchez (1990) published dental metrics that would have been of value in this comparison.

Three more discoveries, to date unpublished, have recently come to light in the same Mousterian level D that produced the fossils just described. They are two isolated teeth, one pubic bone fragment, and another incomplete mandible (the last with traces of combustion—Barroso et al. 1993). The U/Th age of level D is ca. 35,000 B.P., while dates from overlying level I, also with Mousterian industries, seem to indicate a very late survival of this culture after 30,000 B.P.

BIOANTHROPOLOGICAL DATA—THE EARLY UPPER PALEOLITHIC

The early Upper Paleolithic is well represented on the peninsula, especially for the Aurignacian complex, best known from Catalunya and Cantabria in

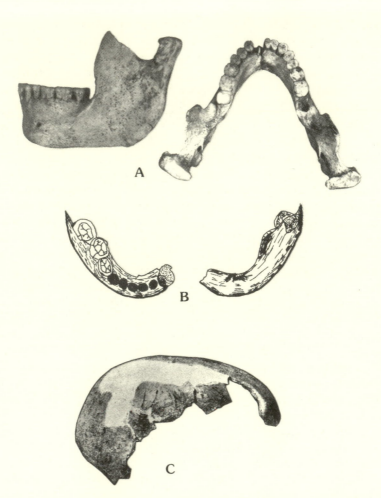

Figure 10.3. (A) Zafarraya mandible (after García Sanchez 1984), (B) Castillo infant mandible fragment (after Basabe 1984), (C) Camargo vault (after Hoyos Sainz 1947). A and C are reduced; B is 1:1.

northern Spain. I refer here to the remarkable AMS ^{14}C dates from La Arbreda cave, in Gerona (Bischoff et al. 1989), and Abric Romaní, in Barcelona (Bischoff et al. 1994), and to the crucial series from level 18 in El Castillo cave (Santander), published by Cabrera and Bischoff (1989). The initial dates from El Castillo have now been confirmed by new determinations from the laboratories at Gif-sur-Yvette and Oxford (Cabrera et al. 1993). An age of ca. 40,000 B.P. for the Archaic Aurignacian in northern Spain appears to be well documented, implying a long period of spatial and temporal overlap there between the late Mousterian, the early Aurignacian, and the Châtelperronian, with some cases of interstratification of the latter two cultures, as documented, for example, at the cave of El Pendo, in Santander (González Echegaray 1993).

El Castillo Cave (Puente Viesgo, Santander)

Archaic Aurignacian level 18 from this important site produced human remains in the course of the 1911–1914 excavations of Breuil, Wernert, and Obermaier (Cabrera 1984). An unpublished study by Henri Vallois (1933) and the short note by Basabe (1974) are actually the only existing documents pertaining to these fossils, which have unfortunately disappeared from the provincial museum in Santander. The Vallois manuscript has recently been discussed by Garralda (Garralda et al. 1992). It indicates the presence of some relatively gracile cranial fragments and one large, relatively unworn, lower molar, corresponding to one or two adult individuals (Table 10.3). In addition were several small skull fragments

Table 10.3. Crown Dimensions of Permanent (M2) and Deciduous (m1, m2) Molars from El Castillo B and C Compared with Other Aurignacian, Middle and Upper Paleolithic Teeth

Tooth		Source Sites	Mesio-distal Diameter	Bucco-lingual Diameter	Crown Index Robusticity
M2		Castillo B	12.0	11.5	126.0
	(2)	European neandertals ($n = 12$)	10.43 ± 0.72	11.0 ± 0.81	
	(3,6)	Qafzeh-Skhūl ($n = 7$)	11.70 ± 0.9	10.7 ± 0.6	
	(4)	Les Rois 1	11.0	11.0	121.0
	(5)	Les Rois 21	10.0	10.0	100.0
	(5)	Les Vachons 1	11.8	11.9	140.4
	(7)	Les Abeilles	10.3(?)	10.5	108.15
	(8)	Early Upper Paleolithic	$\bar{x} = 11.3 \pm 1.00$	10.8 ± 0.82	
		($n = 22$) V $= 9.5 \pm 12.8$	9.8–12		
m1		Castillo C	9.0	7.0	72.0
	(1)	Bacho Kiro 1124	8.8	7.5	66.0
	(2)	European neandertals ($n = 7$)	8.84 ± 0.41	7.48 ± 0.66	
	(3)	Qafzeh-Skhūl ($n = 5$)	8.8–9.3	7.1–8.5	
	(8)	Upper Paleolithic ($n = 12$)	$\bar{x} = 8.1 \pm 0.77$	7.1 ± 0.53	
m2		Castillo C	11.0	9.0	99.0
	(1)	Bacho Kiro 559	10?	9.5?	
	(2)	European neandertals ($n = 9$)	10.43 ± 0.52	9.24 ± 0.42	
	(3)	Qafzeh-Skhūl ($n = 6$)	9.9–11.4	8.8–10.7	
	(4)	Les Rois 1	10.0	9.5	95.0
	(5)	Les Rois 33	11.5	10.0	115.0
	(5)	Fontéchevade 2	10.3	9.1	93.7
	(8)	Upper Paleotlithic ($n = 20$)	$\bar{x} = 10.3 \pm 0.75$	9.1 ± 0.56	

Key: (1) Glen and Kaczanowski 1982; (2) Tillier 1982, 1983; (3) Tillier 1979 and unpublished data; (4) Gambier 1989; (5) Gambier 1993; (6) Vandermeersch 1981; (7) Garralda and Vandermeersch 1993; (8) Frayer 1978
Source: Garralda et al. 1992

and the incomplete mandible of a 3- to 5-year-old child, the symphysis of which was robust and without an accentuated chin (Figure 10.3). The mandible preserved the two right deciduous molars, which were large (the mesio-distal diameter surpasses the corresponding mean for neandertal children—Table 10.3). The human remains from El Castillo confirm the general robusticity of early Upper Paleolithic humans and also show some archaic features, but it is impossible to classify them as "evolved" neandertals or "archaic" modern humans.

Morín and Camargo (Cantabria)

Two other discoveries associated with Aurignacian artifacts have been reported in Cantabria. One is the famous pseudomorph (positive cast of a human body) (Morín Burial No. 1), from the eponymous cave (Villanueva de Villaescusa, Santander), dated to an early phase of this culture (González Echegaray and Freeman 1978b). Because of its uniqueness, and apart from the great archaeological interest of the find, the only anthropological data that can be obtained from the pseudomorph is the estimate, by J. Lawrence Angel and T. Dale Stewart, of the remarkable stature (185–195 cm) of this supposedly male individual, arrived at on the basis of the length of its right "arm" (Freeman and González Echegaray 1973; Freeman 1994). The estimate must be taken with a grain of salt, however, given the very considerable postmortem alterations suffered by the corpse during the fossilization process, and the fact that arm (humeral + radial) length is a less reliable indicator of stature than is femoral length. Given the great inter- and intrapopulational variability characteristic of stature, it would be unwise to try to use it to assign Morín Man to any of the varieties of *Homo sapiens* (Garralda 1992; Garralda and Vandermeersch 1993).

A human cranial vault was also recovered from Aurignacian deposits at Camargo Cave (Santander), during the Sierra excavations of 1908. Thought to be that of a female, it consisted of a long, narrow fragment of a vault, with a curved and prominent occipital and probably an orthometopic forehead (Figure 10.3). Saller (1926) assigned it to *Homo sapiens sapiens*. Because of the cursory published account, and the subsequent destruction of both the fossil and the site during and after the Spanish Civil War (1936–1939), it is impossible to be more specific about the nature of the Aurignacian industry with which the find was evidently associated.

CONCLUSIONS

The anthropological and archaeological data pertaining to the Middle to Upper Paleolithic transition interval in Iberia allow us to draw the following conclusions.

1.   According to various prehistorians, what we think of as the technological transition to the Upper Paleolithic apparently occurred during an episode of relatively mild climatic conditions, rather than during a period of stress. Moreover, it took place over a long period of time, perhaps spanning more than 10,000 years (>400 generations) (Cabrera and Bischoff 1989; Cabrera and Bernaldo de Quirós 1990; Cabrera et al. 1993; González Echegaray 1993; Straus 1992).

2.   The chronology of cultural development was evidently somewhat different in northern and southern Iberia. In the north (especially in Cantabria), the transition is becoming better known not only because of the remarkable ^{14}C dates from Arbreda, Romaní, and Castillo, but also because of the post-1980 excavations at El Castillo, which have yielded rich archaeological material still being analyzed. In the estimation of some workers, these data reveal a clear technological continuity between the late Mousterian and the Archaic or Lower Aurignacian (Cabrera and Bernaldo de Quirós 1990; Cabrera et al. 1993, this volume; González Echegaray 1993). This transitional phase is much less well understood in the central and southern regions. In the latter, and probably also in Portugal, recent work suggests a persistence of Mousterian cultures up to and including the beginning of Würm III, perhaps even after 30,000 B.P. (Barroso et al 1993; Vega Toscano 1990; Zilhão 1994).

3.   The recent discovery of a site with Archaic Aurignacian industries in Andalucia (Bernaldo de Quirós, personal communication 1994) might have implications that would modify previous theories linking the supposed "absence" of early Aurignacian industries in the south to the very late persistence of the Mousterian there.

4.   All Iberian human fossils associated with Mousterian industries have been classified as *H. sapiens neanderthalensis,* although some show a few characteristics that could be considered "evolved" or even "modern" (e.g., the Devil's Tower and Carihuela children, and especially the Zafarraya mandible). They can be explained by either *(a)* claiming that they are part of the normal range of variation of these Würm-age hominids (i.e., by claiming that they fall within the range of variation of the nearly unknown morphological and spatio-temporal variability of these human groups), or *(b)* invoking gene flow, especially in the case of the most recent specimens (i.e., by demonstrating an association between modern humans and the Archaic Aurignacian—something we cannot yet do) (Garralda 1993; Garralda et al. 1992; Vandermeersch and Garralda 1994). If the recent date for the Banyoles mandible is confirmed, it would also fall into the archaeological transition interval. What is interesting about Banyoles is that, if it is really 45,000 years old, its morphological characteristics would force a reconsideration of the variability present in these "recent neandertals."

5.   At present there is, unfortunately, a reasonable doubt about the attribution to *H. sapiens fossilis* of the Carihuela remains associated with the late Mousterian levels in this cave (Garralda 1993). As noted above, they could well date to the Neolithic.

6. Of the extremely scarce, dated Aurignacian fossils, the earliest (those from El Castillo, level 18) exhibit some archaic traits, but they are of uncertain taxonomic affinity (Garralda et al. 1992).

7. The confirmation of the age of the early Aurignacian in northern Spain is an important argument against the notion of an eastern origin for the Aurignacian (Clark 1992; Vandermeersch and Garralda 1994). The new data from Andalucia also call for an alternative explanation.

8. At present, we do not know with certainty whether the cultural variability evident during the transition interval (50,000–30,000 B.P.) is linked to the coexistence of two genetically distinct populations, nor whether the evolutionary biodynamics of these transitional populations can be considered to be independent of their accompanying industries. Of course, the answer to these questions ultimately depends upon the fossils, which are very scarce and fragmentary in the Iberian context. However, the important data furnished by prehistorians are offering new perspectives, questions, and possible answers to our traditional interpretations of the evolution of *Homo sapiens* in the extreme southwestern corner of the European subcontinent.

11

Scenarios for the Middle to Upper Paleolithic Transition

A European Perspective

JEAN-PHILIPPE RIGAUD

CRITERIA FOR IDENTIFYING THE CULTURAL TRANSITION

I analyze here the passage from "Middle" to "Upper Paleolithic" techno-complexes (only one of a number of cultural changes that Europe experienced between 40,000 and 30,000 B.P.) according to certain diagnostic criteria invoked over time by prehistorians on the basis of their particular specializations, their general knowledge of the field, and their more or less direct access to the data base (Bordes 1972; Mellars 1989a; White 1982, 1989b). Among these criteria, the following are most generally employed by European prehistorians.

Technological Change in the Production of Blanks

For Mousterian debitage oriented essentially toward the production of levallois or non-levallois flakes, a more systematically laminar method of production is substituted (the "leptolithization" as used by some European workers). We should ponder the diagnostic character of this technological criterion in light of the fact that in the Middle Paleolithic of the Near East, a blade-dominated Mousterian is well documented long before 40,000 B.P. It gives rise to the Upper Paleolithic Ahmarian, but it disappears before the Aurignacian peopling of the Levant (Meignen and Bar-Yosef 1988). In the north of France and in the Rhone Valley, recent work has shown that Mousterian blade production is equally abundant (Combier 1990; Révillion 1993). A second mitigating factor is that laminar or lamellar debitage is known throughout the later phases of the Upper Paleolithic, which correspond, for example, in southwestern France to the Badegoulian.

More Diversity and Greater Standardization of Tool Forms

The typological repertoire of Middle Paleolithic industries developed around a limited number of themes: sidescrapers, notches and denticulates, backed knives, bifacial tools, etc. Upper Paleolithic industries, on the other hand, offer a more varied typological spectrum in which endscrapers, burins, backed points, shouldered or stemmed points, microliths, etc., appear or reappear. Although some of these Upper Paleolithic themes show up in a tentative way in the Middle Paleolithic, they are always rare and atypical vis-à-vis their Upper Paleolithic counterparts (e.g., endscrapers and burins in certain Upper Acheulean and Mousterian of Acheulean Tradition assemblages; Bordes 1972). In any case, we should always try to determine whether these artifacts are in fact good functional types and not simply the results of our methods of classification.

The greater standardization of tool forms attributed by some (e.g., Mellars 1991) to a growing conceptual normalization appears to me to be indisputable (cf. Clark and Willermet 1995). Morphological variability among Mousterian backed knives, for example, is considerably greater than that observable in their supposed Upper Paleolithic functional equivalents. This apparent trend toward formal standardization is very probably linked to the greater incidence of laminar debitage in the Upper Paleolithic, which tends to produce much more standardized blanks. It is logical, therefore, to suppose that the finished tools made on these blanks would themselves appear more standardized. It is not, however, established that Mousterian backed knives were, in fact, functional equivalents of Châtelperronian backed knives or points, nor that the choice of blanks had the same importance in determining tool form.

Development of an Organic Raw Material Technology

The use of bone, antler, ivory, etc., to make tools seems to be an Upper Paleolithic innovation. Occasional utilization of these organic raw materials in Mousterian contexts has, of course, been noted—expedient or casual bone tools, even symbolic objects (Marshack 1990). Nevertheless, these practices are exceptional and remain very rare. The utilization of organic materials only becomes systematic and common in the Châtelperronian and, above all, in the Aurignacian (White 1982).

Accelerated Rate of Technological Change

Mousterian technocomplexes exhibit relative typological stability over an enormous period of time (ca. 250,000–60,000 B.P.). However, around 40,000 B.P., an acceleration of "creativity" is evident, manifest in the development of new tool forms and new technologies. This phenomenon continues and in fact accelerates still more throughout the entire Upper Paleolithic.

Regional Diversification

On the typological base of the late Middle to early Upper Paleolithic techno-complexes, two attempts at cultural zonation have been proposed: a supposed Châtelperronian "retreat" in the face of Aurignacian "advance" (Leroyer 1987; Leroyer and Leroi-Gourhan 1983) and the existence in Italy of an Uluzzian "province" distinct from that of the Châtelperronian (Gioia 1988). We should not overlook, however, the marked techno-typological variability evident in the Mousterian of the early Würm, which gives rise to distinctive regional facies within the Mousterian itself (e.g., the Pontinian, the Azinipodian).

Personal Adornment and Symbolic Expression

The appearance in the Châtelperronian of traces of personal adornment and their rapid development in the Aurignacian is uncontested by most workers. Discerned in the European Aurignacian since the beginning of the century (e.g., Breuil 1912), and more recently in the Châtelperronian of Arcy-sur-Cure (Taborin 1988), the taste for personal adornment corresponds, in the opinion of some, to a willingness on the part of individuals to differentiate themselves from similar others (White 1982).[1] It is also during this period that manifestations of symbolic expression in the form of mobile and archaic parietal art first appear.[2] That being said, however, we should not rule out the possibility of neandertal symbolic or artistic expression, evidence for which has not been preserved. Funeral practices and other manifestations of neandertal spirituality seem uncontestable in Mousterian sites firmly dated to the early Würm. However, it must also be admitted that such practices are only clearly attested in the initial stages of the Aurignacian.

Economic and Social Changes

Several economic and social changes have been associated with the transition from a Mousterian "stage" to an Upper Paleolithic "stage," among them a systematization (or regularization) of hunting practices. While a certain prudence is urged in this domain in light of results from detailed individual studies, hunting practices developed from those of the Mousterian, when opportunistic strategies played an important role, to the specialized exploitation of game characteristic of the latest phases of the Upper Paleolithic. No significant changes are evident, however, in the Châtelperronian, nor in the initial stages of the Aurignacian.

Population density, somewhat hastily equated with site densities, is another indicator of social and economic change. Compared with the number of sites occupied at the end of the Mousterian, those assigned to the Châtelperronian or to the earliest Aurignacian seem less common. One cannot, then, use growing regional population densities to conclude that there was a discernible demographic increase coincident with, or immediately following, the transition. That only becomes evident, at least in the southwest of France, with the Middle Aurignacian. In the same way, the spatial structuring of early Upper Paleolithic sites, which, one might imagine, would reflect evolved or improved cognitive capabilities on the part of their creators, is not yet well enough established that we can accept it as a socioeconomic marker of the transition.

Language

To finish with this brief review of diagnostic criteria of the Middle to Upper Paleolithic transition, the proposal of Davidson and Noble (1989) can only remain speculative in the absence of any conceivable kind of archaeological or anthropological confirmation. According to these authors, neandertals had only a relatively simple form of language, less structured and less adequate than that of Upper Paleolithic people, which, in consequence, limited them in the domains of cognitive acquisition and in technical and spiritual creativity (Davidson and Noble 1989). Data from St. Césaire and Arcy-sur-Cure do not seem to support this proposition because it might well have been neandertals that made the early Upper Paleolithic industries at these sites, as was the case much earlier in the Near East.

MIDDLE TO UPPER PALEOLITHIC TRANSITION SCENARIOS

Considering these diagnostic criteria and the reservations that go along with them, how was the Middle to Upper Paleolithic transition accomplished in Europe? Several partly conflicting, partly complementary scenarios have been proposed.

The Atlantic European Model

The central tenet of the Atlantic European model is that the Châtelperronian, by virtue of its laminar debitage and tools such as endscrapers, burins, and backed pieces (Châtelperronian points or knives) made on laminar blanks, constitutes an Upper Paleolithic technocomplex in which some Mousterian elements persist. For Bordes (1972), the Châtelperronian represents a logical development from a late Mousterian of Acheulean Tradition base. Leroi-Gourhan (1963), however, saw it as derived from (or more closely related to) the Denticulate Mousterian, and Laplace (1966) preferred to conceptualize it as an evolution proceeding from the Mousterian in general and passing through an undifferentiated phase (on the way to the Aurignacian/Lower Perigordian) that comprised the Châtelperronian as conventionally defined. Since it is interstratified with early Aurignacian levels, the Châtelperronian is pretty clearly contemporaneous with them. After ca. 33,000–32,000 B.P., however, only the Aurignacian is found along the Atlantic facade of Europe (Bordes and Labrot 1967).

The Italian Model

Considered by some to be an Italian equivalent of the Châtelperronian, the Uluzzian exhibits an ensemble of techno-typological characteristics that link it, like the Châtelperronian, to the Upper Paleolithic. In the opinion of Palma di Cesnola (1965–1966), the Uluzzian might have originated in a local, denticulate-rich Mousterian. The Aurignacian follows it without a discernible transition, with the implication of an abrupt break possibly connected with population replacement.

The Central European Model

In eastern Europe, in the Carpathian Mountains, the local Middle Paleolithic is represented by a particular facies called the Szeletian, after the cave of Széléta, in the Bükk Mountains of Hungary. The Szeletian is also found in Moravia, in Upper Silesia in the valley of the Dniestr, and in Moldavia. The Szeletian is characterized by the presence of bifacial foliate (leaf-shaped) points and, in its final phase, exhibits some technological characteristics (e.g., laminar debitage) that will become marked features of the western European Upper Paleolithic. In Germany, Belgium, Poland, and England, other foliate point industries distinct from the Szeletian constitute the Lincombian-Ranisian-Jerzmanovician complex. In central Europe, the initial Upper Paleolithic is represented by the Bohuncian, of probable local origin. This technocomplex combines elements of Middle Paleolithic technology (e.g., levallois reduction) with Upper Paleolithic technologies and tool forms (e.g., laminar debitage, endscrapers, burins, etc.) (Kozłowski and Kozłowski 1979).

The Balkan Model

After a Mousterian sequence in the Balkans that apparently ends abruptly and without issue, an "Aurignacoid" industry—the Bachokirian—develops after ca. 43,000 B.P. It does not seem to be linked or to be derived from the local Mousterian.

The Iberian Model

Following a series of Mousterian industries in the Iberian peninsula of which some, in the south, appear to persist very late (possibly to <30,000 B.P.), the Châtelperronian shows up in the north, in Cantabria, at Cueva Morín (González Echegaray and Freeman 1971, 1973). This pattern is not repeated, however, at l'Arbreda Cave (Catalonia) or at El Castillo (Cantabria), where recent data seem to indicate an Aurignacian occupation considerably older than that of Aquitaine. It is evident that the age of the early Aurignacian at l'Arbreda and Castillo (ca. 40,000 B.P.) poses a problem in respect of models for the Aurignacian peopling of Europe, which generally assume a chronological progression from "early" in the east to "late" in the west (Clark 1992a). Whatever the case, the argument has been made that the l'Arbreda and El Castillo Aurignacian dates do not support cultural continuity with the local Mousterian (Bischoff et al. 1989; Cabrera and Bischoff 1989).

The Middle Eastern Model

The Levantine Mousterian was evidently manufactured by both neandertals and "Proto-Cro-Magnons" (Vandermeersch 1981a). It is characterized by a lot of technological variability, but nothing in it seems to indicate continuity with the "Pre-Aurignacian" of Jabrud rockshelter (Syria). There seems to be a significant gap between the latest laminar Mousterian occupations and the earliest evidence of the Aurignacian, in which heavily laminar industries like the Ahmarian would be placed (Bar-Yosef et al. 1986; Meignen and Bar-Yosef 1988; Vandermeersch 1988).

In sum, the scenarios reviewed here appear to show that, in general, Aurignacian industries are intrusive in the various regional sequences, whether or not the latter are directly derived from the local Mousterian.

CONCLUSIONS

While the origin of the Aurignacian seems to lie outside Europe and the Near East, everything seems to indicate that its development throughout Europe took place while diverse local cultures, already heavily laminar in terms of debitage, were passing into an Upper Paleolithic stage. To underscore the contrast between the Mousterian and the Aurignacian technocomplexes would only contribute to a reductionist view of the problem, masking the profound intervening changes that took place, over a relatively short time interval, at the heart of the late Mousterian industries themselves. Despite the lack of fossils associated with the earliest Aurignacian, the more evolved phases of this culture were clearly the work of *H.*

sapiens sapiens. We have good reasons to think that neandertals were the makers not only of the Châtelperronian but also probably the Uluzzian and the laminar complexes of central Europe. One could inquire, then, into the reasons that might have prevented the neandertals—makers of the earliest laminar assemblages— from achieving the level of technological development attained by the newly arrived Aurignacians. Was it because they were limited in terms of their psycho-functional capabilities? Could they have experienced genetic or demographic "swamping," and how would it be possible to test for this? And, finally, why couldn't they have been the makers of the earliest Aurignacian industries? These are questions that we must put to our bioanthropological colleagues.

ACKNOWLEDGMENTS

I wish to thank the organizers of the International Colloquium *"El Origen del Hombre Moderno en el Suroeste de Europa"* (Madrid 1991) for inviting me to participate. This essay is an updated version of a paper presented at that conference and published by the Universidad Nacional de Educación a Distancia (UNED) in 1993 (Rigaud 1993).

NOTES

1.   It is not clear, however, that all these "elements of adornment" are in fact "person-al" in the strict sense of the term. Why couldn't they have pertained to a collective (group) symbolic repertoire (e.g., emblems) associated with burials or even dispersed accidentally in the domestic context?

2.   The recent dates from Grotte Chauvet seem to indicate a development of symbolic expression and mastery of artistic technique considerably earlier than had generally been supposed.

12

The Concept of "Upper Paleolithic Hunters" in Southwestern Europe

A *Historical Perspective*

J. GONZÁLEZ ECHEGARAY

Over the years, Upper Paleolithic origins have been the subject of various collected works and international meetings, among which the Paris UNESCO Symposium, titled "The Origin of *Homo sapiens*" (Bordes, ed. 1972), is of particular historical significance. The participants in the meetings which resulted in this essay have available to them the latest and most valuable research conclusions, and the special advantage of a southwestern European focus—a region crucial to our understanding of the evolution of prehistoric humans.

On the other hand, since recent work in France and Spain has provided some of the most important new data on Upper Paleolithic origins, some of what I say here might sound excessively repetitive, especially to the nonspecialist reader. I am referring to the widely publicized discoveries at Saint-Césaire or the El Castillo Aurignacian dates. We are all aware of these, of course, and the distinct research foci of different investigators, and their various personal perspectives, cannot fail to enrich our knowledge of the central question of Upper Paleolithic origins.

Having said this by way of a justification for investigating these issues, it seems appropriate now to present an historical overview of this kind of "origins" research, in order to put the hotly debated current issues and problems particular to southwestern Europe into perspective.

THE EUROPEAN CONCEPTION OF THE UPPER PALEOLITHIC: HISTORICIST BIASES

A very generalized conception (and, in my view, an excessively biological one) with respect to the historical development of humankind has sought to link cultural changes in prehistory with the presence of new, and successively more evolved, human morphotypes. In reality, this is nothing more than a more "presentable" version of old racial and invasion theories, often invoked in historical contexts to explain change (see Clark 1994c for a critique of these views). The fact that, during the Middle Paleolithic, the presence of *Homo sapiens neanderthalensis* is well documented, and that remains attributed to *Homo sapiens sapiens* appear in Upper Paleolithic sites, has lent support to the idea that cultural change between these stages coincides with the substitution of one kind of human by another. Underlying it all, we have a concrete example of the old theory at work: the beginning of the Upper Paleolithic is owed to an "invasion" of a new "race" of Cro-Magnons, occupying the territory in which the neandertals formerly held sway.

This line of thinking—essentially historical interpretation—in addition to suffering from confusion owing to a poorly defined and disparate terminology (e.g., culture, biological morphotype, cultural diffusion, human migrations) also lacks any coherence when it is applied to other large-scale changes in prehistory. Following this logic, we would expect that the beginnings of other grand cultural stages would also be accompanied by the presence of new and progressively more "perfect" morphotypes, for which we have no evidence whatsoever. The beginnings of the Mesolithic, the so-called Neolithic Revolution, etc., obviously do not correspond to the appearance of new human morphotypes, so there is the problem of limited generalizability to this explanatory framework. Moreover, in respect of the Lower-Middle Paleolithic transition, we have the paradoxical situation that some human remains from that remote era present a more gracile morphology than Middle Paleolithic neandertals. There is, then, no reason whatsoever in support of the idea of a rigid parallel between cultural stages and the evolution of genus *Homo,* even acknowledging that both follow progressive trajectories.

THE EUROPEAN CONCEPTION OF
THE UPPER PALEOLITHIC:
A HISTORICAL OVERVIEW

The 1920s–1930s

An explicit concern with the nature of the Middle to Upper Paleolithic transition does not show up in the literature until the 1920s, in that it was only some years earlier that the beginnings of the Upper Paleolithic had been defined by a broad consensus. A classic work like *La Préhistoire* of Gabriel and Adrien de Mortillet, in the 1910 edition, still identified the beginning of the Upper Paleolithic with the Solutrean, despite the fact that the Abbé Breuil had been defending the stratigraphic position of the Aurignacian, as interposed between the Mousterian and the Solutrean, since 1906. On the other hand, until the 1930s (and because of the influence of Breuil and Peyrony), the term "Middle Paleolithic" was not in general use, nor had it been established that the Châtelperronian or the Lower Perigordian was something different from the true Aurignacian, in that, until that era, the latter terms were considered "Lower" and "Middle Aurignacian," respectively.

The situation by the mid 1920s is well reflected in the pages of Hugo Obermaier's *El Hombre Fósil* (1925 edition). The beginning of the Upper Paleolithic, he says, "was, without doubt, due to the invasion of a wave of Aurignacian peoples, which completely eliminated the neandertal race in Europe, resulting in a revolution of far-reaching consequences for human culture" (Obermaier 1925:124). Obermaier considered the possibility of separating, on the one hand, the so-called Lower and Upper Aurignacian from, on the other, the true Middle Aurignacian. The latter spread over central and western Europe from an uncertain point of origin. The former, on the contrary, appeared to come from north Africa, where it was represented by the Lower Capsian, that "here (in north Africa) immediately follows the Mousterian, from which it is derived" (Obermaier 1925:126). This culture, later to be called Perigordian, supposedly represented other regional varieties found in Syria-Palestine, in Italy, and even in the south of Russia. In western Europe, and more concretely in Iberia and southern France, then, the Lower as well as the Upper Aurignacian would have had an African origin, while the Middle Aurignacian would have arrived in northern Spain via France.

The 1940s

With the arrival of the 1940s, the idea of what I would call the "Pancapsian" fell out of general use. A broad consensus emerged among Spanish prehistorians (Pericot, Martínez Santa-Olalla, Almagro) that the French and Iberian Upper Paleolithic were basically similar in terms of the gross chronologies, cultures, and stages represented. The notion of a local Capsian was rejected, although it was acknowledged that there were differences between the industries in the north of Spain, more closely linked to French sequences, and those in the central and south parts of the peninsula, which showed regional peculiarities and sequence differences.

At the same time, revisions during the 1940s of the north African site sequences reduced the geographical extent of the Capsian to the Tunisian interior *mesetas* (plateaus) and the Constantine, and led to the conclusion that it never reached the Mediterranean coast or even the Algerian plateaus (Alimen 1955). On the contrary, the ill-named "Ibero-Mauritanian," a supposedly impoverished Capsian with lots of microliths, was extended along the coasts of north Africa as far west as Atlantic Morocco and was seen, at times, to coexist both temporally and spatially with industries of Mousterian or Aterian tradition (Alimen 1955). Both cultures, the Capsian and the Ibero-Mauritanian, are rather late in the Upper

Paleolithic, and it is possible that they had no direct contact with the European shores of the Mediterranean.

The 1950s–1960s

This being the state of affairs at the close of World War II, François Bordes, collecting and systematizing earlier opinions, established in the 1950s his theory on the French origin of the Lower Perigordian, which he derived from the Mousterian of Acheulean Tradition Type B. He also reserved an "oriental" origin for the true Aurignacian, although he did not rule out its partial, in situ development from the Middle Paleolithic in eastern Europe, represented by industries like the Szeletian of Czechoslovakia and Hungary, derived from the *blattspitzen* Mousterian, or the Kostenki I industry, derived from the Mousterian of Starocelie-Volgograd type (Bordes 1958, 1968a, 1968b, 1972). On his part, Georges Laplace (1970) argued for a local and parallel derivation of both the Châtelperronian and the Aurignacian from the Mousterian of the Périgord, while Henri Delporte (1970) presented a version that more closely approximated that of Bordes.

ASIAN ORIGINS

The thesis of an Asian origin for the Aurignacian was particularly debated during the 1950s–1960s because of some discoveries, thought to be relevant to this issue, in the Near East. I am referring here to the so-called Amudian or Pre-Aurignacian culture, identified for the first time by Alfred Rust in Syria at Jabrud Rockshelter II (Rust 1951) and later studied by Bordes (1955, 1960). It was also identified later by Dorothy Garrod and Arthur Jelinek at Mugharet et-Tabūn, on Mount Carmel (Garrod 1956; Jelinek et al. 1973); at Mugharet el-Zuttiyeh, in the Galilee (Garrod 1956); and in the Lebanese rockshelter site of Zumoffen (Garrod and Kirkbride 1961). Its stratigraphic position indicated a very great antiquity, the oldest known culture of Upper Paleolithic affinity in the world. It occurred right in the middle of the full-blown Middle Paleolithic sequence at these sites, interstratified with the Yabrudian and followed by occupations of levallois Mousterian character. Its chronology suggested that it correlated with the Würm I-II interstadial in Europe (Farrand 1972; Zeuner 1961). Its absolute age, despite the fact that archaeological applications of radiocarbon analysis scarcely existed at the time, was estimated to be around 50,000 B.P. (Garrod and Henri-Martin 1961). The industry is very archaic, but it already shows clear resemblances to what would later be the full Aurignacian, with large numbers of blades, keeled and nosed endscrapers, angle burins, etc. (González Echegaray 1978).

If these data were taken at face value, one would be led to think that the first wave of Upper Paleolithic people, coming from an unknown place in Asia, appeared in the eastern Mediterranean (Lebanon, Syria, Palestine) more than 15,000 years earlier than in the rest of western Eurasia, being subsequently absorbed by the local levallois Mousterians, and in advance of a new wave of already generalized moderns which ultimately invaded the whole of the western part of the Old World. The enormous variability, the fluctuations in morphological character states, and even, at times, possible hybridization between neandertals and moderns in Middle and early Upper Paleolithic Levantine populations (as suggested by the fossils from northern Israel, notably those from Mount Carmel and the Galilee) could account for the situation detected in the study of the lithic industries referred to above (Zeuner 1958). But, once again, I want to underscore the necessity for extreme caution in dealing with possible correlations between human (biological) types and culture, since, without going outside the boundaries of the Near East, we have, on the one hand, the human fossils from Qafzeh and Skhūl, practically *H. sapiens sapiens* in morphological terms despite clear associations with Mousterian industries, while on the other, the neandertals from level 25 at K'sar Akil (Lebanon) and those from Amud Cave (Israel) are associated with Upper Paleolithic assemblages (Copeland 1972).

WESTERN EUROPE

With respect to these same issues in western Europe, it is often claimed or asserted or suggested that the earliest Perigordians or Châtelperronians would have been *H. sapiens sapiens,* despite the fact that we cannot evaluate the human skeleton from Combe Capelle (lost during World War II), which was associated with these early Upper Paleolithic industries (Delporte 1970; Lynch 1966), and despite the fact that the very scarce human remains from Arcy-sur-Cure, also recovered from early Upper Paleolithic levels, are clearly very archaic-looking (Leroi-Gourhan 1959). It has even been suggested that the makers of the Mousterian of Acheulean Tradition might have already been *H. sapiens sapiens.* As a synthetic treatment of the "state of the question" in the 1960s, the excellent essay by Louis Pradel, published with extensive commentary in *Current Anthropology,* stands out not only for the scholarship of the author but also for the high quality of the associated commentary (Pradel 1966).

SOUTHWESTERN EUROPE, WITH PARTICULAR
REFERENCE TO SPAIN

Jordá's Thesis

I turn now to southwestern Europe, and more specifically, to the Iberian Peninsula, to see how research during the 1950s and 1960s frames some of the issues and problems of interest to us here. Among the Spanish prehistorians of that era, the most influential was perhaps Francisco Jordá Cerdá, author of *El Solutrense y sus Problemas,* the scope of which is far greater than indicated by its title, and which was perhaps the key publication on Paleolithic research up until that time (Jordá 1955). Jordá's thesis in regard to Upper Paleolithic origins, upon which he expanded in other publications (Jordá 1952, 1963, 1969), departs from a single, salient fact: the lack of confirmation of the existence of the Châtelperronian in Spain at the time he was writing. Its place is occupied in some sites, like Cueva Morín (Vega del Sella 1921), Cueva del Conde, El Otero, and perhaps Cova Negra de Jativa, by an "Aurignaco-Mousterian," which presents elements of both cultures and which suggests, given the appropriate cautions, an in situ transition from the Mousterian to the Aurignacian. For Jordá, only the Gravettian would clearly be of non-Iberian origin, entering the peninsula from France by routes extending around both ends of the Pyrenees, and forming the basis for the Upper Paleolithic cultures of Cantabria, Atlantic coastal Iberia, the inland *mesetas,* and the Spanish Levant.

The Fate of the Aurignaco-Mousterian

With the spate of new excavations in Spain that began at the end of the 1960s, Jordá's recently articulated theories were thrown into a state of crisis. The 1968–1969 excavations at Cueva Morín revealed the existence of an unquestionable Châtelperronian level (10) interstratified between Mousterian and Aurignacian deposits, and the new work showed that what had been considered "transition levels" were nothing more than mixed deposits of various Mousterian and Aurignacian strata (González Echegaray 1969; González Echegaray and Freeman 1971, 1978a). Also, the revision and publication in 1980 (González Echegaray et al. 1980) of the material from the cave of El Pendo, excavated by Martínez Santa-Olalla in the 1950s, seemed to indicate the existence of a second Châtelperronian occupation, as Cheynier had originally suspected (Cheynier 1967). A good summary of all these questions and problems in the north Spanish context, current through the late 1970s, is the work of Federico Bernaldo de Quirós (1982).

Saint-Césaire—The Châtelperronian, an Accommodation Industry

Another event of singular importance to the question of Upper Paleolithic origins was the discovery at Saint-Césaire (Charente-Maritime) in 1979 of a neandertal skeleton in an indubitable Châtelperronian archaeological context (Lévêque and Vandermeersch 1980, 1981). Together with what was already known about the scarce remains from Grotte du Renne and Arcy-sur-Cure (see Gambier, this volume), this find forced a rethinking of the relationship between the human fossil remains and the archaeological data, despite the lack of good samples of either. The most credible hypothesis to emerge from this reassessment was to suggest that the Châtelperronian was not precisely the link between the Mousterian world, identified with *H. sapiens neanderthalensis,* and the world of the Upper Paleolithic, identified with *H. sapiens sapiens,* but rather the material traces of a reaction by neandertals when confronted with the arrival of Aurignacian peoples, presumably *H. sapiens sapiens,* who were beginning to appear in western Europe at the time, and whose presence is documented by interstratified archaic Aurignacian and Châtelperronian levels at sites in southern France (Roc de Combe, Piage) and northern Spain (perhaps El Pendo). The fact that the Châtelperronian seems to disappear thereafter (neither its alleged continuity with the Gravettian, nor the larger question of Perigordian unity, has ever been resolved after more than fifty years of research), lends additional support to the idea that the Châtelperronian phenomenon was simply a transitory response by neandertals to the influx of Aurignacian invaders (Harrold 1981, 1983). However, caution is advised in order to avoid a return to the historicist biases already criticized here.

Implications of Radiocarbon Dates

On the other hand, the latest radiocarbon determinations on the beginnings of the Upper Paleolithic have opened up a number of new avenues for research and discussion. Until the late 1980s, it had been supposed that the earliest Upper Paleolithic industries from central and eastern Europe would have been slightly older than their western counterparts, confirming the notion of an east-west migration of Aurignacian peoples. In this respect, Jermanovician industries give dates of around 40,000 B.P. in Poland (Nieto Perzowa), equivalent to the early Aurignacian of Hungary (Istallosko) (Gábori-Szánk 1970). In the Ukraine, the oldest early Upper Paleolithic date from the open site of Molodova is a rather-uncertain 35,000 B.P. But in Iran and Iraq, the Baradostian appears to be at least 40,000 years old, as does the Dabban of the northeast coast of Africa (Cyrenaica) (McBurney 1972). However, in western Europe, the oldest reliable dates for the Aurignacian (Le Flageolet, Abri Pataud) are not older than ca. 34,000 B.P., and those from the Châtelperronian (Arcy, Les Cottés) are actually somewhat younger (ca. 33,500 B.P.) (Harrold 1983). Châtelperronian level 10 at Cueva Morín (Cantabria) produced a single, rather dubious determination of 35,000 B.P. (González Echegaray and Freeman 1973).

With regard to radiocarbon dates and their implications, things are changing rapidly in recent years. Gábori, for example, after studying central Asian sites with good Middle Paleolithic assemblages, has come to the conclusion that the Upper Paleolithic there is very late (Gábori 1988), thus calling into question a central Asian origin for the Aurignacian. Reliable ^{14}C dates from sites in Russia and Siberia are also generally not very old (Kostenki XVII—23,900 B.P.; Pavlov—26,400 B.P.; Mezin—29,700 B.P.; Eliseevichi—ca. 33,000 B.P.) (Shimkin 1978; Soffer 1985). This finding suggests to Klein that "European Russia has not produced any sites on the basis of which it is possible to argue that there is *in situ* evidence for a transition from the Mousterian to the Upper Paleolithic" (1973:114).

Recent discoveries in the Balkans at the caves of Bacho Kiro and Temnata describe a "transition" industry between the Mousterian and the Upper Paleolithic, dated to ca. 40,000 B.P., but apparently without any direct connection to the underlying local Middle Paleolithic. At Temnata, there is a genuine Aurignacian, albeit later in the sequence. Also, the cave of Couvin, in Belgium, supposedly contains a Mousterian with Upper Paleolithic tendencies, with an age of >40,000 B.P. (Kozłowski 1988b).

Perhaps the most spectacular dates, and those with the most important implications for modern human origins in western Europe, are those recently reported by Victoria Cabrera from the classic cave site of El Castillo (Cantabria). In level 18b, unquestionably attributed to the early Aurignacian, three samples were dated to 40,000, 38,500, and 37,700 B.P., respectively, for a mean age of 38,700 ±1,900 B.P. (Cabrera and Bischoff 1989; Cabrera, this volume). On the other hand, the cave of L'Arbreda (Gerona) has yielded an early Aurignacian, dated at ca. 38,500 B.P. (Bischoff et al. 1989). If Aurignacian level 18b at Castillo dates to the Würm II-III interstadial, it would be earlier than the Châtelperronian of nearby Cueva Morín and contemporary with the denticulate Mousterian levels from this site.

CONCLUSIONS

As Straus (1990) has noted, these new dates call into question both the existence of an east-west migration of Aurignacians in Europe and the idea of a rapid transition between the Middle and Upper Paleolithic. To which I would add that the idea of prolonged contemporaneity (ca. 10,000 years) between the Châtelperronian and Aurignacian cultures would seem to be reinforced; it might even have extended back into the Mousterian time frame. What remains exceptionally murky is the relationship between human racial and cultural groups. Caution is advised here, given that the equation between neandertals and the Middle Paleolithic, on the one hand, and moderns and the Upper Paleolithic, on the other, can now be seen as only a very general, gross approximation. When we look at the hard data, it is very difficult (likely impossible) to substantiate in fine detail an assertion of this kind.

In conclusion, the preceding are what I take to be the major themes and questions that the workers most directly involved in this research appear to have identified over the past few years. I would like to close by expressing the hope that continuing exchanges of scientific observations and interpretive hypotheses will throw more light on this important interval in the history of humankind. The beginnings of the Upper Paleolithic can only be understood by extending the time frame of investigation back into the Middle Paleolithic and by extending the geographic frame of inquiry beyond the relatively small regions with which most of us are concerned.

ACKNOWLEDGMENTS

I am grateful to the organizers of the International Colloquium *"El Origen del Hombre Moderno en el Suroeste de Europa"* (Madrid 1991) for inviting me to present this essay as the Inaugural Address. This chapter is an updated version of a paper presented at that conference and originally published by the Universidad Nacional de Educación a Distancia (UNED) in 1993 (González Echegaray 1993).

13

The Transition from the Middle to the Upper Paleolithic in the Cave of El Castillo (Cantabria, Spain)

VICTORIA CABRERA VALDÉS, MANUEL HOYOS GÓMEZ, and
FEDERICO BERNALDO DE QUIRÓS

The El Castillo sequence is one of the best known on the Iberian Peninsula, and it is unique because it preserves a record of the entire Paleolithic, subdivided into 26 levels comprising sterile deposits alternating with those documenting human occupation. Hugo Obermaier's 1910–1914 excavations, done under the aegis of the Institute of Human Paleontology (Paris), determined that there were 18–20 m of stratified deposits in the cave—one of the longest sequences in Europe (Obermaier 1925). In brief, and taking into account a recent reanalysis of Obermaier's work, there are two early Middle Paleolithic levels, an Upper Acheulean deposit, two Mousterian levels, two Aurignacian levels, two Upper Perigordian levels, one Middle Solutrean deposit, three pertaining to various stages of the Magdalenian, and a single Azilian occupation (Cabrera 1978, 1984).

Three aspects are noteworthy regarding our restudy of the original excavations:

1. As might be expected, the 1910–1914 excavations were deficient in terms of modern methods and standards of data recovery and recording. However, and despite this, Obermaier and his colleagues practiced horizontal exposures (*décapage horizontal*) of at least most of the relatively distinct natural strata. This is clearly indicated by the extensive photographic archives that he created, and the drawings and sketches done during the course of the fieldwork. Each anthropogenic level was treated as a distinct cultural unit (Cabrera 1984).

2. The stratigraphy at El Castillo was very clear, and the black, organic cultural deposits were easily distinguished from the reddish sterile levels with which they were interstratified. As a consequence of the recent work, however, it is now possible to identify distinct occupations within each of Obermaier's archaeological levels.

3. Firsthand accounts from the period indicate that the original excavations emptied almost two-thirds of the site, and subsequent refilling together with the

neglect suffered by the site after 1914 led to the impression that it had been completely dug out. However, reanalysis of the original field notes and the many drawings and sketches made at the time allowed us to identify two substantial witness sections which, without doubt, preserve many of the old levels (certainly levels 13–25, and probably others as well; Cabrera 1984).

COMPARISON OF THE ORIGINAL AND MODERN EXCAVATIONS

The modern excavations used the same level enumeration system as Obermaier, although—as noted—it was possible to distinguish various archaeological and geological sublevels within his strata. The recent fieldwork allowed for a comparison of the results of the reanalysis of the artifacts and documentary evidence from the 1910–1914 excavations. In particular, it became evident that

1. Traces of Upper Paleolithic deposits remained along the south wall of the cave, albeit covered by their erosional debris (slumping of sediments from the 1910–1914 cut face).
2. The existing parking lot covered part of the site up to the edge of Obermaier's transverse section, and up to the elevation of level 13. We cut back the parking lot about 6 m to allow space for a horizontal exposure, recovering thereby a large part of the intact site preserved underneath the overhang of the cave.
3. In the exterior part of the cave, under the parking lot, and after roof fall corresponding to sterile levels 13–17 had been removed, a part of the old sequence corresponding to levels 18–25 was exposed, which also extended back into the cave along what remained of the longitudinal section. Level 18 was identified both by its position between sterile layers 17 and 19 and also by the presence of rocks in the section that are evident in the photoarchives of Obermaier.
4. Geophysical methods (SEV) determined that at least 7 m of intact sediments still existed below level 23.
5. Of Butzer's (1981) stratigraphy, it was only possible to identify with certainty his levels 5a–5e (distinctive because they are stalagmitic deposits), which are equated with level 23. An extended critique of Butzer's stratigraphy lies outside the scope of this chapter.
6. As a consequence of cleaning the existing section and tearing up the parking lot, an 8 m long section, perpendicular to the axis of the cave, was obtained, and an axial section 3–5 m long, depending on depth. Both could be excavated to a depth of ca. 4.5 m.

The extent of the preserved deposits, the constraints imposed by what remained of the parking lot (the site is a popular tourist attraction), and the contours of the cave itself essentially determined the course of the recent excavation. After 1980, the horizontal excavation was centered on a surface some 40 m² in area, of which 24 m² were usable in the case of Level 18. Our intent was to analyze the activities and occupations within level 18, the probable locus of the Middle to Upper Paleolithic transition. Level 18 is Aurignacian in conventional terms, and it is clearly separated stratigraphically from underlying Mousterian level 20 by a sterile deposit (level 19).

STRATIGRAPHY, SEDIMENTOLOGY, AND PALEOCLIMATIC INTERPRETATIONS

The archaeological site of El Castillo is located at one of the entrances of an extensive karstic system of the same name on a conical hill in the town of Puente Viesgo, near the city of Santander, in north-central Cantabria. The cave consists of a large chamber truncated on one side by a cliff, with a roughly rhomboidal entrance, some 18–20 m wide and about 20 m high. Prior to the 1910–1914 excavations, the cave was almost completely filled with sediments. Two passageways lead from the archaeological site in the cave mouth back into the Monte Castillo karstic system, the lower of which is blocked by deposits of level 20 and the upper of which remained partly open after infilling of the cave mouth.

Sediments were introduced via the lower passageway from the cave interior by karstic rejuvenation until it was blocked during the formation of level 20, so that, from a sedimentological point of view, the environment of the archaeological site combines aspects of a true cave and those of a rockshelter, with accumulation of both internally and externally derived deposits. After blockage of the lower passageway during the formation of level 20, the site essentially became a big rockshelter, and it began filling with external debris derived mainly from the south flank of the Castillo hill. These sediments accumulated in the form of a talus cone with its top nearly at the level of the overhang, and extending in a fan-shaped deposit into the interior. The development of the talus cone, together with the retreat of the overhang due to erosion, successively reduced the area available for human occupation over time.

The levels involved in the transition are 20 (Mousterian), 19 (sterile), and 18 (Aurignacian) (Figure 13.1). A more detailed description of the formation processes of these levels is given in Cabrera et al. (1993), together with a sedimentological and paleoclimatic analysis of levels 17–15, important for understanding the overall paleoclimatic sequence. We emphasize here the cultural aspects of level 18, which appears to document the Middle to Upper Paleolithic transition interval.

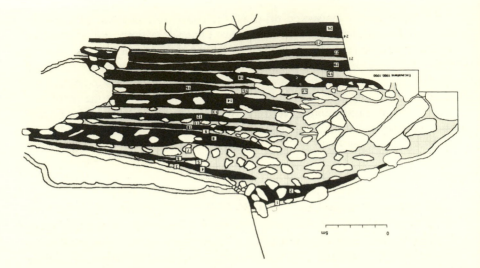

Figure 13.1.   Stratigraphy of El Castillo cave, indicating the areas excavated during the
1980–1990 seasons.

AURIGNACIAN LEVEL 18—CULTURAL INTERPRETATION

Obermaier's 1910–1914 Excavations

In the archaeological collection from the 1910–1914 excavation of level 18,
the following factors are important:

1. There is a high incidence of endscrapers (IG *[indice de grattoir]* =
    34.84), among which carinate (16.75%) and nosed or shouldered forms
    (4.3%) stand out, and a scarcity of burins (IB *[indice de burin]* = 10.32),
    dominated by dihedral forms (7.03%).

2. Among the carinate endscrapers, a type which is characteristic of the site,
    made on half of a small, rounded pebble, is noteworthy. These pieces
    show the same retouching technique as many Quina sidescrapers, also
    made on half-pebbles (Cabrera 1984, 1988), which indicates appropria-
    tion by the makers of the level 18 Aurignacian of the same kind of blank
    as is found in some Quina Mousterian assemblages—a possible element
    of continuity.

3. Aurignacian blades are present in moderate numbers (4.19%), having
    been made on flake-blades (flakes approaching blade proportions), a char-
    acteristic feature of the Cantabrian Aurignacian (Bernaldo de Quirós
    1982a,b).

4. Another characteristic is the high incidence of archaic (Mousterian-like) pieces, among which sidescrapers are especially common (24.03%). There is also an atypical Châtelperron point made of quartzite.
5. The bone/antler assemblage includes ten split-based points with oval cross-sections.

All these characteristics place level 18 squarely within the Cantabrian Typical Aurignacian (Aurignacian I), but the bone spearpoints are especially important elements of this assignment. In general the Cantabrian Aurignacian is characterized by little typological variability; for this reason our approach is based more on stratigraphy than on technology and typology (Bernaldo de Quirós 1981, 1982a,b).

In the original collections from level 18, the presence of split-based bone points was considered more important diagnostically, in the absence of chrono-stratigraphic data, than the abundance of sidescrapers, which were considered a "late survival" (Obermaier 1925). The radiometric dates, together with the sedimentology, allow for a reinterpretation of this level, locating it in the earliest phases of the Aurignacian sequence. In this way, the high incidence of side-scrapers, instead of representing a Mousterian "survival," allow us to consider level 18 as falling in the transition between a Charentian Mousterian (Quina subtype) and the Typical Aurignacian.

As for the faunal inventory from the old excavations, studied by Vaufrey, we have only species attributions, supplemented on occasion by MNI counts. The level 18 fauna is characterized by a dominance of red deer (216 individuals), equids, *Bos* sp. (29 individuals), some remains of rhinoceros originally attributed to *R. merckii* (the old designation, now considered to comprise two species), and some traces of *E. antiquus* (?).

The old excavations indicated that level 18 was an Aurignacian industry that contained increasing numbers of sidescrapers toward the bottom of the deposit. However, Obermaier's field notes at first referred to this level as Mousterian Alpha, and it was only the appearance of the split-based bone points that led him to reclassify it as Aurignacian Delta, reserving the term "Mousterian Alpha" for level 20, which was clearly separated from level 18 by a thick deposit of red sterile silts.

The Post-1980 Excavations

Even taking into account the fact that the areal extent of the modern excavations was much smaller than that of Obermaier, our results are strikingly similar to his (Figure 13.2). In the recent work, level 18 was subdivided into three sublevels (18b1, 18b2, 18c). Throughout these subdivisions, the characteristics of the lithic industry agree remarkably well with Obermaier's (1925) descriptions, if probable functional differences between the two excavated areas are taken into

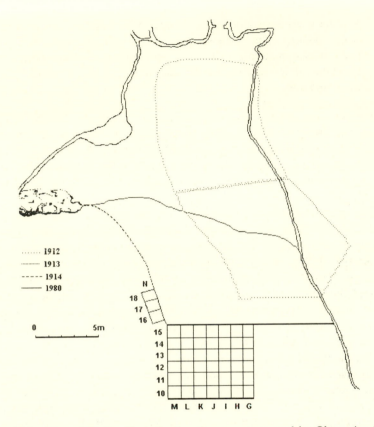

Figure 13.2. Plan of El Castillo indicating the areas excavated by Obermaier (1910–1914) and Cabrera (1980–1990).

account. In this regard, in the outer area of the cave, where only levels 18b1 and 18b2 are represented, a dense concentration of bones is associated with a lithic assemblage with a high incidence of casually retouched limestone pieces, together with quartzite hammerstones and rarer quartzite flakes and debris, and practically no flint. This area also has a scatter of small charcoal flecks and some larger pieces. The faunal inventory by anatomical parts shows abundant cranial fragments, mandibles, and elements of the axial skeleton, which suggests a primary butchering area, done with large, casually worked, heavy-duty cutting tools that could be easily replaced (Cabrera and Bernaldo de Quirós 1984). Limestone is the ideal raw material for this since it is easily worked, and the only raw material in the area from which large tools can be manufactured, since neither flint nor quartzite occur in sufficiently large cobbles or chunks.

Sublevel 18c has a series of rich charcoal concentrations but without clear boundaries. In them, very thin (<1 cm) lenses of carbon are evident, but no burnt bone or fire-cracked rock, nor evidence of features that would allow us to identify them conclusively as hearths. We are inclined to consider them discard areas where hearth debris along with material adjacent to hearths was thrown during successive clearings of the central part of the site. In support of this is the proximity of the "discard zone" to the sloping wall of the cave, which would have precluded use of the area as a living space, and the high incidence of knapping debris in relation to finished tools.

The industry from these sublevels does not vary much from one to the next. There are simple and carinate endscrapers, a few dihedral burins, and abundant side-scrapers, as was the case with the old collections. Sidescrapers are especially common in sublevels 18b2 and 18c. The Aurignacian blades follow the general Cantabrian pattern of being manufactured on laminar flakes, as is also the case with later Aurignacian levels like those at El Pendo and Cueva Morín (Bernaldo de Quirós 1982c). A flint example from the base of 18c is morphologically and technically similar to one made on quartzite from the old excavation (Cabrera 1984).

The predominant raw material is a black Jurassic limestone which does not come from Monte Castillo itself, but which occurs as large, rolled cobbles in the streambeds of the surrounding tributary valleys of the Río Pas, the major drainage in the area (Cabrera and Bernaldo de Quirós 1985). It is heavily altered chemically, especially in the excavated portion of 18c. In the area of 18b2, this limestone almost always occurs in the form of flaking debris, nuclei, and unretouched flakes associated with faunal remains. However, among the less-altered pieces from 18c are a carinate endscraper and a dihedral burin, which offer the possibility that, among the many pieces rendered unrecognizable by alteration, there could be some characteristic Aurignacian tools. Near the base of 18b2, an atypical Châtelperron backed point made of fine-grained quartzite showed up in 1986. Flint is very scarce as medium-sized or larger elements of debitage, but very common in debris less than 1 cm long, especially in the area of the longitudinal section. Most of the flint is gray and could have come from workshops located in other areas of the cave, having been redeposited here in the course of successive "cleanings" of the main zone of occupation.

The distal (working) end of a bone "chisel" was recovered from sublevel 18c; it exhibits a regular series of horizontal lines (hunting tallies) which resemble those from lower Aurignacian levels in other parts of Europe. The distal end of an antler point was also found in 18c, as well as some elongated antler splinters, probable by-products from the making of spearpoints. These artifacts are also present in the old collections, along with the ten split-based bone points mentioned earlier.

The fauna from the recent excavations resembles that from the 1910–1914 excavations, although these data are still preliminary since the definitive study of Altuna and Pike-Tay has not been published. Red deer are especially common. In parts of 18b2, dismembered pieces of the same individual animals (proximal pieces of scapulae, vertebrae, rib fragments) are intermingled with the lithic industry, especially limestone flakes and coarse-grained quartzite hammerstones. The numerous deer mandibles also tend to confirm Vaufrey's MNI count of 216 individual red deer from the old excavation (Cabrera 1984). Big bovids and equids are present but not very numerous. Rhinoceros molars were found, representing two species and including some juveniles. Finally, an ivory tusk fragment from an elephant was recovered on the edge of the main occupation area in level 18b2. It shows teeth marks of scavengers and was evidently worked over by them prior to being collected by humans. Obermaier also noted the presence of mammoth *(E. antiquus)* in this level. In sum, the identification of our level 18 with Obermaier's Aurignacian Delta seems secure. Our analyses present an entirely consistent picture with respect to both the lithic and bone industry, as well as the fauna.

The presence of scavengers is indicated along the outer edge of the 18b2 occupation, always in the upper part of the deposit, and also by the micromammal remains. Their traces are clearly distinct from those that appear on the numerous faunal remains associated with the rest of the archaeological material, interpreted as butchering marks related to muscle stripping of ungulate (especially cervid) carcasses (Pumarejo and Cabrera 1992). The presence of scavengers and regurgitated owl pellets indicates temporal discontinuity in the human use of the cave, especially toward the end of the level 18b2 accumulation. The identification of Obermaier's level 18 with ours (18b1, 18b2, 18c) appears secure on lithostratigraphic grounds, as determined by its position in the sequence, by its composition, as well as by the clear color differences between it and the levels that bracket it.

SUMMARY OF THE ARCHAEOLOGY

From an archaeological point of view, based on the cleaning of the old sections as well as the recent horizontal excavation, it is clear that:

1. in Level 20, in the longitudinal cut, the four recovered cleavers clearly identify this level with Obermaier's Mousterian Alpha, and
2. both the collection from the 1910–1914 excavations and our recent, post-1980 research allow us to establish a concordance between Obermaier's level 18 and ours.

With respect to the cultural characterization of these deposits, we note that, given the concordance of those levels, the presence of bone spearpoints, an engraved antler chisel bit, antler splinters, and other bone/antler elements allow us to identify these deposits as Aurignacian. Level 18 has a total thickness of 15 cm, subdivided into three sublevels, although the exact stratigraphic position of the spearpoints recovered by Obermaier cannot be determined. Most of the bone/antler industry from the recent excavations comes from the base of 18c, which suggests that, if the antler splinters and blanks are part of the process of spearpoint production (as is generally assumed), these points were present right from the beginning of the formation of level 18.

Analysis of both lithic collections indicates that the industry shows, on the one hand, a technological affinity with the underlying Quina Mousterian but, on the other, also has a sufficient number of typical Aurignacian types that it can be considered to pertain to a transitional phase of the Archaic Aurignacian (Cabrera and Bernaldo de Quirós 1990). Therefore, the Middle to Upper Paleolithic transition appears to have taken place in situ at El Castillo over the interval of time represented by the deposition of levels 18–20, and without any external influences.

ISOTOPIC DATES: CHRONOSTRATIGRAPHY AND CORRELATIONS

Ten AMS radiocarbon dates from level 18 from three different laboratories (University of Arizona, Oxford University, and the CNRS facility at Gif-sur-Yvette) are published here for the first time (Table 13.1). They are entirely consistent with the Arizona dates reported in Cabrera and Bischoff (1989). Samples under analysis at Gif-sur-Yvette give an average age for Level 18b2 of 38,500 B.P. and for level 18c of 40,800 B.P. (Cabrera et al. 1996). The Oxford dates are published in Hedges et al. (1994) (Figure 13.3).

The chronostratigraphic position of levels 15–20 has been reasonably well established from paleoclimatic studies and from the dates obtained for level 18 (Cabrera and Bischoff 1989). The pollen chronology has been correlated, taking latitudinal and geographical differences into account, with the Grande Pile sequence, the most complete in western Europe that is also supported by an adequate number of radiocarbon determinations (Woillard 1978; Woillard and Mook 1982). Levels 16–18 evidently formed during a cold phase with a minor warming trend within level 17; dates for level 18 range between 38,000 and 40,000 B.P. The level 17 deposits can be correlated with the inter-Hengelo/Denekamp cold phase in the pollen chronology, dated at Grande Pile to 34,100 ± 600 B.P. Level 15, which follows them, should correspond to the Denekamp Interstadial (Zone 16 of Grande Pile), which has produced five dates ranging between 30,820 ± 210 and 29,090 ± 250 B.P. Level 19, which also looks like an interstadial, and the bottom part of level 18 (sublevel 18c) were probably formed during the Hengelo Interstadial (Zone 15 at Grande Pile), dated to 40,000 ± 600 B.P. Level 20, which exhibits few sedimentary differences from level 19, contains a colder pollen spectrum that could be equated with the cold phase immediately preceding the Hengelo Interstadial.

Table 13.1.   El Castillo—Radiocarbon Dates from the Different Sublayers of Level 18

Level	Lab. No.	Date (B.P.)	σ
18b1	AA-2406	38,500	± 1,800
18b2	OxA 2475	40,700	± 1,600
18b	2OxA 2474	38,500	± 1,300
18b2	AA-2407	37,700	± 1,800
18b2	OxA 2473	37,100	± 2,200
18c	GIFA 89147	42,200	± 2,100
18c	OxA 2477	41,100	± 1,700
18c	OxA 2476	40,700	± 1,500
18c	AA-2405	40,000	± 2,100
18c	OxA 2478	39,800	± 1,400

While these correlations with Grande Pile seem to be reasonable and coherent, it is by no means so easy to make a direct correlation between El Castillo and the sequence known from Aquitaine (Laville et al. 1983). In principle, and taking the level 18 dates and the approximate ages proposed for the Aquitaine phases into account, level 18 would be situated between the end of Würm II and the first part

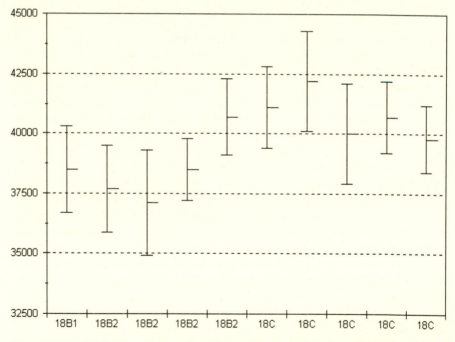

Figure 13.3   El Casillo—Radiocarbon Dates with Error Bars from the Different Sub-layers of Level 18.

of the Würm II-III interstadial (the climatic optimum), and level 15 would be in cold phase IV of Würm III. These correlations are, however, incompatible with the climatic characteristics of these levels, and those of the rest of the analyzed El Castillo levels. So there is a clear discrepancy in the temporal sequence of cold and more moderate phases when the El Castillo data are compared with those of Aquitaine. In our view, this discrepancy rests upon two important factors:

1. The absence in Aquitaine of deposits and dates pertaining to the level 15–18 interval at El Castillo. The Les Cottés Interstadial is then taken as a reference point for a palynological correlation with La Ferrassie as the ceiling of the Würm II-III interstadial, with an upper age limit of 34,500 B.P., and pushing back in time a sedimentary hiatus for which we have evidence of pedogenesis and subsequent erosion (at El Castillo).
2. The hypothetical correlation between Aquitaine and Grande Pile for the Würm II proposed by Laville et al. (1983), where the Würm II-III interstadial, estimated to date between 39,000 and 34,500 B.P., is made to correspond with the sharp peaks that define Zone 15 in the Grande Pile pollen curve. These peaks are thought to correspond to the Denekamp Interstadial, dated between 30,820 ± 210 and 29,090 ± 250 B.P. (Woillard and Mook 1982), and not the Hengelo Interstadial (40,600 ± 600 B.P.), which would be more appropriate given the El Castillo radiometric dates.

Admitting, for the sake of argument, the correlation between Würm II-III and Denekamp (which would tend to weight climatic criteria more heavily than isotopic data), the correlation with El Castillo is problematic because the Würm II-III interstadial would correspond to Castillo level 15, which would perforce put levels 16–20 squarely in Würm II. The correlations of EC level 16 with Aquitaine phase VIII; level 17a with Aquitaine phase VII; levels 17b, 17c and 18 with phase VI; level 19 with phase V; and level 20 with phase IV are feasible in a coarse-grained way, although there are local climatic peculiarities and differences in the degree of expression of the climatic phases between the two sequences. However, this would imply that the two El Castillo Aurignacian levels (16, 18) would fall perforce within the early Würm, and being largely contemporaneous with Mousterian deposits in sites in the Aquitaine (e.g., Combe Grenal, Pech de l'Azé, La Micoque, La Ferrassie) (Laville et al. 1980).

In an effort to resolve these difficulties, taking the sedimentological and climatic data from El Castillo into account, weighting the absolute dates more heavily than the climatic criteria, and acknowledging the rather good correlations between El Castillo and Grande Pile, we propose the following correlations between the El Castillo levels and the Aquitaine chronology.

1. Level 15 (Denekamp) would correspond to the Arcy Interstadial, phase III of Würm III (31.5–30 kyr B.P., according to Laville et al. 1980).
2. Levels 16–18b would correspond to cold phases I and II of the Würm III in Aquitaine, including the minor climatic amelioration of level 17a.
3. Levels 18c and 19 (Hengelo) would be correlated with the Würm II-III interstadial.
4. Level 20 would correspond to cold phase VIII of the Würm II in Aquitaine.

Supporting the above correlations are gross similarities between the sedimentological-climatic characteristics of the level 15–20 sequence at El Castillo and those of the Aquitaine phases proposed to correlate with them. Thus, three phases of sedimentation are recognized in level 19, with the second and third marked by pedogenesis (although not by the development of a true paleosol). The moderately cold character of levels 16–18 agrees well with phases I and II of Würm III, in that intense cold is not documented until phase IV. Wet, cold level 20 can be correlated with phase VIII of Würm II. On the other hand, these correlations force a reconsideration of the age of the onset of the Würm II-III interstadial in Aquitaine, to a date in excess of 40,000 B.P.

Finally, and again hypothetically because of a lack of radiometric dates, the transition levels at El Castillo might be largely contemporaneous with those from the cave of Carigüela (Granada) where, after a temperate humid phase attributed to the Würm II-III interstadial (Unit 6) and possibly correlated with El Castillo level 19, two cold episodes follow (Unit 5). Although indicative of dry climatic conditions, these levels at Carigüela might be the southern equivalents of El Castillo levels 16–18, taking into account the latitudinal differences between the two caves. Unit 5 at Carigüela is followed by a long temperate, humid phase (Unit 4) which, although not as humid as El Castillo level 15, could possibly be correlated with it, again taking the geographic and latitudinal settings of both sites into consideration (Cabrera et al. 1993; Vega et al. 1988).

ACKNOWLEDGMENTS

This chapter was first presented at the International Colloquium *"El Origen del Hombre Moderno en el Suroeste de Europa"* (Madrid, 1991) and is an edited and updated version of a paper presented at that conference and published by the Universidad Nacional de Educación a Distancia (UNED) in 1993 (Cabrera et al. 1993). The transition research was supported by a grant from the Comisión Interministerial de Ciencia y Tecnología (No. PB92-0562).

IV

WESTERN PERSPECTIVES
The Anglo-German Research Traditions

Written by archaeologists and human paleontologists, the five papers in this section represent both replacement and continuity perspectives. Arguing that the current version of the modern human origins debate resembles a "thicket of misreadings, polarization, and biases," the prominent replacement advocates Günter Bräuer and Christopher Stringer identify a number of issues that, if confronted directly, should reveal underlying preconceptions and assumptions of the various workers involved in this research. Among these issues are the adequacy (or lack thereof) of paleontological tests; different construals of the molecular evidence for African "Eve"; the effect of different implicit species definitions; and the nature of evidence for continuity.

Steven Churchill attacks the reductionism implicit in addressing behavioral, cultural, and morphological evolution as independent problem domains, each with its own "package" of methods and theories. In light of the commonality of the problem addressed—the biocultural evolution of *H. sapiens sapiens*—he argues for a process-oriented, holistic approach that would integrate paleoanthropology within the broader discipline of evolutionary biology. This, he suggests, would shift modern human origins research more toward the investigation of the mechanisms of change rather than toward the end products of those mechanisms, with which it is currently concerned.

The chapter by continuity advocate David Frayer focuses on temporally vectored rates of change for craniometric and, especially, dental data from the Middle to Upper Paleolithic transition interval in Europe. Using comparative evolutionary rates measured in darwins *(d)*, he shows that average rates of change between European late neandertals and early Upper Paleolithic hominids fall within the range found in recent *Homo sapiens*. In other words, there was no "accelerated rate of change" between neandertals and early Upper Paleolithic Europeans, as claimed by replacement advocates.

Archaeologist Lawrence Straus focuses on the Iberian evidence over the 40,000–30,000 B.P. transition interval, characterized by very early Aurignacian dates in the north (thus precluding any simplistic east-west "migration" of Aurignacians) and by a very late persistence of neandertals and Mousterian archaeological assemblages in the south. These data, he suggests, have profound implications. They call into question the *ex oriente lux* presuppositions of most replacement advocates, and more generally the normative conceptualizations of both neandertals and the Aurignacian.

The chapter by Cambridge-trained American archaeologist Betsy Schumann is concerned with the effects that disciplinary boundaries have on our construals of pattern and the meaning of pattern in archaeology and human paleontology. She suggests that the way these boundaries are imposed on data, which is in turn a consequence of unwitting allegiance to a particular paradigm, will inevitably lead an analysis in a particular direction, toward a particular conclusion. Awareness of bias factors implicit in boundary definition and in classification might help to eliminate the circular reasoning characteristic of much MHO research.

14

Models, Polarization, and Perspectives on Modern Human Origins

GÜNTER BRÄUER and CHRIS STRINGER

It seems that in the long-lasting controversy on modern human origins little agreement has been reached between the proponents of the multiregional evolution and Out of Africa models. The current situation looks somewhat like a "thicket" of misreadings, polarization, and biases, which instead of clarifying the problems have caused more confusion (cf. Frayer et al. 1994a; Stringer and Bräuer 1994). Thus, a number of basic questions about preconceptions, biases, and assumptions will be addressed in this paper, providing some insights into major causes for such a development. We have not restricted ourselves to our own model because the relevant problems become more evident and clear in comparison with opposite views. In addition, proposals for more efficient research are outlined.

MODELS AND THEIR TESTABILITY

A basic strategy for deciding between hypotheses is to test them. However, what should the hypotheses to be tested look like? Should they represent realistic scenarios or simplified models? These are important questions which have led to basic misunderstandings in the modern human origins debate. A major problem of testability relates to the effects of gene flow proposed by the models. An Out of Africa model that assumes some, or more intensive, mixing of archaic and modern populations during the replacement process cannot be falsified by evidence for some continuity outside of Africa (Bräuer 1989, 1992a, 1992b; Stringer 1992a; Stringer and Bräuer 1994).

A test of the multiregional evolution hypothesis is faced with considerable difficulties as well. Evidence for regional continuity is not an assumption exclusive to this model since, in a replacement scenario, mixing between archaic and modern humans could mimic in situ evolutionary change (Bräuer 1992a; Stringer 1992a; Stringer and Bräuer 1994). Moreover, the unclear effects of assumed gene

flow between the geographic areas might go so far that even regional replacements would be tolerable within the multiregional evolution model (Wolpoff,
personal communication 1994).

A direct test of these complex models is difficult. Thus, one has to simplify the
assumptions to provide testable hypotheses. Yet, we cannot expect that such
simplified models will explain all of the evidence. We rather think that they
might be able to show the predominant mode of the process. Such an attempt has
been made by Stringer and Andrews (1988b) who chose an extreme Recent
African Origin (RAO) model to test whether a complete replacement hypothesis,
assuming no or negligible gene flow outside Africa during the replacement, can
be disproved. Although it was concluded that the complete replacement hypothesis was largely in agreement with the facts, there were some problems, especially
concerning the Australian evidence. Replacement might have been the predominant mode, but a complete replacement hypothesis was not confirmed from the
data. Instead, varying degrees of gene flow or hybridization could be envisaged
as well (Bräuer 1992b; Stringer and Bräuer 1994). However, as the continuity
data are mostly problematic (see below), we think it difficult to show convincingly whether only little or larger gene flow has to be assumed for the various
regions. In addition, a brief hybridization phase could have occurred in some
regions (and could be recognized in the fossil evidence) without there necessarily
being genetic continuity between the hybrids and recent populations (Stringer
1992a).

BIASES AND POLARIZATION

In numbers of papers we are confronted with claims that the Eve Theory (ET)
can be refuted on the basis of fossils alone (e.g., Frayer et al. 1993:19). Yet, what
is the meaning of this Eve Theory? Is it identical to the Out of Africa model? It
seems rather that this Eve Theory is a special construction based on three biased
interpretations.

The first is the assumption that the extreme test hypothesis of Stringer and
Andrews (1988b), which assumes complete replacement, is the only valid global
model. It is ignored here that a realistic Out of Africa model could also allow for
some degree of continuity as a result of mixing. Second, this theory focuses on
only one possible interpretation of the mtDNA data which proposes a complete
replacement of all existing populations by the descendants of a single Eve who
lived in Africa about 200,000 years ago (Frayer et al. 1993). This date is some
100,000 years older than the best-dated fossil evidence for the appearance of
early modern humans (e.g., at Klasies and Qafzeh). It would also preclude
mixing between Eve's descendants and archaic humans in other regions of the
world. However, other interpretations are possible, such as that offered by Cann
(1992), who does not exclude the possibility of mixing but considers the finding

of Neanderthal maternal lineages unlikely in present-day humans. Third, this theory assumes that, for proponents of Recent African Origin (RAO), modern humans spreading out of Africa must have represented a new species which would have made mixing with archaic humans impossible (Frayer et al. 1993:18; Wolpoff et al. 1988). Thus, Smith (1992a:145) uses the term "speciation/replacement." Yet, this again is far from reality. For example, whether we regard the morphological differences between the Neanderthals and modern humans as intraspecific (GB) or as specific (CBS), we are certainly not dealing with a biospecies concept with absolute fertility boundaries, but only with paleospecies (Bräuer 1992b:404; Stringer 1992a:19). As we explained elsewhere (Stringer and Bräuer 1994:416-417), "neither of us feels that taking a RAO position necessarily entails accepting that dispersing modern humans must have represented a new species, contrary to the implications of Frayer et al." Moreover, there is considerable evidence from closely related living mammalian species of hybridization in both natural and captive conditions, whilst separate species identities are maintained.

Thus the Eve Theory, in the sense of multiregionalists, appears to be a very biased extreme model and an unrealistic "straw man" in which neither Frayer et al. nor we believe. Nevertheless, this version of the Eve Theory is proclaimed by practically all proponents of multiregional evolution to be the real Out of Africa model (Frayer et al. 1993; Thorne and Wolpoff 1992), which they can easily refute. Thus, for Smith (1992a:153) "it is only necessary to show that there is some degree of regional continuity across the archaic/modern boundary" outside of Africa to refute this model. G. A. Clark (1992a:193) even assumed that any convincing evidence for continuity would refute *all* replacement models. This is like proclaiming Coon's extreme theory of independent regional evolutionary lineages to be the true multiregional model and then claiming that any evidence for discontinuity refutes all multiregional evolution models. We recently pointed out clearly that equating ET with RAO is inadequate, and we commented critically on the various predictions from ET (Stringer and Bräuer 1994). Yet, focusing on the extreme ET in many publications has caused an artificial polarization over a number of years. This largely unrealistic controversy and polarization, combined with extreme titles such as "The case against Eve" (Wolpoff and Thorne 1991) or "The end of Eve? Fossil evidence from Africa" (Wolpoff and Thorne 1993) over the past few years, has clouded the problems concerning the continuity data (see also Bräuer 1992a, 1992b). We would remind everyone concerned that Out of Africa models existed *before* the Eve Theory was proposed (e.g., Bräuer 1984b; Stringer 1984a).

PROBLEMATIC CONTINUITY DATA

To understand the process of the origin of modern humans, we are convinced that the study of morphological features in archaic and modern humans is extremely relevant. Such analyses give the impression of regional continuities as well as gaps. Thus, no matter which model one favors, the study of such features is essential. In particular, the multiregional evolution model is largely based on data of morphological features that are said to occur in one region more or much more frequently than in others, thus indicating a certain continuity of the gene pools over long time periods. Quite a number of lists of such clade features have been suggested and enlarged over the years (cf. Thorne and Wolpoff 1981; Wolpoff et al. 1984). Many of the respective traits were proposed by Weidenreich (1939b, 1943a), Coon (1962), and others. However, this fact does not make them necessarily reliable or currently useful. The study of such features is by no means simple. A number of problems and pitfalls have to be considered before interpreting the occurrence of such characteristics in archaic and subsequent modern specimens as evidence of regional evolution. Among them:

1. Is the respective feature clearly defined (e.g., facial flatness is quite a variable trait and might need further specification)?
2. Has the variability and present geographic distribution of the features been determined by adequate worldwide studies?
3. Can a feature occurring among "archaic *H. sapiens*" or *H. erectus* be regarded as representing the same trait as in recent humans? Is, for example, a malar eversion in a 700,000-year-old *erectus* specimen with a massive masticatory apparatus comparable to a moderate eversion present in a gracile modern skull?
4. Has the feature been maintained through drift or selection? In the former case, drift effects might have to be assumed for a million years or more. In the case of selection, these features could have continually reevolved in the same region, producing homoplasy rather than phylogenetic continuity (cf. Stringer and Bräuer 1994:416).
5. Is the feature derived with regard to a specific hominid group (as, e.g., the Neanderthals), or is it a primitive retention widespread among archaic humans of various regions?
6. What evidence is needed and available to decide in favor of regional evolution or, alternatively, as evidence of mixing or gene flow during a replacement period?

Because the suggested regional features are critical to the controversy of multiregional evolution versus replacement, a number of specialists have evaluated them in recent years. What do these studies tell us about the relevance of these traits for modern human origins?

Groves (1989a) reassessed a list of sixteen regional traits of the mongoloid lineage, nine of the australoid lineage, plus an additional nine features which are said to link the Ngandong sample to modern australoids. As result of this survey, he found "little evidence for special likeness of modern 'Mongoloids' to *Homo erectus pekinensis*" (1989a:279) and with regard to the Australasian lineage, he also concluded that "from the 'shared characters' point of view, the Regional Continuity model lacks much real substance" (1989a:281).

Habgood (1989) examined the Australasian features as well, considering a list of twelve as proposed by Thorne and Wolpoff (1981) and Wolpoff et al. (1984). He concludes, "it is evident that all of the characters proposed . . . to be 'clade features' linking Indonesian *Homo erectus* material with Australian Aboriginal crania, are retained primitive features present on *Homo erectus* and archaic *Homo sapiens* crania in general. Many are also commonly found on the crania and mandibles of anatomically-modern *Homo sapiens* from other geographical regions, being especially prevalent on the robust Mesolithic skeletal material from North Africa" (1989:259). Habgood (1989:250) even found nine of the twelve Australasian regional features on the Kabwe cranium from Zambia. Furthermore, Stringer and Bräuer (1994) note that every one of the regional features claimed to link the Ngandong and WLH 50 fossils can be found in the Omo 2 calvaria from Ethiopia.

In another study including nine proposed regional features for east Asia (China), Habgood (1992:280) came to the same conclusion: "it is evident that none of the proposed 'regional features' can be said to be documenting 'regional continuity' in east Asia as they are commonly found on modern crania from outside of this region . . . , and are consistently found on archaic *Homo sapiens* and/or *Homo erectus* crania throughout the Old World."

A frequently cited example of the occurrence of continuity features concerns the Sangiran 17 hominid, the only Java *erectus* specimen with a well-preserved face. Although there is an unbridged time period of perhaps 700,000 years up to the earliest faces of modern humans in this region, Thorne and Wolpoff (1981) and Wolpoff et al. (1984) claim a continuity with regard to such features as marked prognathism and the malar tuberosity. However, the presence of the latter feature could not be verified by us or by Habgood (1989:254). Additionally, H. Baba (National Science Museum, Tokyo) has produced a new reconstruction of the Sangiran 17 cranium. Now it no longer appears to exhibit the marked prognathism of the earlier reconstruction (by Wolpoff), and Baba (personal communication, 1994) was also not able to find a malar tuberosity.

One of us (Stringer 1992a) also checked late Pleistocene north African cranial samples with regard to the occurrence of supposed "mongoloid" and "australoid" regional features. It became evident that relevant Chinese features, such as transversely flat faces and shoveled upper incisors, and southeast Asian traits, such as high prognathism, strong supraorbital region, everted malars, and malar tuberosities, all occur with moderate or high frequencies among these North Africans as well. Stringer was criticized by Frayer et al. (1994a) for including in his study a feature (auditory exostoses) abandoned by multiregionalists, yet the paper by Wolpoff (1992) in the *same volume* still mentions this character uncritically!

Wu and Bräuer (1993) checked a number of features commonly seen in China with regard to their occurrence and variability in African "archaic *Homo sapiens*" and early modern material. Nearly all of them also occur in Africa, some showing greater variability there. Moreover, in many features, early modern Chinese specimens show deviations from the pattern of "archaic *H. sapiens*" of this region and exhibit broader spectra similar to those seen in African archaic and early modern *sapiens*.

Probably the most detailed recent study of suggested east Asian and Australasian continuity traits has been carried out by Lahr (1992). Her results indicate that most of these features can no longer be regarded as evidence for regional evolution. Many of the suggested features were too variable to be of use, but of the thirty remaining features studied, nineteen showed no significant regional distribution (e.g., the course of nasofrontal and frontomaxillary sutures) or showed trends other than the ones proposed by the multiregional model. These latter include the profile of the nasal saddle, sagittal keeling, facial flatness/lack of prognathism, and eversion of the lower malar border. The remaining eleven exhibit higher frequencies or a significantly different mean value in the regions, as proposed by the multiregional model, but only two are related to east Asia. Yet, Lahr (1994:49) also points to strong doubts with regard to the "confirmed" Australian features. They are plesiomorphous and their manifestation may be functionally constrained. Moreover, several of these features, such as a strong supraorbital ridge, frontal flatness, and the position of minimum frontal breadth, also occur with high frequencies among the north African Afalou and Taforalt samples. In spite of his other criticisms, Habgood (1989:263) mentioned a combination of frontal and facial features which may constitute evidence of some regional continuity in Australasia. According to Lahr's more recent results, however, the frontal features occur most frequently in the Afalou/Taforalt sample. Only marked prognathism and the malar tuberosity would remain most frequent in Australia, but these characters turned out to be problematic with Sangiran 17 (see above) and occur among north Africans as well (Stringer 1992a). Additionally Lahr (work in progress, personal communication) has shown that such characters were common in early modern humans such as the Skhūl-Qafzeh hominids. Thus, this problem needs further evaluation from a wider comparative perspective than just comparing samples in Java and Australia, or within China.

Shovel-shaped incisors are often regarded as a significant mongoloid clade feature, and indeed they frequently occur on archaic and modern specimens from this region. However, it cannot be ignored that shoveling is a complex character and also occurs in Africa from early *Homo* (e.g., OH 16, OH 39) and *Homo erectus* (e.g., WT 15000) to "archaic *Homo sapiens*" (e.g., Rabat) and modern humans, as from Jebel Sahaba and Wadi Halfa (Anderson 1968; Brown and Walker 1993; Greene et al. 1967; Howell 1960; Tobias 1991). The Neanderthals have high frequencies of shoveling as well. If this character is regarded as a continuity feature in China, what about its occurrence in Africa, and why is its loss in Europe at the Neanderthal/early modern interface not a mark of discontinuity (Stringer and Bräuer 1994:419)?

It has not been our intention to deal with the proposed regional features in detail here, but only to summarize some of the recent evidence. This summary shows that most of these features are highly problematic and inadequate as support for regional evolution. Thus, the current evidence is insufficient to disprove a replacement model that might allow for some degree of gene flow (see also Habgood 1992:283 for the Far East). This point has been missed by Frayer et al. (1994a:427) when they claim, "But what, besides the observations of continuity in many places, could possibly support multiregional evolution? What besides that very continuity, when there is supposed to be discontinuity caused by African replacement, could better refute the Eve theory?" We think that the established evidence for continuity is small and rather could be the result of limited hybridization that mimicked the effects of in situ evolutionary change (Stringer and Bräuer 1994:417).

Nevertheless, as mentioned above, we are ready to consider all well-documented evidence for regional continuity, and we assume that some traits might finally turn out to be reliable evidence for continuity. Among them could be the horizontal-oval (H-O) mandibular foramen which is reported to occur among 53% of Neanderthals and 25% of early modern Europeans, but which became quite rare with time (Frayer et al. 1993:31). However, the early modern occurrences may include a possible Neanderthal specimen from Vindija, and the number of specimens compared from Africa ($n = 1$) and the Near East ($n = 2$) is so small that no one can currently predict what the frequencies would look like for these regions if larger samples were available. Moreover, as Smith (1978: 528) mentions, the border of the foramen might develop during the life of an individual as a response to, say, extensive use of the dentition. This possibility is supported by the fact that the H-O foramen is missing in nearly all juvenile Neanderthal mandibles (exception: the Le Moustier youth). Lieberman (1995:175) also doubts the value of this character, as it occurs outside Europe as well, as for example in Zhoukoudian H1 and OH 22.

In sum, we see increasing evidence from recent studies that there are many problems with most of the suggested regional features. This makes further basic research necessary. And in particular the model of regional continuity must be demonstrated for each grade of human evolution, including the establishment of present-day regionality. Evidence is accumulating that many early modern crania in Europe and China do *not* show clear signals of modern regionality (e.g., Sarich 1995; Schumann 1995a; Stringer 1992b; Wright 1992). To explain this under a recent African origin model would require either that these crania are not ancestral to recent populations in the same region, or that regionality was still evolving—neither of which falsify the model. However, such results cannot be explained under any multiregional model with a long time scale.

MULTIREGIONAL EVOLUTION OR CONTINUITY

Another basic question we want to address is the meaning of "regional evolution" versus "regional continuity." Are these terms identical, or would some degree of regional continuity be possible within an Out of Africa scenario as well? In coping with the increasing difficulties of the continuity data, the proponents of the multiregional evolution hypothesis have obviously moved to a more flexible model, which would even allow for replacements in some regions (Wolpoff, personal communication 1994).

According to previous statements, the multiregional evolution hypothesis proposes "that a balance between gene flow and selection oriented the direction of evolutionary change in human populations of the middle and late Pleistocene" (Wolpoff 1989a:137) and "that *Homo sapiens* evolved from ancient humans gradually in many parts of the world," and thus, "regional continuity in the features of human fossils should be the norm" (Wolpoff and Thorne 1991:37-38). Frayer et al. (1993:17) mention that "multiregional evolution is an explanation about the worldwide pattern of Pleistocene human evolution." Following Weidenreich's suggestion, this model assumes a network of vertical lineages of regional changes and horizontal interregional connections by gene flow between Europe, Africa, and Asia (Frayer et al. 1993:Fig. 1). We have been criticized for ignoring this component of gene flow in the multiregional model (Frayer et al. 1994a), yet the very quotation from Stringer and Andrews (1988b) used to demonstrate this point omitted the words "together with gene flow" from the quotation! This is yet another example of forced polarization in the debate.

However, how intensive were the horizontal connections indicated by Frayer et al.'s figure between Africa and East Asia? Thorne (1993:173)—although one of the authors of this paper—appears to be unconvinced of such connections, claiming that "descendants of the Java and Peking people do not become extinct but give rise, without African influence, to the modern people of their region." A similar view is held by Wolpoff and Thorne (1991:39, 40). In contrast, Pope's (1992a) analysis of the craniofacial morphology of Chinese hominids could not support such views. He sees strong indications of rapid extraregional introductions in both northern and southern China, especially with "archaic *Homo sapiens*" (1992a:289). For modern humans of Asia and Australasia as well, Smith (1992a:154) assumes "that non-indigenous influences, probably ultimately of African origin, were of considerable importance in the emergence of modern human anatomical form." This wide range of views among authors of the same article shows that there is not much agreement on how the process of multiregional evolution might have taken place in the Far East.

Concerning Europe, the views on the regional evolutionary process have changed more obviously (cf. Bräuer 1992a). Some years ago, Smith (1982:685) was convinced that "the simplest and most logical hypothesis supportable on the basis of present knowledge is that the archaic-to-modern-*H. sapiens* transition in

South-Central Europe was indigenous." He saw "no reason to believe that gene flow into this region from the Near East or anywhere else was any more responsible for the archaic-to-modern-*H. sapiens* transition in South-Central Europe than gene flow from South-Central Europe was for the same transition in the Near East or elsewhere." For western Europe, in agreement with Wolpoff (1981:823), he regarded the continuum as less distinct, but new discoveries appeared to be closing the gap (Smith 1982:686). Owing to increasing evidence to the contrary, Smith has more recently favored a replacement model for western Europe (Smith and Trinkaus 1991) and has also strongly modified his former position for south-central Europe. In view of the increasing evidence of an earlier existence for modern humans in Africa and the Levant, he now assumes that a substantial external influence was involved in the emergence of modern Europeans, for the numerous changes from the Neanderthals to early moderns would be difficult to explain without recourse to an extra-European influence (Smith and Trinkaus 1991:284). It also became obvious that interpretation of the central European remains is more difficult than previously thought, forcing Frayer (1992b:180) to admit that for this sequence it is "difficult to ascertain what are true transitional specimens, since age, sex and individual variation (along with problems in chronological placement and the incomplete nature of some of the fossil remains) can prejudice judgments about what is and is not a transitional fossil."

In coping with the increasing need to accept strong extraregional influence and immigrations into Europe and the Far East, the descriptions of the regional evolutionary processes have been more and more restricted to the recognition of some degree of continuity. Thus, for Smith (1992a:146) it is not even necessary to claim "that regional archaic human populations were invariably the major contributors to modern human gene pools in their respective regions. At its most basic level, continuity is only an argument that archaic humans played some role in the origin of regional modern human populations. . . ." Frayer (1992b:187) also proposes, on the basis of the persistence of traits across the Neanderthal/modern boundary, a degree of continuity indicating "some measure of genetic contribution of Neanderthals to subsequent *Homo sapiens* populations." However, as one of us has stated elsewhere (Bräuer 1992a, 1992b), this view no longer means regional evolution. Such continuity does not necessarily reflect regional evolutionary changes primarily caused by selection and/or drift but merely the common presence of certain mostly problematic features, for whatever reasons. As outlined above, such evidence agrees well with a replacement scenario assuming some mixing between archaic and immigrant modern populations (also see Stringer and Bräuer 1994).

RESEARCH PERSPECTIVES

Most recent meetings (e.g., a seminar held in Kyoto in 1994) have revealed that continuing artificial polarization and misrepresentations will not lead to much further progress. Instead, it would be more sensible to make efforts to agree on the real problems and then to look for adequate approaches to examine and perhaps solve them. Moreover, it would be effective if proponents of different views worked jointly on these specific problems. For example, it is difficult to understand why specialists cannot agree after much discussion whether a specific tuberosity exists on a well-known specimen or not. Much work has still to be done on the assumed regional continuity features. Although certain definitions for many of these traits are available, specialists do not always agree on how one or another feature should be determined. This has also led to continuous misunderstandings. As an example of the problems involved, at the beginning of her doctoral research Lahr (personal communication, 1989) was unable to obtain relevant character definitions from one of the main proponents of multiregional evolution, and was advised to create her own definitions. This is hardly a recipe for consistency and testability! Efforts should therefore be made to reach basic widely accepted definitions for data collection. For example, it is difficult to understand why there is no agreement on how facial flatness should be determined. Are there different kinds of flatness which need different study methods?

Moreover, further progress is needed on the question of the occurrence of proposed Australasian features among final Pleistocene North Africans. Why not take casts of relevant Australasian specimens, such as WLH 50, to places where the north African material is available and compare the respective features in detail? (WLH 50 itself requires proper publication so it can be discussed in detail. A metrical study of WLH 50 by Stringer demonstrates that, despite its large size, WLH 50 is similar in cranial shape to recent Australians and a Skhūl-Qafzeh sample and, although much more distinct, is nevertheless closer to a late archaic African sample than to a Ngandong sample.) Besides different views on general models, differences in intellectual traditions, training, and special experience might also lead researchers to look at features in ways very different from those of other researchers. Thus, workshops are needed which offer the possibility for experts to agree on definitions of features and their analysis.

Any commentary on future progress in the debate on modern human origins cannot ignore the extra dimensions added to the discussion by the genetic data. They have had a great impact on the debate, particularly since the Cann, Stoneking, and Wilson article of 1987, and undoubtedly contributed to an increasing polarization. The perceived rise and fall of "mitochondrial Eve" can be tracked by the way different factions in the debate have utilized the growing body of genetic data. While the firm conclusions of those relying on mtDNA analyses for phylogenetic resolution were certainly premature in 1987, some of the original proposals have received considerable support from recent work, such as the complete sequencing of a number of hominoid mtDNA genomes (Horai et al. 1995), supporting an African ancestry and inferred coalescence for humans at 143,000 $\pm$ 18,000 years.

In a completely different approach, Rogers and Jorde (1995a) have used mismatch distributions and coalescence theory on DNA data to surmise that the population size of the original stock of modern humans was probably less than 7,000 breeding females and that "all available genetic evidence is consistent with the proposition that the major human populations separated from a small initial population roughly 100,000 years ago." While these conclusions are compatible with (but do not prove) a recent African origin scenario, they surely falsify that of multiregional evolution.

It will be interesting to see how multiregionalists respond to such results, in the face of comments like that of Gibbons (1995b:1273) who concluded "the evidence coming out of our genes, like the putative African founder population itself, seems to be sweeping the field." These kinds of analyses challenge all those working on fossil or archaeological materials to review our attitudes to such work, whether to become more critical or more receptive, but at minimum we must recognize that such research will greatly increase, not decrease, in its scope and influence in the near future and must be addressed, not ignored.

15

Morphological Evolution, Behavior Change, and the Origins of Modern Humans

STEVEN E. CHURCHILL

A growing number of paleoanthropologists have been sounding the alarm that the archaeological transition from Middle to Upper Paleolithic and the biological shift from archaic to modern humans, traditionally considered to be linked phenomena, may in fact have been independent. The perceived pattern of coupled morphological and behavioral transitions may owe more to historical coincidence than to causal relationships (Chase and Dibble 1990; Clark 1989a, 1992a; Clark and Lindly 1989a, 1989b; Klein 1989a). The association of neandertals with Châtelperronian assemblages indicates that some degree of modern behavioral capabilities were present in late archaic humans, while early modern humans with Middle Paleolithic technologies in South Africa and the Levant suggest morphological modernization preceded behavioral modernization in some places. Attempts to treat modern human origins as a single biobehavioral event, then, are inherently flawed, and the integration of phylogenetic with behavioral or cultural reconstructions leads to specious and tautological reasoning (Foley 1991; Wood 1994). A lack of consensus about the nature and processes of the archaeological, behavioral, and morphological transitions of the Late Pleistocene results in a situation where support for a given phylogenetic model can be mustered by selective use of archaeological and behavioral inferences, many of which are themselves post hoc conjectures heavily weighted by paradigmatic biases. Given the uncertainty surrounding every aspect of the transition, and the heavy reliance on inductive reasoning in the modern human origins debate, we are left with little foundation for evaluating competing models. While conceptual paradigms are inherently self-affirming (Kuhn 1962b, 1970), we have reached a point where phylogenetic models are as well, and positivist science has broken down.[1]

This situation has caused some workers, in accord with a strategy advancing a full or partial reduction of paleoanthropology to biology (Foley 1987a), to advocate a more atomistic approach that considers behavioral and anatomical modernization independently (Chase and Dibble 1990; Foley 1991; Wood 1994).[2]

Failure to do so, they argue, can lead to circular reasoning, whereby modernity in one realm is taken as de facto evidence of modernity in the other (see Wood 1994). Others claim that such an approach is unwarranted and dualistic, that evolutionary processes are complex, and that the various aspects of the transition must be approached with an integrative, holistic methodology (Clark 1992a; Graves 1991; Lindly and Clark 1990a; Mellars 1989b). These calls for, and arguments against, reductionism are redolent of the long-standing struggle in biology between provincialists (who view biology as a province of the physical scientists) and autonomists (who argue for theoretical and methodological auton-omy of biology based on emergent properties of living things), with the dissen-tion centering on the importance of emergent properties of human cultural and adaptive systems.

The historical antecedents of today's phylogenetic models were clearly devel-oped around prevailing ideas of behavioral evolution (Trinkaus 1992a; Trinkaus and Shipman 1993a, 1993b). Since coincident (and hence seemingly parallel) change in human anatomy and archaeological assemblages is one of the over-arching patterns for the Late Pleistocene, phylogenetic models, at least those that attempt to address processes underlying the emergence of modern humans, must accommodate behavior change (see Harrold 1992). Until quite recently, contem-porary workers recognized the importance of behavioral evolution to understand-ing the processes involved in the origins of modern humans but dealt with it only in passing (e.g., Stringer and Andrews 1988b). Behavioral models are now com-ing increasingly to the fore; in some cases the role of behavior has been explicitly incorporated into phylogenetic models, while in other cases it has become more cloudy (see Clark 1989a; Frayer et al. 1994a; Gamble 1993; Stringer 1989a; Stringer and Bräuer 1994; Stringer and Gamble 1993, 1994; Wolpoff 1989a, 1989b). To many, the part of behavioral evolution in modern human origins has become increasingly hazy thanks to the Mousterian hominids of the Levant. There we have seemingly contradictory stories told by the morphological and archaeological evidence for adaptive parity between neandertals and early mod-ern humans.[3]

We might ask if a more reductionist (and/or atomistic) approach is the way out of the horse latitudes in which debate is now adrift, or whether an integrative, holistic orientation is required. Does the clear incorporation of ideas of behavioral evolution into phylogenetic models make these models more testable (by more clearly defining hypotheses about cultural or adaptive evolution that can be tested with archaeological data), or does it only serve to obfuscate simpler, inherently more testable phylogenetic models by adding new layers of complexity? Was behavioral and morphological evolution linked, and do phylogenetic analyses and behavioral models require one another? I contend that the perceived utility that some see in adopting a reductionist approach is a function of larger paradigmatic biases. Proponents of regional continuity have historically argued as autonomists, while adherents of replacement have typically behaved as provincialists.

Regardless of which stance one takes, the problem of modern humans origins specifically, and the field of paleoanthropology generally, is not suitable for reductionism. True reduction of paleoanthropology to biology is not possible because the former deals with subjects that cannot be addressed with theories pertaining to organic evolution, namely the transmission and evolution of culture. Atomistic reduction also is inappropriate. Statements of phylogenetic relationships are themselves statements about morphological, behavioral, and cultural evolution. To say, for example, that strong genetic continuity characterized the archaic to modern human transition in some areas is to say that in those regions archaic morphology and behavior evolved into modern morphology and behavior. Also, the ideological and historical bases of current phylogenies are intimately tied to notions of human behavioral evolution. Because behavioral models are *implicit* in phylogenetic models of the origins of modern humans, a clear separation of research realms is not only infeasible, it is impossible.

BEHAVIORAL PARADIGMS AND MODELS OF MODERN HUMAN ORIGINS

The historical bases of current models of modern human origins have been discussed at length (Brace 1964a; Graves 1991; Spencer 1984; Spencer and Smith 1981; Trinkaus 1982; Trinkaus and Shipman 1993a), and it is well understood that the precursors of these models were the products of the intellectual and social milieu of their day. The presapiens and neandertal phase models developed around the then-known fossil record in the context of implicit ideas about behavioral (primarily social) evolution (Trinkaus 1992a). Boule's (1911–1913, 1921) emphasis on the morphological differences between neandertals and modern humans implied a behavioral gap between the two that was in large part responsible for the ultimate replacement of one by the other. This, coupled with longstanding discomfort with the idea that so bestial a character as a neandertal could be directly and recently ancestral to us, formed the ideological basis of the presapiens theory (Brace 1964a; Smith 1992a; Spencer 1984; Spencer and Smith 1981; Trinkaus 1982). The alternative neandertal phase model was first articulated by Hrdlička (1914, 1927, 1930), whose critique of the presapiens view was phrased in behavioral, not morphological, terms (Hrdlička 1927). Modern incarnations of these models are themselves influenced by the social and political currents of a postwar world grappling with issues of racism and decolonization (Cartmill et al. 1986; Corruccini 1994; Graves 1991 [and associated comments]; Haraway 1988). Our own historical behavior has profoundly shaped our view of human prehistoric behavior.

The recent African origin and multiregional evolution models of modern human origins are embedded in different behavioral paradigms. Recent African origin is a dendritic model, descended from the presapiens theories of Boule and Keith (see Brace 1964a), which postulates a geographically restricted, recent origin of anatomically modern humans with subsequent global expansion and supplantation of indigenous archaic populations. This model predominately adheres to a morpho-typological approach, emphasizes clades in phylogenetic relationships, and stresses major biological and behavioral disconformity between archaic and modern humans (Trinkaus and Shipman 1993b). The multiregional evolution model, on the other hand, follows a population biology approach,[4] relies as heavily on the idea of grades as clades, and emphasizes gradual rather than punctuated shifts in cultural and morphological patterns across time and space. Multiregional evolution is a reticulate model, with its roots in the polycentric schemes of Schwalbe, Hrdlička, and Weidenreich (see Brace 1964a), which sees the emergence of modern humans as a pan–Old World change in behavioral and morphological grade. The behavioral paradigms that intertwine with evolutionary models condition our perception of the phylogenetic and morphological origins of modern humans, as well as determine what behavioral evidence we are likely to see as worthwhile and valid (Trinkaus and Shipman 1993a). Accordingly, study of biocultural evolution is intimately linked to resolution of the problem of modern human origins (Smith 1985, 1991).

NEANDERTAL PHASE, THE NEW PHYSICAL ANTHROPOLOGY, AND MULTIREGIONALISM

Brace's (1964a) resurrection of the neandertal phase model is unequivocally linked to ideas about behavioral evolution. By combining the unilineal, reticulate evolutionary schemes of Weidenreich (1943b, 1947) and Hrdlička (1914, 1927, 1930) with a Washburn-like (1951, 1960; Washburn and Avis 1958) emphasis on tool use as an evolutionary agent, Brace was able to propose a behavioral mechanism that would explain the anatomical shift from archaic to modern humans. Even Brace's definition of neandertals as "the man of the Mousterian culture prior to the reduction in form and dimension of the Middle Pleistocene face" (1964a:18) linked cultural and morphological definitional criteria.

The idea of a connection between human tool use and skeletal morphology antedates Brace and in fact was one of the linchpins in the emerging themes of the "new physical anthropology" and its intellectual issue, "man the hunter," in the 1950s and 1960s (see Haraway 1988). This idea finds its first breath in Darwin's *The Descent of Man* (1871), in which he proposed that the loss of large canine teeth in humans was due to their replacement by technological means of defense. While Darwin clearly recognized that structures reduced in size in animals under domestication (Darwin 1899), he lacked knowledge of the skeletal

gracilization that characterized Pleistocene hominid evolution. Thus Darwin never took the next logical step of arguing that modern human morphology was
attributable to self-domestication. This idea did emerge, however, as the fossil
evidence for an early Quaternary human existence became more abundant. Both
Gorjanović-Kramberger (1904) and Hrdlička (1911) expanded Darwin's hypothesis to argue that improvements in tool use led to a reduction in the size of the
teeth and jaws in humans, which brought about accompanying reduction of the
face and modernization of the cranium.

Washburn reiterated the role of tool use in hominization (1960; Washburn and
Avis 1958; Washburn and Moore 1974) as part of a larger adaptive complex
articulated as "man the hunter" (Haraway 1988). The dominant emergent theme
of Washburn's new physical anthropology was the integration of comparative
anatomy, with its roots in biological structuralism (dating to Aristotle—Moore
1993; Wake 1991), with the neo-Darwinian synthesis, leading to an emphasis on
functionalist and adaptationist explanations (Haraway 1988; Straus 1991). Nowhere was this program more manifest than in Washburn's thinking about cultural evolution. Tool use was not only the defining behavioral characteristic of
humanity, it had *driven* human evolution, so much so that Washburn could assert
that "tools makyth man" (Washburn and Moore 1974:61).

The logical sequel of tools makyth man is that *man makyth himself.* By
making and using tools that then alter the evolutionary forces operating on us,
humans became *participants* in their own evolution. This is an emergent property
that could not have been predicted from the laws of organic evolution. This is
also an idea that has held powerful sway in the minds of several generations of
paleoanthropologists (e.g., Tobias 1981; Gowlett 1987). Haraway sees as the
primary myth of "evolutionary scientific humanism" (i.e., Washburn's new
physical anthropology) the idea that:

> Culture remakes the animal; the persistent western dualism of nature and culture
> is resolved through a self-making productionist dialectic, providing a universal
> foundation of human unity. Man is his own product; that is the meaning of a human
> way of life. Both mind and body become the consequence of a primary adaptive
> shift registered in the bones and muscle, which are signs, literally, of a special way
> of life. (1988:223)

While intuitively pleasing, the "tools makyth man" concept lacked a mechanism that would account for morphological change following behavioral evolution. Brace (1963, 1964b) provided that mechanism in the form of the probable
mutation effect—an idea that has found little support among either anthropologists or geneticists (e.g., Brues 1966; Holloway 1966; Prout 1964; Wright 1964).
Regardless of the veracity of probable mutation, this was a clear effort to account
for patterns of hominid gracilization both in terms of coinciding developments in
lithic technology and with reference to microevolutionary processes important to
the new synthesis (which only then—in the 1950s and 1960s—was starting to

have an impact in physical anthropology (Cartmill et al. 1986). Brace's earlier writings focused on craniofacial reduction, linking increased tool efficiency and cooking in the Upper Paleolithic with relaxed biomechanical loads on the face and cranium, thus allowing an accumulation of mutations in the genes that had been maintaining facial robusticity (Brace 1962a, 1962b, 1963, 1964a). Later he extended the model to the postcranial skeleton (Brace 1991a, 1991b, 1991c), emphasizing Upper Paleolithic innovations in resource extractive technology. Implicit in Brace's view is the idea that morphological grades in hominid evolution are coeval with behavioral grades—that cultural and physical evolution cannot be separated.

Subsequent treatments of the "technology/gracilization" model either make no reference to the underlying biological mechanism leading to skeletal reduction (e.g., Brose and Wolpoff 1971) or shift emphasis to a passive selection model of gracilization (e.g., Smith 1985). Thus most recent treatments invoke a metabolic load effect, whereby relaxation of stabilizing selection that had maintained metabolically costly tissues (bone and muscle) is tantamount to directional selection to reduce those tissues. Reduction of robusticity is then the inevitable outcome of technological development—more efficient tools reduced the need for strength in daily activities, relaxing selection pressures that had maintained muscular hypertrophy and skeletal robusticity. This idea continues to be invoked as the leading candidate for gracilization (Harrold 1992; Jelinek 1994; Smith 1985, 1991; Trinkaus 1983, 1986).

Weidenreich (1947) was hard-pressed to explain the seemingly parallel emergence of the modern human morphotype in disparate regions of the Old World under his polycentric model of human evolution. He fell back on a vaguely defined notion of orthogenesis, one that was more a reiteration of the empirical observation of rectilinear evolution (assuming multiregional evolution) than of predestination. Weidenreich was forced to admit that the mechanism was a mystery (1947:234). Following the metabolic load effect, the emergence of new technology and behaviors could have resulted in rapid morphological change at the archaic to modern human boundary (Smith 1985). If the morphological contrast between archaic and modern humans is merely one of culturally mediated robusticity, then much of evolution in the genus *Homo* was a function of hominids becoming technologically more efficient without necessarily changing behavior patterns (hence an argument for *behavioral continuity* in the context of technological progress can be made). Furthermore, developmental canalization may have resulted in convergence on modern morphology in disparate populations undergoing gracilization, even though selection may have been operating at different genetic loci across the groups (Smith 1985). A shared *Bauplan* in members of the genus *Homo* makes reversion to developmental default states under relaxed selection not only possible, but likely (Wake 1991). Thus the essentially independent origin of modern human morphology in different regions of the Old World (as indicated by the multiregional evolution model) can be explained by technological evolution and diffusion without reference to ortho-genesis (Smith 1985; Wolpoff et al. 1984, but see below).

The behavioral paradigm that embraces the multiregional evolution model is a culturally deterministic one. It emphasizes the emergent properties of human culture, holding that human evolution cannot be understood with reference to evolutionary theory alone. A complete account must include the special effect of culture on the transformation of morphology. Paleoanthropology, by this view, cannot be reduced to biology. This paradigm stresses adaptive, gradual change, with behavioral preceding morphological change. It is a progress view, with an implicit cognizance that the capacity for modern behavior is very ancient, and all that was needed was time for it to develop (perhaps a holdover from the ortho-genic views of Weidenreich—see Bowler 1989). An additional, and important, theme of this paradigm is the idea that archaic and early modern humans were not separated by any substantial behavioral gap, thus allowing for behavioral conti-nuity across the transition (e.g., Agogino 1964; Clark and Lindly 1989a; Lindly and Clark 1990a; Pope 1994).

PRESAPIENS, DETERMINISM, AND REPLACEMENT

The first formulations of presapiens thinking by Schwalbe, Boule, and Keith divulge a tacit notion of behavioral primitiveness in neandertals. The predomi-nant view among late nineteenth and early twentieth century European scholars, including even antievolutionists such as Virchow, was that the neandertals were peculiar, bestial, and not befitting ancestor status for modern Europeans. While overt statements of this nature have fallen out of favor (and in fact were consid-ered "reactionary and essentially racist" in the 1960s: Cartmill et al. 1986:418; see also Corruccini 1994), the conceptual underpinning has persisted in recent incarnations of replacement models.[5] The recent African origin model is based on behavioral inequality between late archaic and early modern humans (e.g., Foley 1987b; Stringer and Andrews 1988b). This superiority could have been something as simple as enhanced long-term memory, leading to greater planning depth, information transfer, and social complexity in early moderns (Stringer 1991a). Despite criticisms of the behavioral superiority pretext of replacement (Clark and Lindly 1989a, 1989b; Lindly and Clark 1990a, 1990b; Marks 1994; Thorne and Wolpoff 1992; White 1990; Wolpoff 1994), this idea has become the focus of recent didactic profferings of the recent African origin model (e.g., Gamble 1993; Stringer and Gamble 1993, 1994).

Brace's (1964a, 1989) assertions that replacement models embodied cata-strophic ideology have been supplanted in recent years by claims, largely from detractors, that they are reliant on colonial metaphors and egregiously see prehis-tory as the extension of recent human sociohistorical interactions into the distant past (Bowler 1991; Clark and Lindly 1991; Graves 1991; Wolpoff 1994b; see also Corruccini 1994). This type of thinking is an outgrowth of the marriage of nineteenth century progressive evolutionary models (see Trinkaus and Shipman 1993a) with the culture history emphasis of European prehistorians (see Binford 1987; Foley 1987b; cf. Clark 1989a). While the impact of cultural and political trends is clearly evident in the parallels between imperialist expansion and the recent African origin model, the influence of recent intellectual currents in biology and paleontology are less often identified. The behavioral models underlying the recent African origin model also originate in part from (or at least are harmonic with) the rising popularity in the 1980s of cladistics as a taxonomic method and punctuated equilibria as an allied evolutionary process (Cartmill et al. 1986). Peripatric speciation (associated with the establishment of new adaptive advantages) followed by dispersal with replacement of the parent species has been argued to be a common theme in evolutionary history (Eldredge and Gould 1972; Stanley 1975). Hence no special pleading is required to understand the evolutionary origins of modern humans. They can be understood in terms of macroevolutionary principles in the domain of zoology, and paleoanthropology can be reduced to evolutionary biology.

The behavioral paradigm associated with recent African origins is thus based on a colonial metaphor coupled with an emphasis on cladogenic origin of new taxa. Historically, imperialist expansion was based on technocultural inequality, and by analogy the recent African origin model demands similar inequality between late archaic and early modern humans. The paradigm sees behavioral change as exaptive and dynamic (Foley 1987b; Groves 1989b; Stringer and Gamble 1993, 1994), rather than adaptive and gradual as seen by proponents of multiregional evolution. Also evident are shades of biological determinism, generally emphasizing the evolution of intelligence as a releasing event that allowed the expansion of moderns out of Africa and their replacement of intellectually inferior indigenous populations. This kind of deterministic thinking about hominid evolution overall was most apparent in the writings of early workers such as G. E. Smith (see Bowler 1989; Landau 1991), but it also emerges in varying potencies in recent works (Foley 1987b, 1989; Groves 1989b [arguably the avatar of this position]; J. D. Clark 1993).

THE MEANING OF MORPHOLOGICAL CONTRASTS

Irrespective of phylogenetic models, there is a growing literature both in archaeology and paleontology claiming significant behavioral differences between late archaic and early modern humans. Arguments for behavioral contrasts have been contested (e.g., Binford 1984, 1985, 1989 vs. Chase 1986, 1988, 1989; Lindly and Clark 1990a [and associated comments]; Trinkaus 1986, 1989 vs.

Tillier 1989), yet these same arguments have been readily accepted as support for replacement scenarios. Some workers have even been accused of manipulating the behavioral evidence for phylogenetic ends (see, e.g., comments associated with Gargett 1989). Nowhere has the interaction between phylogenetic models and behavioral interpretations been more apparent than in discussion concerning early modern humans associated with Middle Paleolithic industries. Modern human fossils have been recovered from Middle Stone Age contexts in South Africa (the most secure case being Klasies River Mouth—Singer and Wymer 1982; Rightmire and Deacon 1991) and from levallois Mousterian contexts in Israel (Skhūl and Qafzeh caves—McCown and Keith 1939; Vandermeersch 1981b). Lithic and faunal similarities suggest broad-scale behavioral resemblance between these Middle Paleolithic moderns and late archaic humans (but see Lieberman 1993; Lieberman and Shea 1994). If there was some inherent behavioral superiority in the earliest modern humans, it is not clearly evident in the archaeological record. However, the important morphological differences between the Levantine early moderns and neandertals in the postcranial skeleton, involving robusticity, body proportions, and joint shapes and orientations, prompt some to argue for important behavioral contrasts between these groups (Ben-Itzhak et al. 1988; Trinkaus 1992b, 1993a). This has led to a situation where multiregionalists, preferring to see behavioral similarity (and hence behavioral continuity) between groups, have emphasized the archaeological similarities (Lindly and Clark 1990a; Marks 1994), with some even rejecting the functional-morphological paradigm underlying the morphological evidence for behavioral contrasts (Wolpoff 1989a; Wolpoff et al. 1991). Others have backed off from the argument that behavior change played a focal role in the origin of modern morphology (Smith 1985 vs. 1994). Supporters of recent African origin, on the other hand, prefer to see behavioral discontinuity, leading them to accept the morphological evidence and reject the archaeological evidence for adaptive equality. In a statement reminiscent of one made by Vince Sarich (1971b) more than two decades ago, Stringer has claimed that, given the veracity of the assumptions behind the functional-morphological paradigm, "then there *must* have been significant behavioral differences between the Skhūl-Qafzeh hominids and the neanderthals, whatever the lithic remains are supposedly saying" (1990b:249; emphasis in original).

Interestingly, while advocates of recent African origins accept the utility and validity of functional-morphological arguments, their behavioral paradigm is not dependent on the form-function paradigm for support. With a biological deterministic mindset, the recent African origin model is comfortable with the idea that modern behavior and morphology evolved independently. As long as these two traits ended up in a single "modern package," gracility could have hitchhiked along with adaptively superior behavior (Stringer 1992a). Since these things didn't necessarily evolve in unison, the gracility of the Skhūl/Qafzeh hominids in the absence of archaeological evidence for modern behavior is not problematic

(Mellars 1990; Stringer and Gamble 1993, 1994). Additionally, arguments have surfaced that the modern advantage may have been demographic (e.g., shorter interbirth intervals or lower mortality) and not behavioral (Zubrow 1989). It is, however, more appealing to proponents of recent African origins to reject the archaeological evidence (or to claim that key early behavioral shifts were archaeologically invisible—Stringer 1991a) and to see the morphological differences as reflecting behavioral differences. Nevertheless, the onus of identifying the adaptively advantageous features of modern humans, whether behavioral or demographic, still falls on adherents of this model (contra Stringer and Gamble 1994:46).

Ironically, the behavioral underpinnings of the multiregional evolution model are dependent upon the form-function paradigm—the very paradigm called into question by some multiregional evolution defenders. Although derived from Brace's technology/gracilization model, the behavior-morphology relationship envisioned by Brose and Wolpoff (1971) differed in that the key behavioral shifts were to be sought in the Middle Paleolithic, not at the transition. They identified this shift as increased production of more-efficient, single-purpose tools—reducing dental loading and robusticity—along with improved climatic buffering in the late Middle Paleolithic. The effect of behavior change on skeletal morphology, largely through natural selection, was gradual. Thus one would expect a considerable lag time between behavior change and its morphological manifestations, such that the emergence of modern form does not relate directly to the archaeological transition from Middle to Upper Paleolithic. Still, behavioral and anatomical evolution are clearly linked. The message here is that the traditional typo-technological transition is not the behaviorally meaningful one (at least with respect to anatomical change). In the Levant we find modern morphology in advance of increased use of task-specific tools, forcing revaluation of the traditional behavioral paradigm of multiregional evolution.

THE SABOT IN THE MECHANISM: SKHŪL-QAFZEH AND THE FORM-FUNCTION PARADIGM

The abandonment of old ideas about hominid-culture relationships (which, incidentally, began in response to the articulation of the technology/gracilization model by Brace [1964a]—e.g., Brothwell 1964; Genovés 1964; Müller-Beck 1964) has both thrown doubt on the validity of the functional-morphological (form-function) paradigm and left the role of behavioral evolution in the origins of modern humans open to question (J. D. Clark 1993; Dibble and Chase 1990; Gowlett 1987; Harrold 1992; Marks 1994; Tillier 1989). Rather than seeing the mismatch of morphology and technology as a dilemma that must be explained away, I prefer to see it as an opportunity to examine our conceptual biases and identify lines of future research.

The multiregional evolution paradigm has, until recently, embraced the form-function paradigm, perhaps overzealously as a deus ex machina, to explain the widespread origins of gracile modern human morphology. Possibly the clearest explication of the relationship of these two paradigms comes from Harrold's (1992) delineation of the Morphology-Behavior-Archaeology (MBA) model. This model holds that both hominid morphology and archaeological remains are monitoring, at some level, hominid behavior, and thus there should be a fundamental concordance between these two lines of evidence. The underlying acceptance that morphology reflects behavior is an expression of faith in the form-function paradigm. The apparent incongruity in these records in the Middle Paleolithic of the Near East and Africa represents a problem either for the MBA model or for the functional-morphological paradigm itself. Discerning which requires a clear idea of the difference between the model and the broader paradigm.

The functional-morphological paradigm derives from what can best be called biological structuralism, with its emphasis on "organism-level phenomena such as the generation, self-organization, and transformation of specific form" (Wake 1991:543). Functional morphology has a great antiquity, beginning with Aristotle's doctrine of final cause, and was greatly popularized in *Form and Function* by E. S. Russell (1916). The paradigm finds expression in the works of Darwin, D'Arcy Thompson, and Keith (see Armelagos et al. 1982). Functional morphology was an integral part of phylogenetic systematics in the early decades of this century (Fleagle and Jungers 1982), but the two areas of inquiry have since become divorced under the assumption that non-functional, "neutral" traits, being less subject to homoplasy, are better suited for phylogenetic reconstruction (an idea that dates to Darwin [1859]). The "new physical anthropology" saw a renewed interest in form-function studies, largely owing to Washburn's focus on behavior, form and function, and functional complexes. The formative work in this paradigm was entirely empiricist, involving the generation of adaptive stories made to accommodate the perceived morphological evidence—an approach rightfully criticized as the "adaptationist programme" by Gould and Lewontin (1979—see also Clutton-Brock and Harvey 1979). In biology, the shortcomings of the adaptationist program have been slowly redressed both by developments in the field of teleonomy (e.g., Curio 1973) and by the increased influence of logical positivism (Schlick 1953; Woodger 1952) and Popperian principles (Popper 1958) on the biological sciences since the 1950s. This has led to an increased amount of work establishing the relationship of behavior to morphology through actualistic studies of uniformitarianist principles, analogous to the "middle range research" of archaeologists (Binford 1983). Exemplars of this work (with respect to the relationship between activity and skeletal robusticity) include Goodship et al. (1979), Lanyon et al. (1979), Woo et al. (1981), and Lanyon and Rubin (1984). At its current state of development, the form-function paradigm relies heavily on the assumption of uniformitarianism in bone remodeling and degenerative processes (Trinkaus 1992a), and on the assumption that natural selection operates to maximize mechanical design efficiency over evolutionary time (e.g., Smith and Savage 1956). While the conceived relationship of form to function varies among practitioners (see Cartmill 1982), there is a unifying assumption that behavioral change or pressure to change functional modalities precedes morphological evolution; hence the latter is a clear reflection of the former.

Some biases and problems inherent in the form-function paradigm include an emphasis on the postcranial skeleton as the most appropriate unit of analysis (and the converse idea, that the cranium is best suited for phylogenetic studies: Armelagos et al. 1982), a near exclusive emphasis on selection as the major agent of evolutionary trends (e.g., Bielicki 1975), and a still too ready tendency to generate inferentially weak post hoc arguments (and in human paleontology especially, ad hoc arguments) that rely entirely on the plausibility of their warranting arguments (Binford 1981; Clark 1988; Clutton-Brock and Harvey 1979) or that, by their very made-to-order nature, cannot fail (see Gould and Lewontin 1979). Post hoc explanations are an unavoidable part of the development of observational sciences (Clutton-Brock and Harvey 1979), and the widespread use of multivariate, pattern-recognition data analysis methods in human paleontology should be enough to convince anyone that we are indeed still in the formative stages of our discipline. However, the practice of coupling a priori expectations (derived from paradigmatic biases) with post hoc reasoning is a deadly one.

The repudiation of the form-function paradigm by some multiregionalists (Wolpoff 1989a; Wolpoff et al. 1991) comes at a time when basic experimental research is strengthening our ability to make meaningful behavioral inferences from skeletal material (see reviews in Ruff 1992; Trinkaus 1992b). This creates the schizophrenic situation of rejecting as untenable a paradigm whose inferential basis is increasingly supported by empirical positivist approaches. The denial of a form-function explanatory framework concerns physical evidence (given new chronological frameworks for Levantine and African early modern humans) that no longer fits with a preconceived paradigm of behavioral evolution. It is not the form-function paradigm that fails (since it has not been firmly established that behavioral differences did *not* exist between the relevant taxonomic groups), rather it is the MBA model that does. This case challenges the notion that both archaeology (especially in the form of artifact typologies) and anatomy are providing comparable information about behavioral evolution—a dubious assumption to begin with. Given that experimental and comparative work strongly supports the form-function paradigm, we might profitably try to identify the operative principles causing the disparity between these two lines of evidence.

Assuming the form-function paradigm is valid, the morphological differences between Middle Paleolithic archaic and modern humans (see Clark and Lindly 1989b for a review) strongly suggest significant behavioral differences between these groups (Trinkaus 1992b, 1993a). The role of developmental plasticity in shaping adult morphology, because it results in morphology that relates to the habitual behavior of the individual (rather than the selection history of the population) and is thus a source of fine-grained behavioral data, is increasingly seen as important to understanding these differences. Cortical bone tissue in the shafts of long bones, for example, is highly adaptable to changing load intensities and patterns. Studies of upper and lower limb bone cross-sections in recent people report a high concordance between morphology and economic behavior patterns (see review in Ruff 1992). Since these plastic aspects of skeletal morphology should be monitoring behavior at the level of habitual subsistence and technological activities, we would expect a concordance between them and archaeological monitors of the same things.

The most obvious explanation for the failure of the MBA model is that the archaeological record is reflecting behavior at a different level. According to Clark (1989a:154), assemblage composition, site layout, settlement patterns, and subsistence practices are "the organizational properties of past systems that monitor adaptation." Artifact typologies alone are likely not monitoring adaptation; thus inferences of behavioral similarities or differences between groups based on typological comparisons are questionable at best (see Chase and Dibble 1990; Clark and Lindly 1989b; J. D. Clark 1993; Harrold 1992; Lindly and Clark 1990b; Pike-Tay 1990; Trinkaus 1986). In addition, given that typologies are based on arbitrary distinctions (arbitrary with respect to the way the tools were conceptualized and used; it is merely assumed that a correlation exists between typology and technology—Clark 1989a), and that Middle and Upper Paleolithic assemblages are understood by reference to different typological systems, typological-based assessments of adaptive evolution are especially susceptible to paradigmatic bias (Clark 1989a; Clark and Lindly 1989a). Thus significant behavioral shifts may have occurred with the origins of modern humans that impacted subsistence efficiency but are not necessarily reflected in the typologically defined artifact record (Harrold 1991a; Mellars 1990). Behavioral innovations that are invisible to traditional modes of artifact analysis may have greatly affected hominid interpopulational dynamics, as well as having had a large impact on the subsequent course of morphological evolution. This suggests that the obituary for the form-function paradigm is premature, and that we may be well advised to redouble our efforts at doing the middle-range research necessary to strengthen our inferential models and to resolve the issue of which line of evidence is most fruitfully informing about Pleistocene hominid adaptation (see Lindly and Clark 1990a vs. Trinkaus 1990). Robusticity and morphological indicators signal a behavioral shift with the appearance of early modern people in the Near East and Europe (Ben-Itzhak et al. 1988; Churchill 1994; Trinkaus 1992b, 1993a). Some archaeologists (e.g., Clark 1992a) claim that a smooth transition from Middle to Upper Paleolithic is evident in archaeological monitors of adaptive behavior. This discrepancy is where we should be focusing our attention.

Harrold's (1992) Morphology-Behavior-Archaeology model fails in the case of the Levantine Middle Paleolithic. Tests of MBA-type relationships in other regions and time periods have also met with equivocal success (e.g., Frayer's [1981, 1989] test of Brues's [1960] "spearman and archer" model; see also Churchill 1994). This is because MBA-type explanations have relied too heavily on empiricist modes of generating post hoc explanations. For example, Harrold (1992:220) provides a list of morphological correlates of suggested cultural developments. These are not, however, deductively derived predictions from the form-function paradigm. They are instead a priori observations from archaeology which, in conjunction with prior knowledge of morphological evolution, gave rise post hoc to the model (see Tillier 1989). Since they are accommodative, they fit the data. Wolpoff's (1989a; Wolpoff et al.'s 1991) rejection of the form-function paradigm is, in the first place, a crisis of accommodation.[6] What is really being rejected is the MBA model, not the overarching functional-morphological paradigm. In the second place, this is not a global rejection of functional morphology, but is instead a qestioning of the applicability of this approach to between-group variation within the genus *Homo* (while it might still be valid across higher taxonomic levels).

Inductive, accommodative modes of model building (as with the MBA model) are problematic not only because we cannot then test the models with the only data at our disposal (those used to construct the model), but also because it eliminates the potential hypothesis-testing value of having two semi-independent data sets (morphology and artifacts) that reflect to some extent the same thing (behavior) (see Gould and Lewontin 1979). Our ability to play the two data sets off of one another in the future will depend, as well, on adequate testing of proposed morpho-behavioral and archaeo-behavioral relationships using data from the modern world; in other words, we need to do the middle-range research necessary to build an inferential base.

PROVINCIALISM VS. AUTONOMISM: DO IDEAS OF BEHAVIORAL EVOLUTION BELONG IN PHYLOGENETIC MODELS?

The question of how to deal with behavioral evolution in phylogenetic models revolves around our perception of the nature of the event itself. Is "the problem" of modern human origins really a series of problems, each capable of being understood in isolation (Wood 1994), or can phylogenetic, adaptive, and cultural evolution be understood only in the context of one another? In this chapter I have made the point that behavioral models, no matter how much people argue they are separate, are inherent parts of phylogenetic schemes. Paleoanthropology cannot be atomized, and it cannot be reduced to biology. Our field started as an area of inquiry clearly within the realm of biology. Darwin advanced a single, unified theory to account for the origins both of humans and of biological diversity generally (Rosenberg 1985), such that all life, human or otherwise, could be explained by the same natural principles:

> The most important item on the early Darwinian agenda was not to elaborate what we would now describe as a theory of human evolution (i.e., a particular historical explanation of human origins), but almost the opposite—to show that human traits required no extraordinary explanation but were predictable and natural results of the general evolutionary process (Cartmill et al. 1986; Bowler 1986). (Cartmill 1990:175)

Paleoanthropology began to distance itself from the theoretical framework of biology in the years after World War II, ironically in large part because of increasing incorporation of Neodarwinian principles into fossil studies. The new synthesis stressed adaptation—a common chorus among practitioners of the new physical anthropology—resulting in the eventual replacement of natural-law explanations with an adaptive-shift model of human emergence (Cartmill 1990). This model sees the human career as beginning with an adaptive shift centered on tool use, such that hominid origins was the result of a singular historical occurrence (Cartmill 1990). This shift of focus forces an autonomistic position on paleoanthropology. An emerging dependency on technology in the genus *Homo* could not have been predicted by natural laws, and thus human origins cannot be a province of biology. As pointed out by Cartmill (1990; Cartmill et al. 1986), the adoption of an adaptive shift model also had the consequence of divorcing paleoanthropology from science (since contingent historical coincidences are outside the purview of science). This situation is the result of an inappropriate emphasis on human uniqueness, a condition that does not automatically preclude a reunion with the field of biology. The idea of unique or emergent properties is dependent upon hierarchical concepts—that entities at higher levels have emergent properties neither exhibited by nor predictable from lower-level entities (Ayala 1985). While hierarchies conceptualized in biology are usually those running from atoms to cells to individuals to species, the autonomist perspective in human paleontology owes to a view of humans, with emergent properties such as culture and intelligence, as occupying the upper levels on the hierarchical great chain of being. Claims of hierarchy and emergence as grounds for epistemological autonomy have been argued to be invalid (Ayala 1985). A better case for autonomy, or for only a partial reduction of paleoanthropology to biology, is that our discipline fails Nagel's (1961) two necessary and sufficient conditions for reduction of one science into another. While the emergence of cultural behavior can be addressed with ecological theory, the transmission and evolution of cultural systems do not obey genetic laws of inheritance or follow microevolutionary processes (see Clark 1989a). Hence paleoanthropological laws and theories cannot be derived as logical consequences of the laws of organic evolution (Nagel's condition of derivability). Likewise the language of archaeology cannot be redefined in the language of population biology (contra Foley 1987b); thus the condition of connectability fails as well (Nagel 1961).

Reduction is not possible, but integration of paleoanthropology and biology certainly is. I would argue, however, that dividing a complex research question into smaller, more manageable questions (the "atomistic reductionism" of Wood 1994) runs counter to this effort. On one level, adaptive evolution can be understood without reference to phylogenetic relationships between the relevant actors. The commonly used taxonomic nomina "archaic" and "modern" human can be used in the sense of grade designations (Brace 1991b). Archaic and early modern human adaptations can then be explored regionally without reference to what came before or after. This is, however, akin to a culture history approach to behavioral evolution (i.e., descriptive without concern for process). If we want to understand the processes leading to the evolution of modern biobehavioral characteristics, and hence make our science relevant to the broader evolutionary sciences, we must sort out phylogenetic relationships. Change in behavior in specific regions can only be understood with knowledge of the relationships of temporal groups to one another (see Wolpoff 1989b). Understanding the evolutionary processes that produced human behavioral shifts is, to me, the more interesting question, and one that requires a holistic approach. The same can be said of the need of phylogenetic studies for adaptive models. At a base level, phylogenetic relationships can be entirely understood by examining nonadaptive, neutral traits with phenetic and cladistic methodologies. But again, the evolutionary processes behind the phylogenies thus produced can only be understood with respect to the behavior of the actors. To paraphrase Clark (1989b:449), since we claim to subscribe to the paradigm of evolutionary biology, with an emphasis on mechanisms of change, we should be concerned with why change was taking place, not simply with the end product of these events. Ironically, to integrate paleoanthropology with evolutionary biology fully—in other words, to achieve the kind of theoretical integration that is the ultimate goal of reductionism—we must embrace a holistic approach. Otherwise, we will continue to generate evolutionary explanations that are unique to the genus *Homo,* defy testing, and are of little interest to the field of biology.

It is time for the different phylogenetic models of the origins of modern humans to make explicit the role of behavioral evolution in that process. I agree with Frayer et al.'s (1994b) claim that the greatest hope for resolution of the origins of modern humans debate lies in how fully contradictory the models are, and that conflicting interpretations of the data should not be minimized, but scrutinized. In this vein, explicit behavioral models, fully imbedded within the phylogenetic models, can serve to deduce theories whose predictions can be tested with the archaeological and functional morphological data sets. This sort of deductive approach has been sorely lacking in paleoanthropology (Clark 1989a, 1989b). Only in this way can evolutionary processes be fully articulated. There is a historical trend for developing sciences to shift from a narrow focus on substances—the end results of natural processes—to a broader concentration on the processes themselves (Brace 1991b). Paleoanthropology has taken only a few tentative steps down this path. For all the interest behind the fate of the neandertals and where modern humans came from, we need to focus on the processes that produced them, and behavioral evolution was undoubtedly a key component in those processes.

ACKNOWLEDGMENTS

I am very grateful to Geoffrey Clark and Cathy Willermet for inviting me to participate in this project. My thanks to Bob Franciscus for suggesting reductionism in modern human origins research as a topic for this chapter. I am also grateful to Trent Holliday, Garnett McMillan, and Erik Trinkaus for helpful comments on an early draft.

NOTES

1.   While logical positivism, Popperianism, and empiricism generally came under attack in the 1950s and 1960s (e.g., Quine 1961), leading to their abandonment among philosophers of science, their influence in the sciences, especially in biology and the human sciences, lives on.

2.   "Reductionism" is used with different senses among paleoanthropologists. Foley (1987a) follows the traditional usage in the sciences (e.g., Ayala 1985; Rosenberg 1985), meaning that the laws and theories of a given field can be subsumed within and explained by reference to (i.e., "reduced to") the laws and theories of another, broader field (e.g., Watson and Crick's discovery of the biochemical structure of DNA rekindled efforts to reduce biology to physics and chemistry). Foley's "methodological reductionism" (1987a:5–6) calls for the reduction of paleoanthropology to biology generally, perhaps with an emphasis on the fields of evolutionary ecology and zoological systematics. Foley's attempt to use hominid cultural behavior, considered by most anthropologists to be an emergent property, as a systematic trait can be construed as a reductive effort to connect archaeology with systematics (Foley 1987b).

Wood (1994) follows Levins and Lewontin (1985) in using reductionism to indicate a strategy of dismantling a subject of study into component parts that are easier to approach analytically or conceptually. This approach can best be characterized as *divide et impera* with the goal of synthesis only after all of the constituent parts have been fully apprehended. While these two uses of reductionism differ, they are not incompatible strategies. The atomistic approach favored by Wood may work to divide a complex subject that is irreducible (sensu Foley) in its entirety to a series of problem areas which can themselves be studies by varied disciplines (e.g., culture change by cultural models, phylogenetic evolution by macroevolutionary models).

3.   Whether or not one accepts specimens from Skhūl and Qafzeh as anatomically modern, their postcranial skeletons do contrast in robusticity and aspects of morphology with those of Near Eastern neandertals (Trinkaus 1992b, 1993a).

4. The recent African origin model emphasizes the macroevolutionary events of origins of a new taxonomic group *(Homo sapiens sapiens)*. Proponents of punctuated equilibria have traditionally argued that macroevolutionary events *cannot* be explained by microevolutionary processes (see Ayala 1985). Champions of multiregional evolution, on the other hand, stress more the microevolutionary processes (mutation, drift and adaptation to local conditions by natural selection [all causing regional differentiation], balanced by gene flow [preventing speciation]) that culminate in macroevolution, the phyletic origination of new taxa.

5. I have purposefully avoided discussing demic diffusion models (such as Bräuer's Afro-European sapiens model—Bräuer 1984b), because the behavioral foundations of these models are harder to recognize. I consider them replacement models, since modern (more adaptive) behavior ultimately "replaces" archaic behavior.

6. And a crisis it is, since it leaves the multiregional evolution model without ties to behavioral mechanisms that help explain the Old World–wide emergence of gracile modern human morphology.

16

Perspectives on Neanderthals as Ancestors

DAVID W. FRAYER

European Neanderthals represent a remarkable series of fossils in the human record—remarkable in their morphology and variation across time and space, in the cultural/spiritual remains they left behind, and in their persistence through some of the coldest (and warmest) parts of the Würm period in Europe. They are also remarkable in maintaining interest and intrigue among the scientific and lay community, given the long-held view of their irrelevance to living humans occupying the Old World west of India. While Neanderthals are probably second only to dinosaurs as fossils recognized in the mass media, they have acquired a completely different "public" image. For example, there has never been a "lovable" Neanderthal on American television like the adorable purple dinosaur, Barney; just the opposite, since in certain contexts there are few things worse than to be called "a Neanderthal" or to be accused of "acting like a Neanderthal." Besides the negative connotations, the popular and professional opinion is that Neanderthals were evolutionary failures leaving no descendants and, therefore, no genes in modern European populations. Thus, while being a "dinosaur" is not necessarily always a compliment, they at least are suspected of having left some descendants in the modern world, while Neanderthals have been shoved to the edge of the human tree like an old bachelor uncle or a spinster aunt. Nearly three decades ago Brace (1964a, 1968) documented much of this thinking, arguing that from the moment of their discovery Neanderthals have been consistently and unjustly ridiculed and rejected. Others on both sides of the replacement vs. continuity issue (Klein 1989b; Radovčić 1985; Spencer 1984; Stringer and Gamble 1993) show this attitude continues to the present. It is my contention that Neanderthals from Europe have been unfairly driven from the human tree, and that, while different from the humans who followed them, Neanderthals represent the most likely ancestors of "modern" Europeans.

Like others who have studied the problem of Neanderthals and modern human origins, I bring to the table my own bias and perspective. Besides my graduate work at the University of Michigan and a long-term collaboration with Milford Wolpoff, two important experiences have influenced my thinking on the phylogenetic place of Neanderthals. The first is my past and continuing work with

European Upper Paleolithic, Mesolithic, and Neolithic human skeletal material. Based on my analysis of dental trends over these periods in western and central Europe, I have been able to document substantial evolutionary changes within the Upper Paleolithic (Frayer 1978, 1984) and more recently from the Mesolithic to the Neolithic (Frayer n.d.). My observations that evolutionary changes have produced greatly reduced human dentitions in the so-called modern humans have influenced my perspectives about the relationship of the Neanderthals and the Upper Paleolithic people who followed them. The second important influence relates to the end of my first trip to Europe in the spring of 1974, when I met Jan Jelínek and studied the collections in the Moravian Museum in Brno. I was especially interested in visiting the former Czechoslovakia, which in those days of harsh socialism was sometimes a difficult, complicated destination. But besides studying the fossils housed in Brno, I wanted to meet Jan Jelínek, who had just published a major paper in *Current Anthropology* (1969) on the central European material. My experiences there, both with Jelínek and with the fossils in the Moravian Museum, had a lasting effect on my thinking about the role of Neanderthals in human evolution.

While much of the Paleolithic skeletal material in the Moravian Museum (especially the human remains from Predmostí) was destroyed near the end of World War II in a fire at Mikulov Castle on the Austrian border, an incomplete cranium from Mladeč had been rescued from the charred remains by Jelínek and his colleagues. This specimen (Mladeč 5) suffered only minor damage from the fire. It was not then widely known in the literature, and seeing it for the first time had an immediate impact on my thinking about the relationship between Mousterian and early Upper Paleolithic people. Mladeč 5 was found in a cave complex in Moravia in 1904. It was associated with early Aurignacian artifacts and, although still not radiometrically dated, undoubtedly derived from the early Upper Paleolithic based on the distinctive stone and bone tools. Mladeč 5 shows some truly distinctive Neanderthal features, such as its low angular skull, lambdoidal flattening, and a distinctive occipital bun. These features are illustrated in Figure 16.1 in which Mladeč 5 is compared with the Neanderthal cranial vault from Belgium, Spy 2. In the same figure is a lateral view of the Qafzeh 9 vault, which many consider to be representative of the likely ancestral population from which Upper Paleolithic Europeans originated. Mladeč 5 looks nothing like the Qafzeh 9, which possesses a high rounded skull with no lambdoidal flattening and not even a hint of occipital bunning. While there is more evidence than just Mladeč 5 and the comparisons shown in Figure 16.1, it is apparent to me in 1995 just as it was in 1974 that Neanderthals must have had some relationship to the early Upper Paleolithic Europeans. Otherwise, specimens like Mladeč 5, if uniquely descended from Qafzeh 9–like populations and unrelated to populations represented by specimens such as Spy 2, would have had to develop many of the same features that are commonly found in European Neanderthals. W. W. Howells (1974) once observed that Upper Paleolithic humans are "instantly recogniz-

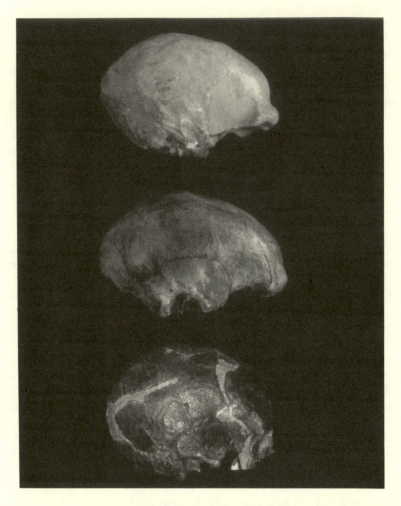

Figure 16.1. Lateral views of Spy 2 (top), Mladeč 5 (middle), and Qafzeh 9 (bottom). All specimens are casts, and the face of Qafzeh 9 has been covered with black cloth to make the views more comparable.

able as anatomically modern." I have always disagreed with this observation and to the contrary concluded in 1974 that Mladeč 5 was "instantly recognizable" as having Neanderthal ancestors. Mladeč 5 and the other material from the site continue to influence my perspective on the origins of the Upper Paleolithic Europeans.

OTHER EVIDENCE, OTHER INTERPRETATIONS

Mladeč 5 is not the only evidence linking the European Mousterian and the European Upper Paleolithic. For the remainder of this essay, two arguments are reviewed which have been used to eliminate Neanderthals from subsequent human evolution. As I read and interpret this evidence, contrary to supporting the separation between Neanderthals and the people who followed them in Europe, the data demonstrate an undeniable connection between European Neanderthal and Upper Paleolithic populations. I review two aspects of this evidence here: (1) morphological links across the European Mousterian to the Upper Paleolithic and (2) rates of change for metric data across the same two periods.

MIDDLE TO UPPER PALEOLITHIC MORPHOLOGICAL LINKS

The first argument relies on a series of features which have been claimed to be uniquely derived (autapomorphic) traits characteristic of the European Neanderthals. From lists compiled by much earlier workers (e.g., Boule 1911–1913; King 1864), Stringer, Hublin, and Vandermeersch (1984) identified about twenty anatomical characteristics (presumably under genetic control) which serve to define Neanderthals as a group apart from earlier or later populations. They argue that these "derived characters . . . are actually unique only to neanderthals . . . [and] provide important evidence for excluding them from the ancestry of modern humans" (Stringer et al. 1984:54). Others have argued that such features in fact designate Neanderthals as a separate species (Rak 1991; Tattersall 1995).

In an earlier paper, I reviewed many of these "unique" traits and showed that while some of the features may be rare in recent or contemporary Europeans, they certainly do not occur in low frequency in the subfossil populations which directly followed the Neanderthals in Europe (Frayer 1992a). One feature is the horizontal-oval (H-O) form of the mandibular foramen. This feature, originally described by Gorjanović-Kramberger (1906) for the Krapina Neanderthals, concerns a presumably genetic trait on the interior aspect of the mandible. In most recent European mandibles, the opening where the mandibular nerve enters the interior aspect of the mandible is "V-shaped," which is considered the normal condition. But in many European Neanderthal mandibles the uppermost part of this area is bridged, so that when viewed from the top the opening takes on a horizontal-oval shape. As suggested by Smith (1978), the bridging across the opening is probably related to an expansion of the sphenomandibular ligament, but given its near absence in modern populations the actual etiology of the H-O foramen is unknown and may be unknowable since there is little chance of finding a pedigree. Yet the trait is most likely of genetic origin since it does not appear to be influenced by environmental factors. The evidence for this is found in hominids with large masticatory structures and/or extensive tooth wear (where one would predict an expansion of the sphenomandibular ligament insertion). In these specimens, the H-O pattern is always absent. Moreover, the trait appears to be a European marker since it is very rare outside Europe, where it is found in only one fossil mandible from Africa, Asia, or Australia. Yet, the H-O mandibular foramen is common in European Neanderthals ($\sim$53%) and in the early Upper Paleolithic people ($\sim$18%) who followed them in Europe. After the glacial maximum (dated about 18,000 B.P.), the frequency of the H-O trait falls to less than 7% in the late Upper Paleolithic populations and in the Holocene populations it drops to the modern European frequency of less than 2%. Other so-called unique traits of Neanderthals are commonly found in the immediately succeeding European hominids (Frayer 1992a). In fact, more than half of the features commonly used to define Neanderthals are not unique to them, but show substantial frequencies in the early Upper Paleolithic populations (Frayer 1992a).

Others have demonstrated that many of these Neanderthal "autapomorphies" are not reliable as a means for distinguishing "moderns" from Neanderthals, or even Neanderthals from their precursors. Trinkaus (1993b) has shown that another mandibular feature, the posterior positioning of the mental foramen on the external face of the mandible, can no longer be viewed as a distinguishing character. This foramen marks one of the terminating points of the mandibular nerve and was generally considered as being located under the first molar in Neanderthals and farther forward on the jaw in modern humans (Stringer et al. 1984:55). According to Trinkaus, the Neanderthal condition is highly variable and "more than half of the European 'classic' neanderthals have their mental foramen mesial to M1," which is typically the "modern" position (Trinkaus 1993b:268). Beyond this, Wolpoff (1994a:Table 14.3) has itemized a series of facial, cranial, and postcranial characters which link—not separate—European Neanderthals from the people who follow them. Thus, considerable evidence points to the persistence of these "neanderthal autapomorphies and common traits" (Stringer et al. 1984:55) into the Upper Paleolithic populations which succeeded the Neanderthals in Europe.

At the same time, these identical features are generally absent in the human fossils from Africa (Omo, Border Cave, and Klasies River Mouth) and the Near East (Skhūl and Qafzeh) who reputedly represent the source populations for the early Upper Paleolithic people of Europe (Bräuer 1989; Stringer and Andrews 1988b; Stringer and Gamble 1993; Tattersall 1986). For example, the H-O trait is absent in the African and Levantine mandibular samples. While these samples are small, the absence of these features has been used to define them as modern, which is generally true except for the early Upper Paleolithic people who possess the "unique" Neanderthal features. The fact that the so-called Neanderthal autapomorphies occur in the early Upper Paleolithic of Europe and not in the known "Eve" populations presents some formidable problems for the advocates of total replacement. These authors discount any interbreeding between the resident European Neanderthals and the invading "moderns," so the Neanderthal unique features could not be due to gene flow. Thus, while Stringer and Gamble argue that the last Neanderthals were able to get close enough to the invading moderns to copy their tools (1993:182), the two groups were apparently not capable of breeding with each other. Some even attempt to sustain the argument that Neanderthals and "moderns" existed "side-by-side for 50,000 years and never [had] sex" (Shreeve 1995:1). Leaving aside the whole question of violation of competitive exclusion, the existence of the Neanderthal features in Upper Paleolithic skeletons undermines the logic of this position and creates the highly unlikely requirement that the identical features evolved independently a second time in European people who followed a supposedly nonancestral population which nonetheless had exactly the same traits. And, following the reappearance of these identical traits, they rapidly decreased again in their incidence in the descendant populations. No amount of genetic analysis or replacement thinking can get around the fact that Neanderthal "unique" features appear in the early Upper Paleolithic. An alternative, more direct conclusion is that Neanderthals contributed to the Europeans who followed them and no abrupt replacement occurred between the Mousterian and the Upper Paleolithic. There is no easy way to dismiss the importance of these persisting features and, unless one simply ignores the presence of these "unique" Neanderthal anatomical traits in the early Upper Paleolithic fossils, there is no reason to question the links between Neanderthals and early Upper Paleolithic Europeans.

RATES OF CHANGE FOR METRIC DATA

A second argument against the inclusion of European Neanderthals as ancestors involves evolutionary rates. It is odd that the same scholars so willing to accept an abrupt punctuational event for the origin of modern humans (as is required in the Eve theory), or for the rapid appearance of racial (geographic) characteristics after the establishment of moderns in replacement models, argue that Neanderthals could not be ancestral to modern humans because there is not enough time for one to evolve into the other. Yet, it is a common, decades-old argument that European Neanderthals differed so profoundly from modern hominids that there was insufficient time to allow them to evolve into Upper Paleolithic humans (Boule 1911–1913; Klein 1992).

To cite a typical argument, Stringer contends that "To convert even the earlier last glaciation neanderthals into the earliest a[natomically] m[odern] Europeans would require a tremendous acceleration in evolutionary change over a few thousand years, preceded and followed by relative (but not absolute) morphological stasis" (1984b:66). Along the same line of reasoning, Bräuer has argued that "The numerous changes throughout the entire skeleton can be shown to have occurred within an extremely short period of time, whereas many tens of millennia would have been necessary for the evolutionary development of these features" (1989:136). Neither of these authors provide (or cite) evidence to support excessive evolutionary rates, the length of time required for "evolutionary development of these features," or estimates of what level of selection would be necessary to account for differences between Neanderthals and early Upper Paleolithic Europeans. It is unclear to me that the last two issues can be satisfactorily resolved. The speed at which traits are developed, lost, or changed is dependent on a number of factors, not the least of which are random forces and the intensity of selection (Lande 1976). For any evolutionary model involving forces of evolution, it is also important to have estimates of heritability. Yet, many of the morphological traits found in Neanderthals and in early Upper Paleolithic fossils occur in low frequency in living populations and, to my knowledge, have not been studied. Consequently, it is not possible to provide anything but a guess as to how long "would have been necessary for the evolutionary development of the traits" (Bräuer 1989:136). However, since evolutionary rates are quantifiable and a tremendous acceleration in them can be evaluated, information about the rate of change is relevant to scenarios concerning the evolutionary fate of European Neanderthals.

In the remainder of this chapter, data for evolutionary rates are reviewed for dental and craniofacial dimensions between Neanderthals and the Upper Paleolithic samples. These are compared to rates calculated for a series of lineally related samples for which no one denies an evolutionary relationship or, much less, considers as separate species. Like the "unique" traits used to define Neanderthals, evolutionary rates provide no support for the contention that Neanderthals cannot be ancestors to subsequent Europeans.

SAMPLES AND METHODS

Two sets of rate comparisons are considered—craniofacial and dental. In a previous publication (Frayer 1992a) other samples from the Levant and Africa were utilized in comparisons, but since the Skhūl/Qafzeh and African Eve samples are so chronologically removed (>50,000 years) from the Upper Paleolithic groups, and because there are no intervening fossils, it seems likely that these populations went extinct without leaving ancestors in late Pleistocene Europe. Consequently, only European samples are considered here. The bulk of the European Neanderthal dental and craniofacial data derive from measurements on the original specimens (Wolpoff, personal communication 1991), but these data have been supplemented with specimens described in the literature and some additional material I have studied. For an average date of this group, I use 70,000 years B.P., corresponding to the midpoint of the span of Mousterian industries in Europe (Mellars 1986). A late Neanderthal sample from western and central Europe has also been compiled, consisting of specimens associated with terminal Mousterian and Châtelperronian levels from Arcy-sur-Cure, Hortus, Kůlna, Le Moustier, La Quina (levels 1–2), Saint-Césaire, Šipka, and Vindija. It is unlikely anyone could argue that the human material from these sites dates to the early Würm or represents anything other than the latest Neanderthals in Europe. An average date of 37,000 B.P. is used for this sample and was calculated from published absolute dates (or estimated dates) for the sites. Since craniofacial measurements are very poorly represented in this group, the late Neanderthal sample is only used in calculating rates of dental change. In those tables, the late Neanderthal group always has a sample size of at least three. All the remaining samples derive from my personal study of the original specimens, except for those that have been destroyed or for which I did not have permission to study. In these cases, I have extracted the measurements from the literature or, where possible, from the cooperative sharing of data by colleagues. Compared with data from an earlier publication (Frayer 1992a), nearly all the samples are slightly modified due to the inclusion of new data. For example, all of the new Dolní Vestoniče materials along with numerous Neolithic remains from the Czech Republic, Slovakia, and Hungary have been added. A few specimens from the Upper Paleolithic samples have been deleted as skeletal material and its context are reevaluated. As in the earlier analysis the modal dates for these samples are 26,500 B.P. for the early Upper Paleolithic, 14,000 B.P. for the late Upper Paleolithic, 8,000 B.P. for the Mesolithic, and 6,500 B.P. for the Neolithic. They are based on published radiocarbon dates and extrapolations from them to sites and levels within Upper Paleolithic, Mesolithic, and Neolithic sites. Because I have very limited craniofacial measurements from the auricular point for Neolithic specimens, rates for craniofacial metrics are only calculated for the earlier two periods. Evolutionary rates were calculated using Haldane's (1949) formula for a darwin *(d)*

$$[\log_e \bar{x}_2 - \log_e \bar{x}_1]/t$$

where $\bar{x}_1$ and $\bar{x}_2$ are the sample means and t is the time interval between the two samples expressed in millions of years.[1]

RATES OF CHANGE

For determining rates of craniofacial change, direct measurements from the auricular point to fourteen landmarks on the face are used. This standard point located above the external auditory meatus is often present even on fragmentary skulls, whereas the more traditionally used *basion* (on the anterior face of the foramen magnum) is often missing or reconstructed. Using only measurements from *basion* to points on the face and cranium substantially reduces sample size in all groups. Consequently, measurements from the auricular point which is on essentially the same coronal plane generate larger sample sizes and provide equivalent biological information. Midline (mid-sagittal plane) and facial measurements lateral to it are used since they are considered to be consistently large in European Neanderthals (Stringer et al. 1984:55).

Despite the contention of those who argue for elevated evolutionary rates, the rates of change in these fourteen measurements for the Neanderthal–Upper Paleolithic comparison are consistently low compared with those of the other samples (Table 16.1). For example, the maximum rate observed for the European Neanderthal–early Upper Paleolithic comparison is 1.8 darwins, which is lower than seven of the fourteen rates for the early to late Upper Paleolithic transition and five of the fourteen rates for the early Upper Paleolithic–Mesolithic comparison. Moreover, the average rate of change between the European Neanderthals and the early Upper Paleolithic is .8 *d*, which is substantially below the average rates for the two post-Mousterian comparisons. The rate between the Neanderthal and early Upper Paleolithic sample is *less than half* the average darwin between the early-late Upper Paleolithic (1.8 *d*) or between the early Upper Paleolithic and Mesolithic (1.6 *d*).

Two unambiguous conclusions can be drawn from the rates for these mea-

Table 16.1. Rates of Change (in Darwins) for Facial Measurements from the Auricular Point

	Neanderthals to Early Upper Paleolithic	Early to Late Upper Paleolithic	Early Upper Paleolithic to Mesolithic
Auricular point to			
prosthion	1.1	1.9	1.6
nasospinale	1.0	1.9	1.8
nasion	.9	1.4	1.3
glabella	.8	1.4	1.3
zygomaxillarae	0.0	1.3	2.0
M1/M2	1.8	1.0	.8
P3/P4	1.2	2.0	1.3
I2/C	1.1	1.8	1.3
inferior nasomaxillary suture	.6	1.6	1.9
jugale	.7	2.8	2.4
frontomalareorbitale	.4	1.9	1.8
alare	.7	1.6	1.6
palatine suture cross	.6	1.8	1.7
post-*orale*	.5	2.4	1.9
Average change	.8	1.8	1.6

sures of craniofacial change: (1) neither a "tremendous acceleration" nor even a rapid evolutionary rate is required for the transition of European Neanderthals into the early Upper Paleolithic, and (2) these rates of change show that anything but stasis characterizes the post-Neanderthal period in Europe, which exhibits substantial reduction in facial projection from the auricular point. Some of these rates are affected by the time interval (see below), but following Gingerich's logarithmic scale, the average rate and highest rate for the Neanderthal–early Upper Paleolithic proposed transition is well within his observed limits and comparable to change within other lineal taxa (Gingerich 1983:160). In short, the Neanderthal– early Upper Paleolithic rates fit comfortably in his "Domain IV" post-Pleistocene rates of change, indicating that they are not excessively high.

Rates for mandibular and maxillary dental changes are presented in Tables 16.2 and 16.3. In both tables, a sample of late Neanderthals is included since, unlike for the craniofacial measurements, more than one or two individuals can be compiled. However, this sample remains small, especially for the incisor and canine dimensions, and means for these individual teeth are likely to vary markedly with the addition of newly discovered, but currently unpublished, specimens. In these dental rate comparisons it is also possible to include a Neolithic sample which consists of specimens from western and central Europe.

For anterior tooth lengths and breadths (Table 16.2), rates expressing change between the total Neanderthal sample and the early Upper Paleolithic *never* represent the highest evolutionary rate. Rather, mean differences between the Mesolithic and Neolithic show the highest darwin values, in each case for individual anterior tooth lengths and breadths. For example, in the mandibular anterior teeth the highest rate between the Neanderthal and early Upper Paleolithic sample is 1.6 *d* for I1 breadth and the mean rate of change of the six dimensions is .8 *d*. In the same six dimensions, the highest rate of change between the Mesolithic and Neolithic is 27.1 *d* and the average rate of change is 16.6 *d*. Rates of change for the anterior maxillary teeth parallel the mandibular rates, with the Neanderthal–early Upper Paleolithic interval consistently showing the lowest rates of change and the Mesolithic to Neolithic transition exhibiting elevated rates. *The latter are nearly twenty times the Neanderthal–early Upper Paleolithic rate.*

In the mandibular and maxillary tooth areas (Table 16.3), the total Neanderthal sample compared with the early Upper Paleolithic specimens consistently has rates of evolution below, sometimes substantially below, virtually all the other sample comparisons. In both jaws, the average rates of evolution between the total Neanderthal and early Upper Paleolithic sample (1.1 *d* for the mandible, 1.2 *d* for the maxilla) represent the lowest average rates for all of the comparative samples. As in the comparisons of the anterior tooth dimensions, the Mesolithic-Neolithic rates for the mandibular and maxillary dental changes are consistently the highest. In this comparison the average rate of change is 23.2 *d* in the mandible and 28.7 *d* in the maxilla. These results are markedly higher than any of

Table 16.2. Rates of Change in Mandibular Incisor and Canine Mesio-Distal Lengths and Labio-Lingual Breadths between Selected Groups, as Measured by Darwins

	I1 Lt	I1 Br	I2 Lt	I2 Br	C Lt	C Br	Mean*
MANDIBLE							
Neanderthals to early Upper Paleolithic	.2	1.6	.3	1.1	.9	.3	.8
Late Neanderthals to early Upper Paleolithic	2.1	6.1	3.1	5.1	4.8	1.5[†]	3.8
Early to late Upper Paleolithic	1.9	.6	2.2	1.5	2.0	2.5	1.8
Mesolithic to Neolithic	27.1	14.6	19.6	13.7	17.3	7.5	16.6
Early Upper Paleolithic to Neolithic	2.3	1.9	2.4	2.2	2.4	2.8	2.3
MAXILLA							
Neanderthals to early Upper Paleolithic	.3	1.0	1.0	1.6	.7	.8	.9
Late Neanderthals to early Upper Paleolithic	.9[†]	1.1	2.7	4.9	1.5	1.8	2.1
Early to late Upper Paleolithic	2.3	.5	2.0	4.1	1.7	1.9	2.1
Mesolithic to Neolithic	29.8	11.9	25.6	0.0	15.0	10.3	15.4
Early Upper Paleolithic to Neolithic	2.6	1.7	2.4	2.4	1.8	2.1	2.2

* of six tooth rates
† Early Upper Paleolithic mean exceeds late Neanderthal mean.

the other sample comparisons. For the mandibular areas the maximum rate involves the M1 (30.1 d) and in the maxilla the greatest rate is 32.9 d for P3. All of them are magnitudes of order higher than any other rates of change, including those for the Neanderthal–early Upper Paleolithic transition.

Although the fact that the rates between the Neanderthal and early Upper Paleolithic are the lowest rates of all the sample comparisons is an important observation, the consistently higher rate of change between the Mesolithic and Neolithic may be in part related to the short time gap between the two groups. Gingerich (1983), for example, has shown that small mean differences can be expressed as high evolutionary rates when t is small. Using 521 rates for a variety of different taxa and widely different time periods, Gingerich divided a logarithmic scale into four domains, each related to the length of time over which the groups were sampled. He was able to demonstrate that rates of change in laboratory experiments of very short duration produced the highest rates (his Domain I) compared with examples of species colonization covering 70–300 years (Domain II) and rates for post-Pleistocene mammals sampled from be-

Table 16.3. Rates of Change in Mandibular and Maxillary Tooth Areas between Selected Groups, as Measured by Darwins (Tooth Areas = Mesio-Distal Length × Bucco-Lingual Breadth)

	Canine	*P3*	*P4*	*M1*	*M2*	*M3*	*Mean**
MANDIBLE							
Neanderthals to early Upper Paleolithic	1.3	1.8	1.1	.1	1.1	1.3	1.1
Late Neanderthals to early Upper Paleolithic	1.9	4.8	3.1	.6†	3.2	5.3	3.2
Early to late Upper Paleolithic	4.1	2.7	3.0	1.1	1.5	1.3	2.3
Mesolithic to Neolithic	23.5	27.1	17.8	30.1	27.8	12.9	23.2
Early Upper Paleolithic to Neolithic	5.2	4.1	2.9	2.7	3.4	3.5	3.6
MAXILLA							
Neanderthals to early Upper Paleolithic	1.6	1.4	1.5	.8	.8	1.3	1.2
Late Neanderthals to early Upper Paleolithic	3.1	1.5	3.4	1.0	2.4	2.9	2.4
Early to late Upper Paleolithic	3.4	4.2	3.2	1.0	2.5	3.0	2.9
Mesolithic to Neolithic	27.7	32.9	26.1	27.0	32.1	26.3	28.7
Early Upper Paleolithic to Neolithic	4.2	4.6	3.7	3.1	4.5	3.9	3.9

* of six tooth rates

† Early Upper Paleolithic mean exceeds late Neanderthal mean.

tween 1,000 and 10,000 years (Domain III). The lowest rates occurred in his Domain IV, which consisted of comparisons of fossil invertebrates and vertebrates sampled from a minimum of 8,000 years to a maximum of 350 million years.

All of the rates presented in Tables 16.2 and 16.3 fall within his observed limits of Domain IV, even the Mesolithic to Neolithic average and maximum rates (Gingerich 1983:160). Nevertheless it is useful to compare rates which are calculated over approximately the same time interval to determine if an excessive amount of change is required to allow for the transformation of Neanderthals into Upper Paleolithic people. In this regard, the late Neanderthal–early Upper Paleolithic rates can be compared with the early to late Upper Paleolithic rates (both sampled over about the same time period of 10,000 years) or to the early Upper Paleolithic–Neolithic rates, which are sampled over about twice that length of time (20,000 years). Both of the latter comparisons involve change within a species (no one questions that Upper Paleolithic and Neolithic humans belong to

Homo sapiens), and it is apparent from the rates that the Neolithic, like the late Upper Paleolithic, has undergone marked dental reduction.

In the mandible (Table 16.2) the average rate of 3.8 *d* for the late Neanderthal–early Upper Paleolithic comparison exceeds the rate from the early to the late Upper Paleolithic (1.8 *d*) and that from the Upper Paleolithic to the Neolithic (2.3 *d*). While some might argue this shows the impossibility of Neanderthals evolving into early Upper Paleolithic humans, the differences in the rates for the mandible dentition are not excessive and represent the only comparison in which the average darwin for the late Neanderthal–early Upper Paleolithic exceeds the other sample comparisons. Also the substantial change in some mandibular dimensions (e.g., I1 breadth) explains why the Neanderthal rates exceed those for the more modern samples. In the maxillary anterior tooth dimensions, the late Neanderthal–early Upper Paleolithic average rate (1.8 *d*) is less than both the early to late Upper Paleolithic (2.1 *d*) and the early Upper Paleolithic–Neolithic (2.2 *d*) comparisons. One maxillary dimension for Neanderthals (I2 breadth) slightly exceeds the highest rate in the early-late Upper Paleolithic comparison. It is apparent from the data on rates of change in the incisor and canine dimensions of both jaws that the transition between the Neanderthals and the early Upper Paleolithic involved relatively high rates of change for some specific dental dimensions, but that the overall or mean rate of change was comparable among all three sampled intervals.

A similar pattern is present in the mandibular and maxillary tooth areas (which are generally represented by larger samples) in that the average rate of change for the late Neanderthal–early Upper Paleolithic interval is not especially different from the average darwins calculated for the early to late Upper Paleolithic or for the early Upper Paleolithic to the Neolithic. Thus for the mandibular tooth areas, the average rate (3.2 *d*) is just less than that for the early Upper Paleolithic–Neolithic (3.6 *d*) and slightly above the early to late Upper Paleolithic mean rate of 2.3 *d*. In the mandibular areas only the M3 shows a greater rate of change (5.3 *d*) than the highest early Upper Paleolithic–Neolithic comparison (5.2 *d*), but this difference is inarguably minor. For maxillary canine and posterior tooth areas, a similar pattern exists in that the late Neanderthal to early Upper Paleolithic average rate of change is consistently less than that found for the other two modern comparisons. The late Neanderthal to early Upper Paleolithic rate is 2.4 *d,* compared with rates of 2.9 *d* between the early and late Upper Paleolithic and 4.0 *d* between the early Upper Paleolithic and Neolithic. As in the mandible, individual rates of change in tooth area between the late Neanderthals and the early Upper Paleolithic samples are consistently within the ranges found for the modern samples.

CONCLUSION

In summary, based on these comparative evolutionary rates, the average and individual rates of evolutionary change between European Neanderthals (or late

Neanderthals) and early Upper Paleolithic hominids are not especially rapid. Rather, rates of change between either European Neanderthal samples and the early Upper Paleolithic are within the magnitude of change found for recent *Homo sapiens*. Thus, contrary to the commonly stated argument that not enough time exists for European Neanderthals to be ancestral to subsequent Europeans, these data clearly demonstrate that there was no "tremendous acceleration" in rates of change between the Neanderthals and the Upper Paleolithic Europeans. For me, these data falsify the argument that European Neanderthals as a group cannot be ancestral to subsequent *Homo sapiens* in Europe (at least with respect to metric features of the face and teeth) because too much change is required over too little time. Moreover, based on the rates of dental evolutionary change, there is nothing to support the contention that European Neanderthals represent a separate species. Such a conclusion would only hold if one is also willing to accept a speciation event between the early and late Upper Paleolithic, between the Mesolithic and Neolithic, or between the early Upper Paleolithic and Neolithic, since all of these comparisons have similar, or in some cases considerably higher, average or individual evolutionary rates.

While rates of dental evolutionary change by themselves do not prove that Neanderthals are ancestral to early Upper Paleolithic Europeans, these results do indicate that European Neanderthals cannot be eliminated as possible ancestors based on speculations which require grossly elevated evolutionary rates. Moreover, the period following the Neanderthals in Europe is not characterized by absolute or relative stasis but by marked change within the Upper Paleolithic and from the Upper Paleolithic to the Neolithic. These observations should put to rest both the contention that differences between the European Neanderthals and the early Upper Paleolithic require an exorbitant rate of change and the unsupported claim that tooth size shows little absolute or relative change after the appearance of the Upper Paleolithic. Those who still maintain that European Neanderthals are unrelated to subsequent European *Homo sapiens* must look to other data; these data do not include the presence of so-called Neanderthal autapomorphic traits or exorbitant rates of change.

ACKNOWLEDGMENTS

This research was made possible through NSF Grant BNS 84-49057 and several awards from the National Academy of Science and the University of Kansas General Research Fund. Without the support of these agencies, the work could not have been accomplished. I thank the curators of European skeletal material for allowing me the privilege of examining their collections and M. H. Wolpoff for access to data and casts in the Paleoanthropology Laboratory, University of Michigan. The manuscript was improved by critical comments of S. Gray.

NOTE

1.   In Frayer (1992a) the evolutionary rates were tabulated incorrectly so that all rates are off by two decimals. While this mistake does not affect the comparisons within the paper, the rates are clearly too low. Here the rates are given in standard darwin units; I apologize for my previous error.

17

The Iberian Situation between 40,000 and 30,000 B.P. in Light of European Models of Migration and Convergence

LAWRENCE GUY STRAUS

In January 1986, James Bischoff of the U.S. Geological Survey wrote me that he was planning to go to Spain and asked me which sites would merit uranium-series dating. I replied with a list that included El Castillo, Abric Romaní, and Banyoles (among several other sites, notably Lezetxiki, Cueva Morín, El Pendo, Carigüela, and Cova Negra). The results of Bischoff's work in northern Spain (and the subsequent involvement of R. Hedges and colleagues at the Oxford Accelerator Lab in duplicating Bischoff's results from El Castillo and from L'Arbreda) are now well known (and indicate a Middle to Upper Paleolithic technological transition at ca. 40,000 B.P.). The use of AMS radiocarbon dating together with uranium-series dating extended the calibration of ^{14}C dates back to 40,000 B.P. (Bischoff et al. 1988, 1989, 1992, 1994; Cabrera and Bischoff 1989; Cabrera et al. 1996; Hedges et al. 1994; Juliá and Bischoff 1991). Suddenly a peripheral region of Europe, the Iberian Peninsula, never before at the center of the debate on the Middle to Upper Paleolithic transition or on Neandertal–Cro-Magnon replacement, was catapulted into the limelight with radiometric determinations associated with Upper Paleolithic materials *as old or older than* the most ancient ones then known from central and eastern Europe. These dates can be interpreted to call into question the traditional ex oriente lux "invasionist" hypothesis to explain the "abrupt" appearance of Upper Paleolithic technologies and lifeways (and, by assumed extension, anatomically modern humans) from the Near East and southeast Europe to western Europe (Straus 1990, 1994; Straus et al. 1993).

At the same time, restudies of Cova Negra and Carigüela and excavations at the newly found sites of Cova Beneito and especially Zafarraya, all in southeastern Spain, and excavation of Gruta da Figueira Brava in southern Portugal converged to suggest that Mousterian technology and Neandertal anatomy had persisted in southern Iberia until at least 30,000 years ago, some 10,000 years after the appearance of Aurignacian technology (and the *presumed* appearance of

Homo sapiens sapiens) in Catalonia and Cantabria (Antunes 1990; Barroso et al. 1984; Hublin 1994; Iturbe et al. 1993; Vega 1990; Villaverde and Fumanal 1990). South of a line transcribed by the Ebro and Tagus rivers, in a vast region separated from the Maghreb by the 10 km wide Strait of Gibraltar, archaic hominids with archaic life ways survived as apparently uninfluenced contemporaries of modern humans with Upper Paleolithic technologies! This startling development—on a par with the 1979 discovery of a Neandertal associated with a Châtelperronian industry at Saint-Césaire in southwestern France (Lévêque et al. 1993)—calls into question the alleged "overwhelming adaptive superiority" of the bearers of the Aurignacian industry and their alleged recent African origin.

INTRODUCTION

It is my intention here to pose a series of questions, to cite a number of facts, and to suggest some alternative explanations for the Iberian situation, but without pretending to a dogmatic certainty that my point of view is absolutely correct outside of that context. Prehistoric archaeology and human paleontology have far too often provided "surprises" like those just mentioned for such hubris. For example, in the few months between the oral presentation of this paper in Capellades (Barcelona Province) in March 1995 and its rewriting in August 1995, the accelerator dating of charcoal drawings in Grotte Chauvet to ca. 30,000 B.P. (Clottes et al. 1995) would seem to call into question my notion that fully developed cave art had appeared fairly late in the Upper Paleolithic. This is a fast-developing field, which is why it is exciting.

Can the biological and cultural changes that occurred ca. 40,000–30,000 years ago be analyzed separately, or are they inextricably linked in a causal relationship (as so frequently and eloquently argued by R. Klein)? For those who believe that biological evolution is the motor of all change—especially in regard to the development of the brain and its cognitive functions—the cultural changes of the Upper Paleolithic follow inevitably, and *the* break with all that had gone before is seen as *abrupt and complete* (e.g., Mellars 1989b; White 1993). These scholars stress such traits as the *sudden* appearance of art, ornaments, and bone technology. And they tend to see a more or less *diametric* opposition between Middle and Upper Paleolithic life ways (subsistence, settlement, technological, social, symbolic, and ideological systems, etc.).

My perspective is a bit more cautious and stems from two sets of observations that should trouble paleoanthropologists more than they seem to when simple biological replacement models are invoked to "explain" the Middle to Upper Paleolithic transition: first, the supposed mismatch between fossils and cultural remains, and second, cultural change is continuous over the long term, not concentrated strictly at 40,000 B.P.

THE "MISMATCH" BETWEEN FOSSILS AND CULTURAL REMAINS

1. Are the Qafzeh and Skhūl fossils really almost "modern" proto–Cro-Magnons (e.g., Stringer 1989b; Vandermeersch 1992) or are they rather more "archaic" (e.g., Arensburg 1991; Kramer et al. n.d.)? Human paleontology often seems, to this archaeologist at least, to engage in considerable semantic play. Whatever the remains are biologically, they are associated with Mousterian assemblages acknowledged to be generally the same as those associated with Near Eastern Neandertals at Kebara, Amud, Shanidar, and Tabūn (Clark and Lindly 1989a). The Qafzeh/Skhūl fossils may be roughly contemporaneous with some or most of the Levantine Neandertals at ca. 100,000 years ago or less (e.g., Bar-Yosef 1992). There is no convincing evidence of "different" or "superior" adaptations on the part of the Qafzeh/Skhūl population(s) (cf. Trinkaus 1992b).

2. The Saint-Césaire and Grotte du Renne Neandertal remains are associated with Châtelperronian assemblages that many authors (e.g., Harrold 1989) consider to be Upper Paleolithic, while others find them "transitional" or highly variable in terms of content and assemblage composition (e.g., Lévêque et al. 1993; Straus and Heller 1988).

3. There is now meticulously published evidence for the association of Neandertal-like human remains with both Mousterian and classic Aurignacian-like assemblages (the latter including multiple bone points and an engraved bone in a level recently AMS dated to 33,000 B.P.) in Vindija Cave, Croatia (Karavanic 1995 and references therein).

4. Numerous cases of robust and some very "transitional" fossils are associated with early Upper Paleolithic assemblages (e.g., Velika Pečina, Mladeč), notably in central and southeastern Europe (Frayer et al. 1993, with references).

5. There are many supposed cases of anatomically transitional or "modern" humans associated with Middle Stone Age or Mousterian assemblages in both sub-Saharan and Saharan Africa, but with no apparent major changes in subsistence or other aspects of life ways (e.g., Klasies River Mouth, Border Cave, Omo, Florisbad, Laetoli, Jebel Irhoud, Haua Fteah; Klein 1989b).

If the biological change was so important, why are there so many of these "mismatches"? And why did it take anatomically modern humans so long to reach Europe (i.e., not at ca. 100,000 B.P., but at ca. 40,000 B.P.)? Indeed, why were Neandertals supposedly able to supplant anatomical moderns in the Levant under harsh glacial conditions, as argued by Bar-Yosef (1989)? Were Neandertals and anatomical moderns really so different as to preclude/exclude mating and thus gene flow since their archaeologically apparent cultural "signatures" are so similar?

CULTURAL ADAPTATION IS CONTINUOUS OVER THE LONG-TERM, NOT CONCENTRATED STRICTLY AT 40,000 B.P.

1. Significant cultural change may have been underway long before 40,000 years ago. Stiner's (1994) research in Latium (west-central Italy) shows more active Neandertal hunting and shellfishing after ca. 55,000 B.P. than before. Unambiguous blade technology (including use of the true crested blade technique on prismatic cores) developed episodically *throughout* the Mousterian and before (Conard 1990; Lamdan and Ronen 1989; Révillion and Tuffreau 1994; Ronen 1992). Excluding well-known examples of pseudo-artifacts (such as the grooved bear teeth from Sclayn Cave; Chase and Dibble 1987), there is limited evidence of art and/or ornamentation in Mousterian contexts, sometimes associated with deliberate human burials (e.g., La Ferrassie, La Quina, Cova Beneito, Quneitra— Marshack 1989, 1996; Iturbe et al. 1993).

2. The early mobile art of southwest Germany (especially Geissenkloesterle) and now perhaps Belgium (Trou Magrite) is spectacular at 36,000–38,000 years ago (J. Hahn, Universität Tubingen, personal communication 1995; Otte and Straus 1995) and the new dates for rupestral art in Grotte Chauvet are, if confirmed, extraordinarily old at ca. 31,000 B.P., but how common and widespread was this art? Certainly there are, at least to date, vast areas of Europe with no evidence of early art. Is it absolutely certain that its appearance was so abrupt? Might it not have had antecedents (perhaps in/on different media than ivory, antler, or bone)? It remains a fact that *most* dated mobile and parietal art is much later (≤20,000 years ago).

3. Change is *mosaic* in character within the Upper Paleolithic, both in terms of traits (some early, others late) and in terms of regions (i.e., change occurred in some regions but not in others). It is not contemporaneous everywhere even throughout so small a continent as Europe; indeed it seems to have often been dependent on regional circumstances. Chief conditioning factors included latitude, topography, location (e.g., coastal vs. interior), available lithic raw material, and the distribution in time and space of food resources and its effects on demography. In short, cultural changes happened in specific contexts, not in a void and not simply as a result of unexplained "migrations."

4. Adaptive cultural change continued throughout the Upper Paleolithic into the Mesolithic and beyond. It seems to have been more or less episodic (as reflected by the traditional culture-stratigraphic units: Aurignacian, Gravettian, Solutrean, early Magdalenian, late Magdalenian, Azilian/Federmesser, and Mesolithic) and correlated with major environmental fluctuations (i.e., the Interpleniglacial with significant interstadials, the Upper Pleniglacial centered on the Last Glacial Maximum, the Tardiglacial, the Late Glacial Interstadial, Dryas III,

and Preboreal). These changes affected all subsystems of the Upper Paleolithic adaptive system (technological, subsistence, settlement/ territoriality/mobility, social, and ideological/cybernetic). Significant material changes well after 40,000 B.P. included the trend toward laminar blanks, invention of new lithic tools and especially weapons, developments in bone/antler technology (especially atlatls, *bâtons de commandement,* eyed needles, and new *sagaie* sizes, types, and hafting systems—mostly after ca. 20,000 B.P.). Situational specialized, and *massive,* hunting of key game species also occurred, sometimes accompanied by overall trends toward diversification (with fishing, shellfishing, birding, and perhaps some systematic use of edible plants) *beginning* in the Solutrean and especially early Magdalenian and accelerating in the late Magdalenian in many regions (such as Cantabrian and Mediterranean Spain, Portugal, southern France). Finally, there was a considerable spread and growth in complexity of mobile and cave art (a second "art boom") in the Magdalenian, 15,000–20,000 years after the so-called start of the Upper Paleolithic. In short, I see a continuum of adaptive change, well underway before 40,000 B.P.—perhaps with several inflection points in the *rate* of change at ca. 40,000, 35,000, 20,000, 15,000, and 10,000 years ago, in apparent correlation with significant environmental fluctuations.

Thus, in my view, the Upper Paleolithic/Mesolithic was a *process,* not a monolithic entity (despite the regrettable tendency to reify hallowed nineteenth-century taxonomic constructs). This view coincides in some respects with the notion of a process of "leptolithization" articulated a generation ago by Georges Laplace (1970), who, in addition, thought that the Upper Paleolithic developed more or less in situ and involved Châtelperronian *and* Aurignacian elements together with Mousterian ones in the formation of the classic laminar technologies of the late Upper Paleolithic. I am skeptical of simplistic invasion hypotheses with their absolutist punctuational character because there is so much evidence to support local changes and inventions; gradual, ongoing adaptations to changing regional environments; and a mosaic pattern in time and space. However, I remain open to the possibilities of diffusion and gene flow (i.e., human cultural and reproductive contacts).

Viewed from the short time scale of the Upper Pleistocene, the Upper Paleolithic was not one macro-punctuational event. Of course, viewed on a *geological* time scale (or even on the time scale of the whole of hominid evolution), the Upper Paleolithic, as a global phenomenon, can be seen as an "event," but spread out over the 40,000–10,000 B.P. interval, and with the *development* of complex adaptive systems (social, ideological, artistic, cybernetic, subsistence, technological, organizational, and ecological). Fundamental changes included demographic increase and radiation to Australia, northern Europe, Siberia, and the Americas, with ever greater reliance on elaborate strategies, tactics, planning, technologies, social organizations, and belief systems.

"ORIGINS" OF THE UPPER PALEOLITHIC?

The invasion scenario depends on two notions: (1) the cultural (i.e., ethnic) identity and unity of the Aurignacian and (2) the existence of an extra-European source for this cultural tradition. This theory is profoundly normative in character and implicitly explains all similarities in artifact form in terms of cultural identity. Without denying that social groups existed or that they might have had some conscious or unconscious material "markers" as long ago as 40,000 years ago, I argue that the data can be interpreted otherwise. What is the evidence for an extra-European origin for the so-called Aurignacian?

The blade-rich Ahmarian industry at Kebara Cave (northern Israel) is accelerator dated between ca. 43,000 ± 2,000 B.P. and 38,000–36,000 B.P., while the latest Mousterian at the same site dates to ca. 48,000 B.P. (Bar-Yosef et al. 1996). However, the Ahmarian does not resemble the European Aurignacian any more than other, early, blade-rich industries in the Near East and north Africa do, and some of these are in fact much older (Clark and Lindly 1989a; Marks 1990; Ronen 1992). Nor do the earliest typologically Upper Paleolithic assemblages from Boker Tachtit in the Negev (southern Israel, which, at 46,000–47,000 B.P., fill the stratigraphic gap that exists in Kebara; Marks 1983). The flake-dominated Levantine Aurignacian at Kebara with some blades, bladelets, El Wad points, and keeled and other endscrapers dates to only 36,000–28,000 B.P. (Bar-Yosef et al. 1996). The early Upper Paleolithic of the Near East has not produced any ornaments or bone tools, the "bottom-line" diagnostics of the European early Upper Paleolithic. Either the Aurignacian is a cultural entity (and if so, it is younger in Israel than in Europe), or it is a phenomenon produced in large part by technological convergence, or it spread from Europe *to* the Near East. The last hypothesis was recently adopted by Gilead and Bar-Yosef (1993), following Garrod. Thus the dating of the Aurignacian and other (related?) Upper Paleolithic industries in southeast Europe is critical to all invasionist arguments.

Unfortunately, despite the magnificent excavations and interdisciplinary analyses led by J. K. Kozłowski, the dating of the critical deposits at Bacho Kiro and Temnata Caves in Bulgaria remains ambiguous, particularly the contradictions among thermoluminescence (TL), conventional, and accelerator radiocarbon dates (Kozłowski 1982; Kozłowski et al. 1992) (Table 17.1). The only indicators of age in excess of 40,000 B.P. from southeast Europe are one old conventional radiocarbon date on charcoal from basal layer 11 at Bacho Kiro (>43,000 B.P.) and two TL dates on burnt flints from basal layer 4 at Temnata: 45,000 ± 7,000 and 46,000 ± 8,000 B.P. The enormous error ranges on these dates could place the true age of the earliest Upper Paleolithic assemblages at these sites at 38,000 B.P. using just one standard deviation. This would be in line with two recently published AMS dates of ca. 38,000 B.P. for Bacho Kiro layer

Table 17.1. Oldest Dates (≥37,000 B.P.) for European Early Upper Paleolithic Assemblages

Country	Site	Culture[a]	Level	Method	Lab No.	Date (B.P.)	s.d.
Bulgaria	Bacho Kiro	Bachokirian	11	AMS	OxA-3183	37,650	1,450
					OxA-3213	38,500	1,700
	Temnata		11 base	¹⁴C	GrN-7545	>43,000	
		Aurignacian	4 base	TL	GdTL-255	46,000	8,000
					GdTL-256	45,000	7,000
			4 mid	AMS	J. Kozlowski lecture 3/27/95	ca. 38,000	
Hungary	Istállóskö	Aurignacian	9 base	¹⁴C	GrN-4659	44,300	1,900
			9 mid	¹⁴C	GrN-4658	39,700	900
	Szeleta	Szeletian	4	¹⁴C	GXO-197	>41,700	
			3	¹⁴C	GrN-6058	43,000	1,100
Moravia (Czech Rep.)	Bohunice	Bohuncian	4a	¹⁴C	Q-1044	40,173	1,200
					GrN-6802	41,400	+1,400/−1,200
					GrN-6165	42,900	+1,700/−1,400
	Stránská Skála IIIa	Bohuncian	3	¹⁴C	GrN-12606	41,300	+3,100/−2,200
	Stránská Skála III	Bohuncian	5	¹⁴C	GrN-12297	38,200	1,100
					GrN-12298	38,500	+1,400/−1,200
	Vedrovice V	Szeletian		¹⁴C	GrN-12375	39,500	1,100
					GrN-12374	37,650	550
Slovakia	Certova Pec	Szeletian		¹⁴C	GrN-2438	38,400	+2,800/−2,100
Poland	Dzierzyslaw	Szeletian	upper	TL	GDT1-349	36,500	5,500
	Nietoperzowa	Jerzmanowician	6	¹⁴C	GrN-2181	38,500	1,240
Austria	Willendorf I	Aurignacian	2	¹⁴C	GrN-11195	41,700	3,100
	Willendorf II	Aurignacian	1-2	¹⁴C	GrN-11190	39,500	1,350
Germany	Geissenklösterle	Aurignacian	IIIa	AMS	OxA-4595	40,200	1,600
			IIa	AMS	OxA-4594	36,800	1,000
Belgium	Trou Magrite	Aurignacian	3	AMS	CAMS-10352	41,300	1,690
Italy	Grotta Fumane	Aurignacian	A2	¹⁴C	UtC-1774	40,000	+4,000/−3,000
					UtC-2048	36,500	600

continued

241

Table 17.1. Continued

Country	Site	Culture[+]	Level	Method	Lab No.	Date (B.P.)	s.d.
France	St. Césaire	Châtelperronian	8	TL	average of six dates	36,300	2,700
	Roc de Combe	Châtelperronain		AMS	J.-Ph. Rigaud and J. Simek, personal communication	ca. 38,000	
	Grotte XVI	Châtelperronian		AMS	ibid.	ca. 38,000	
Spain	El Castillo	Aurignacian	18	AMS	AA-2405	40,000	2,100
					AA-2406	38,500	1,800
					AA-2407	37,700	1,800
					OxA-2477	41,100	1,700
					OxA-2475	40,700	1,600
					OxA-2476	40,700	1,500
					OxA-2478	39,800	1,400
					OxA-2474	38,500	1,300
					OxA-2473	37,100	2,200
					Gif-89147	39,500	2,000
	L'Arbreda	Aurignacian	BE111	AMS	AA-3781	39,900	1,300
					AA-3782	38,700	1,200
					AA-3779	37,700	1,000
					AA-3780	37,700	1,000
	Reclau Viver	Aurignacian	CE103	AMS	OxA-3729	37,340	1,000
	Romaní	Aurignacian	TIII-27	AMS	OxA-3727	40,000	1,400
			2	AMS	AA-8037B·	37,900	1,000
					AA-7395	37,290	990
				U-series	average of 37 dates	43,000	1,000
Portugal	Gato Preto	Aurignacian		TL	average of two dates; Marks et al. 1994	38,100	3,900

N.B.: At most add 5,000–6,000 years to all ^{14}C dates for U-series calibrated ages

242

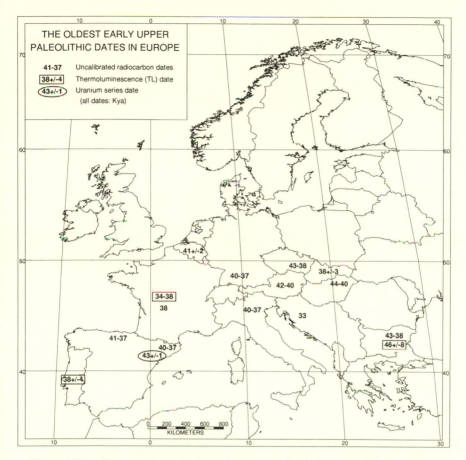

Figure 17.1. Oldest radiometric dates for Early Paleolithic occupations in Europe.

11 (Hedges et al. 1994) and a new AMS date by the Oxford Accelerator team of ca. 38,000 B.P. for Temnata layer 4 announced by R. Hedges and J. K. Kozłowski at a workshop on the Middle to Upper Paleolithic transition in Capellades, Spain.[1] Excluding the three old Bulgarian dates (and one old conventional determination of 44,300 ± 1,900 B.P. on bone collagen from cryoturbated basal layer 9 in Istállóskó Cave in Hungary, which may be less credible than a date from upper layer 9 of 39,700 ± 900 B.P.; Allsworth-Jones 1986), dates for the earliest Aurignacian assemblages *across all of Europe* cluster around 40,000–38,000 B.P. (as do the earliest dates for the Szeletian, Bohunician, and Châtelperronian), with no clear chronological priority to the east (Figure 17.1).

In my opinion, many or most aspects of the Aurignacian can be explained by technological convergence (especially the lithic typological and technological similarities between certain European and Near Eastern assemblages) and *some* diffusion, but *without* invoking migrations or invasions of a scale or nature reminiscent of those of the historical period with its high population densities and land-based food production economies (see Clark 1994b). Perhaps some of the more distinctive diagnostics like split-base antler points were "invented" in limited areas and were subsequently shared among groups (of Cro-Magnons *and* late Neandertals?) using both Aurignacian and Szeletian lithic tools. Nonetheless, it seems to me that we continue to be prisoners of the "cultural" typological categories invented by prehistorians over a century ago, and are thus condemned to look for and "find" gaps and leaps in the record among these units; even if they are not apparent, we seem compelled to invent them. If the "spread" of the early Upper Paleolithic (and, more specifically, the Aurignacian techno-complex) was simultaneous, and if it were simply the result of a westward expansion of ana- tomically modern humans, why did it take so long for "adaptively superior" moderns to move north? This is where the (seemingly contradictory) Iberian evidence makes an important contribution to an attempt at unravelling the mys- tery of the transition.

Finally, is it even clear that the early Upper Paleolithic is all that "advanced"—except, perhaps, in terms of ornaments and art, which do appear *in some sites, in some regions,* early within the Aurignacian and Châtelperronian? Where is the evidence for *radically* superior adaptations (changes in the mode of subsistence strategies, tactics, planning, etc., for the procurement of food and other resources)—other than the admittedly significant development of antler points (Knecht 1993) that were, nonetheless, simply tips for spears thrust or thrown without the assistance of the atlatl? Whether the hypothetical differences in social organization between Middle and Upper Paleolithic systems proposed by Soffer (1992) occurred abruptly and as early as 40,000 years ago in Europe is *completely unknown* and, however intriguing as a possibility, remains merely speculative. And why is there no evidence of complex material culture in Africa until less than 40,000 years ago (with the *possible* exception of harpoons argued to be 90,000 years old from the Upper Semliki Valley in Zaire; Brooks et al. 1995; Yellen et al. 1995)?

WHO WERE THE EARLY AURIGNACIAN PEOPLE?

Regarding "authorship" of early Aurignacian and other early Upper Paleolith- ic artifact assemblages, we have not advanced very much since the Saint-Césaire Châtelperronian Neandertal discovery in 1979. Fragmentary (but supposedly diagnostic) remains characterized as "Neandertal" have been found in association with Aurignacian lithic and osseous artifacts in level G1 of Vindija Cave (Croatia), now dated to 33,000 B.P. (Karavanic 1995; Smith and Ahern 1994).

The late Mousterian levels from this site also yielded Neandertal remains; all have been characterized as evolving in the direction of Upper Paleolithic Europeans (Wolpoff et al. 1981). These finds strongly suggest a significant degree of in situ evolution from Neandertal to Cro-Magnon in southeast-central Europe (see Frayer, this volume).

So far as the Iberian Peninsula is concerned, unfortunately we have neither a firsthand description nor a modern diagnosis of the adult molar and the child mandible found by Obermaier before World War I in the early Aurignacian of El Castillo (Cantabria), and the remains are lost (Cabrera 1984; Cabrera et al. 1993; also this volume). We do not know the real provenience, nor do we have a scientific description, of the so-called Aurignacian (Aurignacian sensu stricto + Gravettian) skull cap from Camargo (Cantabria), found by L. Sierra in the early twentieth century and since lost (Obermaier 1924). The same is true for the possibly deliberately buried (intrusive?) femur found by Amador Romaní in the 1910s in Aurignacian levels in the Catalonian site that bears his name (Bartrolí et al. 1995).

As in France and elsewhere in western Europe, there is still a real absence of well-dated, diagnostic, well-described human fossils unequivocally associated with the Aurignacian before ca. 30,000 B.P. (*pace* Gambier 1989; Gambier and Sacchi 1991). This makes it all the more ironic that, at present, the only well-dated, well-provenienced, well-described hominid remains to be found with both the Châtelperronian and the early Aurignacian are Neandertals (see Garralda 1993, this volume). Whether the late Aurignacian (with unequivocal Cro-Magnons) and the early Aurignacian (with no clear-cut hominid associations) are the same "thing" is anything but clear. This is especially problematic since de Mortillet's and Breuil's notion of "Aurignacian" has already been called into question at least once in the past by Peyrony (1933), and other suggestions for breaking up this "cultural phylum" have been made from time to time (e.g., Djindjian 1993; Pradel 1970). If the Aurignacian is in fact many different things, the association between late manifestations thereof with modern humans is no prima facie evidence that early manifestations must also have been the work of moderns.

THE IBERIAN RECORD

A most striking piece of evidence against the invasionist hypothesis for Upper Paleolithic origins in Europe is the apparent survival of Mousterian technology and Neandertals themselves in southern Iberia as late as 30,000 B.P. As noted above, many independent sources of data from several sites point to this persistence below the Ebro-Tagus line for at least 10,000 years after the Aurignacian (and "modern" humans?) had existed in Catalonia and Cantabria (see reviews by

Table 17.2.   The Latest Mousterian (and Chatelperronian) Dates in Northern Spain

Site	Province	Level	Date (B.P.)	Lab No.
Morín*	Cantabria	10	35,875 ± 6780	SI-951a
Millán	Burgos	1a	37,600 ± 700	GrN-11021
		1b	37,450 ± 650	GrN-1161
Peña Miel	Rioja	C	39,900 ± 10,500	UGRA-128
Ermitons	Girona	IV	36,430 ± 1800	CSIC-197
			33,190 ± 600**	OxA-3725
L'Arbreda	Girona	BE-116	41,400 ± 1600	AA-3778
			39,400 ± 1400	AA-3777
			34,100 ± 750	AA-3776
			44,560 ± 2400	OxA-3731
Romaní	Barcelona	+30 cm	40,000 ± 1500	U-series (USGS)
			42,800 ± 700	U-series (USGS)
			39,200 ± 1500	U-series (USGS)
			41,400 ± 1500	U-series (USGS)
Roca del Bous	Lleida	R3	38,800 ± 1200	AA-6481

* Châtelperronian
** Judged to be ca. 3,000 years too young, owing to error in calculation of background radiation
   (R. Hedges, personal communication, March 28, 1995).

Straus et al. 1993; Straus 1989, 1994) (Table 17.2, Figure 17.2). There was apparently cultural and biological stasis in the Spanish Levant, Andalusia, and southern Portugal, despite the fact that, for 50,000 years, the alleged source of human modernity in Africa lay only 10 km to the south across the Strait of Gibraltar.

At Cova Negra, Carigüela, Cova Beneito, and especially Zafarraya, there are indirect and/or direct indicators that Mousterian technology and even Neandertal anatomy survived into the Würm II-III interstadial and perhaps even into Würm III (i.e., beginning of oxygen isotope stage 2). This would imply that, although somewhat different from those of the early Upper Paleolithic, Middle Paleolithic adaptations were successful, not "fossilized," and at least relatively flexible (witness the "comings and goings" of blade technology throughout the course of the Middle and even late Lower Paleolithic). Neandertals were capable of adopting new cultural solutions, but the differences (at least at first) between Middle and Upper Paleolithic adaptations were not so massive as to force all Neandertals into an Upper Paleolithic mode (or into extinction)—hence the southern Iberian "refuge." On the other hand, *the* Upper Paleolithic developed in fits and starts, unevenly in space, time, and aspect.

**Latest Mousterian and Earliest Upper Paleolithic in
Northern Spain**

In Cantabrian Spain we lack chronometric dates for the end of the long Mousterian sequences at the key sites of El Castillo, Cueva Morín, and El Pendo,

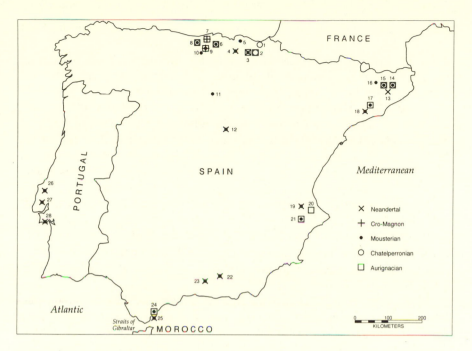

Figure 17.2. Selected sites on the Iberian Peninsula with neandertal or Cro-Magnon remains and Mousterian, Châtelperronian, or Aurignacian assemblages.

1   Ekain
2   Labeko
3   Lezetxiki
4   Axlor
5   Kurtzia
6   Morín
7   Camargo
8   El Pendo
9   El Castillo
10 La Flecha
11 Millán
12 Los Casares
13 Bañoles
14 L'Arbreda
15 Reclau Viver
16 Els Ermitons
17 Romaní
18 Agut
19 Cova Negra
20 Les Mallaetes
21 Beneito
22 Carigüela
23 Zafarraya
24 Gorham's Cave
25 Devil's Tower
26 Columbeira
27 Salemas
28 Figueira Brava

but pollen and sedimentological data suggest that the transition took place under relatively temperate, parkland Interpleniglacial conditions (equivalent to Hengelo; see Straus 1992 and references therein). At Cueva Morín the Châtelperronian, atop a long Mousterian sequence, is radiocarbon dated to 36,000 ± 7,000 B.P. (note the huge standard error). In nearby Burgos Province, there are radiocarbon dates of 37,000–38,000 B.P. for the Mousterian of Cueva Millán (Table 17.2). In Catalonia the situation is far clearer, with several accelerator dates centered on 40,000 B.P. at L'Arbreda (Gerona) and several uranium-series dates also centered on 40,000 B.P. at Abric Romaní (Barcelona). There is a conventional ^{14}C date of 36,400 ± 1,800 B.P. for the Mousterian of Els Ermitons and apparently an AMS date of 38,800 ± 1,200 B.P. for the Mousterian in Roca del Bous (Lérida).

The transition to the Upper Paleolithic seems to take place in both Cantabria and Catalonia under the same moderate interstadial conditions as the latest Mousterian, with no coincident environmental break (Burjachs and Juliá 1994; Butzer 1981; Dupré 1988).

As shown in Table 17.1, there are now ten coherent AMS radiocarbon dates from the Gif, Arizona, and Oxford labs centered tightly on 39,000–40,000 B.P. for basal Aurignacian level 18 at El Castillo. The basal Aurignacian of L'Arbreda now has five AMS dates centered on 38,000–40,000 B.P. from the latter two labs and the Aurignacian of nearby Reclau Viver now has an Oxford AMS date of 40,000 B.P. Aurignacian level 2 of Romaní has been dated by 37 U-series dates to an average of 43,000 B.P. and by two AMS dates to ca. 38,000 B.P. An Aurignacian-like industry at the open-air site of Gato Preto in north-central Portugal has been dated by two TL determinations on burnt flints to 38,100 ± 3,900 B.P. (Marks et al. 1994; but see Zilhão 1993:138).

Despite the chronological indications of an abrupt transition in northern Spain (certainly reinforced by the marked typological and technological contrasts between the Mousterian and Aurignacian lithic assemblages at Romaní; Vaquero 1992), however, there is considerable evidence for technological, settlement, and subsistence continuity between the Mousterian and the Châtelperronian and early Aurignacian in Cantabria (Straus 1990; Straus and Heller 1988). Thus the *nature* of the transition remains unclear and in serious need of detailed investigation. Again, the impression is one of a mosaic of changes.

Late Mousterian in Southeastern and Southern Iberia

In Levantine Spain there are two sites where arguments have been made for a late-surviving Mousterian: Cova Negra (Valencia) and Cova Beneito (Alicante) (Iturbe et al. 1993; Villaverde and Fumanal 1990) (Table 17.3). The uppermost Mousterian levels at Cova Negra, with no Aurignacian-like characteristics, are argued on geological grounds to date to a period of temperate/humid conditions at the end of the Würm II-III interstadial and to the succeeding colder conditions of early Würm III. They would, therefore, by definition postdate 34,500 B.P. Cova

Table 17.3. Dating of the Mousterian (and Neandertal) in Southern Iberia

Site	Chronostratigraphic Placement	Bases of Dating
Cova Negra	Würm II-III and early Würm III	Geochronology: paleosol, followed by *éboulis* (within long sequence back to Eem) ^{14}C dates of 29 ± 6, >29, and >29 kya
Cova Beneito	Würm II-III and early Würm III	Sedimentology, palynology, and macropaleontology: temperate/humid, then cold/dry AMS-date of 30.2 ± 7 kya (or 38.8 ± 1.9 kya)
Cariguëla[†]	Würm II-III and early Würm III	Geochronology: temperate/humid, then dry/cold (within long sequence back to Eem)
Zafarraya[†]	Würm III	Micropaleontology U-series and AMS dates of 32–29 kya
Figueira Brava[†]	Würm III	U-series date of 30.6 ± 11 kya ^{14}C dates of 30.9 ± 0.7 kya and 30.1 ± 0.55 kya
Pedreira das Salemas		^{14}C date of 29.9 ± 1.0 kya
Columbeira[†]		^{14}C date of 28.9 ± 1.0 kya
Foz do Enxarrique		U-Th date of ca. 33 kya
Lapa dos Furos		^{14}C date of 34.6 ± 1.0 kya
Caldeirão		^{14}C date of 27.6 ± 0.6 kya

[†] Neandertal

Beneito (Alicante) contains terminal Mousterian lithic assemblages with some indications of change in an Upper Paleolithic direction plus an engraved bone and a perforated tooth. Sedimentological, palynological, and microfaunal evidence indicates the existence of colder, somewhat drier conditions at the end of the Mousterian than had pertained earlier. This is thought to mark the beginning of Würm III, a conclusion supported by one conventional radiocarbon date of 30,200 ± 700 B.P. but contradicted by an AMS date of 39,000 ± 1,900 B.P. After a depositional hiatus, the overlying late Aurignacian is dated to 33,900 ± 1,100 B.P. by AMS and 26,000 ± 900 B.P. by conventional ^{14}C (Iturbe et al. 1994). The only other so-called Aurignacian levels in southern Spain that have been dated are at Les Mallaetes (Valencia—29,700 ± 600 B.P.) and Gorham's Cave (Gibraltar—two dates of 28,000 B.P.).

It is in Andalusia that we currently have the most powerful evidence for survival of Mousterian-using Neandertals into Würm III (Table 17.3). At Cariguëla (Granada) there is a very long Mousterian sequence, unfortunately with no clearly provenienced radiometric determinations. Geological arguments place the latest typical Mousterian levels in a cold period that followed the Würm II-III interstadial, hence Würm III (Vega 1990). A Neandertal parietal was found in one of these terminal Mousterian levels—roughly correlated with a zone in the nearby Padul pollen core that indicates cold, open environments in the region. Zafarraya (Málaga) has yielded a Mousterian assemblage in carefully controlled recent excavations of a deposit that has microfaunal indicators of cold, Würm III conditions with two series of coherent dates (AMS and U-series) ranging between 32,000 and 29,000 years ago (Barroso et al. 1984; Hublin 1994). The Mousterian level also produced Neandertal remains, notably a magnificently preserved mandible.

Finally, recent excavations in southern Portugal, at Figueira Brava cave, have uncovered a Mousterian industry with an associated Neandertal-like premolar, now dated by two conventional ^{14}C dates and one U-series date ranging between 30,000 and 31,000 years ago (Antunes 1990). This would seem to confirm earlier references of Portuguese Mousterian sites with late dates: Pedreira das Salemas at 29,900 ± 1,000 B.P. and Gruta Nova da Columbeira at 28,900 ± 1,000 B.P. (Antunes et al. 1989). In turn it seems to be confirmed by a recent AMS date of 27,600 ± 600 B.P. for a level capping the Mousterian sequence in Gruta do Caldeirão (Zilhão 1993).

All these indications suggest to me that Mousterian and Aurignacian technological systems overlapped for 10,000 years on the Iberian Peninsula. This calls into question, I would argue, the supposedly overwhelming adaptive superiority of the latter vis-à-vis the former, at least in the environments of the southern half of the peninsula.

Subsistence Evidence

Space limitations do not permit me to detail the subsistence evidence from Mousterian and early Upper Paleolithic contexts in Iberia. The reader is referred to such syntheses as those of Altuna (1972; Altuna and Mariezkurrena 1988), Cabrera (1984), Clark (1987), Clark and Yi (1983), Estevez (1987), Freeman (1973, 1981), Klein and Cruz-Uribe (1994), LAUT (1992), and Straus (1977, 1983b; Straus and Heller 1988).

In Cantabria there is considerable evidence for subsistence continuity between Mousterian, on the one hand, and Châtelperronian and Aurignacian, on the other, with each occupation exhibiting small assemblages with few elements per MNI of bovines and horses plus larger amounts of red deer, but never in massive quantities (as would be the case in the late Upper Paleolithic levels). Ibex, chamois, and boar were not significantly hunted in either period, and there is hardly any evidence for fishing or shellfishing (despite the proximity of the Atlantic coast, especially during Interpleniglacial times). There is some tendency for bears and large carnivores to decrease in early Upper Paleolithic levels relative to Middle Paleolithic ones, as humans seem to have made more frequent use of caves over time (Straus 1982). Klein and Cruz-Uribe (1994) have not found any evidence for large-scale specialized hunting of red deer in the Aurignacian of El Castillo, in contrast to the old report of Vaufrey cited by Cabrera (1984) in which it is unclear as to whether he was referring to MNI or (more likely) to NISP. Caves utilized by humans are essentially the same in the Mousterian, Châtelperronian, and Aurignacian.

We have very few situations for comparing Mousterian and Aurignacian faunas from specific stratigraphic sequences in Mediterranean Spain, where, in general, horses, red deer, bovines, and rabbits were the main fauna in both periods. At L'Arbreda there seems to be an increase in both horses and rabbits in the Aurignacian in relation to the underlying Mousterian horizons, with a notable decrease in cave bears (as in many Cantabrian sites). It is during the late Upper Paleolithic that Mediterranean assemblages become heavily dominated by either red deer or ibex, depending on site location (as in Cantabria). Shellfishing and fishing appear in the record (notably at Nerja in Málaga) much later, in the Magdalenian, again paralleling the Cantabrian situation. Subsistence intensification (with both situational specialization in the massive hunting of either red deer or ibex and overall diversification to include the taking of swift, wary and/or dangerous game; fishing; and shellfish gathering) developed not at the Middle to Upper Paleolithic transition but during the course of the late Upper Paleolithic, some 20,000 years later. Steep, upland and coastal habitats were not systematically exploited by humans until then. There is thus subsistence and settlement continuity across the Middle to Upper Paleolithic transition, whether the technological transition was gradual (as in Cantabria) or more abrupt (as in Catalonia). And in southern Iberia (where admittedly the data are very scarce), there is no indication of adaptive change between earlier and late-surviving Mousterian horizons.

Finally, despite the evidence of early mobile art at Geissenkloesterle in southwestern Germany (ca. 38,000 years ago) (and possibly at le Trou Magrite in Belgium in the same period), the early parietal drawings in Chauvet Cave (ca. 31,000 years ago), the relatively early handprints in Gargas and Cosquer caves (ca. 27,000–28,000 years ago), and the relatively early bas reliefs at La Ferrassie (ca. 28,000 years ago), there is as yet scant evidence of early art or ornament in Iberia. The first beads in Cantabria are from a relatively late Aurignacian level at El Pendo (Barandiarán 1980) and although the simple linear rock engravings at La Viña, El Conde, and Hornos de la Peña (?) may be attributed to the Aurignacian, their age is in fact unknown.

CONCLUSIONS

At least in terms of archaeological evidence, we are a long way from invoking a simple invasion of Europe from the Near East in order to explain the appearance of the Upper Paleolithic. In contrast, it seems reasonable to speak of the *development* of Upper Paleolithic life ways, not at a single date of 40,000 B.P., but over a period of 30,000 years. This would seem ironic, since the Iberian dates have made the early Aurignacian appear to be an instantaneous phenomenon across all of Europe. Yet, deconstructed and leaving aside the unresolved question of the fate of the Neandertals, the Aurignacian is merely an archaeological construct, with a unity more apparent than real and perhaps the result of both technological convergence and specific esoteric inventions combined with group contacts and diffusion. In many cases the new technologies were grafted to one degree or another onto existing ones; in other cases the break was more abrupt. We need to understand "why" in either case. The convergence in terms of lithics between some aspects of the European and Near Eastern "Aurignacians" should serve as a lesson to us: formal similarity does *not* equate with contemporaneity or cultural identity. In any event, there is no compelling evidence that the new technologies conferred such an advantage on their bearers as to immediately supplant the old. This is most spectacularly demonstrated by the case of the Mousterian "refugium" in southern Iberia, so close to Africa with its Aterian techno-complex and to northern Spain with its precocious Aurignacian, and yet so far. . . .

In the murky realm of fossil hominids, who knows what surprises might lie in store for us? Who would have predicted Châtelperronian Neandertals, Aurignacian Neandertals, or 30,000-year-old Neandertals? Yet all are now being accepted, at least by many workers. The facts have a tendency to confound our expectations—or at least they should.

NOTE

1. In both caves, however, the ca. 38,000 B.P. dates come from higher zones of the respective early Upper Paleolithic layers than do the earlier dates in excess of 43,000 years (J. K. Kozłowski, Universytet Jagiellonski, personal communication 1995). It is difficult to compare TL and AMS dates, the former being highly imprecise and the latter systematically too young in this age range.

18

Biological and Archaeological Classifications
Boundaries, Biases, and Paradigms in Upper Paleolithic Research

BETSY SCHUMANN

Classification, the grouping of individual objects (organic and inorganic) into categories or classes, is a necessary part of human perception and thought. In this capacity, boundaries aim to discriminate between and order these objects based on their similarities and differences; the result is the reduction and division of large quantities of variable data into groups of a smaller, more manageable size. The need to order and simplify diversity and the ability to distinguish objects as separate entities forms an intrinsic part of scientific and historical academic discourse. The fields of anthropology and archaeology are no exception in that they have historically relied heavily on systems of classification to provide a framework for more theoretical considerations.

In both archaeology and anthropology, scholars have constructed taxa (groups of distinct organisms or objects) and categories (the ordering of taxa within a hierarchic classification) of like artifacts and hominid specimens to form larger classification schemes. The biologist accomplishes this with clades, populations, species or subspecies, whereas in archaeology, groups such as types, industries and cultures are constructed to identify and describe artifact diversity in the material record. The divisions used by the archaeologist and anthropologist, however, are not always analogous. Classifications in archaeology are created according to criteria devised by the examiner and do not necessarily reflect or rely upon underlying processes; systematics, on the other hand, used more frequently by biologists and anthropologists, orders entities based on an underlying process, such as evolution. The distinction between these two methods for ordering is hazy at best. In this paper, taxonomy (the method for naming and grouping objects, an essentially descriptive science), classification (the ranking and ordering of taxa), and systematics (the ordering of groups based on comparison and an underlying process) are not used interchangeably. A taxon is a group of organisms (or objects) recognized as a formal unit, and each taxon is assigned membership to a category (Simpson 1963). Each taxon with its categorical rank is then superordinated and subordinated to form a hierarchic classification.

It is the division into taxa or groups which requires boundaries, a discriminating grid based on assumptions, to be imposed upon the data. Today, the recognition of meaningful taxa can be determined more easily, although not necessarily more objectively, with the aid of computerized database management and multivariate statistical approaches. This paper is concerned with how these boundaries are formed and ultimately how they affect our interpretation of the various databases in the archaeological and paleoanthropological records of the late Pleistocene.

Although necessary to any analysis, it is the imposition of boundaries (taxa, categories, and systems) upon the data which characterize interpretations of diversity. How scholars draw these boundaries reflects the various paradigms within which they work. It is the paradigm, a collection of biases and preconceptions about the world or a system for understanding and organizing certain aspects of it, that affects the way in which patterns of diversity are interpreted (G. A. Clark 1993a). Paradigms, therefore, have a strong effect on how boundaries between objects (tool types, cultures, etc.) and individuals (species, subspecies, races, etc.) are constructed. Archaeological and paleoanthropological paradigms manifest themselves in the way that they define the relationship (i.e., make assumptions) between the interpreter and that which is interpreted and the way in which the interpreter ultimately defines culture and the relationship between humans and their culture. Classifications ultimately rely on assumptions, and it is those assumptions that form the basis of the paradigm within which the researcher works. It is the more complex, derived paradigms that become methodologically problematic, with large bodies of data and research relying on mutually exclusive systems or assumptions. Current research on the Upper Paleolithic suggests that boundaries are imposed on the archaeological and paleontological databases in order to support the particular paradigm or system that each scholar advocates.

In this chapter, the historical use and construction of biological and archaeological classification schemes and the development of those paradigms that affect our interpretation of the fossil hominid and archaeological records are discussed. I will explain how these boundaries can infuse bias into an analysis and permit the substantiation of those paradigms upon which they are dependent. Finally, I will demonstrate how various paradigms and the hominid and archaeological boundaries which have been defined affect my own work on Late Pleistocene biocultural evolution and the origins of modern humans.

BIOLOGICAL CLASSIFICATION

Since Aristotle's attempt to document the diversity of the animal world (*Historia Animalium,* fourth century b.c.) and order it according to a *scala natura,* classifications have been used to construct a hierarchy of taxa based on the similarities of organisms and the relationships between organisms inferred from their taxonomic characters. Aristotle, through the use of comparative anatomy and reproductive biology, grouped 580 different kinds of animals into general assemblages (Mayr 1982). Since then, the attempt to order the natural world dominated many aspects of scientific work, yet it was not until Carolus Linnaeus (in his *Systema Naturae,* 1758) that a simple but straightforward and methodological classification scheme, based on only four categorical levels, was adopted. Linnaeus employed downward classification based on logical division (in contrast to the upward classification schemes used today) in order to identify and categorize all known plants and animals. The years which followed were marked by a growing interest in taxonomy, with the work of Cuvier, Lamarck, and Buffon and eventually Darwin bringing an end to the use of downward classification (Simpson 1961b). The application of upward classification by grouping specimens together based on common descent revolutionized taxonomic thinking (Mayr 1982).

In his *On the Origin of Species* (1859), Darwin applied the notion of higher and lower categories (intrinsic to an upward classification, whereby a category refers to the rank or position of the different taxa) to his classification scheme. He stated that "from the dawn of life, all organic beings are found to resemble each other in descending degrees, so that they can be classed in groups under groups" (1985:397). Darwin not only developed his theories of evolution and natural selection, but in doing so, he established the importance of discriminating between characters similar because of common descent or through convergence (adaptations acquired independently in several unrelated evolutionary lines). Darwin rejected the belief that the more important a structure is for an organism's survival, the more important it will be for classification (Mayr 1982), supporting his belief in the importance of convergence.

By the 1960s, the need for a method that would systematically quantify similarities and differences between taxa (thereby reducing observer subjectivity) led to the development of numerical phenetics. This methodology allowed for degrees of similarity to be quantified and for a subjective (qualitative) taxonomy to be converted into an objective (numerical) taxonomy (Sokal and Sneath 1973). Phenetics, while concentrating on variation at a more individual level (closely related forms rather than the higher taxa), differs from other classification methodologies in the weighting of characters, the principle through which similarity is established. Once the distance between the operational taxonomic units (clustered taxa) is determined, phenetic classifications, displayed as a phenogram or tree, result from the further clustering of taxa into different hierarchical levels based on the overall percentage of similarity of those taxonomic units to be included (Scotland 1992). The resultant tree, based on distance between the taxa, may have phylogenetic implications although the phylogenetic significance is not the ultimate goal of phenetics. It is merely a methodology for resolving relationships between taxa.

Closely following the work of Darwin, the evolutionary classification scheme frequently used today, cladistics, arose as an alternative to numerical phenetics. Cladistics relies on the presence of autapomorphies (derived characters acquired by one sister group and not the other); the major difference between cladistics and phenetics is that cladistics differentiates between ancestral or derived characters and phenetics does not. Cladistics is based on genealogies (in the form of cladograms): a sequence of branching patterns (dichotomies), each representing the splitting of a parental species into two daughter (or sister) species (Hennig 1966). Cladistic classifications weight the importance of autapomorphies and synapomorphies (shared traits either derived from a common ancestor or acquired through convergence), termed *characters,* which can be either discrete or continuous. Characters are coded, examined for their polarity (the direction of character evolution—Kitching 1992), and then used to construct the most parsimonious (minimum length) tree. While cladistic analysis has been criticized on the grounds that it ignores anagenetic change occurring within the split lines (sister taxa; Scotland 1992) and for the lack of a unified, consistent method for defining, selecting, and discriminating between characters (Clark 1988, 1989a; for an alternative view see Foley 1987b), it remains a powerful and relatively objective tool in modern biological and paleoanthropological classification.

The history and evolution of biological classification is already covered in a vast multitude of publications; the need to explain and document biological diversity, establish taxa, and ultimately order them dominates scientific discourse. Human evolution is no exception. With the emphasis that paleoanthropology has placed on fossil hominid and modern human diversity, greater weight has been given to the building of lesser taxa, below the species level. Species were once thought to represent the lowest level of diversity within a taxonomic classification and were defined on the basis of their distinctness rather than degrees of difference and their relationship to other species (Mayr 1982). Scholars of human evolution ultimately attempt to classify fossil hominids into species based on their attributes. As the fossil record becomes more complete with the discovery of new specimens, classifications both above and below the species level (subspecies, grades, races, etc.) are now used to describe and classify the diversity observed in the hominid record (Rensch 1959).

PALEOANTHROPOLOGICAL CLASSIFICATIONS

The category of subspecies in animal (including human) and plant classifications is used to denote a regional population of a polytypic species, whereby each subspecies occupies a distinct geographical niche without being reproductively isolated. Further attempts to classify variation below the subspecies level are not uncommon in paleoanthropology, and the use of such divisions is a central issue in the debate concerning modern human origins. The fundamental issue, however, in hominid classifications is the ability (or inability) to sort assemblages of fossils into discrete groups (Tattersall 1986). In later human evolution, the classification of fossil *Homo erectus,* and archaic and anatomically modern *Homo sapiens* according to their morphological variability into species, subspecies, races, grades, and/or populations reflects the various paradigms and models within which they are constructed. Following the stimulus of Weidenreich (1943a), who applied grade distinctions to characterize the morphological continuum he observed from the Indonesian *Homo erectus* material to modern Australasians, Coon (1962) categorized modern *Homo sapiens* (below the subspecies level) into five distinct, allopatric races (grades and clades). Coon's application of racial divisions to modern human variation supported his belief in the antiquity of these races, based on independent evolutionary lines and the continuum of certain morphological features from *Homo erectus* through the various archaic stages and their distinct geographical locations. Coon's theory of the parallel evolution of human races, however, is more "extreme" than that of Weidenreich's, which required genetic exchange (gene flow) across regions to prevent speciation from occurring. It is Weidenreich's work (not Coon's, which has been rejected by most modern scholars) that serves as the basis for the current multiregional model of modern human origins (Frayer et al. 1993).

The division of Pleistocene hominids into grades and clades rather than distinct species is frequently used by advocates of the multiregional model of modern human origins (Brace 1964a, 1967; Frayer et al. 1993; Thorne and Wolpoff 1992; Wolpoff 1989b). Multiregionalists emphasize grade and clade distinctions (based on transitional, mosaic fossils) and are in favor of lumping specimens into taxa rather than splitting them (Clark 1991). Contrarily, those scholars who support a single origin for all anatomically modern humans (splitters) separate groups of hominids at a species and subspecies level, without intermediate grades, and assert that modern *H. sapiens* arose as a new species (Groves 1989b; Howells 1976; Stringer 1989b, 1992a; Stringer and Andrews 1988b; Stoneking and Cann 1989). These two models differ fundamentally in their attribution of fossil specimens to "archaic" *Homo sapiens* (Lahr 1994). The boundaries imposed on the hominid record are not only dependent upon the paradigms which dominate the debate over modern human origins, they also reflect the chronological, geographical, and archaeological frameworks within which they are discussed.

Recently, evidence from the fields of genetics and linguistics has been used to classify and "group" late Pleistocene hominids and recent modern human populations (Cann et al. 1987; Cavalli-Sforza 1991; Cavalli-Sforza et al. 1988; S. Hedges et al. 1992; Relethford and Harpending 1994; Rogers and Jorde 1995a; Stoneking 1993; Templeton 1993).

While the expansion of paleoanthropological research into other scientific domains enhances the research conducted on late Pleistocene human evolution, it is not without difficulty. Too often these lines of evidence are *not* kept separate in the compiling and analysis of a database. One cannot replace or fill a void in the biological record with evidence from archaeological or genetic data; fossil hominid taxa can only be defined on the basis of morphological characteristics, not, for example, on the absence or presence of archaeological evidence (Bar-Yosef et al. 1990). However, once the archaeological, skeletal, or genetic database is established, then it can and should be used to test the other lines of evidence which have been assembled in the other fields; biological boundaries (subspecies, populations) can be used to test the validity of, for example, archaeological classifications (industries, cultures, etc.). A link or correlation (either temporal or causal) between these lines of evidence, however, must be established first if the data are to be used to define and confirm taxonomic and behavioral units (Dibble and Chase 1990). The use of these often historically constrained and frequently paradigm-dependent classifications across disciplinary boundaries can be a powerful analytical tool, yet it may require the reevaluation of such classifications as well as the inherent paradigms under which they are constructed.

ARCHAEOLOGICAL CLASSIFICATIONS

Not unlike the biologist, archaeologists have concentrated on the classification of inanimate objects found in the archaeological records of past peoples. Theoretically, archaeologists use classificatory schemes formulated on the basis of biological analogies to sort and identify artifacts in a similar way to that in which animals and plants are classified (Shanks and Tilley 1987).

Archaeological classifications (taxonomies) can be hierarchical in much the same way that biological classifications are. However, when an archaeological classification (typology) is taken to mean a set of mutually exclusive classes (Adams and Adams 1991; Voorrips 1982), it is not always hierarchical. Only when different categories are combined and assumptions are made about their relationships do they become hierarchical (Dunnell 1971). A typology, therefore, is only one kind of archaeological classification. Classifications comprise categories (unlike their use in biological classifications, archaeological categories are taxa and refer to objects grouped together on the basis of similar features) or classes; a typology, then, is a classification which is used to segregate or sort objects into predefined, discrete groups (categories). In other words, the act of classifying is the creation of categories; the formulation of a typology is the method used for sorting artifacts into those categories, all at the same level of abstraction. The term *type* therefore can be used to denote a category with sharp, well-defined boundaries. Types, defined by certain attributes or variables, are in all classifications mutually exclusive and differentiated from each other by their sharp boundaries (cf. Willermet and Hill, this volume).

Both classifications and typologies use types to classify artifacts. Archaeologists describe, define, and sort artifacts into types on the basis of their similar attributes in order to obtain an image of "structure": the order, pattern, and relationship among artifacts (Spaulding 1982). The differential treatment of attributes (and in fact the meaning of what constitutes an attribute) has led to misunderstandings and strong differences of opinion concerning classifications among archaeologists (Hodson 1982). Artifacts can be classified not only by their individual attributes but also by "artifact clustering," whereby certain combinations of attributes will define particular classes or types (Hodson 1982; Adams and Adams 1991). As an alternative to artifact clustering, Spaulding (1977) advocates an approach to artifact classification (known as variable association) whereby the variables of an artifact are examined for their dependence on or independence from other variables of that same artifact. Although frequently used interchangeably, the terms *attribute* and *variable* are defined by Spaulding and Hodson differently. In the simplest of analogies, a variable has a value or property (color, size, material) and an attribute is one of the ways in which a variable is manifested (blue, large, obsidian). It is the search for patterns and relationships between variables and attributes (including the spatial and temporal ordering of the data) which is the central concern of artifact classifications. Nevertheless, archaeological classification in general attempts to group artifacts into assemblages (larger categories), whereby the degree of similarity between assemblages indicates their relatedness; similar assemblages are then grouped into facies, which according to their similar characteristics are classified further as traditions or cultures (Sackett 1991).

While little agreement has been reached on the best method for classifying artifacts (Clarke 1978; Read 1982), artifact classifications ultimately are used to convert artifact variation into culturally meaningful terms (Sackett 1991). Therefore, only those attributes that have cultural "recognition" are used as type-defining attributes (Dunnell 1971; Rouse 1960). Beyond the tenets of artifact and industrial typologies, however, are the larger issues of interassemblage variability and the succession or sequence of cultures. I will now turn my focus to these issues and how they relate to European Upper Paleolithic research.

European Paleolithic research was dominated throughout the first half of this century by French archaeologists. The earliest typological systematics (those of Lartet and Christy 1865 and de Mortillet 1883), based on *fossiles directeurs,* were characterized by the succession of one industry after another in a unilinear, chronological sequence. It was not until the work of Breuil and Peyrony that the linear sequence of cultural change was questioned (Sackett 1991). In their construction of industrial (technological) subdivisions, both archaeologists demonstrated that the progression from one industry to another need not be logical, linear, or evolutionary. It was Peyrony's (1933) reevaluation of the Aurignacian which led to his observation that archaeological subgroups within an industry (as well as the industries themselves) could not only coexist but evolve in parallel

(Sackett 1991, although see Straus 1987b for an alternative view). Peyrony's typological classifications of the Upper Paleolithic and his biocultural interpretation of regional, typological variation laid the groundwork for the further development of what is today known as the Old World paradigm. Both the earlier phylogenetic classifications and Peyrony's parallel phyla schemes (while still dependent on the presence of fossiles directeurs) interpreted regional, temporal, and cultural variation as "behavioral expressions of distinct races of *Homo sapiens*" (Sackett 1991:125-126). Bordes furthered this work in two ways: firstly, he developed a detailed assemblage-based systematics whereby tools could be classified and assemblages described in a quantitative manner *(la méthode Bordes)* and secondly, he reevaluated the significance of the stratigraphic sequence of changes, finding (in concordance with Peyrony) that changes within the archaeological record were neither directional nor gradual (Bordes 1992). This confirmed his belief that, as in the Old World paradigm today, archaeological divisions (between cultures) and variation (within one culture) were characterized by human ethnicity. Cultural diversity was interpreted as the different material expressions of different people (Clark 1991; David 1966; Hayden 1982).

The ideological differences between the Old World paradigm and the New World paradigm have their roots in the traditional typological classifications (and their interpretations) of Peyrony and Bordes. The New World paradigm developed in direct response to the culturally dominated European interpretations of the Paleolithic record. These two paradigms differ in their definition of culture and the relationship between material culture and its makers. While the ideological differences between these paradigms have recently been discussed at length (see Straus 1987b; G. A. Clark 1993a, ed. 1991), they can be summarized (albeit briefly) as follows: in the Old World paradigm, culture is viewed as a material expression of distinct ethnic or social groups with cultural change occurring abruptly, as a result of population migration and replacement (Bordes 1961a, 1972, 1973, 1992; David 1966; de Sonneville-Bordes 1960, 1966); according to the New World paradigm, culture, an adaptive response to various environmental conditions, is independent of any human ethnic or social affiliation and can be characterized either by its spatial and temporal distribution or by functional variability (Binford 1973; G. A. Clark 1993a; Shanks and Tilley 1987; Straus 1987b). Models of Upper Paleolithic adaptive strategies stress either the physical environment and population changes (Bahn 1983; Bailey 1983; Clark and Straus 1986; Straus 1976) or the social environment (Conkey 1978, 1980; Gamble 1982, 1986; Madden 1983; Wobst 1976), in other words, sociocultural networks that linked human groups over vast regions as a result of interactions maintained for the exchange of mating partners, material goods, or information (Mueller-Wille and Dickson 1991). Not only do these paradigms and models reflect interpretations of culture, style, and artifact patterning within the landscape, they also affect the formation of boundaries between and within assemblages, industries, and cultures. Classifications, once thought to be "neutral" devices and independent of theory, have now become as paradigmatic as the interpretations themselves.

Today, however, the debate surrounding the New World/Old World paradigms may not be as polemic as some authors think (Harrold 1991b). Both models recognize the importance of using typologies and systematics in conjunction with other lines of evidence, such as settlement patterns, raw material procurement, and subsistence activities. The current controversy over interpretations of the Paleolithic record may have arisen from a misunderstanding of the terms and concepts used to describe and interpret the material record (Harrold 1991b). It is still unclear what the French scholars meant by "ethnic" or "social" groups. Does this mean that the Aurignacians were an ethnic group who conformed to an identity-conscious social unit, or that they are biologically discernible from their predecessors and successors? It is Harrold (1991b) who feels that the simple definition of such terms and concepts (such as ethnic, abrupt, and gradual) may help to alleviate the problems of much of this historically constrained debate.

The controversy over modern human origins can be characterized by many of the same issues that dominate Upper Paleolithic research. Advocates of a multi-regional origin of modern humans (and most likely advocates of the New World paradigm) postulate that the change from anatomically archaic to modern *Homo sapiens* (intraspecific, grade changes) does not imply concomitant behavioral change (Clark and Lindly 1988, 1989b; Lindly and Clark 1990b). Because the transition from a Middle to Late Stone Age (or Paleolithic) technology does not coincide with the biological transition to anatomically modern humans in Africa, the Levant, and China (for example), the material culture of early anatomically modern humans is continuous and "mosaic" in nature across this transition (Lindly and Clark 1990b). Whereas the Old World paradigm postulates that cultural change indicates population replacement (Klein 1992; Mellars 1989b), the New World paradigm denies any such relationship, based on the lack of concomitant behavioral and biological change. Rather, the differences noted between assemblages are argued to reflect the various typological systematics by which the artifacts are described (Klein 1992). Even in Europe, however, where anatomically modern humans are found only in association with Upper Paleolithic technology, the New World Paradigmatists would claim (1) that the transition to early Upper Paleolithic industries (i.e., the Châtelperronian, Szeletian, Uluzzian) does not coincide with the earliest appearance of anatomically modern humans in Europe, but occurs prior to their appearance, and these industries are found in association with both neandertals and modern humans (Allsworth-Jones 1986, 1990; Kozłowski 1982, 1988a; Lindly and Clark 1990a) and (2) that the true "human revolution" did not occur during the early stages of the Upper Paleolithic but at some time after 20,000 B.P. (Clark and Lindly 1989b; Duff et al. 1992; Lindly and Clark 1990b; Straus and Heller 1988). The characterization of cultural differences (i.e., the boundaries between Mousterian, Châtelperronian, Aurignacian, and Gravettian assemblages) closely follows the polarity of the New

World/Old World paradigms. Advocates of the Old World paradigm see clear-cut, sharp boundaries between these industries (Mellars 1989b; White 1989b), whereas others see these boundaries as indiscernible and obscure, representing gradual and cumulative cultural adaptations (Lindly and Clark 1990a, 1990b; Straus 1990; Svoboda 1988; Karavanic 1995).

BIOCULTURAL CHANGE IN THE EUROPEAN UPPER PALEOLITHIC: BOUNDARIES AND BIASES

Inherent in past studies of hunter-gatherer adaptive strategies is the separation of the biological and archaeological records. The use of more than one avenue of inquiry (be it genetic, archaeological, biological, or linguistic) has been recognized only recently as useful for understanding human biocultural evolution at the intraspecific level during the late Pleistocene. My conviction in the necessity of using more than one avenue of inquiry, albeit a large undertaking for one person, results from a strong interest in human "biocultural" evolution (as well as the inability to decide whether to conduct my doctoral research in biological anthropology or archaeology) and, more important, from the opinion that the biological and archaeological records of this period should be examined in parallel. Certainly, during a period such as the European Upper Paleolithic where we witness a biological and cultural restructuring of human societies, these two records need to be examined in conjunction with each other. Studying the biological changes *without* an in-depth analysis of the archaeological record (or vice versa) only serves to paint a very incomplete picture, certainly if one assumes that there is a direct correlation between the material culture and those who produced it (as in the Old World paradigm). The understanding of modern human origins and their subsequent evolution, be it late Pleistocene, Mesolithic, Neolithic, or in the present (i.e., documenting modern human diversity), requires a multifaceted research protocol: an examination, understanding, and correlation of all facets of all available records.

Research concerned with the Upper Paleolithic is, I feel, ultimately dependent upon the substantiation of various theories of modern human origins and reflects the need to "cut and divide" this period: to form chrono-geographical boundaries in the biological and archaeological data. The imposition of such boundaries results, on a general level, from the historical separation of the disciplines of biological anthropology and archaeology, and on a more specific level from the historical division and classification of the archaeological and fossil evidence by their various cultural associations and their spatial and temporal provenances.

The work in which I am currently engaged deemphasizes traditional European typological classifications by rejecting simplistic claims for the "cultural" or "ethnic" associations of the standard Upper Paleolithic classification units. Because the archaeologist classifies human ethnic groups according to culture (tool type or artifact style), and because no exact definition has been offered to explain what is inferred by human ethnic groups, then perhaps the establishment of a more specific, dependent relationship between human groups and their material culture should be documented rather than assumed, as it has been for the Upper Paleolithic. To date, the only hominids found in direct association with the Aurignacian are anatomically modern *Homo sapiens* (although, see Lindly and Clark 1990a; Karavanic 1995 for an alternative view); the assignment of an archaeological culture to ethnic human groups implies that the Aurignacian is a biological "group" or "population" that differs from, for example, the Gravettian or Solutrean. Defining human boundaries (ethnic, cultural, and even biological) according to typological industries or facies implies an equally distinct biological boundary. Before we can begin to define cultural or ethnic groups during the Upper Paleolithic on the basis of material culture and understand the implications of such a definition (as the Old World paradigm would suggest), it appears not only necessary but logical to determine whether in fact there is any biological pattern that concurs with and conforms to patterning and divisions in the archaeological record.

My research has examined the European Upper Paleolithic for patterns of biological change within three frameworks: chronological, geographical, and archaeological. Unlike other research which concentrates on the pattern of biological change during this period (Frayer 1978, 1984, 1988; Smith and Raynard 1980; Smith 1982, 1984), I chose not to "cut" and "divide" the fossil record into regional or chronological units; my analysis examined the hominid fossils (65 in total) as individual specimens. If there are chronological or regional differences in cranial morphology, then the analysis of individual specimens will indicate any such pattern. I feel very strongly that by imposing a boundary (whether chronological, like Frayer's, or geographical, like Smith's) onto the fossil record unnecessarily interjects bias that ultimately can obscure any natural pattern within in the biological record.

An example of how the use of different boundaries will change the results of an analysis makes this point clear. Using only four metric cranial traits, I examined the Upper Paleolithic hominid data for regional patterns. I conducted this analysis five times, each time using a different geographical boundary. Table 18.1 lists the hominid sites included in this example and their regional placement for each test. Test 1 divided the hominids into one of five groups based on the geographical location of the original finds: southwest France, northern Europe, Grimaldi, Italy, and central Europe. Test 2 divided the sample into two groups, northern and southern, with the boundary set at 44° N latitude. Test 3 also divided the sample into northern and southern groups, but with the boundary set at 48° N latitude, resulting in the hominids from southwest France being considered "southern" as opposed to northern as in Test 2. Test 4 divided the hominids into two groups, eastern and western, with the boundary placed at 18° longitude. And

Table 18.1.   Hominid Specimens Grouped Geographically for the Five Tests

Hominid Site	Test 1: Regions	Test 2: 44° N	Test 3: 48° N	Test 4: 18° E	Test 5: 19° E
Abri Pataud	Southwest France	north	south	west	west
Arene Candide	Italy	south	south	east	west
Barma Grande	Grimaldi	south	south	west	west
Brno	Central Europe	north	north	east	east
Bruniquel	Southwest France	north	south	west	west
Cap Blanc	Southwest France	north	south	west	west
Chancelade	Southwest France	north	south	west	west
Cioclovina	Central Europe	north	north	east	east
Cro-Magnon	Southwest France	north	south	west	west
Dolni Vestoniče	Central Europe	north	north	east	east
Gough's Cave	Northern Europe	north	north	west	west
Grottes des Enfants	Grimaldi	south	south	west	west
Laugerie Basse	Southwest France	north	south	west	west
Maritza	Italy	south	south	east	east
Mladeč	Central Europe	north	north	east	east
Oberkassel	Northern Europe	north	north	west	west
Ortucchio	Italy	south	south	east	east
Pavlov	Central Europe	north	north	east	east
Le Placard	Southwest France	north	south	west	west
Predmôsti	Central Europe	north	north	east	east
Saint Germain	Southwest France	north	south	west	west
San Teodoro	Italy	south	south	east	east
Vogelherd	Central Europe	north	north	east	east

the final test, Test 5, also created an east-west division, with the boundary placed at 19° longitude. In Test 5, the specimens from Arene Candide fall into the western group, whereas in Test 4 Arene Candide is part of the eastern group.

A nonparametric, one-way analysis of variance was performed five times on each of the four variables using the different geographical groups described above to determine whether or not significant differences exist between the regions. Table 18.2 lists the significance values for the cranial variables examined in each test. The specific results of these tests are not relevant to this discussion; what is important is the fact that three different sets of results emerge. The addition or removal of certain specimens from the various groups led to conflicting results.

Of course there are other ways in which these specimens can be grouped. The point is that such boundaries can "lead" the analysis towards an expected (or desired) conclusion, and the results of such analyses are inherently biased. This bias is dependent upon and reflects the paradigm or model of modern human origins which the scholar advocates. Hypotheses and assumptions made about the Pleistocene record are not arbitrary and must be tested; boundaries that are based on these hypotheses and assumptions lead to research which is concep-

Table 18.2. Significance Results (*p* values) of ANOVAs Performed on Five Different Regional Groupings

Variable	Test 1	Test 2	Test 3	Test 4	Test 5
Maximum cranial length	NS	NS	*	NS	*
Fronto-malar breadth	NS	NS	NS	NS	NS
Maximum frontal breadth	*	NS	NS	NS	NS
Basion-prosthion length	**	NS	NS	NS	NS

* $p < 0.05$
** $p < 0.01$
NS = not significant

tualized in order to support a paradigm. A biased analysis only serves to enhance the difference between what a hominid group is in biological terms, what that group is according to the available data, and what it is according to the paradigm. Such research can only serve to reinforce the original paradigm; any information gained is ultimately paradigm bound and leads researchers away from actually testing those original paradigms.

My chronological, geographical, and archaeological analysis of the Upper Paleolithic biological record (Schumann 1995b) was used to test the validity of the correlation between the hominid and archaeological divisions of this period and the assumptions of the Old World paradigm. If we are to assume a direct correlation between the various Upper Paleolithic cultures and human ethnicity, do we also assume that as the culture changes there is a concomitant change in the biological makeup of the hominids? With the recent focus and attention given to late Pleistocene biocultural evolution, there is merit in investigating the nature of the relationship between biological and cultural change and the validity of biological boundaries (either chronological, geographical, or cultural) of the Upper Paleolithic.

Although according to my work the evidence for biological change during the Upper Paleolithic is negligible (Schumann 1995a, 1995b), suppose (for argument's sake) that there were significant changes in cranial height and maximum cranial breadth, suggesting a change through time in general cranial shape. Can this (hypothetical) biological change be correlated with a synchronous change in the archaeological record? If this change occurred in the fossil record concurrently, for example, with the appearance of the Gravettian, then perhaps a correlation between biological and cultural change could be postulated. Only once a chronologically defined relationship between biological and cultural change is identified can we hypothesize that the period between 30,000 and 28,000 B.P. witnessed both a new archaeological industry and a new biological population. If this relationship existed, then from both of these avenues of inquiry a statement about early human ethnicity (in both a cultural and biological sense) could be made. Unless we actually test assumptions of biological and cultural change during the late Pleistocene and incorporate new lines of evidence (i.e., settlement and subsistence patterns, raw material procurement, exchange networks), we will not be able to answer questions about modern human (biological and behavioral) evolution. However, to date, no one has attempted to investigate any such bio-cultural relationship during the European Upper Paleolithic beyond arguments of functional morphology (Trinkaus 1983, 1987; Harrold 1992).

CONCLUSION

One of my objectives is to test the postulates of an existing paradigm, that of the relationship between human ethnicity and biological and behavioral change. The examination of this period therefore requires that I distance myself from the confines of this paradigm of modern human origins and late Pleistocene biological and cultural evolution. My work begins with the appearance of anatomically modern humans in Europe; how, why, and when they appeared is, for this work, not of great consequence. I have approached this subject by constructing both an archaeological and a biological database for the entire Upper Paleolithic. The use of more than one avenue of inquiry allows me to test the relationship between biological and cultural change as well as other models of late Pleistocene evolution. From this it becomes possible to extrapolate into other relevant areas of research, such as the origins of modern humans or more specifically to the notion of multiple dispersals or subsequent population movements into Europe (Lahr and Foley 1994) and within Europe (Gamble 1982, 1986; Madden 1983).

I am not claiming, however, that my work is completely bias-, boundary-, or paradigm-free, but I do feel that my approach to this subject recognizes the effect of drawing boundaries and the biases and paradigms upon which those boundaries and their interpretations depend. These cultural and biological classifications have dominated Paleolithic discourse throughout this century. Rather than remaining within the confines of historically construed paradigms and models and reinforcing the dichotomies of current research, perhaps there should be greater concern for ultimately removing these biases by testing the validity of biological and cultural classifications created in the past and used in the present. The interpretation of pattern and diversity, asking "what it means" and extrapolating these inferences, is crucial to an understanding of prehistoric societies. It is these different interpretations of diversity that are the focus of scientific discourse and debate; it is when the diversity is interpreted solely according to paradigm without an independent assessment of pattern that controversy inhibits scientific progress.

V

ASIAN PERSPECTIVES

Chapters by western and indigenous scholars are grouped together in this section, which begins with a historical overview of research in China and Indonesia by physical anthropologist Geoffrey Pope. Initiated by westerners, paleoanthropology in the Far East is the product of both periodic influence and isolation from the West. Global and regional sociopolitical factors (not least of which are World War II and the Cultural Revolution in China) loom large in this periodicity and resulted in the perpetuation of research methods and theoretical concerns originally introduced by westerners in the 1920s and 1930s. Since the mid-1970s contact between the Far East and the West has increased dramatically, and has led to the emergence of indigenous research traditions that have their own particular characteristics and concerns. However, it is interesting to note that Asian researchers have always favored continuity so far as modern human origins are concerned.

Chinese paleoanthropologist Wu Xinzhi reviews some of the elements of Weidenreich's regional continuity hypothesis, which was based mainly on shared characteristics of Chinese *Homo erectus* and modern Mongolians. He contends that Chinese archaic *Homo sapiens* exhibits a distinctive morphology that overlaps only partly with that of its African counterpart, and that European neandertals possess a different morphology, corresponding to other parts of the African range. Differences between east Asian and European Upper Pleistocene hominids are ascribed to genetic drift during and after the colonization interval.

Susan Keates points to insufficent funding, an exceptionally "coarse-grained" time/space grid, and continued use of obsolete methodologies as important, practical considerations that put obstacles in the path of research on the archaeology of the biological transition interval in China. These factors, the generally conservative nature of Chinese artifacts, and the rare, late manifestations of art and symbolism make the study of hominid behavior a more challenging pursuit than it is in Europe, Africa, and the Near East. She argues that the Asian evidence has often been ignored in modern human origins research, and that until it is given due consideration, a compelling explanation for modern human origins will not be possible.

The essay by replacement advocate Marta Lahr argues that, paradigmatic bias notwithstanding, alternative models for modern human origins do not represent different readings of the data. Instead the biggest problem is different concepts of evolutionary process held by the opposing camps. She suggests that two basic

267

assumptions—the absence of speciation events, and the irrelevance of changes in the rate of gene flow over the past million years—drive the continuity model to the extent that alternative hypotheses are precluded. She argues that only by decoupling the events and processes that produced modern human diversity will more compelling explanations of our origins be possible.

Finally, Australian Colin Groves argues for the priority of human paleontology in any resolution of the modern human origins debate; genetics and archaeology are of secondary importance, he says. Identification of morphoclines in the anatomical features of the relevant fossils is crucial because only shared, derived characters offer evidence of relationship. However, in any actual problem domain, it is by no means easy to distinguish these from primitive or generalized retentions. Inadequacies of the time/space grid exacerbate sampling problems and complicate the picture still further, especially in Asia.

19

Paleoanthropological Research Traditions in the Far East

GEOFFREY G. POPE

Paleoanthropology is largely the creation of westerners working in foreign countries. In Africa, European and American paleoanthropologists have continued to dominate the directions of human evolutionary studies from their very inception. Until recently, research traditions in the Far East have diverged very little from the original foci established by western scholars. However, in contrast with the West, in Asia the question of modern human origins is very nearly not a research topic at all. The infraregional origins of various extant Asian groups have long taken precedence over the origin of *Homo sapiens* in general. The idea that anatomically modern Asians arose as the result of a recent African migration finds no support in the available Asian data nor among the researchers that interpret them.

HISTORICAL OVERVIEW: THE OLD AND NEW TRADITIONS

Paleoanthropological research in Asia has generally been driven by the ideas and agendas of the European and American scientists who have worked there. The Eve controversy represents the most recent interaction between eastern and western approaches to the study of human evolution. Although all of the prominent Asian paleoanthropologists have received their formal training in the western research traditions, this training consists of a thin veneer superimposed on a rich philosophical legacy and turbulent political history which strongly influences the indigenous "worldview" of human evolution in Asia. The combination of western science and oriental culture has resulted in perspectives and methods that have grown out of the "old tradition" of colonial dominance and the "new tradition" of Asian independence.

The "old tradition" in Far Eastern paleoanthropology was one in which foreign researchers interpreted the Asian record almost exclusively in terms of their understanding of European and African evidence. Native Asians were at most "trainees" and not colleagues of the early westerners. As such, some eventually came to adopt the autocratic style which characterized many early western researchers. The "new tradition" involves both westerners and indigenous Asian workers. Western influence remains substantial, but the native workers now conduct their own field and laboratory research. In the new tradition, research is oriented toward making sense of local data, rather than determining how Asian data fit the paleoanthropological "facts" of the rest of the Old World. At the same time the Eve controversy has occasioned a new defense of the model of regional continuity between fossil and extant Far Eastern populations originally detailed by Weidenreich (1943a).

CHINA AND INDONESIA

China and Indonesia have yielded most of the paleoanthropological evidence in the Far East. Questions surrounding the origin(s) of modern humans have become a nexus for debates which not only focus on the direct evidence for the location and timing of the appearance of modern humans, but which have also forced anthropology (and western culture as a whole) to once again confront subjects such as race, intelligence, and culture. This chapter does not address these more profound questions but here notes that a concern about the origins and antiquity of races continues to surface periodically in western culture (Cose 1994; Morganthau 1994). Today, only a handful of westerners have an adequate grasp of Far Eastern paleoanthropological data. Unfortunately, some have nonetheless offered facile pictures of the Asian data in the service of reifying theories of modern human origins developed on the basis of western evidence. This practice has continued the long-standing tradition in which data flow from research efforts in the East, only to be interpreted in frameworks developed in the West.

MODERN SOCIAL AND POLITICAL INFLUENCES

A recounting of the growth of paleoanthropology in China and Indonesia offers both contrasts and parallels in the two nations. The succession of Communist ideologies in China has had very little theoretical impact on human evolutionary studies. Similarly, the postcolonial years in Indonesia, also marked by great social and economic upheaval, have seen a continuation of research directions, orientations, and field methods which have remained largely unchanged from the colonial past. Both nations are only now beginning to emerge from decades of isolationism in which very little joint research was undertaken with westerners.

Two important forces have contributed to the present status of human evolutionary studies in China and Indonesia. The first is the fusion of modern internationalism with traditional ethnocentricism, both of which have been conscientiously cultivated by these emergent national states. This is especially true of China, where pride in the antiquity of culture has always constituted an integral part of the national ethos. The second important influence on paleoanthropological research in Asia has been the "return of the foreigners" and their money. In both countries it was the efforts of foreigners that first defined the methods and objectives of paleoanthropological research, and the frameworks established for the recovery and interpretation of fossils have been adopted and maintained by Asian scientists. Along with a substantial increase in material resources, foreigners have also introduced new techniques, theories, and research objectives which are only now beginning to have an effect. However, in their dealings with scientists of the New Asia, westerners often encounter cultural idiosyncrasies which they fail to understand. One important and as yet unfinished achievement of the new paleoanthropology in China has been to cast off the scientific prejudices of its European founders.

THE MODERN RESEARCH TRADITION

The modern research tradition in Asia is almost wholly a reflection of the western belief that hominid evolution was essentially similar in all parts of the Old World. It is only recently, in the wake of debates such as the "Eve" controversy, that both expatriate and indigenous researchers have begun to formulate hypotheses that are specifically tailored to Asian research questions. The "old tradition" in Asia was of purely western manufacture. It was based on a western desire to reify the evolutionary sequence in Asia as it had been discerned in Africa, and especially Europe. The mammalian biostratigraphic succession (Teilhard and Piveteau 1930), paleoclimatic sequence (Lee 1939; Liu 1991a, 1991b), Paleolithic phases, and the "stages" of hominid evolution were for decades assumed to be the same or very similar to those of the West (Teilhard 1937, 1938, 1941). The old tradition was adopted wholeheartedly by the young Asian "trainees" who have now become the leading paleoanthropologists in Indonesia and China. At the core of the old tradition, however, was the frequently unpublished conviction that hominids had originated in Asia. Demonstrably non-modern morphology in Asia was therefore considered to be older than similar morphology in African and European hominids. When Robinson's arguments for the recognition of *Paranthropus* (1950, 1953) as a hominid became undeniable and as African discoveries began to accumulate, the "old guard" Asian specialists "withdrew" to the position of maintaining an equal antiquity for the oldest comparable Asian finds (Tobias and von Koenigswald 1964; von Koenigswald 1973). Some of the original western workers, especially von Koenigswald, supported this contention in the twilight of their careers and proclaimed that australopithecines were also present in Java (von Koenigswald 1973).

Running through the new tradition is the preconception that many of the evolutionary problems raised by research in Africa and Europe are irrelevant to an understanding of hominid evolution in the Far East. There is, for instance, no Acheulean, no "neandertal problem," no Middle Paleolithic, nor an Upper Paleolithic "revolution." The "problem" of regional continuity has until very recently not been a problem at all, because both the archaeological and hominid paleontological data were most parsimoniously explained as the diachronic persistence of technologies and morphological characters which appeared during the early Pleistocene arrival of hominids in the Far East (cf. Zhang 1991). However, the actual antiquity of most hominid finds in the area has always been a sensitive issue. Workers in the old tradition, without benefit of radiometric techniques, had been content to classify finds as representative of the broad divisions of the Pleistocene; in this respect they differed little from their European colleagues. The new tradition departs from the old not only because it is controlled by native Asians, but also because many of the new research problems are being redefined in directions which stem primarily from the Asian data themselves.

THE EVE CONTROVERSY

The Eve controversy, which originated among western scholars, is beginning to have an impact on the interpretation of the Asian evidence by Asians. In Indonesia, however, the impact has been minimal and the question of the age of the earliest Sangiran hominids easily takes precedence over the question of continuity between fossil Javanese and extant Australians. The Eve controversy is actively influencing phyletic interpretations of early *Homo sapiens* in China (Zhang 1991), but research has been much more concerned with understanding the apparently extreme variability between penecontemporaneous specimens such as Dali, Jinniu Shan, and Yunxian. The new tradition has increasingly recognized that one of the major problems in Chinese Paleolithic archaeology is not the relationship of Chinese assemblages to those from the rest of Eurasia and Africa, but rather the geographic variability within the Far East itself. In Southeast Asia, the peculiarly Asian question of whether *Homo erectus* made stone tools at all has been raised. Like the old tradition, the new one arises primarily out of questions foreigners are asking. The big difference between them is that few of the foreigners raising the questions have any firsthand familiarity with the data and the primary literature. The initial western workers did have a firsthand familiarity with the data, but they examined it according to western conceptual frameworks.

THE COLONIAL PAST

In general western paleoanthropologists brought with them the conviction that the prehistory of the Far East would be similar to that of the West; their conclusions were "pre-cooked." The glacial ages of China and even Indonesia must have been synchronous with those in Europe. Paleolithic archaeology would, therefore, show a succession of stages similar to those of Europe. In both Indonesia and China vaguely defined taxonomic guidelines and only a general concern with provenience (especially in Indonesia) continue to characterize modern work. On a more abstract level the notion that hominids arose first in Asia has continued to influence the interpretation of new discoveries. This is less marked in China, where the prominent paleoanthropologists have been formally trained in anatomy and taxonomy.

While the early western researchers brought their own prejudices and experiences to bear in the Far East, they also unwittingly encouraged the development of an inclination that already existed among indigenous intellectuals. This was, and frequently still is, a decided penchant for antiquarianism. China's long tradition of antiquarianism is more entrenched and far older than that of Europe. The physical possession, or ownership, of a fossil or object continues to be a subcurrent that runs through the work of many Asian researchers. Fossils are still sold on the streets of Beijing and in the villages of central and eastern Java. A predilection toward antiquarianism is logical in the context of indigenous cultures that emphasize the heritage of their ancient past. In Indonesia, Dutch culture brought to their colonies a concerted interest in arts and antiques which fit in well with the *tukang* system, a network of door-to-door tradesmen who specialize in certain goods and services. In China, the organization in charge of what we would call cultural resource management is the Bureau of Cultural Relics Protection. It has increasingly been involved with the regulation of foreign research in China.

HUMAN PALEONTOLOGY IN CHINA

At the end of the nineteenth century and well into this one, Asia was widely held to be the "cradle of mankind" (Nelson 1926; Osborn 1926, 1927). Although subsequent African work has made this a moot point in the West, the idea is still given serious consideration by many Asian scientists. It originated in the West principally because of the belief that the harsh climate of central Asia would have provided the requisite selection pressures which would have produced *Homo sapiens*. Humans were thus born out of the adversity of inhospitable places like the Gobi Desert. Although formulated in the West, the idea also fit well with traditional Asian pride in the antiquity of their own cultures. The American Central Asiatic Expedition and the later and more important excavations at Zhoukoudian (Black 1927; Black et al. 1933) validated the idea that humans were oldest in Asia. Black's successor, Franz Weidenreich, further codified the importance of China in the scheme of human evolution (Weidenreich 1939a, 1939b, 1943a). Andrews, Black, and especially Weidenreich also provided models of interaction between westerners and Asians.

Before the Zhoukoudian excavations, explorers like Swen Hedin and Roy Chapman Andrews provided what were from the indigenous perspective negative images of the paleontologist as adventurer. Sven Hedin's style of exploration was dashing, far-flung, wasteful, and reckless. Though Hedin managed to fill in many "blank spots" on the map, as he called them, his expeditions in central Asia routinely lost men and pack animals in large numbers (Hedin 1925). Davidson Black and John Gunnar Andersson refused to cooperate in a joint expedition with Hedin because Hedin had publicly acknowledged shipping artifacts out of Tibet (Xinjiang Province) (Jia and Huang 1990). This development was to prove prophetic, because money has always been a central point of contention between foreign research teams and the Chinese government.

The work of Roy Chapman Andrews was equally flamboyant. As leader of the Central Asiatic Expedition, Andrews attempted to raise money upon his return from Asia. He presented one of his patrons who had donated $10,000 with a souvenir dinosaur egg he had collected from the Gobi Desert. Consequently, the Chinese government reasoned that each egg was worth that much and that Andrews had taken too many eggs out of the country. He was declared persona non grata, thereby becoming the first paleontologist to be expelled from the Far East (Reader 1989).

If Hedin and Andrews provided negative images, Black and Weidenreich were antithetical and powerful icons of how legitimate international research should be conducted. Their style was far more reserved and organized than that of Hedin or Andrews. They were not explorers, but instead methodical anatomists. Black drew up a carefully worded contract with the Chinese government for work at Zhoukoudian that stipulated prices, fees, and stipends as precisely as any lawyer (see Jia and Huang 1990). Contracts have recently been reinstituted in China in order to provide a framework for international collaboration. While the stipulation of all terms of cooperation is no doubt useful in avoiding misunderstandings, there is also a negative aspect in that, in Chinese culture, contracts are for the purpose of making money.

However, there is little doubt that most of the early western researchers did not view the research skills of the Chinese as on a par with their own. Jia Lanpo related how at first Weidenreich personally oversaw the excavations at Zhoukoudian because, in Jia's interpretation, he did not think the Chinese competent (Jia and Huang 1990). Other Asian nationals had similar experiences.

The colonial era of human paleontology ended with the invasion of the country by Japan (1935) and the subsequent declaration of war (1941). The disappearance of the Zhoukoudian fossils continues as a subject of intense interest among Chinese workers. Few if any Chinese scholars believe that the Americans should be blamed for the loss. In a recent interview, Jia Lanpo, who has retained more of the original Zhoukoudian documents than anyone else, expressed the opinion that the fossils never left China. He believes that they were thrown off the train by drunken Japanese soldiers during a three-day rampage. Because Weidenreich had taken thousands of careful measurements and had made excellent casts of all the hominid material, the loss to the scientific community was mitigated to some extent.

The legacy of the early western workers at Zhoukoudian and their other efforts elsewhere in China was the foundation for the structure of modern research. Until very recently, excavation techniques and recording techniques had changed little from those used in the 1920s at Zhoukoudian. Perhaps the most enduring legacy of the Zhoukoudian work, however, has been the failure to develop specific, problem-oriented research designs. The objective of excavation was, and remains, simply to gain more material. In the West, the "new archaeology" of the 1970s supposedly changed this situation, but in the Far East, excavation is primarily for the purpose of augmenting scarce archaeological and paleontological collections. The hypotheses of the western-dominated "old tradition" continued to be reified through the early 1980s. In the past ten years, however, research strategies have changed far more than they did in the three decades following the exodus of western researchers. Once again it has been the influence of the West that has brought about this change.

HUMAN PALEONTOLOGY IN INDONESIA

As the paragon of modern scientific investigation in prewar Java, G. H. R. von Koenigswald combined some rather unusual characteristics. Because he was trained as a mammalian paleontologist and stratigrapher, he brought with him a desire to divide Javanese stratigraphy into biostratigraphic units comparable in duration and to some degree in content to those of Europe. He established the classic threefold system of Djetis, Trinil, and Notopuro "beds" (actually faunal associations) that has remained in use until very recently. Despite the active resistance of Hooijer, von Koenigswald's stratigraphy, although highly oversimplified, was embraced by most researchers. Much more recently Dutch workers have argued that the Trinil faunas are actually older than those assigned to the

Djetis formation (De Vos et al. 1982, 1994). However, even this newest reassessment is based upon the recognition of subspecific taxa evolved in situ and largely in isolation. From a historical perspective, it is abundantly clear that the expectation of distinct and successive faunal assemblages of the kind known from mid-latitude Pleistocene Europe was an inappropriate model for dealing with the highly conservative endemic fauna of an equatorial island (Pope 1994).

Von Koenigswald's major interest was in the hominids, which he continued to recover until the Japanese invasion of Indonesia. What seems inconsistent with his training is his apparent disregard for stratigraphic provenience, although, in all fairness, the *tukang* system could not have been easily displaced. As the sole purveyors of fossils, the collectors had no interest in revealing the locations of their findspots. The result is that, today, no Javanese fossil has an undoubted provenience. Ironically, it was stratigraphy that von Koenigswald used as a basis for taking issue with Weidenreich's phylogenetic interpretations. The practice of invoking stratigraphy and/or morphology inconsistently in support of various viewpoints continues today among both western and indigenous workers.

The other legacy von Koenigswald left behind was as much a product of the age in which he worked as his own personal perspectives. This was the then-prevalent practice of casually appending new names to each new specimen. Weidenreich was no less guilty of this, but he regarded the taxonomic categories in human paleontology merely as convenient labels (Weidenreich 1945). *Homo modjokertensis* was introduced in order to avoid angering DuBois, *Meganthropus palaeojavanicus* and *Pithecanthropus dubius* were introduced in von Koenigswald's absence. The former takes its name from a suggestion in a letter by von Koenigswald to Weidenreich; the latter expressed Weidenreich's opinion that the specimen was not a hominid. This attitude not only accounts for the numerous taxa recognized before the war, but also those recognized by Indonesian paleoanthropologists today. Indonesian researchers are simply maintaining a tradition of applying new taxonomic appellations in the same way that their western predecessors did. In the 1970s, von Koenigswald's stated belief that australopithecines were present in Asia further fueled a willingness to recognize multiple hominid taxa in Java (1973).

PALEOLITHIC ARCHAEOLOGY IN CHINA

Chinese paleolithic archaeology was created primarily by visiting European prehistorians. With the notable exception of Teilhard, whose interests transcended the boundaries between paleontology, geology, and human paleontology (Teilhard de Chardin 1937, 1938, 1941) and whose opinions were based upon a firsthand exposure to Chinese data, European archaeologists sought to impose

European frameworks on Asian artifacts. Breuil (1927, 1931, 1939), Boule (1927), and later Bordes (1950) brought with them the belief that Far Eastern artifact assemblages would essentially mirror those of the West. Eventually archaeological opinion divided into two camps. The French, again with the exception of Teilhard, felt that lithic traditions like the Acheulean and the Mousterian had simply not been located yet, but soon would be. Breuil (1927, 1931) opined that bone tools definitely occurred at Zhoukoudian, and not just in the Upper Cave (which had produced anatomically modern humans) (Breuil 1939). The other camp, including Movius and Teilhard, felt that the Far East simply lacked these rather sophisticated developments and posited what came to be known as the "cultural retardation theory" (Movius 1944). Movius' field experience in Asia was extremely limited, while Teilhard had come to his conclusion after years of research in China.

Needless to say, the cultural retardation theory did not and does not sit very well with the Chinese archaeologists. The excavations at Anyang and other Neolithic localities in central China hardly suggested to Chinese scholars a heritage that did not extend back into the Paleolithic, and the Upper Cave assemblage indicated an even older indigenous tradition. Some decades later a sort of synthetic research tradition had emerged. At its heart was the belief that the Zhoukoudian industry represented the oldest Asian stone tools, but it was acknowledged that the "intervening" Acheulean and Mousterian had not been found. The search for these traditions continues today. Oddly enough the "Quest for the Acheulean" has recently been echoed in the West by archaeologists who know little or nothing about the Chinese record (Clarke 1990; Klein 1989b, 1992).

One of the most curious developments in Chinese paleolithic archaeology has been the argument over the "big tool" and "small tool" traditions (Yi and Clark 1983). Simply put, it has been postulated that Paleolithic assemblages can be divided into groups that made small tools and those that made large or "heavy-duty" choppers and "chopping" tools (Jia 1985; Jia and Wang 1985b,c; Jia and Wei 1976; Jia et al. 1979). While sites exist that preserve both size classes, they are relatively rare and are "explained away" by invoking a mixing of the two "cultures." Beginning in the late 1980s, Chinese archaeologists began to consider seriously the possibility that different activity facies could result in different modal sizes of stone artifacts. My own research in China strongly suggests that fluvial sorting and selective recovery by Chinese workers has overemphasized this dichotomy, and it is my impression that the twofold division is slowly being abandoned. At the same time, the possibility that archaeological entities such as the Acheulean and Mousterian do exist in China is still being pursued. It is clear that genuine levallois flakes and hand axes (frequently confused in the Chinese literature with bifaces) exist, but are very rare, comprising minute proportions of any Paleolithic assemblage.

EVE AND THE ORIGIN OF MODERN HUMANS

The Eve controversy is only now beginning to have an impact on paleo-anthropology in the Far East, especially in China. In Indonesia, the notion of a recent, African origin for moderns has only been dealt with in an indirect manner, at least in part because early examples of *Homo sapiens* have not been recovered there. A crucial tactical argument for the Eve proponents is that pre-modern Asians are essentially the same as archaic *Homo sapiens* as recognized from Africa and Europe (Pope 1984, 1989c, 1994; Stringer 1984; Stringer and Andrews 1988b; Wilson and Cann 1992). In this regard, Stringer (1989a, 1989b, 1990c; Stringer and Andrews 1988b) has insisted that the impression of a short face (Pope 1989c, 1994) is simply an artifact of postmortem compression and distortion. In fact, after a very brief examination of the original, and inspection of a cast, Stringer (1990c, personal communication 1991) returned from China and admitted that the extreme version of the replacement model may not applicable everywhere in the Old World. Unfortunately this admission has yet to reappear in the scientific literature. This temporary shift I believe to be indicative of what can happen when one has a chance to inspect the actual material.

CHINA

A central morphological issue in China is whether the long sweeping contours of the IZM (infrazygomatic margin) of archaic *Homo sapiens* and neandertals are essentially the same as they are in specimens like Dali and the recently discovered Jinniu Shan material (Pope 1989c). As I have tried to show in a number of publications, they emphatically are not. The typical Far Eastern morphology exhibits a short maxilla, horizontally oriented zygomatics, and a total IZM contour which in no way resembles possible western contemporaries. Furthermore, these and other traits continue virtually uninterrupted from Asian *Homo erectus* through the pre-moderns to extant Far Eastern and Far Eastern–derived populations of *Homo sapiens*. I wish to make clear that this observation in no way argues that Far Eastern morphological features are static. In fact, I and others (Smith, Falsetti, and Donnelly 1989) have indicated that there is substantial gene flow across Eurasia and in and out of Africa. Similarly Bräuer (1992a) has tended to accept the high likelihood of long-range gene flow and admixture. The Jinniu Shan cranium is especially suggestive of this scenario. The only conclusion I can draw from several years of dealing with the fossil evidence on a firsthand basis is that gene flow proceeded in both directions. This is based on the fact that certain morphological features which originated in and were confined to Asia throughout most of the Pleistocene eventually appear in the West. Conversely, certain western features eventually make their appearance in the Far East. I cannot think of a better morphological case for admixture.

From a morphological standpoint, some Chinese (e.g., Wu 1988, 1989, 1990, this volume) and some western workers (Pope 1983, 1984, 1985a, 1985b, 1991, 1992a; Wolpoff et al. 1984) have pointed to a suite of craniofacial features that link *Homo erectus* and modern East Asians. Furthermore, the Xujiayao sample (>10 individuals?) is intermediate to Zhoukoudian *Homo erectus* in a number of features, including overall robusticity (Jia and Wei 1976; Jia et al. 1979). In respect to morphology, most Chinese workers emphasize the mosaic nature of human evolution, Wu (1990) noting that different characters evolve at different rates. A concern with absolute dates in Chinese modern human origins research represents a significant departure from a primary reliance on morphology. The tendency to concentrate almost exclusively on morphology can be traced back directly to Weidenreich, who downplayed the diachronic nature of the Locality 1 hominid fossils, for all practical purposes treating the sample as synchronic. He further believed, often in disagreement with von Koenigswald, that most of the Javanese hominids were contemporaneous with Zhoukoudian. Wu and Zhang have pointed to diachronic changes in *Homo erectus* (Wu, personal communication 1994; Zhang, personal communication 1994). The greater reliance on dates is a very new development in Chinese paleoanthropology, although an assumption of contemporaneity figured in arguments for the presence of australopithecines and *Homo erectus* at the Plio-Pleistocene boundary. Until the 1980s these were the only phylogenetic arguments of any importance in Chinese paleoanthropology.

INDONESIAN CHRONOLOGY

Dating remains problematic in Indonesia, as noted above, and geochronological problems are mainly geological in nature. In both China and Indonesia age estimates based on biostratigraphy are rendered especially imprecise owing to the extremely conservative nature of the archaic mammalian faunas in this part of the world. Despite these difficulties, I suggest that dates for any of the Javanese hominids will not be altered much from what they are thought to be now (i.e., ca. one million years ago). In other words, I do not think either new techniques or more careful studies will substantially alter the conclusion that hominids reached the Far East approximately one million years ago. Whatever the antiquity of the earliest Asian hominids eventually proves to be, it is certain that the vast majority of early Asian hominids are younger than one million years old. In Java, paleomagnetism, fission track, and K/Ar ratios have been the predominant absolute methods for dating fossiliferous Plio-Pleistocene sediments. More recently an ^{40}Ar-^{39}Ar technique has been applied (Swisher et al. 1994). The widespread distribution of volcanic sediments has made absolute dates easier to obtain than in China where suitable igneous materials are very rare. Yet both Java and China suffer from similar provenience problems in that the hominid specimens cannot be reliably associated with the dated materials. This is so far an insurmountable problem in Java. In China, most of the controversy surrounds the age of early *Homo sapiens,* while in Java it is the age of the earliest hominids that impacts directly on the replacement versus regional continuity problem. These issues have been reviewed by Pope (1983, 1988, 1991, 1994).

IMPLICATIONS OF THE NEW JAVANESE DATES

Although linkages are by no means explicit, the new dates from Java have been presented as having a direct bearing on the modern human origins controversy (Swisher et al. 1994). When radiometric dates first became available, it was argued that they were old enough to suggest that australopithecines and habilines could logically be expected there if the dates were proven accurate (Tobias and von Koenigswald 1964; von Koenigswald 1973). At that time, Chinese scientists, believing the Javanese dates to be credible, also reported extremely early *Homo erectus* (Qian et al. 1991) and australopithecines (Gao 1975) from China. In the 1980s these dates came under attack (Mats'ura 1982, 1985; Pope 1983, 1994) and over the next decade it gradually became accepted that both the dates and their relationships to key hominid fossils were unreliable. By the mid-1980s, this resulted in a consensus that the earliest Asians were perhaps only as old as ca. one million years (cf. Zhang 1985).

The new dates suggest an age of ca. 1.8–1.6 million years for the earliest Sangiran hominids (Swisher et al. 1994). These values were obtained on the basis of a new and improved ^{40}Ar-^{39}Ar technique (single crystal dating) which allows the use of a very small mineral sample. They have been widely heralded as a "breakthrough" which supposedly confirms a hominid presence in Java during the Upper Pliocene. The new dates also tend to agree with the original dates obtained by Curtis in the early 1970s (Jacob 1981; Jacob and Curtis 1971). In fact, however, the same problems that undermined the credibility of the first dates also limit the usefulness of the new dates. Simply put, *not a single early Javanese fossil has a reliable provenience.* Taking samples from the alleged Modjokerto site and comparing them with another outcrop of "similar" *(sic)* lithology cannot be used to date the Modjokerto fossil itself.

CONCLUSION

Paleoanthropology in the Far East has been characterized by alternating periods of isolation and contact with western ideas and workers. The initial phases of paleoanthropology in the Far East began with the arrival of western scientists at the end of the nineteenth century (Java) and in the first three decades of the twentieth century (China). In both countries, western interpretations based on work in Europe, and later Africa, have determined the direction of research. During the colonial period some native researchers were trained, if not formally, at least by example. During the isolation of the early postwar period, Far Eastern nations emerged as autonomous political entities. Scientists in these nations continued research in the same mode as their colonial predecessors. The theoretical orientation of Asian researchers strongly supported the view that modern East Asians had evolved directly from Asian *Homo erectus*. Even after the establishment of African australopithecines and habilines as older and better ancestors for later humans, Asian workers continued to promote the view that the Java and China hominids were as old or older than the earliest African specimens. This period can be accurately characterized as one in which research was used to reify the notion of early Asian australopithecines. The recent dates from Java have caused a resurgence of this idea. The most disconcerting aspects of this new contention are the phylogenetic pronouncements of westerners with very little training in taxonomy or knowledge of the history of paleoanthropological research in the Far East. Taking the Swisher et al. (1994) dates at face value, Chinese paleontologists have already begun to accord their own early sites a similar antiquity—this despite the fact that no new Chinese data have been forthcoming (Tang 1991). Once again westerners have returned with new techniques which have been applied irresponsibly to old contexts.

In China, an indigenous tradition of antiquarianism was mingled with the conclusions drawn from research at Zhoukoudian. All fossil discoveries continue to be chronometrically and morphologically evaluated with reference to the Zhoukoudian sequence. Furthermore, most Chinese paleoanthropologists continue to believe that Chinese *Homo erectus* is the direct, lineal ancestor of modern Chinese. It is fair to say that until very recently most Chinese research was engaged in simply filling in the details of the overall pattern of hominid evolution in Asia. This pattern was one in which *Homo erectus* evolved anagenically into modern Asians.

The western supposition that Chinese paleoanthropology is not up to the task of providing even broad conclusions about the remote past brings to mind O. J. Simpson's defense attorneys' contention of "sloppy" and grossly incompetent police work in the case. A similar set of preconceptions is at work in western critiques of Far Eastern methods. It is obvious that in its isolation, Chinese archaeology has been less than rigorous in many instances. However, it is also clear that despite the lack of gridding, standardized measurement, and many other methodologies, there is no Asian Acheulean. Hand axes have been discovered, but they are exceedingly rare. Similarly, paleontologists do not expect to discover australopithecines, although some of the principal Javanese researchers believe they already have. More important, research in both Java and China, but especially China, is increasingly being driven by specific East Asian problems. In both areas indigenous researchers continue to be influenced by both western money and western ideas. Finally, the idea that *Homo erectus* did not evolve into modern East Asians has absolutely no support in the data generated over the past one hundred years.

20

On the Descent of Modern Humans in East Asia

WU XINZHI

This chapter tries to show that Chinese morphologically modern humans *(Homo sapiens sapiens)* are similar to Chinese archaic *H. sapiens* in regard to a suite of features originally defined by Franz Weidenreich during the war years (1939, 1943a). No adaptive value for these shared features has been demonstrated, so environmentally based selective forces cannot be invoked to explain their persistence over an interval of time in excess of 300,000 years. If Upper Pleistocene Chinese morphologically modern humans are the descendants of the African "Eve" postulated by replacement advocates (e.g., Groves and Lahr 1994; Stringer and Andrews 1988b), these features should be explained by genetic drift during the migration of Eve's descendants from Africa to east Asia, a process which, we are led to believe, took place relatively recently, between 200,000 and 100,000 years ago.

The replacement scenarios postulate an "early" migration or radiation, corresponding to the initial colonization of Asia by hominids in the Lower Pleistocene, and a "later" migration, and subsequent replacement, corresponding to the appearance of modern humans in the late Middle or early Upper Pleistocene. Since the features involved are believed to be adaptively neutral, how can we explain the apparent "fact" that the results of genetic drift after both the early and the late migrations could be so similar? In default of any argument for selective forces operating on these features, the probability that this convergence is due to chance alone is virtually nil, especially given that similarity is manifest in a complex of features instead of only one or a few of them. On the contrary, it is much more reasonable to suppose that features shared in archaic and modern Chinese *Homo sapiens* are the result of their genetic connection. Chinese Upper Pleistocene hominids inherited these features from their local Middle Pleistocene ancestors, instead of from the African Eve.

Archaeology is also relevant here. Continuity in Middle and Upper Paleolithic assemblages in China is well-documented, instead of the disjunction implied by replacement scenarios that link fossils to types of archaeological assemblages. If replacement had, in fact, occurred, certainly one might expect some evidence of it in the increasingly well documented archaeological record (changes in the characteristics of lithic assemblages, settlement patterns, resource procurement, etc.).

Based on the application of the coefficient of divergence, I have confirmed the assertion of Suzuki (1982) that morphologically modern hominids from Japan probably originated in south China. Nevertheless, application of this coefficient does not support Brothwell's (1960) assessment of the affinities of the Niah Cave hominids, and in fact it indicates a closer relationship between Niah Cave and the Upper Pleistocene populations to the north of Borneo (rather than to the south of it). My analyses suggest a closer affinity of the Tabon fossils (Philippines) to the Pleistocene hominids of China.

ASIAN UPPER PLEISTOCENE EVOLUTION IN HISTORICAL CONTEXT

Morphologically modern human skulls of Pleistocene age considered to pertain to morphological moderns have been recovered in eastern Asia; most of them come from the Chinese mainland. Where they originated is a much-debated topic. More than fifty years ago, Weidenreich opined that

> there are clear evidences that *Sinanthropus* is a direct ancestor of *Homo sapiens* with closer relation to certain Mongolian groups than to other races. . . . This statement, however, does not mean that modern Mongols derived exclusively from *Sinanthropus* nor that *Sinanthropus* did not also give rise to other races. (1943a:276, 277)

Although representatives of morphologically modern *Homo sapiens* had been found in the Ordos region (Mongolia) in association with Upper Paleolithic industries, and in the Upper Cave at Zhoukoudian, a continuous series comprising many, dated fossils leading from *Homo erectus (Sinanthropus) pekinensis* to *Homo sapiens* was unknown at the time he was writing.

So Weidenreich (1939, 1943a) proposed an early form of what was to become the multiregional continuity model (e.g., Wolpoff 1989b; Wolpoff et al. 1984) based mainly on the fact that some common morphological characters are shared by *Homo erectus pekinensis* and modern Mongolian groups. Those he identified are given in Table 20.1.

Since 1949, many human fossils of different ages have been recovered in China. Although some of Weidenreich's shared features are not evident in all of these new finds, most of them are. In my opinion, they indicate a continuous, in situ pattern of human evolution over the Middle-Upper Pleistocene interval in China, extending back in time to at least 400,000 to 500,000 years ago (Wu 1990).

Table 20.1. Weidenreich's (1939b, 1943a) Common Morphological Characters Shared by *Homo erectus pekinensis* and Modern Mongolians

mid-sagittal crest and parasagittal depression
metopic suture
Inca bone
horizontal disposition of naso-frontal and frontal-maxillary sutures
rounded infraorbital margin
certain "Mongolian" features of the nasal bridge and malar region
maxillary, mandibular and auditory meatus exostoses
small frontal sinuses
shovel-shaped upper incisors
reduced posterior dentition
high frequency of third molar agenesis

COMMON FEATURES OF CHINESE HUMAN FOSSILS AND EAST ASIAN–EUROPEAN COMPARISONS

Morphological features shared by all or most of the Chinese specimens found before and after World War II are enumerated below.

1. The antero-lateral surface of the fronto-sphenoidal process of the zygomatic is positioned more anteriorly than it is in the neandertals. I drew an imaginary horizontal line which represents the orientation of the middle part of the above-mentioned surface (Wu 1988). The lines thus drawn for the process constitute an angle of 85°—more obtuse than the 80° angles found in the archaic *Homo sapiens* skulls from Dali and Maba. The angle thus formed in Peking Man and other *Homo sapiens* skulls is more or less similar to that found in Dali and Maba.

2. The Chinese specimens exhibit a less protruding mid-facial region. This feature is measured by the zygo-maxillary angle, 125° in the Dali cranium and 125–138° in Chinese morphologically modern skulls.

3. The zygomatic process of the maxilla shows certain characteristic features. The demarcation between the anterior surface of this process and the antero-lateral surface of the maxillary corpus is quite distinct and forms a more or less deep depression. The junction between the lower margin of the zygomatic process and the maxillary corpus is rather high. In other words, this junction is rather distant from the alveolar margin and is closer to the orbital margin than in the neandertals. The lower margin of the zygomatic process is oriented superolaterally initially, and then veers laterally and runs horizontally until it joins the maxillary process. These characteristics are also found in other *Homo sapiens* and *Homo erectus pekinensis* skulls in China, while a much deeper notch, the malar notch, is found in Peking Man.

4.   The nasal saddle is flatter in all Chinese *Homo erectus* and *Homo sapiens* skulls than it is in neandertals.

5.   The Chinese specimens have a low upper facial region. The upper facial index (distance between nasion and supradentale × 100 divided by distance between zygions) is 50.1 in the archaic *H. sapiens* skull from Jinniushan and 51.7 in the reconstructed Dali skull. This index falls within the range of morphologically modern skulls in China (48.5–53.8). The archaic skull from Maba is probably the only exception; its high orbit indicates a probable higher upper face.

6.   The orbital margin in the Chinese fossils is angular. The skull from Maba is exceptional; it exhibits a roughly circular orbital margin.

7.   The infero-lateral margin of the orbit is rounded. The Maba specimen is again the only exception.

8.   The general contour formed by the fronto-nasal and fronto-maxillary sutures is a more or less horizontal curve.

9.   Mid-sagittal ridge. The mid-sagittal ridge of Peking Man is very strongly developed and extends down to the posterior part of the frontal and the superior margin of the parietals. It also appears on the upper three-quarters of the frontal squama of the Hexian *Homo erectus* skull-cap; there is also a trace of it on the parietal bone. This ridge also exists in the skulls of early *Homo sapiens* from Dali, Jinniushan, and Maba but it is more weakly developed and stops at the mid-frontal. In the late *Homo sapiens* skulls from China, this ridge is present in the specimens from Ziyang, Upper Cave 103, and Liujiang. Although the ridge becomes less pronounced through time, it is nevertheless present throughout the whole evolutionary lineage from *Homo erectus* down through recent *Homo sapiens* in China.

10.   Shovel-shaped upper incisors are a characteristic morphology that exists without exception in all Chinese fossil human dentition.

In sum, it seems pretty clear that the Chinese Pleistocene human skulls are quite different from their European contemporaries with regard to the above-mentioned features (Wu 1988). In most if not all of the European neandertals and earlier European fossils, the antero-lateral surface of the fronto-sphenoidal process of the zygomatic is disposed more laterally, the mid-facial region is higher and more protruding, the juncture between the lower margin of the zygomatic process and the maxillary corpus is closer to the alveolar margin, the lower margin of the zygomatic process runs obliquely and is continuous with the lower margin of the zygomatic bone without curving, a canine fossa is absent, and the nasal saddle is quite protruding in comparison with the Asian specimens. Although certain features prevalent in east Asia do show up in a few European specimens (e.g., upper facial flatness in Steinheim, shovel-shaped upper incisors in some neandertals), the complex of east Asian features just outlined has not been found in any European Pleistocene hominids. It is clear that the complexes of morphological features characteristic of these two continents are different in general. The difference is the result of basically separate evolutionary trajectories maintained over very long periods of time. The occasional appearance of "east Asian" features in Europe, and the occasional appearance of "European" features in east Asia (e.g., the high upper face, spherical orbit, etc., in the Maba skull) can probably be attributed to the fact that (1) Pleistocene hominids living in these two widely separated continents were derived from a common ancestral stock (i.e., Lower Pleistocene *Homo erectus*), and (2) there was some, probably minimal, degree of genetic exchange between these two lineages during the course of their evolution.

AFRICAN-ASIAN COMPARISONS

To compare early *Homo sapiens* specimens from China with those from Africa is a much more complicated endeavor than to compare the Asian and European samples (Wu and Bräuer 1993). In the orientation of the antero-lateral surface of the fronto-sphenoidal process of the zygomatic bone, the angle formed by the horizontal lines representing the orientation of this surface is 70° and 69° in Bodo and in Broken Hill 1, respectively. The angle in Florisbad is roughly 110° (i.e., this skull shows an even flatter upper face than the Chinese specimens). On this variable, the African skulls exhibit morphologies that resemble both European and east Asian ones. The African specimens show a broader range in regard to this feature than that shown in either the east Asian or the European samples.

The ratio of the glabella subtense fraction to the glabella-bregma chord varies between 0.45 (Jebel Irhoud 1) and 0.58 (Broken Hill 1) in African early *Homo sapiens,* while this index in the Dali and Maba skulls is 0.43 and 0.45, respectively. They thus appear to fall within (or just outside) the lower part of the range of the African skulls. The values for this index indicate that the most protruding part of the frontal squama of archaic *Homo sapiens* is at a lower position in the Chinese specimens than in most of the African ones. Chinese specimens show a narrower range.

The Broken Hill 1 and Bodo skulls have high upper faces; their upper facial indices are estimated at 61.5 and 55.7, respectively. Jebel Irhoud, however, is ca. 49.7—close to that of all early *Homo sapiens* from China. Again, the African samples are more variable.

The morphology of the lower margin of the zygomatic process of the maxilla also shows a broader range in Africa than in east Asia or Europe. In Bodo and Broken Hill 1, this margin runs obliquely upwards and back and is continuous with the lower margin of the zygomatic bone. There is no change in direction at the junction between these two bones. The Eliye Springs skull probably resembles Bodo and Broken Hill 1. Broken Hill 2 and Florisbad are, however, different.

In these latter two specimens, the lateral part of the lower margin of the zygomatic process of the maxilla turns slightly downward just before it joins with the zygomatic bone. The medial end of the lower margin of the zygomatic process (its juncture with the maxillary corpus) is close to the alveolar margin in the Bodo skull, and is situated at a higher position in Broken Hill 1 and in LH 18. This feature is less variable in both the Chinese and the European samples.

The alveolar process in the Bodo and Broken Hill 1 skulls is very protruding, whereas it is only slightly protruding or flat in Florisbad in Broken Hill 2. LH 18 is intermediate in regard to this feature. All the Chinese specimens are much less protruding than Bodo and Broken Hill 1. Again, in regard to this feature, the Chinese specimens are much less variable than those from Africa.

The braincase is broadest in the midregion in all Chinese early *Homo sapiens* skulls, and in some African skulls (e.g., Hopefield, Ndutu). Whereas in LH 18 and Eliye Springs the broadest part is toward the back, as in neandertals, in Broken Hill 1 and Omo 2 the broadest part is located in a more anterior position than that seen in Hopefield and Ndutu. Again, the African specimens are most variable.

Although the skulls from both regions have mid-sagittal ridges on the frontal squama, the ratio of the height of the ridge to the width of its base on the transverse section perpendicular to the outer surface of the bone is larger in the Chinese specimens than the corresponding ratio in the African ones. In other words, the whole prominence is narrower and "steeper" in China than it is in the African skulls.

In sum, and acknowledging that the samples are exceptionally poor and are very widely distributed in space and time, it seems pretty clear that, in most features, Chinese archaic *H. sapiens* exhibit a narrower range of morphological variability—a range that corresponds to only a part of that typical of their African contemporaries. European archaic *H. sapiens* also exhibit a narrower morphological range, but it constitutes a different part of the spectrum from that in China and corresponds to the other part of the range found in Africa. To explain this phenomenon, I suggest that the common African origin of hominids led to basic morphological similarities among the later inhabitants of the diverse continents (a kind of a morphological substrate, as it were), yet genetic drift made later east Asians and Europeans somewhat different from each other, and from the original African populations, in the frequencies of certain genes in their respective gene pools. Natural selection in different very broad geographical regions could also account for some of the morphological differentiation, but, as noted, the features expressing the differentiation can hardly be considered to have had adaptive significance.

A survey of morphologically modern Chinese humans underscores marked similarities to archaic Chinese specimens in regard to the features discussed above. Specifically, the most protruding part of the frontal squama is at a lower position in both time-sequent samples. The broadest part of the calvarium is at the middle third of the maximum cranial length. The upper face is low. The lower border of the zygomatic process of the maxilla is curved, with the juncture between this border and the maxillary corpus relatively distant from the alveolar margin. The alveolar process is moderately protruding. The upper incisors are shovel-shaped. These shared, derived characters probably indicate continuity from morphologically archaic to morphologically modern humans in China.

IMPLICATIONS FOR REPLACEMENT MODELS

According to replacement models, Chinese morphologically modern humans are supposed to be descended from African archaic *Homo sapiens;* the morphological features of Chinese Upper Pleistocene populations should be inherited from the latter. From a replacement point of view, the narrower range of morphological variability seen in Chinese morphologically modern humans can be explained by genetic drift during the later *Homo sapiens* migration out of Africa, but, as we have seen, it is difficult to square this expectation with patterns observed over time in the Chinese data. It is significant that the result of genetic drift during the alleged "later" migration replicates that which supposedly occurred during the initial and much earlier *Homo erectus* radiation. This is, to say the least, problematic. If, in fact, the later migration was a large-scale event or process, how could the results of genetic drift during the "early" and "later" migrations possibly be so similar? Further, the similarity is manifest in a complex of features rather than being limited to one or a few of them. I suggest that such a coincidence is highly unlikely, and am compelled to conclude that there is little evidence for a large-scale migration of the supposed ancestor of east Asian morphologically modern humans from Africa, or from anywhere else. The shared features of archaic and modern *Homo sapiens* in China are unique to that area, and they are infinitely more likely to be the result of genetic connection than any other reason thus far proposed. Chinese Upper Pleistocene hominids inherited these features from the local ancestral stock, rather than from the African Eve.

In addition, if the replacement model were an accurate description of what actually transpired in China over the transition interval, the Late Paleolithic assemblages of China should show evidence indicating Middle Eastern connections. In the Middle East, the Middle Paleolithic pertains to the Mousterian industrial complex (sensu lato). Both archaic and modern human fossils are associated with Mousterian stone artifacts in the Middle East (Clark 1992a; Clark and Lindly 1989a, 1989b). However, the Late Paleolithic of China shows clear evidence of continuity with the Middle Paleolithic of that region, without any indications whatsoever of either accelerated change or disjuncture corresponding to the transition interval. Moreover, there is no evidence of a "Mousterian," as conventionally defined (e.g., Bordes 1968a), in the Middle or Upper Paleolithic of China. This fact has been convincingly demonstrated by all Chinese and some foreign archaeologists (e.g., Li 1993; Pope 1992b; Zhang 1990). It indicates that if the supposed ancestor of morphologically modern humans in China did come from Africa via the Middle East, and if that hypothetical ancestor replaced indigenous Chinese hominids, the "invaders" must have abandoned their own cultural traditions completely and adopted wholesale the cultural traditions of the "vanquished" East Asians. This is so utterly unlikely as to require no further comment. The archaeological evidence gives no support to the replacement model for modern human origins in China.

EAST ASIAN EVIDENCE OUTSIDE CHINA

In addition to the Chinese data, several other morphologically modern human skulls of Pleistocene age have been recovered from various parts of island east Asia. Various suggestions have been proposed for the affinities amongst them, and for their relationships with the Chinese sample. The Pleistocene morphologically modern remains geographically closest to China are those unearthed from Minatogawa, on Okinawa Island, and on Miyako Island, Japan. In respect of Minatogawa, Suzuki has argued

> Insofar as available skeletal material shows, about 18,000 years or more ago, some generalized proto-Mongoloids from south China as well as from north Indochina, having racial ties with the Liukiang Man, the Zhenpiyan Man and Lang Cuom Man, migrated to Okinawa and the mainland of Japan over a landbridge existing at that time. The Minatogawa Man and the Mikkabi-Hamakita Man from the mainland of Japan are probably the immigrants themselves, or the descendants of these immigrants. (1982:45)

In an effort to assess the credibility of this assertion, I applied the coefficient of divergence as defined by Brothwell (1960:337) to make comparisons between the Minatogawa male skull and Late Paleolithic skulls from neighboring areas. The coefficient of divergence (CD) is calculated as follows:

$$CD = \sqrt{\frac{\Sigma_k (a_i - b_i)^2}{k}}$$

where $a_i = A_i/(A_i + B_i)$; $b_i = B_i/(A_i + B_i)$; A_i = the measurement of skull A; B_i = the measurement of skull B; and k = the number of measurements (cited and modified from Brothwell 1960:337).

The results of my comparisons are summarized in Table 20.2 (from Wu 1992b). These figures confirm the assertion of Suzuki (1982) quoted above. The Liujiang cranium is closer to the Minatogawa adult male skull than to any other contemporary skulls found in neighboring regions.

To investigate the degree of relationship between the Minatogawa skulls and

Table 20.2. Coefficients of Divergence between the Minatogawa Male Skull and Other Skulls

	CD
Liujiang (Liukiang)	0.029
Zhoukoudian Upper Cave	0.054
Niah Cave	0.046
Wadjak	0.071
Keilor	0.053

Liujiang further, I calculated coefficients of divergence between two female skulls from the Upper Cave at Zhoukoudian (102, 103) and two female skulls from Minatogawa (II, IV). The results are 0.030 and 0.033, respectively, for these two Late Paleolithic populations. If these results could be taken to imply that the normal intrapopulational difference of east Asian Late Paleolithic populations is something around 0.03 as measured by the CD statistic, then the divergence between Minatogawa and Liujiang populations is so small that it is not unreasonable to conclude that they might have had a common origin.

The human fossils found on Miyako Island include a large piece of occipital squama, and a small triangular fragment probably representing about half of an Inca bone (Sakura 1985). Inca bones have not been found in African populations prior to archaic *Homo sapiens,* but they are fairly common in skulls of Middle Pleistocene age in China. They show up in three or four (of six) *Homo erectus* skulls from Zhoukoudian, and they are probably present in archaic *H. sapiens* from Dingcun, Dali, and Xujiayao. So, the presence of this small bone in the Miyako sample would seem to suggest a closer genetic relationship with indigenous archaic *H. sapiens* in China than with any African population.

Brothwell (1960) applied the coefficient of divergence to investigate the affinities between the Niah Cave hominid from Borneo and penecontemporaneous human fossils from neighboring (and, in some cases, distant) areas. His results are summarized in Table 20.3 (from Brothwell 1960:337). The lowest CD is that

Table 20.3. Coefficients of Divergence between the Niah Cave Hominid and Crania from Neighboring Areas (from Brothwell 1960)

	CD
Talgai subadult	0.057
Keilor adult	0.061
European neandertals	0.105
Le Moustier subadult	0.082
Chinese prehistoric	0.070
Borneo adults (modern)	0.059
Australian adults (modern)	0.042
Tasmanian adults (modern)	0.028
Japanese adults (modern)	0.050

Table 20.4. Coefficients of Divergence between the Niah Cranium and Crania from Other East Asian Pleistocene Sites (from Wu 1992a)

	CD
Liujiang	0.033
Minatogawa	0.046
Upper Cave 101	0.078
Wadjak	0.068
Keilor	0.077

between Niah and modern Tasmanians, so Brothwell concluded that "The Tasmanian and Australian groups are closest to the Niah skull, followed by the Japanese and Borneo groups. The prehistoric Asian types are next" (1960:339). Note that the "Chinese prehistoric" sample used by Brothwell (0.070) is a pooled series of mid-to-late Holocene rather than Pleistocene age.

Because there is clearly vectored morphological change from the Upper Pleistocene through the mid-Holocene in east Asian skulls, comparisons between Pleistocene and modern or Holocene samples are perhaps less compelling than comparisons between two Pleistocene specimens, which would hold temporal change at least approximately constant. So, I applied Brothwell's method to calculate the coefficient of divergence between the Niah skull and Pleistocene crania from neighboring areas. The results are summarized in Table 20.4 (from Wu 1992a).

It is also noteworthy that the cranial index of the Niah skull is 77.8, substantially higher than that of Talgai (73.4), Keilor (72.6) and Wadjak (74.5), as well as those of all other Australian Pleistocene or early Holocene skulls. The cranial indices of the latter do not exceed 73, and most of them are less than 70. On the other hand, the cranial index of the Niah skull is much closer to those from the Pleistocene of China (Liujiang, 75.1; Ziyang, 77.8). The shape of the upper alveolar arch of the Niah skull also resembles that of Liujiang, and it stands apart from those of Pleistocene or early Holocene Australian skulls. Finally, palatal length and breadth (47.5 mm, 37.5 mm) and the cranial index of the Niah skull (78.9) all closely recall the corresponding measures from Liujiang (45.0 mm, 36.0 mm, and 80.0 mm, respectively). I conclude from this that the Niah skull is overall morphologically much more closely allied to specimens from the region to the north of it than to those from the south. The closest resemblance is clearly with the skull from Liujiang, south China.

Finally, some Late Pleistocene human fossils have been recovered recently from Tabon Cave, in the Philippines. Third molar agenesis in a mandible from this cave recalls the similar condition found in the Lantian *Homo erectus* mandible, Peking Man, and the Liujiang skull, and its higher frequency of occurrence in modern Mongolians than in Australian aborigines. On these dental characters, it is reasonable to conclude that the Tabon remains are probably closer morphologically—hence genetically—to humans of Pleistocene China than they are to Pleistocene Australians. A narrow, longitudinal ridge on the outer surface of the nasal bone recalls a similar structure in Dali, Maba, and Upper Cave 101. Together, these resemblances strengthen the affinity of the Tabon population to that of the Chinese mainland.

21

Analyzing Modern Human Origins in China

SUSAN G. KEATES

Differences in archaeological research traditions between the People's Republic of China and the West are substantial. The conceptual framework in which lithic assemblages are studied in China is based on outdated methods and perceptions of hominid behavioral evolution. In consequence, comparisons between Chinese hominid behavior and that in other regions are in several respects limited. Modern methods of lithic and site analyses and a more precise chronology need to be established in Chinese archaeology.

The Middle to Upper Paleolithic "transition" cannot be understood without examining the archaeological record of pre–anatomically modern *Homo sapiens.* Attempts to identify and define the characteristics of the transition in China are especially highly dependent on the pretransition record because of the uniqueness of the lithic assemblages in their generally conservative informality. This characteristic makes it a more difficult and complex task to identify what may characterize the transition. Changes in lithic technology seem to appear in the Late Pleistocene. The archaeological and hominid anatomical evidences indicate behavioral and genetic continuity in China (Aigner 1981; Keates 1994b; Pope 1988, 1992a; Pope and Keates 1994).

Discussions on modern human origins have usually either ignored or marginalized the artifactual evidence from China. This has effectively constrained our understanding of the regional characteristics of early modern humans in a worldwide perspective (see also Keates and Bartstra 1994) and how they evolved.

TRADITIONS AND CONDITIONS OF RESEARCH

China's "Open Door" policy has since the early 1980s encouraged foreign workers to carry out research (e.g., Binford and Stone 1986; Clark and Schick 1988; Huang et al. 1988; Pope 1989b). Although the political disturbances in 1989 proved to be a setback to international research cooperation (such as the cancellation of arrangements with their Chinese colleagues by the majority of western researchers), the past few years have seen an increase in cooperative research (museum studies, surveys, and excavations; see Keates 1995; Pope et al. 1990; Schick et al. 1991; also Deborah Bakken, University of Illinois at Urbana-Champaign, personal communication 1990).

Paleolithic research in China was founded in the 1920s by western researchers whose training was usually in geology and paleontology (Freeman 1977; Keates 1995; Trigger 1989). Thus, archaeological studies were strongly influenced by then-current European perceptions of technological and faunal evolution (Breuil 1927; Pei 1937; Teilhard and Licent 1924). Although they were not always compatible with the evidence (Movius 1944; Pei 1937; Teilhard 1941), many of the principles established in research (e.g., stone tool typology) continued to be followed after westerners ceased to work in China in the late 1930s. This continuity is possibly associated with China's cultural conservatism and its political, economic, and scientific isolation. The ideas of a "culture type" (e.g., Teilhard and Pei 1932) and of three distinct "cultural blocks" (Black et al. 1933) may have been influential in forming the cultural tradition model, the dominant explanation of lithic assemblage variability in China. According to this model, assemblages with a predominance of either small or large tools represent different technological complexes and reflect a gradual change from the oldest to the youngest (Chia et al. 1972; Jia 1980; Zhang 1990). These assemblages are considered diagnostic and representative of largely distinct and independently evolved cultures (Jia 1980; Jia and Huang 1985b; Zhang 1990; and see Chia et al. 1972). A third tradition may have developed in the Upper Paleolithic (Jia and Huang 1985b). This approach has possibly influenced how Paleolithic studies are conducted, and it may have been a factor in continuing some of the lithic analytical methodologies established in the early period of research (see below). A linear and progressive evolution of hominid morphology, resource exploitation, lithic (more complex and diverse tools) and hunting techniques, and the use and eventual control of fire are thought to characterize the Chinese Paleolithic (Jia 1980; Jia and Huang 1985a; Zhang 1985).

Attempts to decipher the archaeology of the transition in China are probably among the most difficult of any Paleolithic region. Chinese regulations mean that setting up an excavation project is a complicated, long, and very political process. More serious is the fact that the archaeological record has suffered from inadequate financial resources available to Chinese researchers for excavation and postexcavation work. This has undoubtedly affected excavation methodology, documentation, curation, data analyses, and publication. For instance, lack of room for curation led to the discard of numerous lithic artifacts excavated from the early Late Pleistocene Xujiayao site before detailed study (Keates n.d.c).

Restricted access to most Chinese collections (in general only selected artifacts and fossils are made available for study) means that detailed analyses are not feasible (e.g., Aigner 1981; Binford and Stone 1986; Freeman 1977; Keates 1995). A request by this author to study the lithic assemblages from Xujiayao was refused "because the specimens have not been published internationally" [i.e., in the non-Chinese literature] (Li Chaorong, Institute of Vertebrate Paleontology and Paleoanthropology, personal communication 1989). Access is also difficult because some collections are dispersed in museums, and in some cases their location cannot be established (Keates 1995). Published studies are very brief (for example, of the important and artifact-rich Zhoukoudian Locality 1, Xujiayao, and Shiyu collections; Chia et al. 1972, 1979; Zhang 1985) and cannot be independently evaluated to a satisfactory extent. Analyses and interpretations focus mainly on tool typology, usually treating waste materials as marginal (Aigner 1981; Keates 1995; Pope, personal communication 1990). Much of the voluminous body of Chinese data lies "buried" and untranslated in journals published at the numerous research institutions across China (Keates 1995; Olsen and Miller-Antonio 1992; Pope and Keates 1994). All of these factors bias interpretations of assemblage variability. They limit attempts to identify what technological features may indicate early modern human behavior, and influence what kinds of generalizations can be made.

An understanding of the archaeological record of the transition has also been skewed by the concentration of research in the northern provinces. This situation is compounded by the continuing removal of dragon bones and fertilizer (guano) from the many karst caves in southern China. Furthermore, southern Chinese prehistory has traditionally been interpreted as less advanced compared with that of the northern provinces (see, e.g., Jia 1980; Wei 1981; Zhang 1990), an interpretation that seems to stem from a historical bias (Trigger 1989).

Archaeologists who have no working knowledge of China tend to represent the stone artifact and other cultural evidences from this region as of minor importance with regard to the behavioral evolution of modern humans (see, e.g., papers in Mellars and Stringer, ed. 1989, including Mellars and Stringer 1989; also Clarke 1990; Mellars 1989b:376–377), seemingly ignorant of the English literature (for instance, Aigner 1981; Jia 1980; Wu and Olsen 1985). Thus, the issue and problematics of the transition have been largely limited to Europe, Africa, and the Near East (Pope and Keates 1994). The influence this bias may have on securing funding for research in China may be significant (Keates 1994a). For example, a proposal to conduct multidisciplinary research at an Early Pleistocene hominid activity site (Pope 1989b) was rejected largely on the basis of the criticisms voiced by some reviewers whose comments demonstrate ignorance of Chinese paleoanthropology and misconceptions about field conditions in China. According to one reviewer, only methods of lithic analysis used in Africa should be applied in China (National Science Foundation 1991).

Unfortunately, unethical behavior has also complicated archaeological investigations. For example, in an apparently misplaced and unscientific sense of competition, some non-Chinese researchers attempted to terminate the fieldwork by a joint Sino-American team at an Early Pleistocene site in northern China, although both teams were investigating two different sites (Keates n.d.a). This has not inspired Chinese confidence in the positive outcome of future international collaborative research.

DEPOSITIONAL AND CHRONOLOGICAL CONTEXTS

One of the principal objectives in the study of hominid sites from any period of time must be the examination of the integrity of the recovered archaeological materials. A major problem in the study of Chinese hominid behavior is the frequently very inadequate excavation and recording methods used. Three-dimensional coordinates and screening are usually not employed. The preservational condition of archaeological materials (natural wear of artifacts, extent of fragmentation of bones, etc.) is often unpublished. Moreover, descriptions of how and where prehistoric specimens were found, and in some cases what exactly was found, can be unclear and ambiguous. For example, frequently it cannot be ascertained whether a specimen was excavated or recovered from the surface. Exhaustive reviews of the primary literature and, where possible, first-hand communication with the excavator(s) can, however, promote a better understanding of a site's depositional history and of other features. Reference to long-term studies of archaeological deposits conducted in non-Chinese regions, for instance, in Africa (Isaac 1977; Potts 1988), is often indispensable in site analyses because of the near absence of such studies in China (Keates 1995).

Although Paleolithic bone and antler tool technologies continue to be widely accepted in China (e.g., Tao and Wang 1987), the evidence from China appears to be rare and late (Binford and Ho 1985:428; Keates 1995; and see below). However, so-called bone tools can give important clues in identifying both the presence of nonhominid bone-modifying agents at archaeological sites, such as hyenas, and the extent of their activities in relation to site formation (Keates 1995). Such studies have been shown to be significant in non-Chinese (Binford 1981) and Chinese (Binford and Stone 1986) archaeological contexts. Taphonomy should be introduced as an integral part of prehistoric research in China.

Natural processes, such as floods and repeated shifts in river courses (especially of the Huanghe) may have substantially affected what we know about the geographical distribution of hominids and the kinds of localities that were occupied. The recovery of archaeological materials is probably also impeded and biased by the considerable depth of loess and river deposits in the Huanghe Basin (Teilhard 1941:63). However, very thick deposits have proved to be areas with good to excellent conditions for in situ preservation where archaeological occurrences have been exposed but not severely disturbed by natural erosion (e.g., Chia et al. 1979; Pope et al. 1990).

Chronometric dating is now slowly replacing the chronotypological approach (Pope 1988; Wu and Olsen 1985). However, too much reliance is still placed on geological and faunal dating, and too little attention is given to considering factors such as redeposition of materials.

ARTIFACTS

Chinese archaeologists recognize a sequence of Lower, Middle, and Upper Paleolithic industries based largely on lithologic, biostratigraphic, and typological criteria (e.g., Chia et al. 1972; Jia 1985; Jia and Huang 1985b; Wu et al. 1989; cf. Black et al. 1933). The Upper Paleolithic is thought to date to about 40,000–10,000 years ago (e.g., Jia and Huang 1985a). The classification into a succession of Paleolithic industries has been interpreted as Eurocentric, and possibly inappropriate for northern (Aigner 1983; Pope 1989a; Watanabe 1985; Yi and Clark 1983, 1985) and southern (Olsen and Miller-Antonio 1992) China. Until substantive critical evaluations of the Chinese Paleolithic have been made, using a geochronological terminology (i.e., Early, Middle, and Late Pleistocene) seems to be more appropriate (Olsen and Miller-Antonio 1992).

The generally unstandardized and conservative Chinese lithic record differs markedly from that of Europe, Africa, and the Near East (Aigner 1981; Pope and Keates 1994). Other important characteristics are the predominance of flake tool assemblages and the low frequencies of core tools and of artifacts with secondary retouch. Late Pleistocene assemblages appear to be relatively standardized, but publications are not detailed enough for independent analyses (see, e.g., Xie and Cheng 1989). The later Late Pleistocene assemblages appear with microlithic artifacts (e.g., Chen 1984; Gai 1985), but the definition of microlithic assemblages by different authors is still rather vague. Many of these assemblages include artifacts that are common in the Early and Middle Pleistocene of China (see, e.g., Dong 1989; Xie and Cheng 1989), while artifacts with a greater amount of secondary retouch modification (compared with previous periods) have been recovered from other Late Pleistocene sites (see, e.g., Yuan et al. 1989). What significance this, and an apparently greater regional diversity of assemblages in the Late Pleistocene, may have for interpretations of the transition needs to be addressed. Some of these assemblages may have been collected from secondary contexts. Certainly, the chronological contexts should be more precisely determined.

Research strategies in China need to change substantially to facilitate understanding of the characteristics of lithic assemblages and to enable informed comparisons between different assemblages. The screening of soils and the curation, recording, and analysis of all archaeological materials, including all micro-debitage, should become a standard. Similarly, the range and actual frequencies of artifact categories (cores, flakes, hammerstones, tools, etc.), raw materials, specimens with secondary retouch, etc., should be documented and published, as should artifact measurements (length, width, thickness, etc.). Metric data can, for example, be used to test the validity of the Chinese small and large tool model in terms of size distribution (Keates 1995), and to assess the claim that tool size decreases steadily in the Paleolithic (see, e.g., Institute of Vertebrate Paleontology and Paleoanthropology [IVPP] 1980; Zhang 1985).

There is also an urgent need for the standardization of artifact and raw material terminology to avoid ambiguities and help communications both among Chinese and between Chinese and foreign archaeologists. A tool classification terminology has been devised that may encourage a more objective analytical framework, one not hindered by assumptions rooted in European and African research contexts (Keates 1994b, 1995; Pope and Keates 1994). Studies of a number of lithic assemblages show that the published record has to be examined carefully. For example, the Xujiayao (see above) and the late Middle Pleistocene Dali tools have been classified into a multitude of categories (see, e.g., Chia et al. 1979; Wu and You 1980). This gives an impression of complexity which in some cases cannot be verified (Keates 1995, n.d.b). Although the facilities to conduct microwear analyses exist in China, they have yet to be integrated into research programs.

Core reduction experiments and refitting of artifacts to examine reduction strategies and fracturing characteristics of different materials may also assist in identifying changes in behavior. Previous attempts at conjoining have not been successful, probably to some extent because whole collections were not available for study (Keates 1995; and see Schick et al. 1991). Previous lithic analyses (of the late Early Pleistocene Donggutuo and Xiaochangliang assemblages) strongly indicate that studies of the raw materials selected for tool manufacture and knapping experiments may be very significant variables to interpret the general informality (cf. Movius 1978) and possibly the size of Chinese artifacts. These analyses also suggest certain commonalities in Chinese and East African hominid behavior (Keates 1995). While a correlation has been suggested between the informality of tool assemblages and an emphasis on nonlithic, especially bamboo, tool technology (Pope 1989a), it has thus far not been substantiated with direct evidence. A large body of information may, however, have been lost.

The constraints that exist in studying Chinese prehistory in exhaustive detail (see above) make it difficult to monitor changes of the pretransition lithic record. A computer database of Chinese Paleolithic sites incorporating published and unpublished data (Keates n.d.d) could facilitate more detailed studies and comparisons of the evidence (for example, the range of technologies in China before, during, and after the transition). An expansive photographic record of artifacts, directly accessible for consultation, would allow both Chinese and foreign researchers to achieve a more informed and hopefully objective impression of lithic assemblage variability instead of having to rely on selected artifacts. It could also assist non-Chinese workers when dealing with materials that are only directly accessible in China and when limited resources constrain the time for study.

Primary questions that could advance our understanding of the transition have become apparent through the direct study of Chinese assemblages and the literature. They have been influenced by what is known about hominid behavior in non-Chinese contexts. This author's European, and to some extent North American, academic background has certainly been influential. Comparing cultural materials from both Chinese and non-Chinese regions has the potential for identifying human universals, and for advancing interpretations of early *Homo sapiens sapiens* behavior from a sufficiently wide geographical perspective.

Some of the questions that could be addressed are: What characterizes the lithic record before, at, and after the apparent time of the emergence of modern *Homo sapiens*? Can a transition be recognized in China? In what respects does it differ from the record previous to the emergence? Are there diagnostic traits? Was there a change in how lithic technology was organized? To what extent can the Chinese lithic record be characterized as informal? Does informality mean that tools were manufactured and used in an expedient manner? Is there a correlation between small tool size or informality and raw material (e.g., tractability, size)? What role may the availability or tractability of raw materials have had on the technologies used, on tool morphology, etc.? What can informal assemblages tell us about technological and cultural capacity? Can changes in core reduction, raw material exploitation, and tool manufacture be recognized that suggest a change in hominid behavior? Is there an increase in the complexity of stone tool manufacture in the Late Pleistocene? Are Late Pleistocene tools smaller and "better" manufactured as claimed by Chinese archaeologists? Did small tool size possibly limit the extent to which tools could be modified by secondary retouch? Could this explain the apparent rarity of tools exhibiting secondary retouch? What kind of patterning can be recognized in Late Pleistocene assemblages that may differ from Early and Middle Pleistocene assemblages?

Detailed study of Chinese lithic assemblages will probably lead to the recognition of greater technological and stylistic variation. This assumption is based on the preliminary study of a number of collections, surveys of sites, and a review of the literature. In particular, the occurrence of very similar small retouched points throughout the Paleolithic (one Mesolithic/Neolithic specimen has also been recognized) of northern, eastern, and southern China had until recently not been noted (Keates 1991, 1994b, 1995). Two of these points were "discovered" in the spoil heap at Xujiayao. The manufacture of these artifacts to a standard pattern, in a variety of raw materials, and with indications that the sites at which these points are found increase in number during the Late Pleistocene, would appear to be of potential significance for the study of the evolution of hominid cognition. It is also intriguing, and requires explanation, that these specimens are found in different environmental zones. Standardization is also apparent in the tool collection excavated from Xujiayao. A study of a small sample of tools by this author found at least twenty "thumbnail scrapers" of a standardized pattern, all made on very fine-grained chert. The extent of standardization at this site cannot presently be assessed because of restricted access to this collection (see above), but it may indicate behavioral change at the beginning of the Late Pleistocene.

An understanding of hominid behavioral evolution from the perspective of the paleoneurological evidence (Pope and Keates 1994) combined with in depth studies of the lithic record could further enhance our knowledge of how certain brain functions may have changed over time. A study of Chinese hominid brain casts is planned (Dominique Grimaud-Hervé, Institut de Paléontologie Humaine, personal communication 1994).

SPATIAL ANALYSES AND SUBSISTENCE TECHNOLOGY

The extent to which hominid activities can be reconstructed is limited. Discrete activity areas cannot be identified for the majority of sites (but see Gai and Wei's 1989 work on the terminal Late Pleistocene Hutouliang locality) because the use of three-dimensional coordinates and sieving of materials is the exception rather than the rule. There is also a bias in the recovery of small fauna. The commercial utilization of fossils (subsequently prepared as medicine), and thus loss of fossils, including excavated specimens, is not uncommon in China. The available faunal data are frequently too poor for purposes of detailed reconstructions (and see Bettinger et al. 1994). For instance, it is often difficult or impossible to establish the number of fossils (NISP and MNI) recovered from sites (the exception is hominids). The very rare taphonomic studies (Lü 1985; Pei 1932, 1938; Pope 1988; Schick et al. 1991; Lü Zun'E, Beijing University, personal communication 1989) may also to some extent be a consequence of the implicit assumption by Chinese workers (see, e.g., Jia 1980; Wei 1985) that the association of artifacts with fossils represents evidence of hominid activity (Keates 1995). Thus, it is difficult to make informed and constructive interpretations of hominid site activities, and to gauge evidence for any changes in subsistence technology across the transition.

It has yet to be established whether the predominance of certain species at Xujiayao and the late Late Pleistocene Shiyu open-air sites may mark some kind of fundamental shift in hominid subsistence strategy. These faunal occurrences may be critical to advance our understanding of hominid resource and subsistence technology. The predominance of two species in the Xujiayao fauna (*Equus przewalskyi* and *Coelodonta antiquitatis* with 91 and 11 individuals, respectively; Chia et al. 1979) and in the Shiyu fauna (*E. przewalskyi* and *E. hemionus* with 130 and 88 individuals, respectively) has been interpreted as evidence of specialized faunal exploitation and "the appearance of 'professional' hunters" (Jia and Huang 1985a:221). Although burnt bones were recovered from both sites, studies do not seem to have been carried out to substantiate these claims with, for example, evidence of butchery marks. However, cut-marked horse mandibles have apparently been identified at Xujiayao, with cuts positioned for tongue removal (Pope, personal communication 1990). The stratigraphic data are not sufficiently detailed to investigate whether the animal fossils recovered from several meters of soil thickness represent discrete episodes of exploitation or "massive kills" (cf. Chase 1989).

Evidence from the earlier Zhoukoudian Locality 1 may be instructive to examine the antiquity of selective faunal exploitation. Horse was identified as the most cut-marked and least gnawed fossil (Binford and Stone 1986). Most tool-marked mandibles are represented by horse, and cutmarks on horse and cervid mandibles suggest, among other activities, tongue removal (Binford and Stone 1986). These data may indicate possible commonalities between late Middle Pleistocene and Late Pleistocene hominid behavior. Although the low frequency of identified tool marks at Locality 1 suggests limited hominid exploitation of animal resources (Binford and Stone 1986), it may alternatively indicate limited use of stone tools and the use of other techniques of carcass utilization (and see Pope 1989a). The representativeness of Binford and Stone's (1986) sample has also been questioned (Behrensmeyer 1986; Olsen 1986; Pope 1988), while the poor preservation of most bones from this site would probably constrain clear identification of any hominid modification.

SYMBOLIC BEHAVIOR AND ART

Symbolic behavior and art are thought by some to be critical indicators of advanced human cognition and behavior (e.g., Mellars 1989a; White 1993a). Two important characteristics of Chinese archaeology are that symbolism and art are rare and appear late. Considering the extensive investigations so far conducted (especially in northern China), the lack of these features appears to be a characteristic of early modern humans in this part of the world. This does not mean that perishable materials may not have been utilized. If culture is defined "as a system of socially transmitted information" (Barkow 1991:140), then culture is rarely expressed in the symbols and art of China.

The very late and very rare evidence for hominid burial relative to other Paleolithic regions seems to be one of the characteristics of modern humans in China. Burials associated with apparent grave goods (bored carnivore canines, perforated seashells, etc.) have been recovered from the late (R. E. M. Hedges et al. 1992) or terminal (Wu and Wang 1985) Late Pleistocene Upper Cave at Zhoukoudian (Jia and Huang 1985a; Pei 1939). Lack of preservation may to some extent explain the late evidence for this behavior; however, the small amount of evidence thus far recovered should not prevent detailed investigations of those materials. Aigner's (1981:128) suggestion that the Zhoukoudian Locality 1 hominids disposed of the dead in parts of the cave separate from the places of activity may infer intentional burial; it would represent the earliest evidence for this behavior in China (Keates 1995). Similarly, it might be profitable to examine whether the presence at Xujiayao of at least eleven adult and juvenile hominids could indicate intentional burial.

It has recently been suggested that an engraved antler fragment indicates a relatively great time depth for Chinese Paleolithic art (Bednarik 1992, 1995). This interpretation is difficult to reconcile with the uniqueness and very late age (terminal Paleolithic) of this artifact (Keates 1994b). The "engraving" on an equid bone (You 1984; Yang 1988:pl. 26) seems to be a natural artifact, possibly root etchings (Keates 1994b).

CONCLUSION

A consensus on modern human origins that has substance cannot be reached if the archaeological evidence of China (and for that matter of other regions of the Far East) continues to be either ignored or minimized. Even though actualistic analyses of the artifactual evidence and its contexts need to be established, there is a substantial body of data which can be usefully employed to make more than superficial comparisons with data from Europe, Africa, and the Near East. Since behavioral scientists are concerned with identifying and understanding the characteristics of the Middle to Upper Paleolithic transition, they must surely also recognize that substantially more resources need to be invested soon in both field and museum research in China.

22

The Evolution of Modern Human Cranial Diversity
Interpreting the Patterns and Processes

MARTA MIRAZÓN LAHR

The idea behind this book is to produce an introspective analysis of the paradigmatic biases that influence current research into the origins of modern humans. I should say at this point that I will not claim, as some do, to be "paradigm-free." I work within a strong paradigm that defines how I have come to believe evolution to work. In Clark's (1991) terms, I'm clearly what he calls a *replacement advocate* (as opposed to a *continuity* advocate). However, I also have a strong background in evolutionary biology which takes me towards hypothesis testing. This means that rather than interpreting information solely on the expectations of a theoretical framework, I try to formalize these expectations into hypotheses that can then be statistically confirmed or refuted. Some will see this as pure empiricism, but I will argue below that what I am applying to this research domain is not solely a reading of data, but the premises and perspectives from evolutionary biology.

Given the amount of recent and fossil data relating to population relationships and characterization, it is clear that the discussions arising from this research should not be at the level they are. The differences in the theoretical explanatory frameworks used by different researchers are of such magnitude that completely opposite, yet internally coherent, interpretations of a single evolutionary event are possible.

The question to be asked is, are such alternative explanations an unavoidable consequence of working with what are, ultimately, unprovable events? It is certainly true that alternative interpretations will always exist, for that is how science proceeds. However, science searches for answers. As is true of all sciences, evolutionary biology progresses by asking questions, formulating hypotheses to test those questions, obtaining relevant data as objectively as possible, examining those data for patterns, testing the hypotheses mathematically, and answering the questions by an interpretation of the results within a theoretical framework. The researchers' paradigms will influence both the questions asked and the answers obtained, but not the actual data or results. If the data are clearly defined and the methods of analysis clearly specified, the results should be evident to all. Discussions and disagreements are thus brought down to the level of differences in interpreting data and not personal theoretical expectations.

THE THEORY BEHIND THE FACTS

In order to study the evolutionary events related to modern human origins and diversity, several different sources of data exist. Each requires different assumptions and inferences, and both the choice of data and the assumptions as to the evolutionary meaning reflect the researcher's educational background and theoretical framework.

Some researchers will use only recent data (biocultural parameters of recent groups of people), from which an inference as to evolutionary patterns and processes is made based on a characterization of recent human populations and their affinities. This has been the case with many geneticists, who argue strongly that genetic data are "better" (i.e., more reliable) than any skeletal or archaeological information. However, recent data can only estimate the affinities of the "survivors" in the evolutionary process and, even here, assumptions as to the character of past processes are necessary to infer past events. Most of those involved in the debate work with data that represent direct measures of past biocultural traits derived from the fossil and archaeological records. Although the problem of inferring past processes from present parameters is avoided, other major assumptions are necessary: diversity in the fossils is assumed to reflect biological processes, and the archaeological assemblages are assumed to represent sociocultural characteristics. Therefore, recent data require a shift from real to inferential processes, past data require a shift from real to inferential parameters. Furthermore, most researchers tend to use a single data set, or build on one data set and use patchy confirmatory information from others. Only rarely are the different data sets involved in the actual formulation and testing of hypotheses, and the reasons are largely pragmatic ones.

I have stated above that these two sources of data, past and present, represent different kinds of data owing to their necessary assumptions. However, in practice they are separated along lines that reflect theoretical analytical affinities: archaeological assemblages and contemporary sociocultural features are studied under similar theoretical frameworks, whereas genetic and skeletal data are studied using concepts derived from evolutionary biology. In analytical terms this division makes sense, for biological and social features have different modes of transmission. However, applying these different analytical techniques to past data precludes their integration at the level of hypotheses. This, in turn, results in incomparable interpretations, which preclude the use of both biological and sociocultural archaeological information within the phylogenetically oriented evolutionary hypotheses. The archaeological and fossil records become independent by force of analysis.

Both multiregional and single origin models are derived from these various sources of data. The problem with defining data certainly plays a role in the current conflict, but it is not the source of the argument. The argument is about the character of the evolutionary process.

MULTIREGIONALISM: A MODEL OR AN EXPECTATION?

A large part of my work has been related to testing the premises of morphological continuity from archaic to modern *Homo sapiens* in east and southeast Asia. In doing so, I have come to believe that such continuity cannot be documented. In the absence of an empirical basis for multiregional evolution, the other assumptions leading to a view of anagenesis as the evolutionary mode can also be questioned. I, like others, believe that independent of the source, the data relevant to the origins of modern humans can be interpreted as indicating a recent single origin for the group. Furthermore, the evolutionary processes leading to such an event are consistent with the diversity of forms and adaptive variability that characterized earlier human evolution. This does not mean that single origin models are problem-free. A number of morphological, behavioral, and chronological paradoxes exist in the record, and these have to be examined in terms of the biases affecting our view of a single origin as a single event. Not least, the problem of deriving modern diversity from a single ancestral source remains largely untackled. However, because the debate is stranded between arguments of multiregionalism and those of localized origin, most of the effort has been directed towards refuting or confirming either view and not towards developing the model of a single origin further.

I will address some of the issues related to interpreting the pattern and process of modern human differentiation in the late Pleistocene. However, in my view, these issues are problems within the broader paradigm of cladogenesis. As Clark (1991, 1992a) does, I consider that the polarized extreme theories of modern human origins (described at length elsewhere) reflect two different paradigms held by the researchers involved. The paradigms behind these models concern the evolutionary process (in terms of gradualism and universality of process) and, in some cases, sociopolitical concerns. I will try to show why I believe that, although historically the formulation of a model based on the premise of anagenesis as the evolutionary mode makes sense, the maintenance of this view today is largely, if not wholly, paradigmatic.

BIASED PREMISES OF CONTINUITY

Views on anagenesis on a multiregional scale call into question biological premises of the scale of selection and gene flow acting to maintain not only widespread clines, but the evolution of polytypic species within such clines. Selective forces acting on individuals within a population and the genetic and demographic effects on that population (in terms of shifting gene frequencies and expansions/extinctions) are thus scaled to the level of the worldwide polytypic species observable today. This universal view of evolutionary events is diametrically opposed to that of many researchers (including myself), who view not only the effects of selection as localized, but the actual benefits of selective processes as well. Consequently, evolutionary change is not seen as a necessary outcome in all populations of a species, resulting in a process of subspecific differentiation that is strongly influenced by the magnitude of gene flow and is nonuniversal and diachronic in character. In paleontological terms, these alternative models of evolution are evidenced in the spatio-temporal pattern of change (i.e., morphological continuity or discontinuity).

I consider that the answer to whether there is evidence for morphological continuity across an archaic-modern boundary should actually be very simple. Either there is evidence of continuity, as defined by those who defend it (I might disagree as to its definition, but that is a separate question) or not. A large part of my research has been concerned with testing the assumptions of morphological continuity in the late Pleistocene, and not trying to demonstrate a single origin for modern humans. I am theoretically inclined to view evolutionary events as cladogenetic, multiple, and nonuniversal. Have these theoretical inclinations biased my attempts at testing continuity? Although paradigm biases affect the formulation of hypotheses, in this case the hypotheses I tested were those formulated by continuity advocates, using their assumptions, choice of data points, and populations. Below I will show why I believe the three premises of continuity (chronological, genetical, and temporal) reflect personal theoretical expectations and not actual interpretations of evolutionary events in the Upper Pleistocene.

Regional Synchrony of Evolutionary Change

An anagenetic evolutionary mode in the genus *Homo* is inferred from the worldwide maintenance of "grade" similarities over one million years, and the view that these "grades" superseded each other concomitantly in several regions of the world. This hypothesis has an empirical component—the chronological data available today. Clark (1991) has argued that dates do not provide a pattern, and that to assert they do represents naivete of interpretation. In Clark's view, the pattern is not a quality of the record per se, but rather that what constitutes a pattern is determined or influenced by paradigmatic biases. But the question of whether we can achieve an objective pattern in the chronology of events is a different one. If such evidence is used at all, it should be valid for either view— independent of whether the result is synchrony or asynchrony.

The question of synchrony, of course, has a direct evolutionary significance when dealing with clinal change. In several works outlining and defending the multiregional model (especially Wolpoff 1989b; Wolpoff et al. 1984), it is argued that clinal theory explains the balance between selection and gene flow, the latter not only maintaining a polytypic species but its evolution as a cline. In order to achieve such an explanation, it is necessary to assume that no speciation events occurred in the period concerned. This is a very large assumption indeed. However, if it is made, then the mechanisms that explain the process of evolution of clines are consistent.

The primary assumption of anagenesis (i.e., absence of clear temporal species boundaries and branching events) is built into the multiregional theory. The maintenance of species cohesion, phenetic similarities, and stability is explained in terms of the evolution of coadaptive genetic systems that do not require gene flow to be maintained (Carson 1982; Mayr 1970). While such coadaptive genetic systems could account for the appearance of each of the "grades" mentioned above and of regional subspecies creating a polytypic species configuration, they could not account for the relationship between local adaptation and differentiation of the regional populations. This is because subspecific differentiation is not believed to affect the balance of adaptation of the new system. This problem is recognized by Wolpoff et al. (1984), who argue that the mechanisms maintaining species cohesion and stability are different from those maintaining phenetic similarities, and that the coadaptive genetic systems are different from one population to the next within a polytypic species. Therefore, it is argued that species cohesion can be maintained through gene flow, even if phenetically the regional populations are responding to local selective forces. Gene flow would further act to provide specieswide evolutionary change. These views, together with the basic assumption of absence of speciation, are clearly stated: "we contend that a special explanation for phenetic similarity is hardly required in the face of the extremely small number of genetic differences distinguishing populations enclosed *within the protected pool of a biological species*" (Wolpoff et al. 1984:460; emphasis added); . . . "The shifting balance model with its multiple adaptive peaks better fits what is known about polytypic species evolution *in the absence of particular speciation events*" (Wolpoff et al. 1984:460–461; emphasis added).

On the other hand, if the appearance of "grades" is seen as the emergence of new coadapted genetic systems that represent new species with specieswide phenetic similarities, the process radically changes in character. The appearance of these new forms would be related to the response to local parameters, and its success evidenced by geographical expansion and diffusion. The establishment of regional populations would therefore give rise to a process of subspecific differentiation, which would maintain species integrity until a new adaptive system emerges within one local population. Pronounced evolutionary change would thus be directly linked to the amounts of gene flow, and in the absence of such gene flow the constraints in altering the coadaptive genetic systems would tend to favor conservatism in form. If an anagenetic, single-species view is not imposed on the data, such an alternative view of evolutionary change, which allows for diachronic and nonuniversal change to take place, is a better model for interpreting the chronological record of the past one million years (Aitken et al. 1993; Grün and Stringer 1991; Grün et al. 1990; Schwarcz et al. 1988, 1991; Stringer et al. 1989; Valladas et al. 1987, 1988).

The World Population of *Homo erectus*

It is clear that anagenic and cladogenic views of evolutionary change are modeling different magnitudes of past gene flow. Large amounts of genetic exchange between widely separated areas would have been necessary in order to maintain the specific unity of regional populations over such long periods of time (one million years) as argued by multiregional advolcates.

Traditionally, two sources of inferential information have been used to infer past levels of gene flows. The first of these is current levels of genetic exchange between foraging groups. It should be clear, however, that this is not actual evidence. Using these data to model past gene flow requires modeling parameters of past demography and geography, and especially, rates of inter- rather than intra-group genetic exchange. Unfortunately, such models are not common.

The second source of inferential data on past inter-regional gene flow is the fossil and archaeological records. The appearance in one area of fossils or archaeological remains that have strong affinities toward those from another region is taken to reflect inter-regional gene flow: "There is a growing body of *direct evidence* for gene flow along this eastern end of the human range" (Wolpoff 1985:361, on the similarities between Wajak, Keilor, and Liujiang; emphasis added). Therefore, gene flow is regarded as the agent maintaining multiregional change through time, rather than the observed change being interpreted as reflecting the introduction of a new form into the area.

It is part of the premises of the multiregional model that the maintenance of clinal equilibria with regional adaptive peaks within a species is independent of the magnitude of gene flow. In Wolpoff's own words: "I reject the notions that selection or drift is likely to 'override' the influence of gene flow, or that gene flow is likely to 'swamp' the effects of selection. The idea that gene flow magnitudes can be too small or too large makes no sense in terms of the balance model of clines" (Wolpoff 1989b:89). However, the paleoanthropological record of the past one million years shows that the balance of clines was not maintained, and that important discontinuities in morphology and occupation occurred. In these circumstances, the magnitude of gene flow becomes not only relevant but necessary to explain the lack of synchrony of events.

While arguing both that the magnitude of gene flow is unimportant and that gene flow was widespread, proponents of the multiregional model largely ignore genetic assessments of paleodemography. Many years ago, Cavalli-Sforza modeled population growth across the Paleolithic-Neolithic boundary (Ammerman and Cavalli-Sforza 1984). This model shows pre-Holocene population sizes that, in terms of world ranges, would be minute. In the past few years, mtDNA data have also been translated into measures of paleodemography by Harpending, Rogers, and coworkers (Harpending 1994; Harpending et al. 1993; Rogers n.d.a, n.d.b; Rogers and Harpending 1992; Rogers and Jorde 1995a; Sherry et al. 1994). These reconstructions are clear: the world population of *Homo erectus* could not maintain the continuous levels of multidirectional genetic exchange necessary for multiregional evolution to occur. The concept that population sizes and magnitude and direction of gene flow varied markedly through time can no longer be ignored.

Temporal Continuity

Temporal continuity is the crux of the multiregional model and should be reflected both archaeologically and morphologically. In the case of archaeology, the gamut of arguments is much greater than those involving fossils because of the problems of data definition, although still finally expressed in terms of arguments against or in favor of temporal continuity. On the other hand, the morphological perspective on multiregional continuity is far clearer, for the data that are supposed to prove such continuity have been defined.

EVIDENCE FOR REPLACEMENT

Arguments for archaeological continuity follow two lines. On the one hand, using theoretical expectations that reject phylogenetic readings of archaeological data, some archaeologists have come to play an important role in maintaining anagenetic views of evolution. Thus, researchers like Clark argue that the distinctions between Neanderthal and modern technologies are imposed on the data: "there is no correlation whatsoever with archaeological industrial configurations, nor did important changes in technology, subsistence and settlement patterns coincide with the biological transition (between archaic and modern humans) in either area" (Clark 1989a:157). On the other hand, there are those who see the interpretation of non-European remains (mainly Asian) as suffering from Eurocentrically imposed biases, and defend "indigenous" interpretations that reinforce the view of local developments (Pope and Keates 1994:543).

A large number of studies have counter-argued Clark's views. As he says, these studies are mostly those of European scholars (e.g., Mellars 1989a, 1993), but not exclusively (e.g., Klein 1992). However, what transpires from continuity views are unclear hypotheses. The question "Is there a biological transition during the Upper Pleistocene in Europe?" is answered in the affirmative, as is the question "Is there an archaeological transition during the Upper Pleistocene in Europe?" The argument actually turns on whether these two events coincide and reflect a replacement of one population by the other, and the significance of these transitions. Showing that the archaeological data do not prove a replacement event does not prove continuity, and similarly, arguing that archaeological change reflects function, process, conditions, etc., and not the identity of the makers, only tells about the applicability of archaeological information to phylogenetic questions. Therefore, if archaeological data cannot be assessed from a phylogenetic perspective, then the archaeological evidence cannot be used to support either view.

However, these arguments are at variance with European evidence, where contemporaneous Middle and Upper Paleolithic assemblages have been found (Bischoff et al. 1989; Cabrera and Bischoff 1989; Mellars 1989a, 1993; Straus 1989, 1994). The "Neanderthal" character of the fossils associated with terminal Mousterian sites (Hublin 1994; Stringer et al. 1984) is a strong justification for attributing these archaeological remains to "Neanderthal people." So, does evidence of contemporaneity attest to a replacement event? The concept of a replacement event needs to be formalized so further controversy as to its character can be avoided. In order to recognize a replacement event (by which a new form, locally evolved or migrant, will outcompete and replace a resident group), it would be necessary to identify a four-stage sequence: (1) the presence of a single population in an area, the original resident form (in the case of Europe, the Neanderthals); (2) the appearance of a new form, or an increase in numbers of an already resident minor group or variant (in our example, modern humans and the Upper Paleolithic); (3) a period of contemporaneity of the two groups (the period between approximately 40,000 and 30,000 years ago); and (4) the disappearance or significant reduction of the resident population (the disappearance of Neanderthals and Mousterian assemblages) would have to be demonstrated. In cases where replacement is incomplete (i.e., the new population becomes the great majority but not the only group in the area), small isolated pockets of evolutionary relics of the earlier form could be observed (e.g., the south Spanish site of Zafarraya). Such pockets may result from either geographical refugia (areas where the new group does not reach—geographical barriers) or ecological refugia (areas where conditions are so stable that the existing adaptations of the early form are not outcompeted by the newly expanding group—stability of niche).

As archaeologists like Klein (1992) and Mellars (1989a) and paleoanthropologists like Stringer (1989b) have argued, the pattern of the transition from archaic to modern in western Europe during the second half of the Upper Pleistocene is consistent with a replacement event, and not with a process of in situ continuous change. A replacement in evolutionary terms does not imply total absence of admixture, just a change in evolutionary trajectories owing to the introduction of a new form. Nevertheless, even in the case of extensive hybridization as argued by Smith (Smith, Simek, and Harrill 1989), the basic concept is an alteration in the magnitude of gene flow with the appearance of modern humans in the area, so that a morphological shift results when gene flow overrides the local pattern of adaptation. It should be noticed that this view is not consistent with the theoretical bases of multiregional evolution (Wolpoff et al. 1984).

MORPHOLOGICAL EVIDENCE FOR DISCONTINUITY

A large part of the argument over the origins of modern humans rests on evidence for morphological continuity from archaic to modern in more than one region of the world (Frayer et al. 1993; Wolpoff 1989b; Wolpoff et al. 1984). If such morphological continuity could be proved, a single regional ancestry for all modern humans to the exclusion of all other archaic populations can be rejected. It is therefore crucial to establish whether there is evidence of regional archaic traits in certain recent populations, so that a regional archaic ancestry, or at least regional archaic genetic contribution, can be confirmed. The design of a test for this hypothesis is relatively simple, for the data proposed to reflect morphological continuity in east and southeast Asia has been specified by Wolpoff et al. (1984) and several researchers before them (Macintosh and Larnach 1976; Weidenreich 1943a; and others). What had not been clearly determined were the methods for objectively quantifying these data. This lack of standardization and quantification has played its own role in maintaining the debate on the incidence and distribution of the features of continuity (Groves 1989a; Wolpoff 1989a), but as paleoanthropological research abandons subjective measures ("larger than," "more rounded," "more or less straight," etc.), this problem will certainly disappear (cf. Clark and Willermet 1995; Willermet and Clark 1995 for a less optimistic view).

I tested the hypothesis of morphological continuity in eastern Asian and Australian populations by quantifying and comparing the incidence of the features of continuity in several fossil, recent, and sub-recent modern populations (Lahr 1992, 1994, 1996). Was this test biased by my theoretical expectations and paradigms? I argue that it was not. Neither the questions asked nor the data used were chosen to demonstrate my theoretical expectations, while the quantification of data gives a particular strength to the inter-regional comparisons. The results of these tests are so categorical that it is difficult to comprehend how the view that this specific set of features reflects multiregional continuity in east and southeast Asia persisted for as long as it has. I understand that earlier researchers did not use statistical analyses and may have used local rather than regional populations to represent the wider range of regional variation, but the persistent use of these features today is unjustifiable.

I believe that sometimes certain concepts are perpetuated because it is impossible for everyone to have firsthand knowledge of the relevant data. With time, these concepts and viewpoints become part of the status quo. The "mark of Java" in Australian skulls is one of those concepts. Several studies have shown that, besides the persistence of archaic features of robusticity shared by archaics and early modern crania from elsewhere (Groves 1989a; Groves and Lahr 1994; Habgood 1985, 1989; Lahr 1992, 1994, 1996; Lahr and Wright 1996; Stringer 1992a), Australian crania are modern in size and shape and more closely related to other modern skulls than to any archaic population of the world (Howells 1989; Stringer 1991b, 1992b). Nevertheless, it is a commonly held view, probably born of hearing it repeated over the years, that Australian crania have features absent in other modern populations that link them to the Ngandong fossils. This is just not supported by the data.

DISCONTINUITY OF OCCUPATION

We have seen that the question of hybridization or admixture between Neanderthal and modern humans is justified given the contemporaneity of these two populations. However, the bases for assuming continuity in Asia are very different and, again, reflect preconceptions about the character of the record.

It is usually assumed that after its original colonization, each area was continuously occupied by hominid populations. However, the issue of local population extinction or continuous occupation must be addressed in terms of the interaction between small population sizes and unstable environmental conditions. As mentioned before, estimates of population size in the Pleistocene point towards very small groups, and the relative genetic contact or isolation of a group would depend on specific ecological and cultural circumstances. Both the biological and nonbiological circumstances of all environments change continuously, and these changes were even more pronounced throughout the phases of glacial cycles. The changes in environmental conditions must have required changes in subsistence strategies by the different hominids, resulting in periods in which certain populations were more densely concentrated and localized, and others more thinly and widely distributed. The genetic processes acting upon groups with these different demographic parameters must have varied markedly, increasing the circumstantial chances of particular populations expanding or going extinct. These demographic fluctuations in terms of population numbers, density, and range are at the root of the problem of continuity of occupation.

So, what is the basis for assuming that there was continuous occupation? Only archaeological data can provide positive evidence for such an assumption. In some areas the archaeological record is sufficiently rich as to allow some measure of stability of occupation. However, other areas lack such rich archaeological records, and hominid presence is inferred from the sporadic occurrence of fossils. Whether these fossils represent a genetic continuum throughout any period of time or temporally independent periods of occupation can only be ascertained morphologically if all fossils within a region show closer links to each other than to other fossils outside the area. Only if this is the case can temporal continuity of occupation be accepted.

Proponents of the multiregional model again betray their biases here. Notwithstanding the lack of archaeological data, morphological analyses of Javanese *Homo erectus* point towards continuity of occupation (Rightmire 1990; Santa-Luca 1980). These morphological similarities are contrasted, however, with the pronounced differences between the last *Homo erectus* (Ngandong) and the subsequent modern populations in the region (either southeast Asian or Australian) (Lahr 1996; Santa-Luca 1980; Stringer 1992a, 1992b). In the absence of archaeological information, morphological data of discontinuity must be used. We cannot use negative evidence to demonstrate that the archaic population of Java went extinct, but we have to accept the positive evidence of discontinuity in the process of population change.

At this point, we should consider what the chances are of a local population going extinct. The distribution and density of archaeological sites at the last glacial maximum shows that many areas suffered important reductions in either population density or human geographical range (Gamble and Soffer 1990; Soffer and Gamble 1990 and papers therein; O'Connor et al. 1993; Smith et al. 1993; Veth 1993), with the possibility that certain areas were virtually uninhabited for considerable periods of time (Klein 1992; Mussi 1990). Reconstructions of late Pleistocene demographic fluctuations show the dynamics of the evolution of human populations, whereby the conditions of specific geographical localities, specific peoples, and specific population densities will determine the response to major environmental fluctuations. Therefore, it is impossible to assume or generalize the parameters that determine the effects of environmental change on populations throughout the world. The circumstantial aspects of the evolutionary processes that make a population expand, contract, or change in response to a particular environment at a particular time are characteristics of all complex dynamic systems and the source of the anachronic spatial patterns in the paleoanthropological record.

BUILDING MODELS FROM THE RECORD: MULTIPLE EVENTS AND THE NONUNIVERSALITY OF PROCESS

Over the course of the past few decades, paleoanthropologists have come to accept that the appearance of several features in human evolution occurred independently. Bipedalism, stone-tool technology, brain expansion, life-history changes, language, personal and artistic ornamentation, trade, agriculture, etc., all appeared at different times in the record (Foley 1996). It is also widely accepted that for most of human evolution more than one hominid form was in existence, with more than one hominid adaptive niche. These broad evolutionary patterns suggest that human evolution did not proceed anagenetically towards a single form, and that certain features we consider "human," like the use of technology, our pattern of postnatal growth, language, and sociality, might not have been part of the make-up of many hominid groups. The same principle of independence of events applies to the late Middle and Upper Pleistocene. Only by decoupling the events will it be possible to establish the patterns and interpret the processes.

The events that are at issue are as many as may be inferred from the data. A few of the most commonly disputed ones are the age of the last common genetic ancestor, the appearance of morphologically modern humans, the development of the Upper Paleolithic, and the establishment of regional populations of modern character. These events have commonly been treated as part of a single anatomical and behavioral "package" associated with becoming modern, especially in Europe. However, it is the forced association of these events that creates the inconsistencies in the interpretation of the patterns and processes in the paleo-anthropological record.

Such misconstruals of associations are among the most important biases influencing interpretation. The expectation of an association between modern humans and Upper Paleolithic tools has created inconsistent arguments on either side of the debate—on one hand, there are those who use the fact that early modern humans do not have "blades" to refute the association of the Upper Paleolithic in Europe with modern people (Clark 1991); on the other hand, the same fact is used to argue that all populations before the Upper Paleolithic were not behaviorally modern (Klein 1992). The advent of the Upper Paleolithic was indeed an extraordinary event, and in the case of Europe, some have suggested that it is associated with the appearance of modern humans. However, in other areas of the world modern people were already present before the Upper Paleolithic or Late Stone Age industries appear in the record, while in Australia and southern South America, these traditions never developed. It is clear that the appearance of modern humans was not associated with the Upper Paleolithic. A cladist looking at this problem would say that the origin of the Upper Paleolithic (not its later extension for the horizontal transmission of behavioral traits) is an apomorphy of European and western Asian populations, and not modern humans. Thus, the Upper Paleolithic does not characterize modernity.

I will mention three other such forced associations that confuse the patterns in the record. Firstly, it is sometimes assumed that the age of the last common genetic ancestor should translate into the age of the last common morphological ancestor. However, that is not the case. Genetic events shared by all modern people may precede the appearance and establishment of a particular morphological pattern for different periods of time. As far as the date of the genetic ancestral population is concerned, estimates range markedly. However, all these estimates are consistent with a point of coalescence of a diverse array of human genes in the Middle Pleistocene and, therefore, are compatible with a recent single ancestry of modern humans (Stoneking et al. 1992). The actual "age" or point of divergence from the ancestral source of recent populations remains unknown.

Secondly, the association of morphological modernity and skeletal gracility has also created arguments that certain specimens are not "fully modern" in morphology. By "fully modern" it is usually implied that they should present a level of gracilization as observed in recent skeletal remains. However, studies of cranial robusticity of early and recent modern populations show that the assumption that "modernity" equates with "gracility" is wrong (Lahr 1992, 1996; Lahr and Wright 1996). The loss of robusticity in the majority of recent populations is a later, nonuniversal event in the history of modern human evolution. Disclosing the actual temporal pattern and functional character of the gracilization process in different modern groups remains a challenge.

Finally, the association between a modern morphology and recent patterns of regional diversity is possibly the most misused, for it represents the empirical differences between the multiregional and single origin models. Those who consider the evolution of regional diversity to have preceded the evolution of a modern morphology argue that the early moderns throughout the world cannot derive from an African source for they lack African characteristics. However, a single origin model is based on the premise that regional diversity is secondary to a modern morphology, and therefore, early modern humans are not associated with any regional pattern as reflected in recent populations (African or otherwise).

It is possible to argue that by decoupling these events, interpreting the patterns and processes becomes less constrained by theoretical expectations. Reconstructing the evolutionary pattern of several processes such as those described above (the development and transmission of technological innovations, genetic diversification from an ancestral source, skeletal gracilization, morphological diversification of peoples, etc.) will bring us closer to an understanding of the evolution of modern humans than we are at present.

PARADIGMS, THEORIES, AND BIASES

To conclude, I reiterate my belief that the exposition of paradigms and biases will not help to achieve a consensus view. Those who firmly believe that cultural beings evolved anagenetically without the genetic effects of small populations affecting the broader picture will continue to do so. I have come to believe that they do so in spite of a contradictory record by not following the scientific procedures that link data to hypotheses to theories. Can theory be decoupled from data? To a certain extent yes, although I am not arguing for a purely empiricist scientific approach. It is clear that theory reaches beyond what data can provide and achieves a level of prediction, interpretation, and integration that transcends data. This notwithstanding, theory cannot be totally independent of data, for that would reduce it to mere speculation or personal opinion. Multiregionalism has become independent of the data it is supposed to be modeling (the synchrony of events, continuous gene flow, continuity of regional occupation and descent) and has come to reflect the personal biases about evolution of those who defend it.

I believe the paleoanthropological record shows that the history of each population is the consequence of the specific chain of events within each region of the world. Therefore, the processes that created modern regional diversity are likely to have been unique to each case. In order to reconstruct this evolution, all sources of past and present evidence should be used, for they convey different aspects of a single process. A multidisciplinary approach has been defended for a long time; however, most work (including my own) builds ideas on a specific data set and uses suitable information from another data set to back it up. That is very different from multidisciplinary integration. The real challenge for those prepared to abandon the preconceptions towards other disciplines will be the definition of common assumptions for data from different sources, so that they may take part in the formulation of hypotheses.

I started this chapter by saying that I work within a replacement paradigm. However, I also stated that I do not proceed in my work by interpreting data in terms of this paradigm, but in terms of hypotheses about pattern and process in the evolutionary record. I am aware that the paradigm may actually bias the choice of questions I ask, but this should not alter the results I obtain. In other words, I know that the particular hypotheses I try to test reflect my theoretical concerns, which in turn reflect paradigms. However, the hypotheses, the data used, the way they are used, and the results obtained are all transparent for anyone to check, contest, and refute. I would argue that the problem should not be stated in terms of paradigmatic versus empiricist perspectives, for in science theory cannot exist without its substantiation, nor data without a theoretical context. Facts should form the source from which modifications of existing paradigms derive.

In an earlier work, Wolpoff stated, "It is the process of refutation, rather than verification, which leads to the replacement of one hypothesis by another, and the purpose of hypothesis testing is to attempt refutations" (1978:461). I believe that multiregional anagenetic views of human evolution have been tested and consistently refuted by studies over the past few years, and that those involved in researching the evolution of hominids in the later Pleistocene are in the process of moving towards paradigms shared with evolutionary biology.

ACKNOWLEDGMENTS

I would like to thank G. A. Clark and C. M. Willermet for inviting me to contribute to this book. R. Foley, R. Haydenblit, W. W. Howells, and F. Lahr commented on earlier versions of this manuscript, and their suggestions and criticisms were much appreciated.

23

Thinking about Evolutionary Change
The Polarity of Our Ancestors

COLIN P. GROVES

The debate over the origin of modern humans polarizes into two camps: multi-regional (regional continuity) and replacement (restricted origin). Despite the acrimony which has all too frequently attended this debate, it is curious that the differences between the two camps have not always been clearly spelled out; it has at times been easy for upholders of one or the other position to claim that the other camp has misrepresented their position. Each of the two models has a more and a less extreme version, and it is necessary to examine the statements of the protagonists in order to determine just where along the spectrum they really do lie.

THE MULTIREGIONAL MODEL

The most extreme expression of multiregionalism would maintain that the earliest population of the genus *Homo* in any given region of the world is exclusively ancestral to all later populations in that region. To my knowledge, no one who claims to be a proponent of multiregionalism would accept this formulation in either its theoretical or its practical aspects. Theoretically—because even Coon, perhaps the most extreme proponent of the model, did not deny that evolutionary change in a given population could have been initiated by gene flow from other regions (Coon 1962:36–37), indeed in all probability there always had been some genetic contact between regional human populations (Coon 1962:369). Practically—because no one seems to regard Lower/Middle Pleistocene Javanese as ancestral to modern *Javanese,* but rather to modern aboriginal Australians and Melanesians (i.e., there have been at least some distributional shifts and "regional" continuity is to be taken rather loosely).

What the multiregional model's exponents propound, then, is some less extreme form of "weak multiregionalism." As expressed by Wolpoff et al. (1984) and Frayer et al. (1993), in what may be taken as the definitive expositions, it involves the maintenance through time of particular morphological complexes in given geographic regions, despite the continuous operation of specieswide evolutionary trends. Gene flow operates, therefore, across both space and time.

But how extreme, or not, is the multiregionalism of this school? When did interregional differentiation start, and were *all* non-sapient *Homo* populations involved in modern human origins, or did some become extinct without issue? Wolpoff and colleagues are explicit: it began with the initial expansion of *Homo* out of Africa (1984:450–451), and all known (morphologically distinct) non-sapient populations left a detectable genetic heritage in their modern vicars— even the Neandertals (1984:470–471).

THE REPLACEMENT MODEL

Stringer does not exclude the possibility of hybridization; he merely relegates to it a minor role, except perhaps for Australasia (1992a:14–16, 20). Bräuer goes further: hybridization certainly did occur, and he identifies certain European specimens as showing its influence but makes it clear that this is in no sense regional continuity (1992a:93); he names his version the "African Hybridization and Replacement model." On the other hand Bräuer is more extreme than Stringer in that he is sure that modern humans evolved in Africa, whereas Stringer leaves open the Middle East as a possible locus, and even draws attention to certain Chinese fossils as intriguing in this respect (Stringer 1993b). Perhaps the most extreme version is that of Groves and Lahr (1994), who are as certain as Bräuer is that it was Africa, and as certain as Stringer that we do not have any evidence for hybridization (in fact, not even in the Australasian case).

WHAT DOES *NOT* COUNT AS EVIDENCE

We are discussing the origin of modern humans (*Homo sapiens,* as that binomial is increasingly restricted). That category is, and can only be, defined anatomically. Certain technologies, and certain behaviors vis-à-vis neighboring groups, may well have been exclusive to *Homo sapiens* throughout its history, but this is not the heart of the matter. Neither are genetic data central—if human mitochondrial DNA has less variation than that of chimpanzees, orang-utans, or gorillas, this may be because the human species really does have less time depth and really did not incorporate genes from other taxa of *Homo.* But it may not— there may have been a recent selective sweep that got rid of the more divergent variants (S. Easteal, School of Medical Research, Australian National University, personal communication 1996), or the molecular clock might be much sloppier than we thought, or the mitochondria of all the neandertal and *Homo erectus* women who contributed to the modern gene pool might by chance have been lost. It is a common misapprehension that human origins models depend, in whole or in part, on genetic and behavioral (including technological) data, but this is not the case.

Frayer et al. (1993) fall into this error when they list a behavioral criterion as the very first of their "predictions" for what they dub the "Eve theory."[1] They consider that there should have been a technological advantage in order for replacement of one population by another to have occurred, that archaeology does not demonstrate this, and that the replacement model is therefore refuted.

Such an argument is not relevant: it is about *how* it occurred, not *whether* it did. If replacement of other species by *Homo sapiens* occurred, only the fossil evidence itself can demonstrate it. Clearly there was some "how" involved in the replacement process. If archaeology fails to document a "spreading technological advantage," then new stone tool types were evidently not part of it. Mechanisms of replacement might or might not include extermination, global competitive superiority, local competitive advantages, greater fertility and/or longer reproductive span, increase in the population of a key resource, some disease affecting the non-*sapiens* taxa, or even just pure chance. Or *Homo sapiens* might really have been more sapient, and the other people couldn't stand it and died of shame.

Why one anatomical type should replace another is a very interesting question, but that one cannot think of *how* it could have happened is not evidence that it did not happen.

WHAT CONSTITUTES EVIDENCE FOR REGIONAL CONTINUITY?

Support for the claim that there has been genetic continuity, from then to now, in a given region is based on features (in cladistic terms, character states) that occur within that region throughout that period. In what sense the features invoked in support of continuity are "regional" is not made clear by Wolpoff et al. (1984) but is spelled out in more detail by Frayer and colleagues (1993): the features do not have to be exclusive to the region (1993:21), merely in their highest frequency (1993:17) and found in distinctive combinations (1993:21–22) there.

These two aspects—highest frequency and combination—will be briefly considered in turn.

Highest frequency. Samples of modern crania are large enough that one can calculate accurate frequencies of occurrence of particular features, and so claim confidently that in some regions a certain feature occurs at significantly higher frequencies than elsewhere. The fossil record, though by now respectable in many respects, is still not sufficient to enable us to give accurate frequencies for the occurrence of features, yet one can certainly say that a particular feature is "commonly" seen in one regional sample but "rarely" in others. If the two features in the same region are the same, ancient and modern, then multiregionalists may have a point. Maybe. But first calculate the frequencies on the modern samples and make sure they are correct. Lahr (1994) claims that they have not always been correctly assessed, and if she is right the consequences for multiregionalism are rather serious.

Combinations, not individual features. Why do clusters of features occur in combination? There are several possible reasons: *(a)* they are consequences of a single pleiotropic gene, or tightly linked genetically, *(b)* they form part of a developmental complex, *(c)* they form part of a functional complex, *(d)* they are selected for by common environmental factors, and *(e)* they occur together by chance. As far as I can see, in the first three cases there would be no option but for the features to co-occur, so the fact that they are found in combination is of no special significance. If they are selected for by common environmental factors, there would be some significance in their co-occurrence only if at least one of the features was *exclusive* to that region at that time (i.e., it was not part of the specieswide genetic heritage). If they co-occur by chance there is much more likely to be some special significance (in terms of regional continuity) to the combination. In other words, common high frequency could well be an indicator of regional genetic continuity; co-occurrence may or may not be.

THE IMPORTANCE OF POLARITY

If two taxa share a certain character state absent from a third, what can we deduce about their relationship? It depends on the *polarity* of that character state: whether it is *apomorphic* (derived) or *plesiomorphic* (primitive).

Suppose we have three groups (or taxa), A, B, and C (Figure 23.1), and they differ from one another in various character states. Let us call them lungfish (A), lizard (B), and dog (C). In one character, integumentary covering, A and B are alike, C differs (Figure 23.1, right): lungfish and lizard both possess scales, dog has hair. Is this evidence that lungfish and lizard are more closely related to each other than to dog? No, because scales are plesiomorphic, hair is apomorphic. We know this by taking into consideration an *outgroup,* a taxon known to be, phylogenetically, not a part of the group under consideration. In this case, say, salmon. Salmon, like lungfish and lizard, are scaly; scales, therefore, simply persist from a more remote common ancestor, while a change has occurred during the evolution of dog to produce hair.

In another character, limb type, B and C are alike while A differs (Figure 23.1, left): lizard and dog have legs while lungfish has fins. Is this evidence that

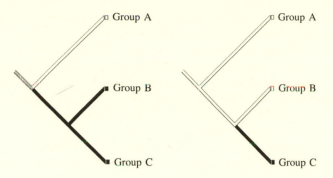

Figure 23.1. Character polarity and its consequences. The apomorphic (derived) character state is shown in black.

lizard and dog are more closely related to each other than to lungfish? Yes, because legs are apomorphic, fins are plesiomorphic. Our outgroup comparison shows this clearly; lizard and dog share a common ancestor (which had evolved legs) subsequent to the separation of the line leading to lungfish (which retained [primitive] fins).

Now, of course, these two respective explanations are not absolutely certain: they are the most probable ones, because they are the most parsimonious. As Felsenstein (1983) puts it, evolution is relatively improbable; therefore it is more likely that any given evolutionary change has happened only once. It might have happened that legs evolved twice, once in the ancestors of lizard, once in those of dog, in which case their common possession of legs would not, after all, indicate a more recent common ancestry, but this is unparsimonious. It might also have happened that scales had evolved three times (separately in the ancestors of salmon, lungfish, and lizard), but this is even less parsimonious. Despite Felsenstein's claim, cladistic analyses have shown that homoplasy (parallel and reversed evolution) *is* a common event; the popularity of cladistic analysis in recent years has shown this quite clearly, as reviewed by Groves (1989b; cf. Clark 1988). Parsimony is not an iron law of nature, it is a rule of thumb (Cartmill 1981), a "most probable" scenario.

The cladistic method does suffer from this drawback: parsimony is assumed, but it is merely the most probable condition, rather than the inevitable one. (Biological relevance does not have to be demonstrated for the employed traits, contra Trinkaus 1992c; neutral characters, uncontaminated by selective advantage, will do just as well.) But in the present case the implication is clear and has nothing to do with parsimony: plesiomorphic character states are absolutely not evidence for common ancestry.

An example will show the significance of this drawback. Kramer (1991) examined modern human mandibles from East Africa and from Australia, and Pleistocene *(Homo erectus)* mandibles from Java. In seven character states the

Australians, but not the Africans, resembled *Homo erectus.* Kramer deduced regional continuity from this—the Australians had inherited their mandibular features from Pleistocene Javanese ancestors—but it does not follow, because the polarity of the characters was not tested. East African pre-*sapiens* mandibles might show either the character states of the modern East Africans, in which case regional continuity would be strongly supported, or they might show the same states as the Australians and Javanese *Homo erectus,* which would thereby be shown to be plesiomorphic and so of no value for the question at hand. But since no such investigation has been made, no conclusion can be drawn either way.

Groves (1989b) made the case that, when polarities are known, most (all?) of the features that have been cited as evidence of regional continuity—those that are truly comparable—are found to be plesiomorphic, and thus not relevant. The question arises, of course, whether outgroups were chosen judiciously. If the earliest non-African *Homo* are on the order of a million years old (Hyodo et al. 1993), then *Homo ergaster,* as represented by such well-preserved specimens as ER-3733 or WT-15000, would be a suitable outgroup. If, however, the earliest Javanese *Homo* (marking the initiation of the Australoid lineage) are ca. two million years old (Swisher et al. 1994), then the *Homo ergaster* fossils could be part of an already separated African lineage. In that case, we could be rather stuck for an outgroup!

I conclude that the case for regional continuity must be much more tightly and logically presented than has so far been the case. To carry conviction the case should be put somewhat as follows:

1. The character states defining the australoid lineage (or mongoloid, etc.) are A, B, and C.
2. They are defined *thus.*
3. They occur at frequency *x* in aboriginal Australians, at frequency *y* in New Guinea Highlanders (and so on), but at no higher frequency than *z* in non-australoid populations (always provided that sample sizes are adequate, and frequencies are statistically meaningful).
4. The characters can be observed in the following presapiens fossils [list] from the region, of which the following specimens [list] possess the "australoid lineage" states of the characters.
5. The characters can be observed in [list] from elsewhere, of which only a very few [list] possess the "australoid lineage" states.
6. And finally—absolutely critically—the character states in question are apomorphic, and here is the evidence that they are.

WHAT WOULD BE EVIDENCE FOR REPLACEMENT?

The same stipulations demanded of the multiregional model must also apply to the replacement model: archaeology and genetics offer at best indirect evidence, and character state polarity must be demonstrated. I would present the case somewhat as follows:

1. The apomorphic character states of *Homo sapiens,* relative to (say) *Homo ergaster,* are as follows [list].
2. The distribution of these states in Middle and Lower Pleistocene *Homo* is as follows [list those that possess the states in question, and those that do not even though the characters can be observed].
3. The greatest number of these states occurs in fossil group A (regional and chronologically circumscribed), which is thereby identified as the cladistic sister-group of *Homo sapiens.*
4. The next greatest number occurs in fossil group B, which is therefore the sister-group of the A + *sapiens* clade—and so on, until we have constructed a cladogram that is as complete as the evidence permits.
5. Next, examine the non-*sapiens* clades to see if any apomorphic states can be found in them.
6. Finally, focus on *Homo sapiens* itself, identify character states within modern populations that are of regional significance (exactly as for the multiregional procedure, steps 1–3), and document their distribution in fossils assigned to the species.

My version of the replacement model predicts that the Middle Pleistocene African fossil group is group A, the sister-group of *Homo sapiens;* that the Neandertals and Middle Pleistocene Europeans are group B, the sister-group of *sapiens* + A; and that East Asian *Homo erectus* is the sister-group of all of these. It further predicts that group A and group B have their own apomorphic states, which can be tracked in the same way as those of *Homo sapiens.* (A taxonomy at species level can then be drawn up as it is felt appropriate.) It predicts finally that the earliest fossils of *Homo sapiens* in Africa will have none of the character states that are of regional significance, that the earliest non-Africans will have few or none, and that regional character states will develop within each region once it is settled (by *Homo sapiens*), such that the earliest fossils of a given region show them incipiently, the later ones more definitively.

CONCLUSIONS

The multiregional/replacement debate has generally been carried out with more heat than light up to now—shall we say, with a low "signal-to-noise" ratio. This chapter examines the premises and finds that they are often insubstantial or irrelevant. The crucial criterion, which must not be lost sight of, is polarity; using this, both parties to the debate can draw up a research program which has every prospect of finally resolving the problem.

ACKNOWLEDGMENTS

First let me thank Geoff Clark and Cathy Willermet for their invitation to contribute to this unusual, intriguing, and worthwhile enterprise. I am presently involved in a research project to study original fossils with continuity advocate, Alan Thorne, and would like to go on record that it is possible for colleagues who differ profoundly in their views to travel and study together with pleasure and without rancor. Despite extensive discussions with Alan and others, notably Chris Stringer, Marta Lahr, Jonathan Kingdon, David Dean, and Dan Lieberman, the ideas put forth in this paper, and especially the way I have expressed them, are all my own.

NOTE

1. "Eve theory" is an inaccurate designation for the replacement model for two reasons: First, in science a theory is a highly corroborated, robust model of the universe (theory of evolution, quantum theory, etc.). I do not think that many multiregionalists would concur that the replacement model has this character. Secondly, "Eve" is a construct—a publicity stunt, perhaps—of geneticists to designate the coalescence point of the human mitochondrial tree. It does not truly describe the entire replacement model.

VI

MOLECULAR BIOLOGY AND ITS IMPLICATIONS

The current furor surrounding the question of our origins was sparked in 1987 by mtDNA research and a compelling biblical analogy that captured the public imagination—the notion of an African Eve, the biomolecular mother of us all. As originally articulated, however, this scenario lasted only until 1992, when Washington University geneticist Alan Templeton exposed fatal mistakes in the interpretation of the computer algorithm used to "root" the tree of mtDNA relationships. Here Templeton presents a statistical test that has the ability to discriminate between the Out of Africa replacement hypothesis and a hypothesis of recurrent gene flow linking Old World populations throughout the Upper Pleistocene. The results falsify the Out of Africa hypothesis and are consistent with both a diffuse origin for moderns in which different modern traits first evolved at diverse locations and were subsequently brought together by gene flow, and with a single origin model in which all modern traits evolved first at a single location and then spread throughout the world by gene flow amplified by selection.

In the next chapter, population geneticists Henry Harpending and John Relethford come to almost the opposite conclusion. Taking a single origin hypothesis as reasonably well demonstrated, they discuss how racial (subspecific) differences might have come about and how different kinds of genetic evidence might suggest resolution of this issue. Different evolutionary dynamics characterize different kinds of genetic systems, they argue. DNA sequence data and multivariate metric trait data typically have high mutation rates and high levels of resolution of group differences, so these systems provide more information about our evolution during the Upper Pleistocene than do other kinds of genetic markers.

Using a simulation approach, Charles Oxnard outlines the parameters of three models that assess, respectively, (1) the effects on species' divergence dates of small numbers of randomly chosen fossils being representative of a much larger number of species that resulted from anagenesis, cladogenesis, and extinction (This model asks the question: Are the current divergence dates realistic?); (2) the effects on subspecies of interbreeding and migration (Are simple notions of one set of major migrations realistic?); and (3) the effects of different models of unilineal genetic inheritance (Is it realistic to derive time and place of origin from the analysis of single-sex inheritance systems?). Many construals of pattern and the meaning of pattern in the genetic evidence are naive, he argues. These simulation approaches allow us to define and generate quantitative test implications for many operational notions (e.g., migration, gene flow) left undefined by archaeologists and human paleontologists.

Vincent Sarich explores the consequences of the recurrent episodes of glaciation on human demography over the course of the Upper Pleistocene. He suggests that each episode of glaciation corresponded to an episode of population displacement that would have increased gene flow and decreased racial variation throughout the middle latitudes of the Old World. Only after climates stabilized following a glacial pulse would human populations settle down again, to begin another round of greater genetic separation, adaptation to local conditions, and increased regional differentiation. The formation of races, then, is a cyclical process that took place repeatedly over the course of the Pleistocene, and one highly correlated with episodes of continental glaciation.

24

Testing the Out of Africa Replacement Hypothesis with Mitochondrial DNA Data

ALAN R. TEMPLETON

The Out of Africa replacement hypothesis presents a dramatic view of the origins of *Homo sapiens* in which anatomically modern humans evolved in Africa about 100,000 years ago, spread out over the Old World, and drove to genetic extinction all earlier human populations found in Europe and Asia (Cann et al. 1987; Vigilant et al. 1991). Much of the data claimed to support this hypothesis have been derived from studies on human mitochondrial DNA (mtDNA), a maternally inherited DNA found in the cellular organelle that makes aerobic metabolism possible. This molecule was chosen to study recent human evolution for two reasons. First, mitochondrial DNA does not recombine like nuclear DNA. This means that mutations accumulate in any mitochondrial DNA lineage in a manner that reflects their evolutionary sequence. Accordingly, evolutionary history is written more cleanly in mtDNA than in nuclear DNA. Second, the rate of mutational substitution of mtDNA is greater than that for nuclear DNA in mammals (Brown et al. 1979). This means that recent evolutionary events have the potential of being marked by mutational change. Hence, mtDNA is an appropriate molecular tool for studying recent human evolution.

Having an appropriate molecular tool is only the first step in testing a hypothesis: the tool must be coupled with a sampling design that is informative about the hypothesis and the data must be analyzed in such a way that discriminates among alternatives using statistically rigorous and objective criteria. Unfortunately, the original papers that put forward the idea that the mtDNA data support the Out of Africa replacement hypothesis (Cann et al. 1987; Stoneking et al. 1986; Vigilant et al. 1991) are flawed with respect to both sampling design and analytical methodology (Templeton 1993, 1994a). The flaw in the sampling design stemmed from a focus not on the Out of Africa replacement hypothesis, but rather upon subsidiary hypotheses that are necessary but not sufficient for the veracity of the replacement hypothesis (Templeton 1994a). In particular, much of the past effort (Harpending et al. 1993; Nei 1992; Stoneking 1994; Stoneking et al. 1992) has concentrated upon demonstrating that the common mitochondrial ancestor of all humanity resided in Africa, that she lived approximately 200,000 years ago—or at least considerably less than the time of the expansion of *Homo erectus* out of Africa—and that human population size has greatly expanded in the recent past. All of these hypotheses are necessary for the replacement hypothesis to be true, but none of them discriminate between the replacement hypothesis and an alternative in which all Old World human populations were genetically interconnected to one another by gene flow, albeit at low and restricted levels, throughout the entire evolutionary time period marked by the coalescence of all mtDNA diversity back to a common female ancestor (Templeton 1993, 1994a). Under the gene flow hypothesis, the common ancestor could have lived anywhere in the Old World, including Africa. Hence, demonstrating that the root of the mtDNA gene tree is in Africa (which has not yet been done: Maddison et al. 1992; Ronquist 1994; Templeton 1993) offers no discrimination of the relative merits of the replacement versus gene flow hypotheses. Second, when populations are genetically interconnected, a haplotype (the type of DNA molecule as determined by the simultaneous state of all the nucleotides or restriction site markers in the study) has the potential of spreading geographically at any time and does not require population range expansion in order to spread. In particular, there is no necessity for the date of coalescence to be older than one million years under a multiregional gene flow model (that is, spreading with *Homo erectus*), as has erroneously been implied (Stoneking 1994). Accordingly, even if the date of the common mtDNA ancestor were conclusively shown to be 200,000 years ago (which it has not—Templeton 1993), this date is equally compatible with a gene flow and a replacement model. Finally, there is little doubt that human population size has expanded greatly in the recent past, and this fact is consistent with both the replacement and gene flow models. It should never be forgotten that the replacement hypothesis is one of expansion of *geographical range,* and statistics showing expansion of *size* are irrelevant to discriminating between the gene flow and replacement models.

The two models do differ in how mtDNA haplotypes are supposed to have spread geographically over the Old World. Thus, a geographical analysis of the mtDNA haplotypes and their gene tree is necessary in order to test the replacement hypothesis versus the gene flow hypothesis. The idea that a geographical hypothesis requires geographical sampling in order to be tested should be trivially obvious, but the original mtDNA data sets used to "support" the Eve hypothesis (Cann et al. 1987; Vigilant et al. 1991) were sampled in such an inaccurate or sparse way geographically that they were useless for testing the replacement hypothesis (Templeton 1993). Thus, in order to test the replacement hypothesis, it is essential to turn to other data sets that have sampled human populations in a more accurate and thorough fashion throughout the Old World. In Templeton (1993), the data of Excoffier and Langaney (1989) were used for this purpose, and this chapter will present some additional analyses of these data.

The second major failing of the original papers supporting the replacement hypothesis was a flawed analytical methodology. First, the estimated mtDNA, maximum parsimony haplotype trees from the restriction site data (Cann et al. 1987) were estimated incompletely (Maddison 1991), and from the DNA sequence data (Vigilant et al. 1991) incorrectly (Templeton 1992). This shows the great importance of assessing the statistical confidence one has in an estimated haplotype tree before proceeding with any additional analysis. In my previous analysis (Templeton 1993), I used the haplotype tree estimated by Excoffier and Langaney (1989) without a formal assessment of statistical confidence, in part because Excoffier and Langaney made explicit their criteria for resolving ambiguities and these criteria had a sound theoretical base with considerable empirical support (Crandall and Templeton 1993; Templeton 1994a). Moreover, the statistics used to test the replacement hypothesis versus the gene flow hypothesis are based upon procedures that are robust to ambiguities in the cladogram (Templeton and Sing 1993). Nevertheless, my previous analysis (Templeton 1993) is still subject to the criticism of no explicit assessment of statistical confidence in the mtDNA haplotype tree (Long 1993; Stoneking 1994), so this deficiency will be corrected here.

Once the haplotype tree (or a set of plausible haplotype trees) has been estimated, rigorous statistical analyses must be performed of the overlay of geography upon the haplotype tree. An appropriate test statistic must have the ability to discriminate between population range expansion and gene flow. The tests given in Vigilant et al. (1991) lack discriminatory ability (Templeton 1993) and therefore are irrelevant to testing the replacement hypothesis. Fortunately, an alternative testing procedure is available that does have the ability to discriminate between geographical associations caused by historical events versus the operation of recurrent evolutionary forces such as gene flow (Templeton 1993, 1994a). Since the original analysis, this statistical methodology has been expanded to include additional statistics that help discriminate among alternatives (Templeton 1994b; Templeton and Georgiadis 1995; Templeton et al. 1995), so this chapter will present an expanded, more powerful analysis of the human mtDNA. In addition, using recent developments in coalescent theory (Castelloe and Templeton 1994), a new method of analysis both of the geographical location of the root of the human mtDNA haplotype tree and of the replacement versus the gene flow hypotheses will be presented. Finally, the implications of these analyses for the evolution of anatomically modern humans are discussed, as well as some broader issues about this controversy and the nature of scientific inference.

ASSESSING CONFIDENCE IN THE mtDNA HAPLOTYPE TREE

The mtDNA phylogenies of Cann et al. (1987), Excoffier and Langaney (1989), and Vigilant et al. (1991) were all estimated using the principle of maximum parsimony. The first step in assessing confidence in an estimated tree

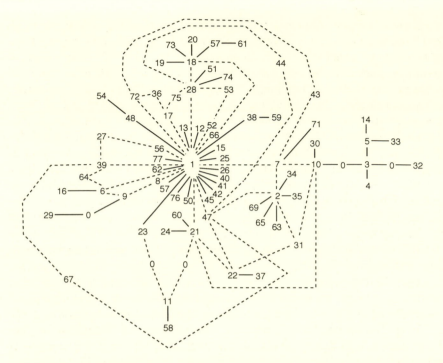

Figure 24.1. The maximum parsimony mtDNA cladograms. The numbers refer to the haplotype designations given in Excoffier and Langaney (1989) and Excoffier and Smouse (1994). Zeros indicate an intermediate haplotype state under maximum parsimony that was not found in the sample. Solid lines indicate a single mutational change that is unambiguous under the principle of maximum parsimony. Dashed lines indicate single mutational changes that may or may not have occurred during the evolution of mtDNA, thus indicating the positions of cladogram ambiguity under maximum parsimony.

is to see if maximum parsimony is an appropriate criterion for the data set to be analyzed. Templeton et al. (1992) give an algorithm for determining the bounds of validity of maximum parsimony. Applying this algorithm to the Old World data given in Excoffier (1990) and Excoffier and Langaney (1989), the probability that two haplotypes separated by only a single mutational connection are related to one another in a parsimonious fashion is 0.99, and the probability that haplotypes separated by two mutational steps are related in a parsimonious fashion is 0.98. As there are no connections among haplotypes that involve more than two mutational changes, no additional analyses are required. Hence, maximum parsimony is well justified as an appropriate criterion in this case.

Thousands of equally parsimonious cladograms are compatible with these data (Excoffier 1988; Excoffier and Smouse 1994). These equally parsimonious alternatives appear as "loops" in the haplotype networks that show all the maximum parsimony connections (Figure 24.1). The equally parsimonious alternatives are defined by all the ways in which these loops can be broken. Excoffier and Langaney (1989) used two criteria to break loops: (1) connections involving two rare (<5%) haplotypes were eliminated in favor of those involving a rare-common connection, and (2) connections involving two haplotypes found in different populations were eliminated in favor of those between geographically overlapping haplotypes. All but one loop was eliminated by these criteria (Templeton 1994a). These criteria are justified by coalescent theory and by an empirical investigation of 29 other haplotype trees (Crandall and Templeton 1993). However, a quantitative assessment of the likelihood of these alternative connections is possible using the techniques and results given in Crandall and Templeton (1993). Moreover, because geographical data will be analyzed with the estimated haplotype tree, it is best to avoid using geographical data to estimate the tree itself. Hence, the second criterion of Excoffier and Langaney (1989) will not be used in the present analysis.

The results reported by Crandall and Templeton (1993) support the first criterion of favoring rare-common connections over rare-rare ones, but they also indicate that this discrimination cannot always be made with 95% confidence. More rigorous criteria will therefore be used. Both coalescent theory and empirical results indicate that a rare haplotype is highly likely to be a tip; that is, it has only one mutational connection to the remainder of the haplotype network. Hence, connections will be eliminated that simultaneously favor a rare-common connection over a rare-rare connection and that also make a rare haplotype a tip in the haplotype network. Moreover, the 5% criterion for rarity is reduced to 1.5%. These combined criteria and the more rigorous definition of rarity ensure that the connection being eliminated is exceedingly unlikely (Crandall and Templeton 1993). Loops are also broken by eliminating connections that simultaneously make a rare haplotype a tip and that favor a rare-common connection over a second rare-common connection, but where the alternative common haplotypes differ by at least an order of magnitude in frequency. The latter component of this criterion stems from the prediction and observation that the probability that rare haplotypes are connected to alternative common haplotypes is proportional to the frequency of that common haplotype (Crandall and Templeton 1993).

Figure 24.2 shows the haplotype network after the connections satisfying the above criteria are eliminated. As can be seen, many loops and alternatives still remain. Hence, there is much more ambiguity in this haplotype tree than indicated by Excoffier and Langaney (1989) when the stricter frequency criteria are used and when geographical criteria are excluded. In particular, three loops and one set of two alternative connections remain; Figure 24.3 portrays these ambiguities in isolation so they can be seen more easily. Figure 24.4 shows the fifteen ways of breaking the loops shown in Figures 24.3a and 24.3b. There are also six ways of breaking the loop shown in Figure 24.3c (each of the six steps in the loop), and two alternatives in Figure 24.3d, for a total of 2,700 possible unrooted trees.

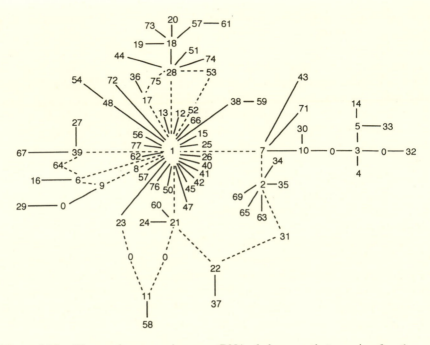

Figure 24.2. The maximum parsimony mtDNA cladograms that remain after the co-alescent criteria described in the text were used to eliminate some of the possible connections.

Using a methodology similar to that given in Crandall et al. (1994), proba-bilities are assigned to all possible ways of breaking a loop. In addition to coalescent criteria, the method of Crandall et al. (1994) uses asymmetries in the probabilities of gains and losses of restriction sites (Templeton 1983) so in some cases additional connections can be eliminated with statistical confidence at least at the 95% level. The loop shown in Figure 24.3a is actually a double loop, so the restriction site asymmetries were incorporated according to the method given in Gerber (1994). As shown by Gerber (1994) for double loops and Crandall et al. (1994) for single loops, the relative cladogram probabilities resulting from these asymmetries are invariant to the root of the cladogram, so no rooting is needed in order to calculate these probabilities. The double loops in Figures 24.3a and 24.3b were drawn so they are comparable in terms of site gains and losses. The probabilities of the various ways of breaking them (Figure 24.4) based solely on gain/loss information, and assuming a strong bias in favor of transitions over transversions (Templeton 1983; Templeton et al. 1992), are 0.18947 for clado-grams I, III, VII, and XV; 0.03158 for II, IV, IX, X, XII, XIII, and XIV; and 0.00526 for V, VII, and XI. Two statistically independent measures—the proba-bility of haplotype 8 being a tip (0.828 using the equations given in Crandall and

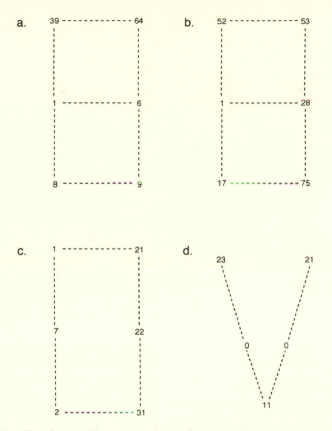

Figure 24.3. The four ambiguous portions of the set of cladograms illustrated in Figure 24.2.

Templeton 1993:Fig. 6) and the probability of haplotype 64 being a tip (0.882)— were incorporated as follows. Eight of the fifteen possible resolutions of this double loop have haplotype 8 as a tip (I, II, VII, IX, XI, XII, XIV, and XV), so the probability mass of 0.828 is evenly distributed among them, and the probability mass of 0.172 (the probability that haplotype 8 is an interior) is evenly distributed among the remaining seven cladograms. Hence, the marginal probability across all possible resolutions of 8 being a tip remains at 0.828. The tip probability for haplotype 64 is used in a similar fashion.

These three sources of independent information (based on nonoverlapping subsets of the data) are combined into a single probability by multiplying them times one another for each respective cladogram and restandardizing so that the sum over all cladograms is one. After this step, additional information is incorporated that is applicable only to a subset of the possible resolutions. In that subset

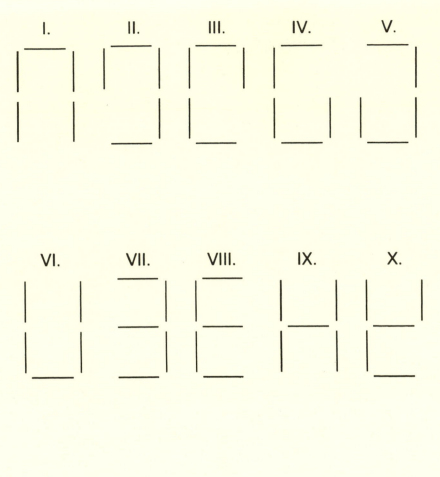

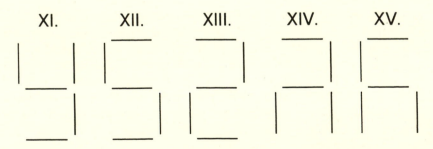

Figure 24.4.   The fifteen possible resolutions of a double loop of cladogram ambiguity.

of cladograms in which haplotype 8 is a tip, the probability of 8 being connected to 1 is 0.986 versus a probability of 0.014 for the 1 to 9 connection, using equation (4) of Crandall and Templeton (1993) and the frequencies of haplotypes 1 and 9 (data that have not been used previously), and the same probabilities also apply to that subset of cladograms in which haplotype 6 is connected either to 1 or to 9, but not both. This information is incorporated into the analysis by using Bayes' Theorem to redistribute the previously calculated probability mass among the pairs of alternatives that differ only in whether or not the haplotype of interest is connected to 1 or 9. For example, cladograms I and II are a pair of resolutions that differ only in 8 being connected to 1 (resolution I in Figure 24.4) or to 9 (II). The previously calculated probability of I is 0.07899, and the probability of II is 0.01316. After using Bayes' Theorem and this new piece of coalescent information, the probability of I is 0.09194 and that of II is 0.00021. The total probability of these two alternatives combined is unchanged, but the coalescent information has altered their relative weights. Combining all these sources of information leads to the probabilities shown in Table 24.1. As is readily evident, several of the alternatives are extremely unlikely. Five of the fifteen possible alternatives (I, VIII, IX, X, and XV) account for 96% of the probability and hence constitute a 95% confidence set. All the cladograms in this set share the connections between haplotypes 1 to 39 and between haplotypes 1 to 8, so these connections are now regarded as resolved at the 95% level.

For the loop shown in Figure 24.3b, in addition to gain/loss asymmetries, there is statistically independent information in the tip probabilities of haplotypes 52 (0.872), 53 (0.892), and 75 (0.892). The combined probabilities are given in

Table 24.1. Probabilities for the Fifteen Alternatives (Figure 24.4) for Breaking the Double Loops Defined by Haplotype Sets {1,6,9,8,39,64} and {1,17,28,52,53,75}

Alternative	Loop {1,6,9,8,39,64}	Loop {1,17,28,52,53,75}
I	0.09194	0.08722
II	0.00022	0.00201
III	0.01875	0.01457
IV	0.00034	0.00243
V	0.00001	0.00006
VI	0.00006	0.00040
VII	0.00004	0.00033
VIII	0.14275	0.10539
IX	0.10021	0.10516
X	0.02379	0.01757
XI	0.00024	0.00242
XII	0.00143	0.01454
XIII	0.00364	0.00243
XIV	0.01532	0.01454
XV	0.60127	0.63095

Table 24.2.   Probabilities for the Six
Alternatives for Breaking the Loop Defined
by Haplotype Set {1,2,7,21,22,31} (see
Figure 24.3c for the haplotype locations in
the loop)

Alternative		Probability
I		0.43379
II		0.06262
III		0.43379
IV		0.06348
V		0.00359
VI		0.00273

Table 24.1, and cladograms I, VIII, IX, X, and XV define a 95% confidence set. The 1 to 17 and 1 to 52 connections are common to all five of these cladograms, so these connections are also regarded as resolved at the 95% confidence level.

Asymmetries in gains and losses do not discriminate among any of the six ways of breaking the loop in Figure 24.3c as all resolutions and possible rooting directions yield the same proportion of gains to losses. The probability of haplotype 31 being a tip is 0.868, and in those resolutions that have haplotype 21 as a tip in this loop (IV and VI in Table 24.2), the probability of the 21 to 1 connection is 0.981 versus 0.019 for the 21 to 22 connection (using equation 4 in Crandall and Templeton 1993). Similarly for alternatives II and V, the probability of a 1 to 7 connection is 0.946 and the probability of a 2 to 7 connection is 0.056. The combined probabilities for the six resolutions are given in Table 24.2, and resolutions I through IV collectively account for 99.3% of the probability. These four resolutions share the 1 to 7 and 1 to 21 connections, which are therefore regarded as resolved at the 95% level.

Finally, there are two ways of connecting haplotype 11 to the remainder of the cladogram (to haplotype 21 or haplotype 23). Haplotype 23 is much rarer than 21, so its tip probability gives the best discrimination, yielding a probability of 0.887 for 11 being connected to haplotype 21 versus 0.113 to haplotype 23.

Figure 24.5 shows the end result of this assessment of cladogram uncertainty. The 95% confidence sets reduce the number of cladograms from 2700 to 200,

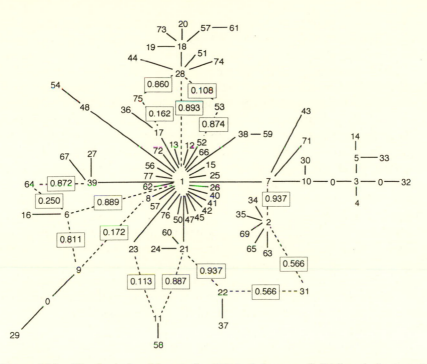

Figure 24.5. The final plausible set of mtDNA cladograms. Solid lines indicate mutational changes that have a probability ≥ 0.95. Dashed lines indicate possible mutational changes that may or may not have occurred, with the boxed number superimposed upon the dashed line being the estimated probability that that particular mutational event actually occurred.

still a substantial amount of ambiguity. The possible connections that are not resolved at the 95% level are shown with a probability attached to them that is the sum of the probabilities of all resolutions that have that linkage. In summary, the degree of ambiguity suggested by the present analysis is far greater than that indicated by Excoffier (1990) and Excoffier and Langaney (1989) or used by Templeton (1993). However, the unequal probabilities indicate that much information is available even though not resolved at the 95% level, and this information will be used in the subsequent analyses.

A CLADISTIC ANALYSIS OF GEOGRAPHICAL DISTANCES

The Out of Africa replacement hypothesis is a geographical/evolutionary historical hypothesis that can only be tested by analyzing geographical data in an evolutionary historical context. The importance of estimating the mtDNA haplotype tree is that it provides the evolutionary historical context for the analysis of geographical data. When the genetic variation is organized into an evolutionary genealogy, the resulting analysis of how geography overlays upon genealogy has been called "intraspecific phylogeography" (Avise 1989, 1994). Such analyses commonly find a strong association between the geographical location of haplotypes and their evolutionary position within a haplotype tree, but the demonstration of such an association per se does not reveal the causes of the association.

Fortunately, more information than mere association can be gathered from a geographical overlay upon haplotype trees; different causes of geographical association can yield qualitatively different patterns that can be assessed through rigorous statistical testing (Templeton et al. 1995). The first step in such an analysis is to translate the estimated set of plausible cladograms into a hierarchical nested design. Haplotypes separated by a single mutation are grouped together into "one-step clades" proceeding from the tips to the interior of the cladogram, then one-step clades separated by a single mutation are grouped together in "two-step clades," etc., until the next level of nesting would encompass the entire tree cladogram. The basic nesting rules are described in Templeton et al. (1987), and additional rules that deal with the complications caused by ambiguities within the plausible set of haplotype cladograms are given in Templeton and Sing (1993).

The standard nesting rules are first applied to that subset of the haplotype connections that are regarded as resolved. Figure 24.6 shows the resulting one-step clades. As can be seen, despite the plausible set including 200 different trees, the nesting state of only four haplotypes is in question. This shows that the basic nesting rules are robust to even large amounts of cladogram ambiguity (Templeton and Sing 1993). Because so few haplotypes have ambiguous nesting states, all possible combinations will be considered to obtain a thorough documentation of robustness of conclusions to cladogram ambiguity. These possible combinations are shown in Table 24.3.

Figure 24.7 shows how the one-step clades are nested into two-step clades. Once again, four one-step clades are left in an ambiguous state, all of which involve loops to the central and most abundant one-step clade, 1-1. The number of ways of breaking these loops combined with the alternative haplotype memberships in some of these one-step clades (Table 24.3) preclude an exhaustive analysis at this level (each run of the program on a data set this large requires about 12 hours on a VAX workstation). Instead, the nesting rules for loops described in Templeton and Sing (1993) are used to nest all these clades of ambiguous status with the central clade in the loop, clade 1-1. This causes a reduction of power but insures robustness to cladogram ambiguity (Templeton and Sing 1993). The analysis at this level is still affected by the alternative one-step clades given in Table 24.3, yielding sixteen different 2-1 clades (Table 24.4) and four different 2-2 clades (alternatives 2-2a, b, c, and d, corresponding to including one-step clade 17a, b, c, or d within it). All these alternatives will be

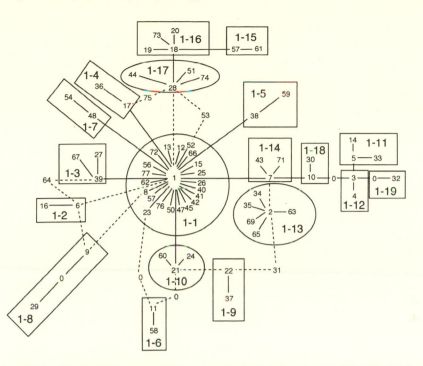

Figure 24.6. The haplotypes nested into one-step clades using the plausible set of clado-grams given in Figure 24.5. Each one-step clade is indicated by an enclosed oval or box and designated by 1-*n* where the 1 refers to the clade level and *n* represents the number assigned to a particular one-step clade.

investigated to insure robustness of conclusions. Figure 24.7 also shows that the nesting stops at the two-step level since all two-step clades belong to a single three-step clade: namely the entire cladogram. The evolutionary relationships among the two-step clades at the three-step level are not affected by ambiguity at all, with clade 2-1 being interior to three tips (2-2, 2-3, and 2-4). Nevertheless, there are still four alternatives at the three-step level because of ambiguities at the one-step level (Table 24.5). Once again, all alternatives will be considered. In this manner, a statistical design that is completely robust to all 200 cladogram possi-bilities has been defined.

Now that the statistical design has been defined, statistics that encode geo-graphical information must be calculated. The analysis of Templeton (1993) used only two such statistics; the average distance (great circle geographical distance) of all members of a clade from their geographical center (which measures how widespread a clade is geographically and is hereafter called the clade distance and symbolized by D_c) and the difference between the average clade distance for interior minus tip clades within the same nesting category at the next higher step

Table 24.3.   The Probable Alternative Ways of Nesting Haplotypes into One-Step Clades

Nesting Clade	Alternative	Haplotype Membership
1-1	a	without haplotype 53
	b	with haplotype 53
1-2	a	without haplotype 64
	b	with haplotype 64
1-3	a	with haplotype 64
	b	without haplotype 64
1-4	a	without haplotype 75
	b	with haplotype 75
1-9	a	without haplotype 31
	b	with haplotype 31
1-13	a	with haplotype 31
	b	without haplotype 31
1-17	a	without haplotypes 53 and 75
	b	without haplotype 53 and with haplotype 75
	c	with haplotype 53 and without haplotype 75
	d	with haplotypes 53 and 75

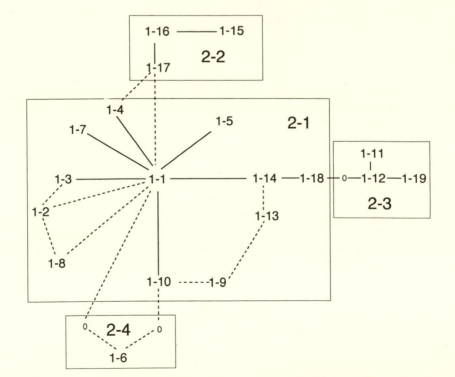

Figure 24.7.   The one-step clades nested into two-step clades using the plausible set of cladograms given in Figure 24.5. Each two-step clade is indicated by an enclosed box and designated by 2-*n* where the 2 refers to the clade level and *n* represents the number assigned to a particular two-step clade.

Table 24.4. The Probable Alternative Ways of Nesting One-Step
Clades into Two-Step Clade 2-1

Nesting Clade	Alternative	One-Step Clade Membership
2-1	a	1-1a, 1-2a, 1-3a, 1-4a, 1-9a, 1-13b
	b	1-1a, 1-2a, 1-3a, 1-4a, 1-9b, 1-13a
	c	1-1a, 1-2a, 1-3a, 1-4b, 1-9a, 1-13b
	d	1-1a, 1-2a, 1-3a, 1-4b, 1-9b, 1-13a
	e	1-1a, 1-2b, 1-3b, 1-4a, 1-9a, 1-13b
	f	1-1a, 1-2b, 1-3b, 1-4a, 1-9b, 1-13a
	g	1-1a, 1-2b, 1-3b, 1-4b, 1-9a, 1-13b
	h	1-1a, 1-2b, 1-3b, 1-4b, 1-9b, 1-13a
	i	1-1b, 1-2a, 1-3a, 1-4a, 1-9a, 1-13b
	j	1-1b, 1-2a, 1-3a, 1-4a, 1-9b, 1-13a
	k	1-1b, 1-2a, 1-3a, 1-4b, 1-9a, 1-13b
	l	1-1b, 1-2a, 1-3a, 1-4b, 1-9b, 1-13a
	m	1-1b, 1-2b, 1-3b, 1-4a, 1-9a, 1-13b
	n	1-1b, 1-2b, 1-3b, 1-4a, 1-9b, 1-13a
	o	1-1b, 1-2b, 1-3b, 1-4b, 1-9a, 1-13b
	p	1-1b, 1-2b, 1-3b, 1-4b, 1-9b, 1-13a

level. Obtaining the average interior minus tip difference is complicated in this
case by cladogram ambiguity. Many clades could be either tips or interiors.
Consequently, in this new analysis, the average difference is obtained by weighting the tip/interior status of any ambiguous clade by the probabilities of it being a
tip or interior. These probabilities in turn are obtained by using the cladogram
probabilities given in the previous section, and summing them for all resolutions
that make the clade of interest a tip, or alternatively an interior. Hence, all
statistics used in this analysis that could be affected by cladogram ambiguity
explicitly take that ambiguity into account.

A second set of distance statistics will also be used in the present analysis. The
nested clade distance (D_n) is the average great circle distance of all members of a
clade from the geographical center of the clade at the next higher step level
within which the clade is nested (Templeton et al. 1995). This distance measures

Table 24.5. The Probable Alternative Ways of Defining the
Membership of Two-Step Clades into a Three-Step Clade

Nesting Clade	Alternative	One-Step Clade Membership
3	a	2-1a, 2-2d
	b	2-1c, 2-2b
	c	2-1i, 2-2c
	d	2-1d, 2-2a

Note: Only four of the sixteen alternatives for clade 2-1 need to be
considered at this level, as all the other alternatives yield identical
numbers at this level.

how far away clade members are from individuals that bear haplotypes from the same or evolutionarily closely related clades. The average difference between interior clades minus tip clades is also calculated for the nested clade distances, once again by incorporating cladogram ambiguity explicitly. The advantage of using these additional statistics is that they provide more power in rejecting the null hypothesis of no geographical association and in discriminating among alternatives when the null hypothesis is rejected (Templeton et al. 1995).

The distributions of these distance measures under the null hypothesis of no geographical associations within the nesting clade are determined by recalculating the distances after a random permutation of clades against sampling location within a nesting clade. One thousand random permutations are used to insure accurate statistical inference at the 5% level of significance (Edgington 1986).

Rejecting the null hypothesis for clade and/or nested clade distances merely tells one that there is a significant association between haplotypes and geography. To discriminate among alternatives, Templeton et al. (1995) give in detail the differing expectations for population range expansion, population fragmentation, and restricted but recurrent gene flow. The expected patterns are summarized here in Table 24.6. Because the patterns can be complex and because the expectations given in Table 24.6 assume a geographically complete sampling design, Templeton et al. (1995) gives an explicit inference key to derive inferences from rejections of the null hypothesis and to warn the user when the sampling design is inadequate. All conclusions will therefore be made in a completely explicit and objective fashion.

Because this inference methodology is discussed in detail in Templeton et al. (1995), only a few additional points will be discussed here that pertain directly to testing the Out of Africa replacement hypothesis. First, Rogers and Jorde (1995a) have criticized the criterion of tips tending to be younger than their interiors under restricted gene flow (Table 24.6). They claim, without supporting modeling or mathematics, that the tip/interior polarity is inapplicable to cladograms in which few mutations distinguish the haplotypes, as is the case for the human mtDNA. The tip/interior polarity was justified both by exact coalescent theory and by computer simulations (Castelloe and Templeton 1994). In performing these calculations and simulations and in defining our original haplotype tree estimation algorithm (Templeton et al. 1992), the *only* types of cladograms considered were those in which all or almost all haplotypes are separated from one another by only a single mutation. Almost all intraspecific haplotype trees have properties similar to that of human mtDNA, and therefore we wanted to develop a statistical methodology that deals precisely with this case for which traditional phylogenetic approaches have little to no statistical power (Crandall 1994; Templeton et al. 1992). Rogers and Jorde (1995a) were apparently unaware that this indicator of temporal polarity was justified for cladograms of the type that they argued would undermine it.

Second, Rogers and Jorde (1995a) confused the criteria used in Table 24.6 for

Table 24.6. Expected Patterns under the Different Models of Population Structure and Historical Events

Pattern I. Restricted Gene Flow

    a. Significantly small D_c values, primarily for tip clades. Some interior clades with significantly large D_c values.

    b. $\bar{D}_c(I)-\bar{D}_c(T)$ is significantly large, where $\bar{D}_c(I)$ is the average clade distance of interior clades within the nested category, and $\bar{D}_c(T)$ is the average clade distance of tip clades.

    c. Average D_c values should increase (and occasionally level off) with increasing clade level in a nested series of clades. If the distances level off, the null hypothesis of no geographical association should no longer be rejected even though rejected at lower clade levels.

    d. The above patterns also hold for the D_n values unless some gene flow is due to long-distance dispersal events; then significant reversals of the above pattern can occur with the D_n values.

Pattern II. Range Expansion

    a. Significantly large D_c and D_n values for tip clades, and sometimes significantly small for interior clades under contiguous range expansion, but some tip clades should show significantly small D_c values under long-distance colonization.

    b. $\bar{D}_i(I)-\bar{D}_i(T)$ is significantly small for $i = c, n$ for contiguous range expansion, and for $i = n$ for long-distance colonization.

    c. The above patterns are not recurrent in the cladogram or are geographically congruent.

Pattern III. Allopatric Fragmentation

    a. Significantly small D_c values, primarily at the higher clade levels. The D_n values at this clade level may suddenly increase rapidly while the D_c values remain restricted, depending upon the geographical configuration of the isolates.

    b. The pattern of distances described in (a) should represent a break or a reversal of the distance pattern established by the lower level nested clades.

    c. Clades showing pattern (a) and (b) should tend to be connected to the remainder of the cladogram by a larger-than-average number of mutational steps.

    d. The above patterns are not recurrent in the cladogram or are geographically congruent.

a population range expansion with those described by Cann et al. (1987) and Vigilant et al. (1991). These two papers argue that range expansion causes many of the older haplotypes to be confined to the ancestral, pre-expansion area whereas many of the younger haplotypes are more widespread geographically. However, in Table 24.6, all that is required is that the younger tip haplotypes be significantly more widespread than expected under the null distribution. Because tip haplotypes tend to be rare, they inherently tend to have a more restricted distribution than interior haplotypes. Hence, a tip haplotype can be significantly widespread and still be more geographically restricted than its interior. Rogers and Jorde (1995a:18) unfortunately equated my criterion to that of Cann et al. (1987) and Vigilant et al. (1991) when they stated that my criterion for inferring expansion events was that some younger clades "occupy more territory than some older clades." Although this certainly would satisfy the criteria given in Table 24.6, it is much more restrictive than the one outlined above. Indeed, neither of the inferred human expansion events made in the original analysis (Templeton 1993) satisfy Rogers and Jorde's criteria, nor do either of the two

expansion events into formerly glaciated areas that were inferred for tiger salamanders (Templeton et al. 1995).

RESULTS OF THE CLADISTIC ANALYSIS FOR HUMAN
mtDNA

Table 24.7 presents the results of the permutation testing for haplotypes nested
within one-step clades and the inferences drawn from them using the interpreta-

Table 24.7. Significant Results of the Permutation Tests of the Null Hypothesis of No
Geographical Association along with the Inference Chain Derived from the Key
Given in Appendix (pg. 357).

Clade	Haplotype	Distance Type	P-Levels under the Alternative Cladograms a	b	Inference Chain (Numbers refer to questions in Appendix (pg. 357))
1-1	1	D_c	0.000^S	0.000^S	1-2-3-5-6-7-8(NO):
		D_n	0.002^S	0.001^S	Restricted Gene Flow, but the
	8	D_c	0.020^L	0.019^L	sampling design is inadequate
		D_n	0.013^L	0.010^L	to eliminate the possibility of
	12	D_n	0.002^L	0.000^L	some long distance gene flow
	13	D_c	0.001^S	0.000^S	along with gene flow limited by
		D_n	0.002^L	0.005^L	isolation by distance (valid for
	40	D_n	0.034^S	0.033^S	both alternatives a and b)
	41	D_n	0.035^S	0.027^S	
	42	D_n	0.038^S	0.048^S	
	45	D_n	0.028^S	0.026^S	
	I-T	D_c	0.005^L	0.057	
1-2	64	D_c	n.a.	0.050^L	1-2-11-17(NO):
					Inconclusive (b only)
1-4	75	D_c	n.a.	0.018^S	1-2-3-4(NO):
		D_n	n.a.	0.018^S	Restricted Gene Flow, Isolation by
	I-T	D_c	n.a.	0.012^L	Distance (b only)
1-13	2	D_c	0.027^S	0.118	1-2-11-12(NO):
		D_n	0.026^S	0.118	Contiguous Range Expansion (a
	31	D_n	0.036^L	n.a.	only);
	I-T	D_c	0.020^S	0.118	1-2-11-17(NO):
		D_n	0.033^S	0.028^S	Inconclusive (b only)
1-16	18	D_c	0.009^S		1-2-11-12(NO):
		D_n	0.013^S		Contiguous Range Expansion
	I-T	D_c	0.009^S		

A superscript S designates a significantly small distance; L, a significantly large distance. If a
haplotype is not present in a one-step clade under an alternative, the P-level is marked "n.a."
for "not applicable."

tive key given in Templeton et al. (1995). Only those results that are significant at the 5% level are presented. Because several alternative clade memberships were considered, the probability levels are given for all alternatives if at least one alternative yields a significant result. In this way, the robustness of the results to cladogram uncertainty can be judged. Despite the additional statistics and different nesting designs from that used in Templeton (1993), the biological conclusions are virtually identical. The pattern within clade 1-1 is that of restricted gene flow, as in the original analysis. Likewise, a significant range expansion is detected within Europe in clade 1-16, as in the original analysis. A novel result is that clade 1-2 produces significant deviations from the null hypothesis under one alternative, but the result is inconclusive. Another novel result is that clade 1-4 produces significant deviations under one alternative, and the results indicate restricted gene flow through isolation by distance. Finally, clade 1-13 produces significant deviations under both of its alternatives, but the conclusions are not robust to cladogram uncertainty. Under one alternative, an African population expansion is indicated (as in the original analysis of Templeton 1993), but under the other probable cladogram the result is inconclusive.

Table 24.8 presents the results for the clades nested within two- and three-step clades. There are many more alternatives in these cases, but as the inference chain obtained by using the key in Templeton et al. (1995) was not changed by any of the alternatives, only the range of *p*-values over the alternative nesting designs is indicated in this table. Many significant results are encountered at these levels, and all of them lead to the inference of restricted gene flow.

In summary, the analysis indicates restricted gene flow at all levels of analysis. Thus, there was limited genetic interchange among Old World human populations throughout the entire evolutionary period marked by the coalescence of mtDNA to a common female ancestor. This inference is incompatible with both the multiregional hypothesis with no gene flow (Howells 1942) and with the Out of Africa replacement hypothesis. Hence, both of these hypotheses about recent human evolution are strongly rejected by the mtDNA data. The rejection of the replacement hypothesis does not mean that population range expansions did not occur in recent human evolution; indeed, the present analysis indicates quite the opposite. A recent range expansion within Europe is detected with this analysis, as with the original analysis (Templeton 1993). However, a second recent intracontinental range expansion detected in the first analysis (Templeton 1993) is not robust to cladogram uncertainty. Hence, by explicitly incorporating cladogram uncertainty into the analysis, the paramount role of restricted gene flow in recent human evolution is actually strengthened (e.g., the new inference of restricted gene flow observed among the haplotypes found in clade 1-4 under one alternative) while the role of recent, intracontinental range expansions is somewhat diminished but not eliminated.

Table 24.8. Significant Results of the Permutation Tests of the Null Hypothesis of No Geographical Association along with the Inference Chain Derived from the Key in Appendix (pg. 357)

Nesting Clade	Nested Clade	Distance Type	Range of P-Levels under the Alternative Cladograms	Inference Chain (Numbers refer to questions in Appendix (pg. 357))
2-1	1-2	D_c	0.000^S–0.001^S	1-2-3-5-6-7-8(NO):
		D_n	0.000^S–0.001^S	Restricted Gene Flow, but the
	1-4	D_n	0.024^S–0.108	sampling design is inadequate to
	1-8	D_c	0.038^S–0.061	eliminate the possibility of some
		D_n	0.000^L–0.001^L	long distance gene flow along with
	1-9	D_c	0.000^S–0.009^S	gene flow limited by isolation by
		D_n	0.000^S–0.000^S	distance
	1-10	D_c	0.000^S–0.007^S	
	1-13	D_c	0.006^S–0.083	
		D_n	0.000^L–0.011^L	
	1-18	D_c	0.040^S–0.064	
		D_n	0.011^L–0.034^L	
2-2	1-16	D_c	0.000^S–0.001^S	1-2-3-4(NO):
		D_n	0.000^S–0.000^S	Restricted Gene Flow Through
	1-17	D_n	0.002^L–0.007^L	Isolation by Distance
	I-T	D_n	0.006^L–0.013^L	
3	2-1	D_c	0.000^S–0.000^S	1-2-3-5-6-2-3-4(NO):
		D_n	0.000^S–0.000^S	Restricted Gene Flow Through
	2-2	D_c	0.000^S–0.000^S	Isolation by Distance
		D_n	0.047^S–0.071	
	2-3	D_c	0.000^S–0.000^S	
		D_n	0.000^L–0.000^L	
	I-T	D_c	0.000^L–0.000^L	

A superscript S designates a significantly small distance; an L, a significantly large distance.

A CORRELATIONAL COALESCENT ANALYSIS FOR HUMAN mtDNA

In the cladistic analysis, time polarity was used in a categorical fashion to define contrasts between younger and older clades (tips vs. interiors; lower nesting levels vs. upper nesting levels). However, the algorithm of Castelloe and Templeton (1994) produces quantitative "outgroup weights" for each haplotype that are highly correlated with the age rank of the haplotypes. As explained in Templeton et al. (1995), if restricted gene flow dominates in importance in explaining the present-day associations between haplotypes and geography, a positive correlation is expected between age and geographical extent of a haplotype, and likewise between outgroup weight and clade distance. On the other hand, Cann et al. (1987), Stoneking et al. (1986), and Vigilant et al. (1991) have argued that the oldest haplotypes are found exclusively in Africa, while evolutionarily younger haplotypes are found throughout the world. Such a pattern would yield a negative correlation between outgroup weight and clade distance. Hence, this correlation provides an excellent means of discriminating between these two hypotheses.

The first step in such an analysis is to calculate the outgroup weights. This is done using the algorithm of Castelloe and Templeton (1994), which assumes that the topology of the haplotype tree is known. In this case, there are 200 probable topologies. Accordingly, the outgroup weight for each haplotype was calculated for all 200 probable topologies and then averaged over topologies using the probabilities for each cladogram resolution that were calculated earlier. The correlation coefficient between the resulting outgroup weights and the clade distances is 0.45, which is significantly positive at the 1% level. If the original haplotype tree of Excoffier and Langaney (1989) is used, this correlation rises to 0.60. These results are clearly compatible with the restricted gene flow model and are incompatible with the predictions of Cann et al. (1987), Stoneking et al. (1986), and Vigilant et al. (1991). Thus, both the cladistic analysis and this correlational analysis strongly demonstrate that the Out of Africa replacement hypothesis is incompatible with the mtDNA data.

The cladistic and correlational analyses can be combined to some extent. Instead of looking at the overall correlation between haplotype outgroup weight and clade distance, one can examine this correlation among clades nested together into a higher level nesting clade. The outgroup weight for a higher level clade is calculated by summing the outgroup weights of all its component haplotypes. These correlations are asymptotically independent from one another. However, the sample sizes will be small, particularly at the level of haplotypes nested within one-step clades, so the usual large-sample tests for significance of a correlation coefficient are not valid. To avoid this problem, the distribution of these cladistic correlation coefficients is calculated by the same random permutation procedure used previously. The null hypothesis is then tested to look for significantly large and significantly small correlation coefficients relative to the mean value generated under random permutation (note: the null hypothesis is no longer that the correlation is zero). The gene flow model predicts significantly large correlations, the population expansion model predicts significantly small correlations. All significant deviations at the 5% level from the permutational null hypothesis are given in Table 24.9. As can be seen, with one exception, all significant deviations from the null hypothesis are large, positive correlation coefficients, once again found at all levels of the analysis. This reinforces the earlier conclusion that Old World human populations have experienced restricted gene flow throughout the entire evolutionary history marked by the coalescence of mtDNA. The one exception is the correlation for the haplotypes found in clade 1-13a, which was significantly small. This is a clade that led to an inference of

Table 24.9.   Significant Correlations between a Clade's
Outgroup Weight and Its Clade Distance as Determined
by 1,000 Random Permutations

Nesting Clade	Correlation (or range of correlations if all were significant over all probable cladograms)
1-4b	0.99^L
1-13a	0.25^S
1-17b	1.00^L
1-17d	1.00^L
2-1(a-p)	0.59^L–0.68^L
2-2(a-d)	1.00^L
3(a-d)	0.84^L–0.88^L

A superscript S designates a significantly small distance; an L, a
significantly large distance.

population expansion in the cladistic analysis of geographical distance (Table 24.7). Hence, the two analyses are concordant whenever they both reject the null hypothesis and never display any significant discordancies. Both imply that the dominant pattern in humans is due to restricted gene flow with the possibility of recent and geographically restricted population range expansions.

Although not germane to discriminating between the Out of Africa replacement hypothesis and the gene flow model, the outgroup probabilities can also be used to see if information about the geographical origin of the root exists. This is accomplished by calculating the relative probabilities across the geographical regions of Europe, the Middle East, Africa, and Asia of the product of the outgroup probabilities of each haplotype (which have already been weighted by cladogram uncertainty) times its regional frequency and summing over all haplotypes. When this calculation is performed, the probability that the root was located in Europe is 0.254; in the Middle East, 0.276; in Africa, 0.195; and in Asia, 0.274. As is evident, there is virtually no information about the geographical location of the root. This nearly uniform geographical distribution is expected under a gene flow model because the older the haplotype, the more time it has had to spread out evenly geographically and hence the more difficult it is to determine its geographical origin.

IMPLICATIONS FOR THE ORIGIN OF ANATOMICALLY MODERN HUMANS

Given the consistent and strong evidence for gene flow among Old World human populations, the multiregional hypothesis with no gene flow and the Out of Africa replacement hypothesis can be rejected as models for the evolution of modern humans. The gene flow model is compatible with two other scenarios for the evolution of modern humans: the multiregional model with gene flow, and the single geographical origin model with subsequent spread via gene flow and selection.

Under the multiregional model with gene flow, there is no single geographical origin of modern humans. Instead, different modern traits evolved first in different geographical regions. Those traits that were adaptive then spread throughout all of humanity by the combined effects of restricted gene flow amplified by natural selection. Modern, adaptive traits that originally arose in different regions would be put together through gene flow/selection interactions, and anatomically modern humans arose from this amalgamation of genetic traits of diverse geographical origin.

Under the single geographical origin model with subsequent spread via gene flow, the genetic complex responsible for the features associated with anatomically modern humans arose first in a single geographical location in the range of ancient humans, which includes Africa. (Note: to the extent that more and more of the traits associated with modern humans had a single geographical origin, this hypothesis blends smoothly into the previous, multiregional hypothesis. Hence, discriminating between the two may prove difficult.) The genetic complex underlying this modern morphology then spread throughout the rest of humanity through gene flow amplified by selection, thereby resulting in the *morphological* replacement of the older human type by the modern type. Although the gene flow is restricted, this complex of modern traits could still spread rapidly and globally if it were selectively favored. Theory (Barton and Rouhani 1993; Crow et al. 1990), experiments (Carson 1961; Wade and Goodnight 1991), and natural examples (Johnston and Templeton 1982; Templeton et al. 1989) show that even low rates of gene flow can create drastic genetic changes in the geographical distribution of traits when those traits are favored by selection. Because the gene flow is restricted throughout the entire period marked by mtDNA coalescence, this means that at any given time during this period regional genetic differentiation among human populations would be expected. This regional genetic differentiation, as long as it was nuclear and thereby subject to recombination with the genes responsible for modern morphology, could persist even as the genes for anatomically modern traits were spreading via a combination of gene flow and natural selection. It is well established that recombination coupled with gene flow and selection allows for highly variable levels of genetic differentiation among subpopulations for different genetic systems, with some genetic systems being homogenized and others remaining well differentiated (Barton 1979; Barton and Clark 1990; DeSalle et al. 1987; Templeton et al. 1989).

The highly variable impact of restricted gene flow upon differentiation of various genetic systems has long been known in humans (Workman et al. 1963), but this feature of restricted gene flow has not been much appreciated in the literature concerning the Out of Africa replacement hypothesis. Nevertheless, a deeper appreciation of the genetic role of gene flow is critical in interpreting the different scenarios of recent human evolution. For example, fossil evidence exists favoring an interpretation of multiregional continuity for many anatomical traits (Frayer et al. 1993; Spuhler 1993), while other evidence exists favoring a single geographical origin of modern humans for other traits (Stringer and Andrews 1988a). Although these interpretations have been perceived as contradictory, they are both compatible with the single geographical origin model with subsequent spread via gene flow (or with a multiregional gene flow model in which the traits on which Stringer and Andrews focused had a single geographical origin). With restricted gene flow, some traits can spread by selection across populations that simultaneously display regional differentiation and continuity for other traits, much in the same manner that sickle-cell alleles of African origin have spread in certain European populations that are otherwise genetically and morphologically differentiated from Africans (Miller 1994; Monteiro et al. 1989; Schiliro et al. 1990). Hence, there is no justification for treating these two fossil data sets and respective analyses as contradictory: both are equally and simultaneously valid under a restricted gene flow hypothesis of the evolution of modern humans. Hence, from a population genetic perspective, the debates about the fossil data are simply an artifact of the erroneous assumption that replacement in a fossil series can only be accomplished by one population replacing another, resulting in one population's complete extinction. However, replacement in a fossil series can also be accomplished by a replacement of the responsible genes at specific loci via interbreeding among the populations followed by selection. Morphological replacement should never be equated with population replacement when those populations have the capacity to interbreed.

The mtDNA analysis clearly shows that gene flow occurred in Old World human populations throughout recent human evolution, but it will never by itself discriminate between the multiregional and single regional hypothesis outlined above. This can only be done with fossil data. Genetic and fossil data are complementary; the fossils reveal the patterns of evolution, and the genetic data can discriminate among the different processes that can yield a pattern (e.g., a replacement of genes vs. a replacement of populations). Together, genetic and fossil data give a rich picture of both the process and pattern of human evolution. The most important process information that the genetic data have yielded in this analysis is that there is only one kind of modern human that has evolved as a single evolutionary unit despite past and current regional differentiation for some traits.

BROADER ISSUES

The debate about the Out of Africa replacement hypothesis has been plagued by several issues which have aroused much argument and passion but little illumination and which deal with fundamental aspects of the nature of scientific inference that are by no means unique to this specific debate. Following is a brief discussion of some of these issues.

Confusing Hypothesis Consistency with Hypothesis Testing

If one has a strong inclination in favor of a particular hypothesis, it is all too easy to examine data sets and analyses for their consistency with respect to the favored hypothesis. If many or all such data sets and analyses are consistent, the favored hypothesis seems to grow in strength and support. However, this is all an illusion when the data sets and analyses are not just consistent with the favored hypothesis, but are also consistent with its alternatives. What are needed for scientific inference are not just consistent data and analyses, but data and analyses that have the potential for *discriminating* among the alternative hypotheses. Data sets or analyses that do not discriminate among alternatives contribute nothing to the scientific validity of the favored hypothesis.

For example, much of the current debate focuses on whether or not there is an African root, with many participants in this debate still behaving as if the resolution of an African mtDNA root would "prove" the Out of Africa replacement hypothesis (Gibbons 1995b). As pointed out in the introduction to this chapter and in Templeton (1994a), only the inference of a non-African root is informative in this case and all alternatives are consistent with an African mtDNA root. There is even less potential discriminatory power for nuclear DNA. Expected coalescence time (that is, the time from the present back to the common ancestral sequence) is a function of the inbreeding effective size for the genetic element under consideration, and this size is innately fourfold larger for nuclear DNA than for mtDNA (because of diploidy vs. haploidy, and bisexual inheritance vs. unisexual inheritance) (DeSalle et al. 1987). Moreover, as will be discussed shortly, molecules that do not recombine (such as mtDNA but not nuclear DNA) show a strong bias towards even shorter coalescence times. If the coalescence time of mtDNA is truly about 200,000 years ago, then the expected coalescence times of almost all nuclear genes are going to be commonly greater than one or two million years. This places the expected coalescence times of much nuclear DNA into a period in which all humans probably lived in Africa. Hence, studies on nuclear DNA are expected to have an African root under *all* hypotheses of modern human evolution.

Dating the age of the mtDNA coalescence to around 200,000 years ago is also irrelevant to discriminating the Out of Africa replacement hypothesis, as mentioned in the introduction and in Templeton (1994a). Acknowledging the irrelevancy of this date, Rogers and Jorde (1995a) have nevertheless argued that one aspect of this estimated coalescence time is relevant: namely, its implications for the inbreeding effective size (Crow and Denniston 1988) of human females. Coalescence times are proportional to the inbreeding effective size, and the inference of a relatively recent mtDNA coalescent time therefore implies a relatively small inbreeding effective size of females. Using standard neutral theory, Rogers and Jorde (1995a:30) estimated a world population of about "3500 breeding females" at the time of coalescence. This number, they argue, is incompatible with the multiregional hypothesis with gene flow because "knowledge that Eve lived recently would imply that the human population was small—too small to have populated three continents." A similar argument has been made for Y-DNA by Hammer (1995).

These size estimates are questionable for three reasons. First, Rogers and Jorde (1995a) ignore the full range of uncertainty in this coalescence time as discussed in Templeton (1993). Hammer (1995) uses the technique described in Templeton (1993) to estimate a confidence interval for the coalescence of Y-DNA, but then ignores it in estimating effective size. Moreover, Hammer treats the mutation rate and the calibration point of the human/chimp split as being estimated without error. Rogers and Jorde at one point allow a twofold range of error in the mutation rate—a gross underestimate of the full range of uncertainty as discussed in Templeton (1993), but in most of their subsequent size estimates they use only one value. All of these sources of error affect the estimate of inbreeding effective size and, when put together, imply that the data of Rogers and Jorde (1995a) and Hammer (1995) cannot eliminate inbreeding effective sizes on the order of 10^5 as possibilities. Second, Rogers and Jorde (1995a) and Hammer (1995) treat the inbreeding effective size as a reasonable indicator of the actual population size. It is well known that inbreeding effective sizes can be orders of magnitude larger or smaller than actual size under realistic biological circumstances (Templeton 1994b; Templeton and Read 1994), so this premise is of dubious validity. When this uncertainty is coupled with the measurement error in the underlying parameters, one could have had population sizes at the time of coalescence in the range of hundreds of thousands to millions.

Third, a pure neutral model had to be invoked to estimate these effective sizes. However, molecules such as mtDNA and Y-DNA are extremely sensitive to hitchhiking effects. For example, suppose that a favorable mutation occurs in the mtDNA only once in a million years—a figure so low that it would be acceptable to even the most ardent neutralists. In such a case, there is a 20% chance of such a mutation having occurred between 400,000 and 200,000 years ago, thereby greatly shortening the observed coalescence time of current human mtDNA. Moreover, such a rate of favorable mutations would insure great stochasticity among species, with some species' current array of mtDNA variation having been greatly affected and others not at all by such hitchhiking effects. (Such hitchhiking has no effect on the interspecific molecular clock. Although hitchhiking speeds up the rate of fixation of variants intraspecifically, it also temporarily reduces the neutral variation in the species, thereby slowing down subsequent neutral fixation. These two effects cancel one another, so hitchhiking does not affect interspecific divergence rates.) Thus, theory predicts that DNA molecules that do not recombine behave very erratically intraspecifically, with a consistent and strong bias to shorter than expected coalescence times. This in turn means that there is a strong bias to underestimate inbreeding effective size. All of these theoretical predictions are supported empirically by an extensive *Drosophila* literature (see references given in Templeton 1994a).

By treating these effective size estimates as simple numbers, a strong case for compatibility with the Out of Africa replacement hypothesis is seemingly made. By treating these effective size estimates as being estimated with many sources of error and as underestimating census size, it is obvious that the range of uncertainty is so extreme as to preclude discrimination among alternatives. Hence, this is yet another issue that can be debated endlessly without resolving any of the underlying hypotheses. Indeed, it is surprising how much of the literature on this debate has focused on issues that have no or little potential for ever discriminating among the alternative hypotheses.

Using Hidden and Questionable Premises

Using an analysis that depends upon some hidden premise is common in scientific inference. All scientists bring to their work a particular worldview that includes premises about how nature works which are so deeply imbedded that they are often not questioned or made explicit. This is why it is so important to publish methods of analyses and inference criteria in sufficient detail, as was attempted here, so that other workers, who may not share the same worldview as the author, can point out these hidden and questionable premises. This often leads to a point/counter-point that is beneficial for all parties as the premises are clarified.

One of the most common hidden premises in the "Eve" literature is that of a split between Africans and non-Africans. Again and again, one encounters papers that date this "split" with different data sets and analytical techniques (many computer programs will obligingly give you a pictorial representation of a tree with a split and estimate a date, regardless of whether or not a split actually occurred). However, under any hypothesis of recurrent gene flow between Africans and non-Africans (such as that indicated by the analysis presented in this chapter), there never was a split. Obviously, the premise of a split needs to be tested instead of merely invoked.

The cladistic procedures outlined here provide a rigorous method for inferring whether the geographical pattern of variation is consistent with an historical split (fragmentation) or no split (recurrent gene flow) using criteria that are completely explicit (Templeton et al. 1995). For example, in analyzing the mtDNA of tiger salamanders, a clear split into eastern and western lineages was detected for mtDNA (Templeton et al. 1995). Using the same explicit criteria, there was no split among any human populations. Quite the contrary, the present analysis documents recurrent and continual genetic interchange among all Old World human populations throughout the entire time period marked by mtDNA. Accordingly, estimating a date for a "split" of Africans from non-Africans based on evidence from mtDNA is certainly allowed by many computer programs, but the results are meaningless because a date is being assigned to an "event" that never occurred.

Most studies using nuclear DNA are based upon some multilocus genetic distance analysis rather than locus-specific haplotype trees. Such genetic distance data can also be input into a variety of computer programs and forced to yield a bifurcating tree of "splits" whether or not such splits actually occurred. Hence, having a computer output showing a split (even with a date) does not constitute proof that a split had occurred. Fortunately, there are distinct expectations for genetic distance patterns under the hypothesis of an African/non-African split versus that of populations interchanging genes as constrained by isolation by distance. In the "split" hypothesis, Europeans and Asians should be equally distant from Africans. Under an "isolation by distance" hypothesis, Africans and Asians should be the most distant, with Europeans being intermediate between the two. The intermediacy of Europeans also means that any branch leading to Europeans from the main African/Asian trunk should be short under an isolation by distance model. Under the split hypothesis, the European and Asian branch lengths should be approximately the same.

Since the 1970s a large number of these genetic distance studies have been performed on human populations using a wide variety of genetic screening techniques (see Templeton 1993 for references). Every one of these studies generates the distance pattern expected under an isolation by distance model. There is not a single nuclear genetic distance study that corresponds to the expectations of the African/non-African split hypothesis, and whenever the distance data were explicitly used to test for goodness of fit to the split hypothesis, it has always been rejected (Templeton 1993). Moreover, the only study to date that directly compares mitochondrial and nuclear genetic data in the same samples of major continental populations found inconsistencies in the genetic distance trees that "undermine the genetic evidence for an African origin of modern humans" (Jorde et al. 1995:523). Ironically, despite this overwhelming evidence that there never has been a genetic split between Africans and non-Africans but rather genetic interchange, the premise of a split is so ingrained that many authors of these studies still talk about an African/non-African split and date this "event" that their own data indicates never occurred. In order to make the obviously discordant genetic distance data fit the split hypothesis, extraordinary scenarios are constructed, such as a group of Africans and Asians (the most geographically distant populations) migrating independently into the Middle East where they engage in "an ancient admixture *event*" (Poloni et al. 1995; emphasis added). Yet, the genetic distance pattern given by Poloni and colleagues (1995:Fig. 4) corresponds to the expectations of the isolation by distance model. Hence, instead of having two human populations independently move thousands of kilometers to a common point, all that is needed is to have humans distributed throughout the

Old World and able to disperse a little bit in any direction every generation, resulting in recurrent gene flow restricted by isolation by distance.

The mtDNA and nuclear DNA data are therefore strongly concordant: there never was a split between Africans and non-Africans; rather, Africans, Asians, and Europeans are genetically interconnected populations that define a single evolutionary lineage with some regional differentiation, as expected under isolation by distance (Templeton 1994a, 1994b). If objective statistical criteria had been used from the outset, there probably never would have been a mitochondrial Eve hypothesis because none of the original arguments (an African root for the mitochondrial tree, a split between Africans and non-Africans, greater diversity in Africans) is supported by the data presented. The more rapidly that all parties begin to focus on data that can test alternative hypothesis, justify their premises, and perform all of these tests with rigorous statistics and objective inference criteria, the sooner this debate will be put to rest.

ACKNOWLEDGMENTS

This work was supported by NIH grant R01-GM31571.

APPENDIX

Start with haplotypes nested within a one-step clade:

1. Are there any significant values for D_c, D_n, or I-T within the clade?
- NO—the null hypothesis of no geographical association of haplotypes cannot be rejected (either panmixia in sexual populations, extensive dispersal in nonsexual populations, small sample size, or inadequate geographical sampling). Move on to another clade at the same or higher level.
- YES—Go to step 2.

2. Are the D_c values for tip or some (but not all) interior clades significantly small or is the I-T D_c distance significantly large?
- NO—Go to step 11.
- YES—Go to step 3.
- Tip/Interior status cannot be determined—**Inconclusive outcome.**

3. Are any D_n and/or I-T D_n values significantly reversed from the D_c values, and/or do one or more tip clades show significantly large D_n values, or interior clades significantly small D_n values, or is I-T significantly small D_n with the corresponding D_c values being nonsignificant?
- NO—Go to step 4.
- YES—Go to step 5.

4.   Do the clades (or two or more subsets of them) with restricted geographical distributions have ranges that are completely or mostly nonoverlapping with the other clades in the nested group (particularly interiors), and does the pattern of restricted ranges represent a break or reversal from lower-level trends within the nested series (applicable to higher-level clades only)?

- NO—**Restricted gene flow with isolation by distance (restricted dispersal by distance in nonsexual species).** This inference is strengthened if the clades with restricted distributions are found in diverse locations, if the union of their ranges roughly corresponds to the range of one or more clades (usually interiors) within the same nested group (applicable only to nesting clades with many clade members or to the highest level clades regardless of number), and if the D_c values increase and become more geographically widespread with increasing clade level within a nested series (applicable to lower-level clades only).
- YES—Go to step 9.

5.   Do the clades (or two or more subsets of them) with restricted geographical distributions have ranges that are completely or mostly nonoverlapping with the other clades in the nested group (particularly interiors), and does the pattern of restricted ranges represent a break or reversal from lower-level trends within the nested series (applicable to higher-level clades only)?

- NO—Go to step 6.
- YES—Go to step 15.

6.   Do clades (or haplotypes within them) with significant reversals or significant D_n values without significant D_c values define geographically concordant subsets, or are they geographically concordant with other haplotypes/clades showing similar distance patterns?

- NO—Go to step 7.
- YES—Go to step 13.
- Too few clades (≤ 2) to determine concordance—**Insufficient genetic resolution to discriminate between range expansion/colonization and restricted dispersal/gene flow**—Proceed to step 7 to determine if the geographical sampling is sufficient to discriminate between short- and long-distance movement.

7.   Are the clades with significantly large D_n values (or tip clades in general when D_n for I-T is significantly small) separated from the other clades by intermediate geographical areas that were sampled?

- NO—Go to step 8.
- YES—**Restricted gene flow/dispersal but with some long distance dispersal.**

8.   Is the species absent in the nonsampled areas?

- NO—**Sampling design inadequate to discriminate between isolation by distance (short-distance movements) versus long-distance dispersal.**
- YES—**Restricted gene flow/dispersal but with some long-distance dispersal over intermediate areas not occupied by the species.**

9. Are the different geographically concordant clade ranges separated by areas that have not been sampled?

- NO—**Past fragmentation.** (If inferred at a high clade level, additional confirmation occurs if the clades displaying restricted ranges by at least partially nonoverlapping distributions are mutationally connected to one another by a larger than average number of steps.)
- YES—Go to step 10.

10. Is the species absent in the nonsampled areas?

- NO—**Geographical sampling scheme inadequate to discriminate between fragmentation and isolation by distance.**
- YES—**Allopatric fragmentation.** (If inferred at a high clade level, additional confirmation occurs if the clades displaying restricted ranges by at least partially nonoverlapping distributions are mutationally connected to one another by a larger than average number of steps.)

11. Are the D_c values for some tip clades significantly large, and/or the D_c values for all interiors significantly small, and/or is the I-T D_c significantly small?

- NO—Go to step 17
- YES—**Range expansion,** go to step 12.

12. Are the D_n and/or I-T D_n values significantly reversed from the D_c values?

- NO—**Contiguous range expansion.**
- YES—Go to step 13.

13. Are the clades with significantly large D_n values (or tip clades in general when D_n for I-T is significantly small) separated from the other clades by intermediate geographical areas that were sampled?

- NO—Go to step 14.
- YES—**Long distance colonization.**

14. Is the species absent in the nonsampled areas?

- NO—**Sampling design inadequate to discriminate between contiguous range expansion and long-distance colonization.**
- YES—**Long distance colonization.**

15. Are the different geographically concordant areas separated by areas that have not been sampled?

- NO—**Past fragmentation.** (If inferred at a high clade level, additional confirmation occurs if the clades displaying restricted ranges by at least partially nonoverlapping distributions are mutationally connected to one another by a larger than average number of steps.)
- YES—Go to step 16.

16. Is the species absent in the nonsampled areas?

- NO—Go to step 18.
- YES—**Allopatric fragmentation.** (If inferred at a high clade level, additional confirmation occurs if the clades displaying restricted ranges by at least partially nonoverlapping distributions are mutationally connected to one another by a larger than average number of steps.)

17. Are the D_n values for tip or some (but not all) interior clades significantly small, or the D_n for one or more interior clades significantly large, or is the I-T D_n value significantly large?

- NO—**Inconclusive outcome.**
- YES—Go to step 4.

18. Are the clades found in the different geographical locations separated by a branch length with a larger than average number of mutational steps.

- NO—**Geographical sampling scheme inadequate to discriminate between fragmentation, range expansion, and isolation by distance.**
- YES—**Geographical sampling scheme inadequate to discriminate between fragmentation and isolation by distance.**

25

Population Perspectives on Human Origins Research

HENRY HARPENDING and JOHN RELETHFORD

We are relative latecomers to research on modern human origins. Harpending became involved through helping obtain samples for Linda Vigilant and Mark Stoneking, both of whom later moved to Pennsylvania State University and persisted in forcing his attention to problems of using genetic data to understand our species' history. Later Relethford, a specialist in quantitative genetics, began working with Harpending and others on using craniometric data to examine models of early human differentiation.

Our perspective on modern human origins is that the issues are really demographic, not phylogenetic. The field suffers from a conspicuous lack of explicit testable population genetic models that address the kinds of data that are available. Instead there has been a lot of "seat of the pants" population genetics that has left us open to the rather destructive controversies that characterize the field. We have tried to proceed by constructing explicitly demographic models of the early history of our species, and then examining whether the models can be falsified by extant genetic data about humans.

In demographic terms the continuity model proposes that our species had a large number of ancestors throughout the Middle and Upper Pleistocene. If archaic human populations worldwide underwent in situ transformation to modern humans, then the agent of transformation was a complex of new, selectively advantageous genes. This complex must have spread by gene flow through our ancestral species. Since these ancestors are supposed to have occupied much of the temperate Old World they must have numbered in the millions or tens of millions.

The replacement model suggests that our species had a very small number of ancestors during the Middle and Upper Pleistocene. If modern humans were not a new species by all accepted criteria of speciation, they were ecologically equivalent to a new species that spread from some restricted point of origin and outfought or outcompeted existing archaic populations. In this model the ancestry of today's genetic material was reduced to several hundreds to several thousands of individuals for an unknown duration during the Pleistocene.

Between these two models, which are really two ends of a continuum, is a model which proposes that modern humans were a new race, spreading by population growth but also incorporating genetic material from extant archaic populations. The number of ancestors, in this case, must number in the tens of thousands to hundreds of thousands.

These special models of the origins of our species have been the subject of much controversy that is amply reviewed and discussed in this volume. Our view, briefly, is that the issue is essentially settled in favor of the replacement model or perhaps something between the replacement model and the partial incorporation model, because of the small effective size of humanity as discussed below.

We are ourselves interested more in two other topics related to the demography of modern human origins. The first topic is the genesis of differences between major human races; that is, how did the differences among major regional human populations develop? (We use race in the sense of subspecific variability—there is no proper technical definition of race in anthropology.) The second topic is the genesis of race differences in within-group diversity. In several genetic systems African populations are more diverse than populations from other regional groups. By this we mean that the genetic difference between two individuals picked at random from the same population is greater if the population is African than it is if the two individuals are sampled from, say, Europe. What is the demographic history of African populations that led to this greater diversity within the region? Models have been proposed to understand these problems, and below we show how genetic data suggest resolution of conflicts between different models.

Since modern humans are anatomically different from our archaic precursors, and since modern humans are associated with a distinctive new technology in some regions, the fossil and archaeological records ought to provide evidence to support or refute the different models. Unfortunately no one seems to have worked out the quantitative genetic consequences of the models for what a patchy fossil record should look like under each hypothesis. Because there are no models, arguments from the fossil record seem so far to be curiously lacking in power. Our approach has been to generate testable predictions about contemporary variation and test these instead.

POPULATION GENETIC ESSENTIALS

We will discuss five categories of genetic traits in this chapter. The properties important for inference about the past are different for each system, so we will sketch the important concepts from population genetics theory first, then discuss how the five different marker systems behave, then use this information to evaluate models of human origins and differences. We will assume that we are dealing in each case with selectively neutral traits, although we do not mean to deny the importance of selection in shaping gene distributions. For example, the pattern that we discuss in mitochondrial DNA could very well be due to a selective episode and a particular kind of population history rather than to the population explosion that we have inferred (Harpending et al. 1993). It is only the combination of evidence from many systems that lets us be confident of our assertions.

The important concepts from population genetics are genetic diversity in populations, genetic difference between populations, the rate at which neutral mutations occur at a locus, and the resolution (see below) of the mutational mechanism at a locus. *Diversity* is a measure of how different two individuals from the same population are from each other, while *difference* is a measure of how much more different two individuals are if they come from two distinct populations.

Diversity in a population is increased by mutation and it is removed by genetic drift. If a long-term equilibrium is reached, diversity is proportional to the product of the population size and the mutation rate. Diversity changes, in response to changing population size, at a rate proportional to the inverse of the population size, that is to $1/N$, and to the mutation rate μ. If population size suddenly increases and N is very large, $1/N$ is small and diversity accumulates slowly. If a population crashes to a small size, on the other hand, diversity is lost rapidly since $1/N$ is relatively large if N is small. This means that the effective size of a population that fluctuates in size is closer to the minimum size over time than it is to the average, since diversity is lost faster than it is regained.

Difference between populations accumulates because of genetic drift and mutation, and it is decreased by gene exchange or migration between populations. The greatest difference should occur (1) at a locus with a high mutation rate between pairs of populations that (2) are small and (3) exchange no or very few genes. At equilibrium the difference between populations is proportional to the reciprocal of population size and the sum of the migration and mutation rates, except for high resolution DNA sequence data where equilibrium difference is independent of the mutation rate. Various models of the origin of race differences give greater weight to one or the other of these mechanisms. For example the famous reconstructions of racial differentiation of Nei and Roychoudhury (1974) assumed that all differences between races were the result of the accumulation of mutations in large populations that did not exchange migrants. In this circumstance Nei's measure of genetic distance is an appropriate statistic, and it can be used to infer separation times of races. We suggest below, however, that this model of the origin of races does not fit the available data.

The *mutation rate* is the rate, usually given as a rate per locus per generation, at which mutations occur. While this concept is familiar, the resolution of the mutational mechanism is not as familiar. *Resolution* refers to the probability that a mutation that does occur is to a new and different form. A low-resolution system would be one in which there is backward and forward mutation between two alternative alleles in a population. In mitochondrial DNA sequences most mutations at each nucleotide site are transitions, changes from one purine to another or one pyrimidine to another. These back-and-forth transitions represent the ideal low-resolution mutational system. At the other extreme are systems in which almost every mutation results a new and different allele. The textbook abstraction of this is the "infinite alleles model" or, in the case of sequences, the "infinite sites model" where it is supposed that every mutation to an evolving DNA sequence occurs at a previously unmutated site.

While the infinite alleles and infinite sites models are probably unreachable abstractions, they may be good models for the process of change in long DNA sequences where there has not been sufficient time for repeat mutations to have occurred. They may also be reasonable models for the process of mutation in genes determining quantitative traits, like the craniometric traits we discuss below. Sequences, regarded as single alleles, preserve more information and have higher resolution (in our sense) than loci that follow the infinite alleles model because a sequence can preserve the whole history of mutations that occurred on the path up to a common ancestor and down to another sequence. With ordinary allelic data, on the other hand, the only datum about two alleles is whether they are the same or different. Multivariate quantitative traits may fall somewhere between simple alleles and sequences, since mutations can occur at any of a supposedly large number of loci that affect traits and it may be unlikely that any two mutations have exactly equivalent effects on all traits.

GENETIC SYSTEMS

Blood groups and proteins are the classical markers that were ascertained in many population surveys before their recent displacement by technologies that work directly with DNA. The most familiar of these markers is probably the ABO blood groups, but several dozens of them were routinely assayed before they went out of fashion. These systems have mutation rates that are low compared with those of the other systems we use, perhaps on the order of one in ten million or so.

Nuclear restriction fragment length polymorphisms (RFLPs) are sites scattered throughout the nuclear genome that happen to be cut by one or another of a set of restriction enzymes. The sequences near some of these polymorphisms have been worked out so that the sites can be assayed rapidly using the polymerase chain reaction (PCR). These are sometimes called nuclear DNA polymorphisms. The mutation rate for these sites is again low, perhaps one in ten to one in a hundred million.

Mitochondrial DNA (mtDNA) is a haploid system that is maternally transmitted. Early studies of mtDNA variation used restriction enzymes, but the popular technology today is to study DNA sequences. For comparing humans most people sequence the control region where the mutation rate seems to be higher than elsewhere in the molecule. The mutation rate for a sequence is the sum of the rates for each site along the sequence: for 700 or so sites of the control region the overall rate may be two in one thousand. Most differences within humans are transitions, and there is apparently substantial back and forth mutation at some sites, so the infinite sites assumption certainly fails for individual mtDNA sites although it might not be so bad for whole sequences.

STRs (short tandem repeats) are loci in which variable numbers of repeats of short sequences occur. These are easily ascertained and have very high mutation rates. The mechanisms of mutation are not very well known but a substantial part of the mutational process involves a site with N repeats mutating to one with either $N + 1$ or $N - 1$ repeats. The infinite sites assumption fails for these loci, and the mutation rate may often be of the order of one in a thousand or even higher.

Quantitative traits are modeled a rather traditional Fisherian scheme in which the observed values are supposed to represent a contribution from both genes and the environment, the relative contribution of the two being determined by heritability. Relethford and Harpending (1994) describe the extension of classical population genetic models to quantitative traits. The analog of the mutation rate for quantitative traits is the rate of appearance of new heritability from mutation. Experimental studies show that this rate varies widely. A figure of one in a thousand is often used (Lynch 1988).

APPLICATION TO PALEOANTHROPOLOGY

The amount of genetic diversity in our species in classical marker systems is what would be expected in a species whose total effective size is 10,000 breeding individuals (Nei 1987). A recent survey of nuclear DNA polymorphisms discussed by Gibbons (1995a) arrives at the same number. The diversity of mtDNA sequences in our species suggests that the number of females is 5,000, in strikingly good agreement with the estimate from other systems. This low genetic diversity in humans means that the total size of our ancestral population must have been very small. More precisely, it must have been very small recently enough that mutation has not had time to erase the signature of the small size. The specific pattern of mtDNA sequence differences among individuals led Harpending et al. (1993) to propose that the recovery from the low size of 10,000 began late in the previous interglacial or early during the most recent glaciation, that is, 100,000 to 40,000 years ago.

Since the multiregional model proposes that the ancestors of our species were spread over most of the temperate Old World, it implies that the number of ancestors must have been more than a million, and certainly not 10,000. The genetic evidence, then, denies the multiregional model of the origin of modern humans. There may be some particular set of historical events that could salvage the multiregional model, such as separate severe regional bottlenecks, but no one so far has come up with such a model that works.

The view from anthropological genetics is essentially that the original debate about multiregional and single origin hypotheses is resolved in favor of the single origin hypothesis. Current research is directed instead at understanding the demographic details of the origin.

AFRICAN DIVERSITY

Genetic diversity within Africa is much greater than diversity within other regions for mtDNA sequences, for other long sequences like globin genes, for craniometrics, and for short repeat polymorphisms. It is no greater for classical markers and nuclear RFLP polymorphisms. One possibility is that classical markers were ascertained mostly in Europeans so that they are a biased sample of markers from the genome, the bias being in favor of those most diverse in Europeans. However, a recent model of ascertainment bias (Rogers and Jorde 1995b) suggests that it cannot account for all of the difference.

There are two simple demographic models that can explain higher genetic diversity within Africa. One is that the population ancestral to contemporary Africans was larger during most of the later Pleistocene. Since a larger population maintains a higher level of genetic diversity, a larger ancestral African population might explain all contemporary race differences in within-group diversity.

The other explanation for greater African diversity is that the ancestors of Africans experienced the great demographic expansion first, long before the ancestors of the other major races. In other words, our population history is funnel shaped, but there are several connected funnels rather than just one. The African funnel opened earlier. Since genetic diversity today is roughly the sum of diversity that made it through the neck of the funnel and diversity that has accumulated since the opening of the funnel, the earlier expansion of Africans predicts that African diversity should be greatest in those genetic systems with high mutation rates, those that have had time to accumulate mutations since the funnel opened. Low mutation systems, on the other hand, have not had time to accumulate diversity since the funnel opened. In these systems contemporary diversity reflects diversity that made it though the neck of the funnel, and there are no particular race differences.

The model of earlier African expansion neatly accounts for greater African diversity in just those systems with high mutation rates (mtDNA sequences, craniometrics, VNTRs [variable number of tandem repeats]) but not in those systems with low mutation rates (classical markers, nuclear RFLPs and DNA markers), just as we observe in the data. We have always favored the alternate model, that African populations were simply always larger in the neck of the funnel (Relethford and Harpending 1994), but we now think that there must have been earlier African expansion.

RACE DIFFERENCES

Two kinds of models have been proposed to account for differences between major regional races of humans. One model is that the history of races is a history of successive bifurcations of expanding populations, with differences accumulating between daughter populations because of the accumulation of mutations and the absence of gene flow. According to this model differences between large populations have accumulated since the origin, and the ultimate origin of these differences has been mutation.

The other model of race differences is that the differences are ancient, reflecting differences in a geographically structured population of our founders in the neck of the funnel. This model suggests that the founding population of modern humans was not a single gene pool but a highly subdivided population consisting of two or more isolated demes. Chimpanzees are a contemporary example of this kind of population structure, with the western variant of the common chimpanzee being so different from the central and eastern variants that some would call it a different species. Under this model race differences are ancient, having accumulated in the neck of the funnel because of isolation and genetic drift, and then were "frozen" by the great demographic expansions of the later Pleistocene.

These two models make different predictions about race differences in different kinds of genetic systems. Under the branching model most race differences reflect mutation, so race differences should be greater for high mutation systems and low or minimal for low mutation systems. Under the frozen ancient structure model, on the other hand, race differences should reflect differences in the ancient subdivided population as well as the effects of mutations that have accumulated since the demographic expansions. In general the effects of mutation would be to obscure the ancient differences at most genetic systems, but not for mitochondrial DNA or for other sequences where it is possible to count differences between genes that have not undergone recombination.

With data available so far, the frozen ancient structure model is the clear winner. Race differences are minimal for mitochondrial DNA and for short repeat polymorphisms, the high mutation systems. They are greatest for classical markers and nuclear RFLP polymorphisms. This is precisely the pattern expected under the frozen ancient structure model and precisely the opposite of the pattern expected under the bifurcating model. Craniometric data ought to act like a high mutation rate system, but they are apparently a high-resolution system in which mutational differences accumulate rather than obscure differences. Thus, craniometrics are different in their response to large numbers of mutations from repeat polymorphisms and mtDNA sequences where a lot of back and forth mutation occurs. Race differences in craniometrics are of the same magnitude as race differences in classical markers and nuclear RFLPs.

CONCLUSIONS

Genetic research into modern human origins has begun to focus on questions and models of the demographic recovery from a bottleneck that occurred when modern humans expanded. The older debate between multiregional evolution and a single origin of modern humans is not of much concern any more, since all the genetic evidence seems to favor the single origin hypothesis. We foresee a new kind of paleodemography in anthropology that uses gene distributions in extant populations to understand population dynamics in the remote past. Paleo-anthropologists ought to be as familiar with the genetic evidence as they are with the osteological and archaeological evidence of early human history. If they become more familiar with genetic data, the artificial divisions between geneticists, anatomists, and archaeologists should at last begin to break down.

26

Time and Place of Human Origins
Implications from Modeling

CHARLES OXNARD

Paradigmatic shifts have occurred a number of times in the history of research in human evolution, often through development of new investigative areas. The most important development during the latter half of the twentieth century has been molecular biology. The beginnings of molecular biology were as long ago as the turn of the century (e.g., studies of blood groups—Friedenthal 1900; Nuttall 1904), but the larger picture was clearly drawn only three decades later (Zuckerman 1933). Perhaps the earliest glimpse of the promise of molecular biology (although not, of course, described as such) was evident in the words of Darwin himself: "Nevertheless, all living things have much in common, in their chemical composition, their cellular structure, . . . and their liability to injurious influences" (1884:424, 425). How much closer could Darwin have come, given the language of his day, to the concepts of the molecular, ultrastructural, and immunological evolution of our times?

The shift has been from concepts developed largely from whole-organism morphology, especially from fossils (including the time dimension), to concepts derived from molecular studies of living species which, as end products, involve time only through extrapolation. When a new science produces a shift of this type, there is usually also some new development of the older science that confirms the new view, which has also been true in this case.

A PARTICULAR PARADIGMATIC SHIFT: THE
RELATIONSHIPS OF THE HOMINOIDEA

One well-known example relates to the relationships of the hominoids. Early whole-organism studies, especially during the latter part of the nineteenth century (Huxley 1872; Mivart 1867, 1873; Owen 1866) and the earlier part of this century (Elliot 1913; Le Gros Clark 1934; Schultz 1936; Wood Jones 1929;

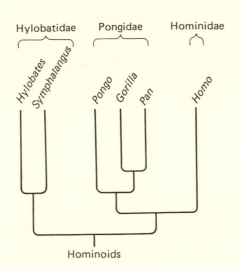

Groupings inherent in
classical morphology

Figure 26.1. General topology of the classical morphological view of the relationships of the Hominoidea.

Zuckerman 1932, 1933), assessed humans (hominids) as evolutionarily separate from apes (pongids—Figure 26.1). A major shift in this view came from the molecular investigations in the 1950s and 1960s, which showed that humans and African apes are a closely knit group, quite separate from Asian great apes (Barnicott 1969; Goodman 1962; Sarich and Wilson 1967) (Figure 26.2).

Many whole-organism investigators (comparative anatomists, paleontologists) arrayed themselves against the molecular findings. Eventually, however, the molecular picture, despite difficulties and argument over interpretation, came to be accepted. Much of this acceptance depended upon the careful and quiet way in which Morris Goodman went about the problem, not only describing the molecular data so that they could be understood by whole-organism biologists but also making the effort to understand the nature of the comparative anatomical and fossil evidence.

Acceptance of the new ideas was also due to changes in the older discipline. Thus, application of a new technology, cladistic analysis (e.g., Andrews and Martin 1987; Martin 1986), seemed to confirm the molecular assessment. Such studies generally placed humans with African apes and separated them from Asian great apes. Unfortunately, cladistics has frequently provided alternative results. Thus, some cladistic investigations still conclude that humans are separate from all apes (Kluge 1983), and others that humans are most closely related to orang-utans (Schwartz 1984).

Groupings explicit in
biomolecular studies

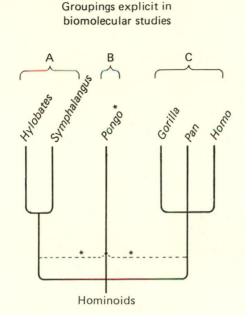

Figure 26.2. General topology of the consensus molecular view of the relationships of the Hominoidea. Asterisk indicates possible variation.

In part controversies occur within cladistics because, though theoretically appropriate, there are problems in its practice (Oxnard 1983–1984, 1987, 1990). One example will suffice. Cladistics depends upon the decision that a character is primitive or derived, which in theory is reasonable enough. However, most real characters are not single underlying characters; rather they are the complex features that we observe. Such complex observable features are likely to be compounds of several (perhaps many) underlying characters. With rare exceptions, then, the features that we actually observe will be some compound of several underlying primitive and derived characters. The outcome required by current practice (that a character be either primitive or derived) is, therefore, unlikely to be correct.

Although it has been little noticed, another method for investigating morphological data has also provided the same answers as the molecules. That is, when anatomical studies attempt to cover as much of the entire body as possible, and when the data so obtained are quantified and analyzed so that redundant information is eliminated mathematically, then results that mirror those of the molecular investigations are obtained (Figure 26.3). These results now stem from several independent investigations—two employing measures of the overall proportions of the body, two summarizing measures from many smaller anatomical

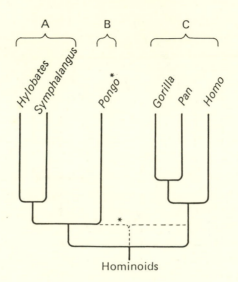

Groupings explicit in
morphometric studies

Figure 26.3. Morphometric view of the relationships of the Hominoidea. Compare with
Figures 26.1 and 26.2. Asterisk indicates possible variation.

regions, and two involving measures of teeth (Oxnard 1978, 1981, 1983, 1983–
1984, 1990). The grouping of humans with gorillas, chimpanzees, and bonobos
has not received further challenge. Thus the application of the molecular revolution in biology to evolutionary studies of the Hominoidea and the coincident
development of new tools in whole-organism studies has been one of those
exciting stages that sometimes occur in the development of science when separate types of investigations and different lines of evidence start to converge.

A NEW PARADIGMATIC SHIFT: THE EVOLUTION OF THE
SPECIES *HOMO SAPIENS*

The prior paradigm shift involved analysis of relationships within a super-
family. We are now in the middle of another paradigm shift that applies at the
species level: the time and place of origin of modern humans. This story, never
very clear in the past, has become much more controversial in recent years
despite the fact that knowledge of human diversity, whether from population
studies of living peoples or from paleontological investigations of a greatly
increased number of fossils, has grown enormously since ca. 1960.

The fossil record of the parent genus *Homo* has now been extended back, depending upon definitions and investigators, from the few hundred thousand years or so of three decades ago to as much as two million and more years ago at the present time. Various forms of *Homo* are now known, including "anatomically modern" humans, various "archaic" humans (all *Homo sapiens*), and even older groups usually designated with alternative specific names like *erectus* and *habilis*. The new controversies depend upon which of two theories is the more likely to be correct.

CONTINUITY VS. REPLACEMENT

The first theory posits early migrations of both *H. erectus* and early *H. sapiens* throughout the world a million or more years ago, together with transition everywhere, at different but still fairly ancient times, of *erectus* into *sapiens*. This theory, with some intermediate genetic admixture, is known as multiregional continuity (Figure 26.4). The second theory envisages a very recent migration of anatomically modern forms from Africa throughout the world only a few hundred thousand years ago together with replacement everywhere of all previously existing species and subspecies. It has been termed the recent replacement theory (Figure 26.4). The first idea requires that the fossils provide evidence of continuity within the major regions of the world, and a degree of discontinuity between them. The second idea requires that the fossils provide evidence of discontinuity within the major regions of the world at approximately 150,000 years ago. The problem arises because investigators supporting each of these theories claim exactly these findings *from the same fossils!*

As with the earlier morphological studies at the superfamily level, these concepts have been illuminated by new findings from today's molecular biology: in this case, studies of mitochondrial DNA (mtDNA). Again, some of the findings seem to be contrary to some of those from morphology, and especially from the morphology of the fossils. One exposition of the work on mtDNA implies that, indeed, human origins were in Africa, and at a relatively recent time estimated to be about 150,000 years ago (Cann et al. 1987). Such a result supports the recent replacement theory. It is based upon two underlying ideas.

The first is a study of patterns of similarity in the mtDNA sequences, attempting to discover how living populations are related to each other through maternal ancestors. This seems to show that all living humans are related (which is not at issue), and that they first appeared in Africa (not necessarily so obvious). The second depends upon an analysis of the rates at which these differences may have been acquired. If it is assumed that the rate of change was reasonably constant, then the amount of difference is proportional to time. The timing can then be set by extrapolating from other divergence times. For instance, if a difference of

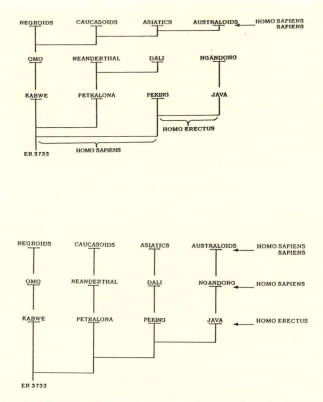

Figure 26.4. Simplified versions of the recent replacement hypothesis (upper frame) and the regional continuity hypothesis (lower frame).

more than 40% exists between the mtDNA of chimpanzees and humans, and if the common ancestor of humans and chimps is set at about five million years ago, then the <2% separation within living humans implies a common ancestor of modern humans at ca. 150,000 years ago.

Of course, these assessments are imprecise. First, the assumption is made that there has been no parallelism or random acquisition of mtDNA sequences. If there had been, the case for a non-African origin would be almost as probable as an African one. Second, other analyses of the same data actually imply that a non-African origin is about as equally probable, and that the date of separation might have been between 500,000 and 900,000 years ago (Long 1993; Maddison et al. 1992). Humans must have had common ancestors somewhere in space and time, but 150,000–900,000 years ago with no certainty as to continent would seem to set pretty wide limits.

TESTING THE EVIDENCE OF MOLECULAR BIOLOGY

This most recent paradigmatic shift is based mainly on molecular biology. Can we look towards new whole-organism studies for further evidence to test both of these theories? Are these indeed the *only* theories to be tested?

The first level of testing refers to species divergence times. Because of the implications of the chimpanzee/human divergence time for human mtDNA studies, times of divergence at the species level are critical. This is especially important because, though the generally accepted time of the chimpanzee/human divergence is about five million years ago, the various molecular studies have actually provided times that vary enormously. Thus, a very short divergence time for humans and chimpanzees of about one million years ago was suggested not very long ago by Hasegawa, Yano, and Kishino (1984). A very long divergence time of thirteen million years ago was estimated by Gingerich (1985). In between, almost every possible time has been suggested (e.g., 3 mya—Cronin 1983; Sarich and Wilson 1967; 6.3 mya—Goodman 1976; 7.5 mya—Goodman et al. 1983; 8.0 mya—Cronin 1983; 10 mya—Bauer 1973; Read 1975; Sibley and Ahlquist 1984). These estimates are, in turn, dependent upon other estimations of the times of other common ancestors (e.g., the common ancestors of all apes) that are even more problematical. Values from one to thirteen million years ago surely imply a level of uncertainty.

Some of the estimates are almost three times the figure used in mtDNA calculations. They would set the original migration time at more than one million years ago. Estimates like this are more an approximation of the time of existence of the genus *Homo,* certainly the species *Homo sapiens,* rather than the subspecies *Homo sapiens sapiens,* not the least its recent variant, "humans of modern aspect." Thus, the need to use a species divergence time raises the question, is it possible to model the evolution of non-interbreeding groups (species) in order to throw light on the process and timing of group divergences?

Because of the implications of migration and interbreeding for human mtDNA studies, the pattern of divergence at the interbreeding subgroup level (subspecies or geographic variants) is a second critical factor. Though the two original theories imply migrations at particular times (early—multiregional continuity; late—recent replacement), the true situation must surely include the complexities of multiple migrations together with reciprocal remigrations, multiple splittings, and reciprocal fusions (interbreedings). A second question is therefore raised by this discussion: is it possible to model the evolution of interbreeding subgroups (subspecies, variants) in order to throw light on the effects of migrations and interbreeding?

Finally, one special element of the mtDNA investigations is the fact that they depend upon inheritance that is carried out through a single sex (but see below).

Does this really provide a straightforward estimate of time and place of population origins, or are there other biological (perhaps sociobiological) factors that could impact upon these assessments? A third question is therefore raised by this discussion: is it possible to model the evolutionary processes of single-sex inheritance patterns in order to throw light on the degree to which time and place of origin may be modified by other factors? It is already known that population size is critical; corrections for small populations are usually employed in mtDNA calculations.

THE NON-INTERBREEDING GROUP SIMULATION

Many studies of non-interbreeding group (species) diversity have demonstrated some of the patterns that can occur in evolution. For example, in any given higher taxonomic group, there may occur: *(a)* a gradual increase in the numbers of non-interbreeding groups (species) over time that results in a bowl-shaped distribution of their numbers; *(b)* early increase followed by later decrease in numbers of groups, producing a vase-shaped distribution; *(c)* a period of greatly reduced numbers of groups after a gradual decrease following a gradual increase, resulting (if we retain the ceramic analogy) in an amphora-shaped distribution (Figure 26.5).

The bowl-shaped distribution of expanding numbers of groups has been simulated a number of times (e.g., O'Hara 1993; work by Tavare cited in Martin 1993). The vase-shaped distribution applies to many creatures (e.g., to a super-family such as the Old World monkeys) for which there has probably been a recent reduction in the numbers of species. The amphora-shaped distribution may apply more specifically to groups like the hominoids which had quite large numbers of species at earlier times but a major reduction (as in the thin neck of an amphora) more recently.

The simulations all assume that the non-interbreeding groups (species) continued unchanged for a period of one million years. The new species obtained at each time level are produced by a process described below. Within each of these distributions, fossils were "discovered" on a random basis at the rate of 3% of all species. The fossil species are shown in bold type for each distribution in Figure 26.5. The extinction rate of 3% (as with all the other parameters) can obviously be changed in further simulations.

Given the fossils that are identified by the random process, and the fact that the remaining (nonfossilized) species cannot be "seen" by the paleontologist, then a phylogeny of the fossils can be "inferred" in a way not too dissimilar to what is done by paleontologists (Figure 26.6). Groups that are close horizontally are more likely to be descended from a common group at a nearer time than groups that are horizontally far apart. Groups that are close vertically are more

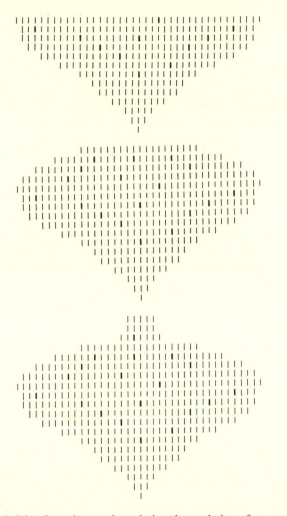

Figure 26.5. Models of species numbers during the evolution of groups of organisms. Each species (vertical line) is assumed to last for 1 million years. Known fossil species, determined on a random basis as 3% of all species, are marked in bold. Upper frame: gradual increase in numbers of species produced, giving a bowl-shaped distribution. Middle frame: gradual increase followed by gradual decrease in numbers of species produced, giving a vase-shaped distribution. Lower frame: gradual increase, then gradual decrease, followed by severe decrease in species diversity, an amphora-shaped distribution.

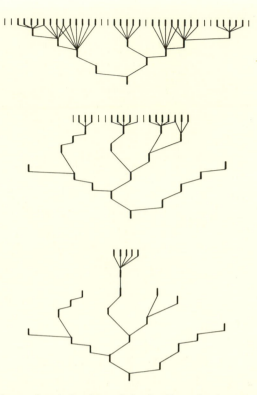

Figure 26.6. Phylogenetic relationships of the fossil species from the models shown in Figure 26.5. In the lowest frame, the amphora-shaped model most closely resembles the situation in the Hominoidea; the most recent common ancestor of the living species is around 3–5 million years old. Six of the nineteen fossils are on the phylogeny leading to the living species.

likely to be related in a lineage than groups that are vertically far apart. The key elements of phylogenies determined in this way are that the fossils are generally related to one another, that earlier fossils are generally ancestors of later fossils, and that later fossils generally give rise to extant forms.

Perhaps as a result, the phylogenies determined in this way look rather similar to many phylogenies available in the current literature for actual fossils. For example, the inferred phylogenies for the bowl- or vase-shaped models are not dissimilar in general form from those that have been put forward for various primate groups that have reasonably large numbers of species at the present day (e.g., Fleagle 1988; Martin 1993). Likewise, for the amphora-shaped model, the inferred phylogeny is rather similar to many phylogenies in the literature for

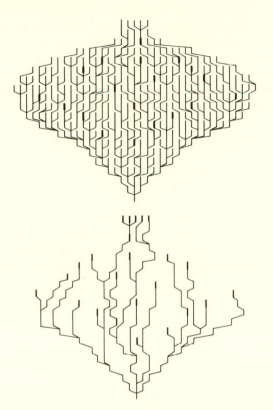

Figure 26.7. The "true" evolutionary relationships for the amphora-shaped model of the lower frame of Figure 26.5. Upper frame: the phylogenetic relationships of all species allowing for both dichotomous and trichotomous splittings and extinctions at specified rates. Fossils (at 3% of all species) are again indicated in bold. Lower frame: phylogenetic relationships of the fossil species alone as given by the model. The most recent common ancestor of living species is at 19 million years ago. Only one of the nineteen fossils is on the phylogeny leading to the living species.

groups exhibiting marked reductions in numbers of species. Thus, the fossil phylogeny of the amphora-shaped model contains a fairly recent common ancestor for all the living species, and most of the fossils lie on the lineages associated with extant species. This is similar to current ideas about human evolution whereby most of the fossils are assessed as being on or close to the human lineage. Even the robust australopithecines, now usually relegated to a side branch, are still assessed as closely linked to the human lineage, and were, when they were first discovered, placed quite firmly on the direct human trunk.

However, the three distributions shown here were actually achieved using a modeling process that involved splitting, continuation, or extinction of groups (see upper frame, Figure 26.7). Splittings and continuations are indicated by oblique lines in the diagrams as occurring at the end of each one-million-year period to give rise to new non-interbreeding groups in the next million years. The first three "generations" of groups were allowed to change and split without other interference in order to get the process started. After that point, a series of rates for splittings and extinctions was assumed to affect all subsequent descendant groups. However, precisely which groups were allowed to split, which to continue unchanged, and which to become extinct was determined using random numbers. This resulted in the evolution of a distribution of descendant groups that show, depending upon the parameters assumed, a variety of distribution shapes.

As a result of this process the total phylogenies for the models are known. One such phylogeny is shown in the upper frame of Figure 26.7 for the amphora-shaped distribution. From it, we can isolate that portion of the phylogeny that contains the "fossils" (lower frame, Figure 26.7). This immediately highlights some of the problems of inferring phylogenies from fossils.

First, it is usual in phylogenies for most fossils to be placed on the evolutionary pathways to the modern forms. This is exactly so for humans; for example, almost all the fossils thus far identified have been placed on or close to the human lineage. This was so even in our own inferred (model) phylogeny—as many as six of the nineteen fossils are on branches that lead to extant species (lowest frame, Figure 26.6). But the reverse is the case for the "true" (model) phylogeny. Only one of the nineteen fossils is on a branch leading to extant species (lower frame, Figure 26.7).

Second, in most higher primate fossil phylogenies, the dating of common ancestry of extant species is generally placed close to that of identifiable fossils. This is certainly so for human evolution. For instance, the time of separation of humans from African great apes (the time of their most recent common ancestor) is generally placed at about five million years ago (i.e., it is placed just earlier than the earliest of the known pre-human fossils—the australopithecines). There are no fossils for chimpanzees—and this alone should give us pause! A date of this order is also true of our inferred model phylogeny in which the overall most recent common ancestor of the extant species is placed at three to five million years ago (given that each species is assumed to last one million years—for the amphora-shaped model—Figure 26.6). Again, however, the reverse is actually the case for the "true" model phylogeny. The most recent common ancestor of all extant species is at nineteen million years; none of the randomly chosen fossils lie on lineages leading to the living forms (lower frame, Figure 26.7).

Third, in most higher primate fossil phylogenies, very few of the fossils are recorded as going extinct. Most fossils are placed on lineages that lead to something. This is certainly the case for hominid evolution. Almost all the fossils that are known are placed on or close to lineages leading to living species (humans).

This is similarly true of the inferred phylogeny from the model; only five of the total of nineteen fossils are recorded as going extinct (for the amphora-shaped model, see lower frame, Figure 26.6). Once again, however, the opposite is found in the "true" model phylogeny. Only one of twenty fossils is on a lineage leading to living species. Nineteen of the fossil species are on lines that go extinct (lower frame, Figure 26.7).

It is thus clear that the trees obtained from knowing only the fossils do not correctly predict the "true" relationships. This is critical for the interpretation of mtDNA data (and most other molecular investigations). Thus, for mtDNA studies, placing recent human origins at 150,000 B.P. depends upon the human/African ape common ancestor existing at about five million years ago (although there is no fossil). The model suggests that a much earlier date is more likely (see also Clark and Lindly 1989a, 1989b).

What evidence is there from real fossils? One fossil that could be used in such an estimate is the earliest australopithecine. If *A. afarensis* is dated at, say, 3.5 million years ago and is not itself an ancestor of both African apes and humans, then ca. 5 million years ago for the assumed common ancestor might well be right. However, that date has already been pushed back by the discovery of *Ardipithecus ramidus* and *A. anamensis*. If they, too, are human ancestors but not ape ancestors, then, at 4–4.5 million years, they come perilously close to predating the supposed human/African ape common ancestor. Should even earlier non-ape australopithecines be discovered, then the five million year estimate would be completely shot.

A second reason for placing the common ancestor at five million years ago are the findings from molecular studies of relationships between extant humans and African apes. However, to go from those molecular data to actual dates also requires assessments of common ancestry—e.g., the common ancestor of humans and apes. The model indicates that similar caveats may apply. And when we look at the range of dates that are posited by various molecular estimates (1.5–13.0 mya for the chimp/human ancestor), then that five million year estimate is highly suspicious. If the five-million-year estimate for the human/chimpanzee common ancestor yields a date of 150,000 years ago for the origins of modern humans, then earlier dates for the human/chimpanzee common ancestor could very easily give 700,000 or 900,000 years or even more. Indeed, another set of studies using data from a Y chromosome analysis, even when based upon the 5 mya common ancestor figure, gives a time of origin of modern humans at 270,000 years, but with 95% confidence intervals that reach much further back (John Mitchell, La Trobe University, Melbourne, personal communication 1996) to as much as 411,000 years (Hammer 1995) or even 800,000 years (Dorit, Akashi, and Gilbert 1995; all reviewed in Jobling and Tyler-Smith 1995). The models thus suggest caution in adopting estimates from molecular data when they are calibrated on the basis of estimations about common ancestors from a fossil record that does not usually include common ancestors.

At this point it must be emphasized that only a small number of simulations have been carried out. Work is proceeding on computational programs that will allow many simulations to be carried out simultaneously. This will provide for *(a)* a large number of runs and sample statistics, *(b)* varying the model parameters, and *(c)* modeling more complex evolutionary mechanisms. Logistically this is not a simple matter. It may be a considerable time before the programs are complete and large numbers of simulations can be undertaken.

THE INTERBREEDING GROUP SIMULATION

The aim of the interbreeding group model of evolutionary change was to study effects of interbreeding and migration. Again, a number of groups were determined. In this case, however, it is assumed that migrations occur from one continent to another so there are four separate sets of interbreeding subgroups (i.e., four "continents") (upper frame, Figure 26.8). One "continent" was chosen as the originating continent and the others arose as a result of these migrations. Once the continents and their contained groups were determined, change was allowed to proceed according to rules not unlike those used before. In this case, however, for the sake of simplicity, only five subgroups were permitted on each continent, and the rates were adjusted so that this number did not change by more than one or two at any time. A "true" model phylogeny was then constructed within and between continents using the following principles.

First, as with the non-interbreeding group simulation, the interbreeding groups eventuating from the "originating" groups were linked by assuming that these groups have the ability, on a random basis, to continue unchanged, to split (only dichotomies were allowed), or to become extinct at each set time period. Second, because this is an interbreeding subgroup model, it must also allow for the opposite of splitting (i.e., fusion—fertile hybridizations). This was also permitted. Third, given that we are modeling migration from one continent to another, the model must include back-migration because if the one can occur then the other can. And if a migration can occur once, it can occur many times. Thus, several migrations and back-migrations were allowed, also chosen randomly, over the complete time scale of the model. Obviously, many runs can be made with different values for the various parameters, though this too awaits further computer development of the models. However, a small number of runs was carried out manually, and one phylogeny showing the full system of branches is shown in the lower frame, Figure 26.8. From this figure, restricted patterns of branching can be derived that would result from comparisons of samples of subgroups from each continent. Figure 26.9 shows four such branching patterns for comparing four different present-day subgroups, one from each continent.

Figure 26.9 shows that, although splittings and fusions, and migrations and

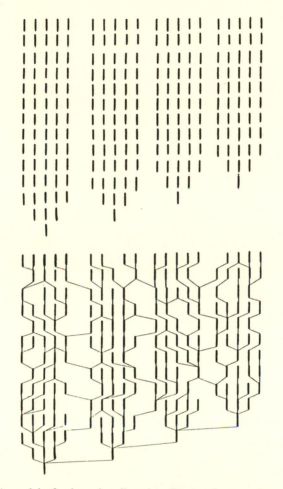

Figure 26.8. A model of subspecies diversity allowing for spread to separate "continents." Upper frame: the actual subspecies groups assuming that each subspecies lasts 0.25 million years. There are five subspecies on each of four "continents." Lower frame: the actual phylogenetic relationships of these subspecies allowing for *(a)* splitting and fusion of groups, *(b)* continuation and extinction of groups, and *(c)* migrations and remigrations of groups.

back migrations, are all allowed many times throughout the time scale of the model, the actual trees for these particular subgroups "find" only links between the subgroups that are far back in time (first and second frames). Even the runs that display the most recent linkages (third and fourth frames) nevertheless have mostly early links. All of the more recent splittings, fusions, migrations, and back migrations go unrecognized in all trees.

It is of particular interest that this is exactly what is posited in both the recent

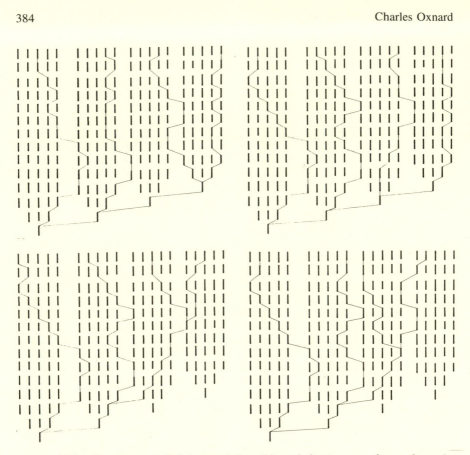

Figure 26.9. Four patterns of phylogenetic branching relating to comparisons of samples of living subspecies, one from each "continent." Despite all the variations introduced by fusions, splitting, extinctions, continuities, remigrations, and migrations at many different times, the overall branching is closely similar to the original branching pattern in the first migrations in the various continents.

replacement and regional continuity hypotheses (Figure 26.4). All that is needed to differentiate the two is the time depth of the model. And that, as we have seen from the non-interbreeding group model, is not easily determined.

THE INDIVIDUAL LINEAGE SIMULATION

These first two models were discussed at the International Institute for Advanced Studies in Kyoto in late 1993 and the Centre for Asian Studies in Hong Kong in 1995, both subsequent to the Australasian Human Biology Society meetings in Adelaide in 1993 (where these ideas were first presented). As a result, a third set of models, of lineages of individuals within an interbreeding subgroup, was worked out.

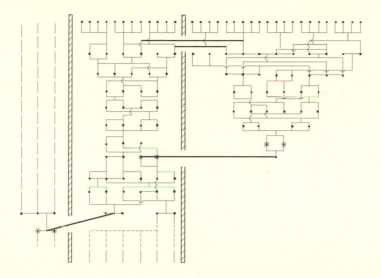

Figure 26.10. A small simulated genealogy allowing for three groups deriving from an initial pairing with several offspring. The model starts with a single pair migrating from a first "continent" to a second which already contains a population. Somewhat later, a further migration follows to a new, empty continent. Much later there is a second migration from the second to the (now occupied) third continent. The genealogy for the first continent is indicated by dashed lines only; the second and third continents' genealogies are shown in full. The vertical hatched lines indicate the separations between the continents. The resultant gene pools of the present-day population overlap as a result of the migrations.

Figure 26.10 shows a genealogy producing three populations after eleven generations. In a first run, two relatively early migrations near the base of Figure 26.10 resulted in three relatively early populations on three separate "continents." Two later migrations near the top of Figure 26.10 resulted in further admixture from the second to the third "continent." The resulting "present-day progeny" at the top of the figure demonstrates what would be expected: an overlap in the gene pools of the second and third populations as a result of the late migration. This model has been made small enough to allow the reader to see the actual individuals concerned. Because it indicates fathers as squares and mothers as circles (standard genealogical convention), it allows us to follow backwards from the present day to the primordial mothers and fathers.

Mitochondrial DNA Relationships

Though very limited, this small model allows us to isolate the mtDNA portion of this genealogy. Thus, for the same lineage as Figure 26.10, Figure 26.11

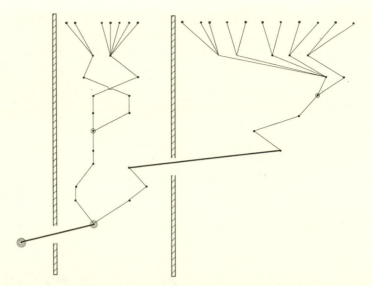

Figure 26.11. The mtDNA relationships of selected individuals of both sexes in the present generation traced back through the mothers to the original mother of the tree defined in Figure 26.10. The "first mother" on each continent is located relatively recently (circled) in the third and fifth generations. The two geographic areas remain separate back to the original migration at the tenth generation (double circle) on the second continent. The original mother (three circles) is, of course, located on the first continent. The most recent migration, deliberately involving only males, cannot be recognized by mtDNA analyses.

shows the tree of relationships of the mothers of mothers of a sample of present-day individuals. It indicates, as we would expect, linkages back through "mothers" to the initial mother in the model.[1]

It is of interest to realize how much (or how little) of the total genealogy of this model is represented in the mtDNA tree. First, the mtDNA tree "finds" only the two early migrations. The reason for this is, of course, that only males were "allowed" to participate in the most recent migration. Another simulation allowing females to migrate might provide a different picture. This finding is especially important because, in many primates with extended social groups (e.g., chimpanzees), females may stay with their natal groups, it being mostly the males that migrate. This aspect of the model in fact fits very well with findings from mtDNA in Old World monkeys such as rhesus where interpopulational mtDNA divergences are much larger than might be expected (Melnick et al. 1993).

In a similar way, primarily male migrations are likely to have been especially frequent in human gene admixture. Even when female migration does occur, if the number of migrating females relative to the size of the new pool is very small, then the effect on the mtDNA tree may be rather similar to there having been no females migrating at all. This is because, even when females do migrate, their influence on the new population may be very much less than that of the males assuming that a few migrant males may impregnate many of the new females, whereas a few migrant females produce relatively few offspring from the new males.

A second feature of this particular mtDNA divergence tree is that it is powerfully affected by population size. This was actually an accidental finding that came out because, in order to keep the model small, population sizes had to be very small at several points in time. As a result it became evident that the divergence trees readily found times in the past when the populations, especially of females, were very small. This is highly likely to have been a major feature of human group diversity in the past, as indeed it is today for some aboriginal populations.

A third feature of this mtDNA tree also became obvious as a result of modeling a very small system by hand. With these very small populations, consanguineous and polygamous matings were necessary to prevent the groups from dying out. These mating phenomena have strongly affected the mtDNA trees. Other mating patterns and different adult sex ratios would also enormously impact the trees. Such phenomena, though often not overtly recognized by human societies today, may have been forced upon many early human populations through the exigencies of environmental harshness and resultant small population size. The tree finally reduces to the "mother of us all" who is located on the originating continent.

Y Chromosomes

It is also possible to model the obverse situation—Y chromosomal inheritance along the patrilineage. Of course, the conventional anthropological wisdom is that this is not possible with mtDNA; the mtDNA in the sperm is said to be discarded at fertilization. This is not, in fact, true (James Cummins, Murdoch University, Western Australia, personal communication 1995). It is, however, true that there is an enormous dilution factor when the small amount of mtDNA in the sperm merges with the large amount of mtDNA in the egg. This dilution factor might have prevented the earlier, less refined mtDNA studies from recognizing the paternal component. Newer techniques now allow it to be measured and, at least in theory, this paternal mtDNA might provide the basis for a patrilineal tree.

There is another structure that might allow us to look for a patrilineal tree—the non-X-homologous portion of the Y chromosome. At the time this work was started (1993), such studies had not been published for humans, although preliminary investigations had been reported (Gibbons 1991). More recent publications show that it has now been achieved (Dorit et al. 1995; Hammer 1995; and commentary in Pääbo 1995). The pattern of relationships of the fathers of the

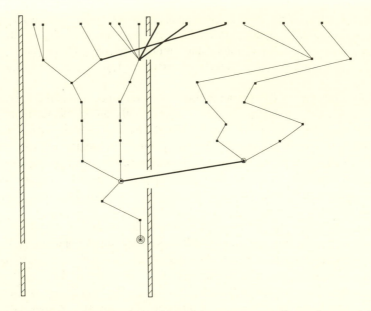

Figure 26.12. The reciprocal relationships of the same present-day males traced through the small amount of paternal mtDNA, or through portions of the Y chromosome of the model shown in Figure 26.11, followed back through the fathers. The "original father" (single circle) on each continent is located much further back in time than the "original mothers" of Figure 26.11 (at the sixth and eighth generations, respectively). The overall original father (two circles) is actually located on the second continent, not the first (cf. Figure 26.11). In this case, the very recent migration is also easily identified.

fathers was isolated from the model shown in Figure 26.10. The resulting tree (Y chromosome tree—Figure 26.12) differs from both the maternal (Figure 26.11) and paternal mtDNA trees in that, although paternal and maternal mtDNA can be included for both females and males in the present generation, only males have the Y chromosomal markers.

Figure 26.12 shows that the Y chromosome tree, although including the same males in the present-day population as were included in the maternal mtDNA tree, is quite different from that tree (Figure 26.11). First, the male-inheritance tree highlights the most recent migration "missed" by the mtDNA tree; of course it would find it because the most recent migration was a migration of males. But other phenomena are also apparent. The clusterings resulting from the "fathers of the fathers" extend more deeply in time than those from the "mothers of the fathers" obtained from mtDNA divergence tree. Perhaps most interesting of all, the "father of us all" exists on a different continent than the "mother of us all"!

It is evident, then, that *(a)* a tree of maternal linkages (mtDNA relationships), *(b)* a tree of paternal relationships (Y chromosomes or paternal mtDNA), and *(c)* the entire genealogical tree of nuclear DNA relationships each expose different aspects of the phylogenetic story. Each is likely to be affected by different parameters (e.g., population size, mating system, level of consanguinity, nature of migration) in different ways. An assessment based solely upon one of these methods is likely to be misleading.

CONCLUSIONS: IMPLICATIONS FOR HUMAN EVOLUTION

The recent time for modern human ancestry derived from *some* mtDNA studies (e.g., Cann et al. 1987; Vigilant et al. 1991) but *not necessarily all* such studies (e.g., Maddison et al. 1992), and now also from Y chromosomal studies (e.g., Dorit et al. 1995), depend ultimately upon times of common ancestry that are plugged into the analyses. For example, it is common to use five million years for the chimpanzee/human common ancestor and fourteen million years for the orang-utan/human common ancestor. The non-interbreeding group model suggests that these times might easily be wrong by a large factor. Recalibrated data for any of these studies might actually point to a date several times longer than those mooted. The mtDNA data may actually be indicating a split within the species *Homo sapiens* into subgroups as long ago as 500,000–900,000 years, or even a split within genus *Homo* into species as far back as 1.8–2.4 million years.

Even with acceptance of the five and fourteen million year figures just cited, the Y chromosome study, though it gave an expected date of 270,000 B.P., also provides a 95% percent confidence interval extending as far back as 800,000 B.P. If confidence limits are to be respected, this means that these data imply at least a reasonable possibility of a figure as high as 800,000 B.P. If the earlier times for common ancestors are not unreasonable possibilities, then the Y chromosome analysis could also easily extend the time of human origins back to ca. one million years ago. It is of interest to note that the mtDNA and Y chromosome studies are internally consistent whether they indicate very recent or very old dates. *It all depends upon the common ancestral estimates.* The model suggests that these may be much earlier than usually assumed.

The interbreeding subgroup model suggests that, whatever the nature of splittings, migrations, and extinctions (and unions, remigrations, and continuities), the linkages among extant forms from the various continents mirror best the oldest of these phenomena, and this despite the likelihood that they have actually been occurring throughout the entire time period concerned. Again, this means that the molecular studies are providing information about the earliest migrations alone; many others may have been "missed."

The sex-linked individual mating models imply that the most useful information is not likely to be about time and place of origin at all. The "mother of us all" on one continent is scarcely compatible with the "father of us all" on another. It's a long way to go for sex! Rather, the sex-linked individual mating models suggest that change in population sizes, differences in mating patterns, adult sex ratios, infanticide, degrees of consanguinity, types of migrations, and much more that may have occurred during our evolutionary history are far more important determinants of patterns in mtDNA and Y- chromosomes studies. Individual populations must frequently have undergone extreme reductions owing to environmental and other hazards. Though small populations can usually be expected to go extinct, this will not always have been the case. Female sex-linked phylogenies may be much more acutely affected by such phenomena than male sex-linked phylogenies.

It must also have frequently been the case that mating patterns were altered by phenomena like monogamy, serial monogamy, polygyny, polyandry, first-cousin matings, uncle-niece matings, and even sibling matings, all of which are found in some human groups today and in recorded history, and many of which may have been obligate in the past in human groups with severe population reductions.

Other social phenomena also affect mating patterns. One of these is (especially preferential) infanticide—well known in both human and nonhuman primates. This practice may be severe enough to result in markedly unbalanced adult sex ratios, even though the sex ratios at birth are likely to be approximately equal. An opposite phenomenon is parental favoritism towards daughters (a result, in some nonhuman primates, of networking amongst mothers in a polygynous situation, and among humans, recorded for the Kenyan Mukogodo—Cronk 1993). All such factors would affect the results of mtDNA and Y-chromosome studies.

Finally, different "migration" patterns can be important bias factors: explorers of virgin territory, marauding "armies" (perhaps primarily male), marauding armies with camp-followers, slave-taking parties (with preference for males as heavy workers, females for housework or pleasure), itinerant merchants and/or craftsmen (of both sexes), and perhaps most important of all, the continuous stream of travelers of all kinds—all these aspects of "migration" would have affected the mtDNA and Y-chromosome studies in different ways (see Clark 1994a).

The overall conclusion from these three models suggests that we are not yet at the point in molecular research where we can leap directly to the time and place of origin. Although a recent origin in Africa may have been "the way it was," there are other more likely possibilities. (1) The time of origin might have been earlier than we expect, (2) originating regions might have been different than we expect, (3) more complex relationships among human populations might have obtained, and (4) different evolutionary processes might have been involved. The whole set of ideas rests upon the modeling processes. It is essential that these models be replicated many times, with many different parameters, and with determination of statistics derived from many simulations. That work is now in progress but it is clearly a major and difficult undertaking.

ACKNOWLEDGMENTS

This work benefited from discussions with colleagues in the Centre for Human Biology. The initial stimulus was a request from Dr. Bill Brede to proffer a paper at the Adelaide meeting of the Australasian Society of Human Biology, and discussion with colleagues at the workshop on mtDNA-Theoretical Approaches to Human Origins held at the International Institute for Advanced Studies, Kyoto (both in 1993). The work was carried further with a presentation at the Fourth International Conference of the Centre of Asian Studies at the University of Hong Kong (1995). I am grateful to many colleagues and students in the Centre for Human Biology, University of Western Australia (especially Profs. Nina Jablonski and George Chaplin) with whom these results have been discussed as they developed, and I am especially grateful to Prof. F. Peter Lisowski (University of Tasmania) who provided much commentary on the text.

NOTE

1. Incidentally, no one should assume that this means that mtDNA studies indicate that there is only a single "mother" of us all (see also Ayala 1995).

27

Race and Language in Prehistory

VINCENT M. SARICH

Concern with our ancestry is a human universal. It is then hardly surprising that the recent debates concerning the origin of our species and of races within it join almost seamlessly with the pre-1859 concerns about monogenesis (currently known as Garden of Eden or Out of Africa models) or polygenesis (currently regional continuity or multiregional evolution). In pre-evolutionary biology times it was the question of whether our species had been created only once, with subsequent racial differentiation taking place "naturally" (monogenesis), or whether each major race had been created separately (polygenesis). Today it is whether our species has a single, geographically and temporally restricted origin (thus, the Garden of Eden), or whether racial lineages can be traced well back into *Homo erectus* times, with "sapiensation" then being a much more gradual process (regional continuity). A concern and fascination with the questions is, however, no guarantee of answers—indeed, questions about ourselves tend to be among those least likely to yield realistic answers. That has certainly been the case here. As we shall see, it is very likely that neither of these scenarios comes close to approaching the probable reality.

My own recent concern with these issues was triggered by a sentence in a letter to *Science* by a group from the Smithsonian commenting on an article by Lewin (1988c) concerning the proposal of Cavalli-Sforza et al. (1988) of a general congruence between language-based and gene-based trees linking extant human populations.

> Any attempt to reconstruct global human history must deal with evidence that linguistic relationships reflect a much later period in human history than the genetic relationships among human populations. (O'Grady et al. 1989:1651)

The response to this sentence has to be, What evidence? The Smithsonian group provided no documentary support for their position either in this letter or in their later article (Bateman et al. 1990, esp. pp. 8–11), and it is difficult to imagine how there could be any. That is, given that any differentiation among populations (genetic, linguistic, cultural) implies actual physical separation, there is going to have to be an appreciable congruence among the pattern of relationships implied by each variable. Even today two populations more similar in anatomy to one another than either is to a third are also much more likely to be more similar genetically, culturally, and linguistically, and this would have been the case even more often in the past. Thus the position of Cavalli-Sforza and colleagues—that there is a general isomorphy of gene and language trees—must be seen as something close to a null hypothesis; that is, as being much closer to a reasonable working assumption than a data-based conclusion. This has long been evident.

Thus the Smithsonian group's argument would have to be regarded as untenable as well as undocumented, and any attempt to validate it as very likely an exercise in futility. But such a judgment tends to deflect attention from what is very likely the real source of these objections. What has passed apparently unnoticed about the Cavalli-Sforza et al. scenario is that it requires an enormous, and completely undocumented, linguistic extrapolation—from the roughly 7,000 or so years over which linguists tend to agree that language relationships can be traced, to the 100,000 or so years which dates the root of their tree. While the extrapolation is enormous and undocumented (and, very likely, undocumentable even if correct), it does at first glance seem to be required by the apparent congruence of the gene and language trees in the context of the dates currently being provided by biochemists, paleontologists, and archaeologists.

The question, then, as is the case so often, is one of the time frame involved; specifically, how far back in time can we trace any of the lineages at issue here?

POPULATION AFFINITIES OF FOSSIL *HOMO SAPIENS*

The only direct evidence as to the antiquity of populational lineages resides in the fossil record, and what we want to know here is when individuals found in a given area begin to have a reasonable probability of being more closely related to modern individuals found there than to those found in other areas. We might, for example, attempt to place early Upper Paleolithic fossils found in areas populated today by "Caucasians" on a tree of extant human populations derived from the data of Howells (1973, 1989). One would expect that if the 100,000-year time scale of Cavalli-Sforza et al. were correct, then these 12,000–30,000+-year-old fossils should fall on the Caucasian clade of the *Homo sapiens* tree, and the literature has long implied that this is actually the case; in other words, that these are anatomically, as well as geographically, European—more closely related to modern Europeans than to modern Asians or modern Africans. That, as will be demonstrated, is simply not the case, though the demonstration is neither easy nor straightforward.

It seems to me that at least three criteria must be satisfied before one can address questions of this sort with any degree of justifiable confidence. First, the algorithm to be used must be able to place known individuals into their appropriate populations or areas with some reasonable degree of reliability. This is obvious, for if it doesn't work reliably for knowns, there can be no rationale for using it on unknowns. Second, it should be able to take a random sample of individuals from known groups and reconstitute those groups without previous knowledge of their number or characteristics. In other words, our algorithm should have a reasonable robustness with respect to assessing the affinities of individuals when compared with other individuals, and not simply with known populations. This is necessary because human fossils are almost always found as individuals, and they obviously do not belong to extant populations. This latter point leads to the third, and more subtle, requirement. If a fossil cannot be viewed as a member of an extant population, then it can only be tested for placement on, or proximity to, a lineage leading to one or more of the latter. But this means that a simple similarity criterion will not do, as showing that fossil X is "most similar" to extant population Y (or to individuals from Y) is without phylogenetic significance until the question of amounts of change along the various extant lineages is taken into account. That is, fossil X may be more similar to population Y than to population Z simply because less change has taken place along the Y lineage, and not because X was part of the Y lineage more recently than it was part of the Z lineage. This last problem does not appear to have ever been recognized, never mind addressed, in the relevant literature.

Any such effort today necessarily begins with the unique and invaluable body of data gathered by Howells (1973, 1989), which he has generously made available on disk. Thus interested workers can now work directly from a large number of individual measurements made by one man on almost 3,000 skulls from, in the main, known populations. I have developed the following approach for satisfying the three criteria just noted. I make no claim here that this is the best possible "algorithm," nor am I especially satisfied with its elegance. It does, however, have the virtue of working—that is, of satisfying the three criteria mentioned above.

The measurements used here (nos. 2, 4–6, 8, 9, 11, 13, 14, 17, 18, 22–25, 32, 33, 35, 37, 39–47, 51, and 54 from Howells 1973) were chosen as *(a)* relatively uncorrelated with one another, *(b)* giving substantial attention to the face as well as the cranium, *(c)* less likely to be missing in the fossil specimens, and *(d)* more objectively measurable. The measurements are identified in Table 27.1. Size-correction was achieved by dividing the breadth measures (nos. 5, 6, 8, 9, 11, 17, 18, 22, 24) by the mean of cranial length and height; the length measures (nos. 2, 23, 25, 32, 33, 51, 54) by the mean of cranial height and breadth; the height measures (nos. 4, 13, 14, 35) by the mean of cranial length and breadth; and the cranial vault measures (nos. 39–47) by cranial length. Then the variation in the contributions to be made by each of the various measures to the overall distance was eliminated by converting each size-corrected measurement to a z-score (using a panel of 105 recent specimens drawn from 21 populations to provide the

Table 27.1. Craniometric Variables (from Howells 1973)

2.	Nasio-occipital length
4.	Basion-bregma height
5.	Maximum cranial breadth
6.	Maximum frontal breadth
8.	Bizygomatic breadth
9.	Biauricular breadth
11.	Biasterionic breadth
13.	Nasion prosthion height
14.	Nasal height
17.	Bijugal breadth
18.	Nasal breadth
22.	Bimaxillary breadth
23.	Zygomaxillary subtense
24.	Bifrontal breadth
25.	Nasio-frontal subtense
32.	Inferior malar length
33.	Maximum malar length
35.	Cheek height
37.	Glabella projection
39.	Nasion-bregma chord (frontal chord)
40.	Nasion-bregma subtense (frontal subtense)
41.	Nasion-subtense fraction
42.	Bregma-lambda chord (parietal chord)
43.	Bregma-lambda subtense (parietal subtense)
44.	Bregma-subtense fraction
45.	Lambda-opisthion chord (occipital chord)
46.	Lambda-opisthion subtense (occipital subtense)
47.	Lambda-subtense fraction
51.	Prosthion radius
54.	Frontomalare radius

reference means and standard deviations). Each individual or population thus became a column of thirty *z*-scores. Distances (inter-individual, individual to group, or inter-group) were then calculated as the average *z*-score per size-corrected measurement. For this study, all thirty measurements, treated in exactly the same manner, with each thus contributing equally to the final result, were used for all comparisons (except, of course, when some were missing for certain of the fossils). I am not making the argument that each measurement would be, in some ideal world, contributing equally to the task at hand of sorting individuals into their respective populations. What I am saying is that neither I nor anyone else knows how to weight characters in this realm. So I recognize and accept that I am trading some precision for a good deal more in the way of objectivity.

The next step is to gain some measure of affinity from these size-corrected and range-adjusted data—and here, rather than relying on one or another clustering program whose internal workings I cannot understand, I used the following more transparent procedure. As a test I chose ten individuals from each of five widely separated populations: Norse, Santa Cruz (Amerind), Zulu, Anyang (Bronze Age Chinese), and Tolai (New Britain); calculated the distance between each pair of individuals; and then calculated the correlations among the columns of distances. Those correlations resulted in five obvious groupings containing 37 of the 50 individuals. Each of the five groups was then averaged into a single composite individual, and the remaining thirteen individuals were compared with those five, with the result that a further eight fell cleanly into their correct units. There then remained two misplacements—one Anyang with Santa Cruz, one Tolai with Zulu—and three ambiguities: one Norse between Norse and Santa Cruz, and two Zulu between Zulu and Tolai. This is already quite good, and any major effort in this direction would rapidly note not only a subgrouping of the African (Teita, Dogon, Zulu) and Melanesian (South Australia, Tasmania, Tolai) samples using this choice of characters and metrics, but also that one character (no. 37, glabella projection) showed a strong bimodality with respect to the African and Melanesian units—such that it alone would sort the Africans from the Melanesians with better than 90% accuracy, which would very likely clear up the Zulu-Tolai confusion. Thus even this very simple approach provides reliable results and gives us some confidence in regard to placing fossils.

Now to criterion three: how do you compensate for the fact that a morphologically "primitive" population (i.e., one closer to the base of the *Homo sapiens* tree) would show greater similarities on the basis of shared primitive conditions, a result having no phylogenetic relevance? The importance of this matter can be assessed by calculating the distances among Howells's 1989 sample of 29 human populations (plus the Tierra del Fuegans I measured in London and Vienna) and then seeing if any of them shows a significantly smaller average distance to the others, thus indicating much less change along its lineage. We find that the Norse, Zalavar, Hokkaido, and Ainu samples are, by this criterion, least derived, showing about 25–35% less change than the average. The Buriats (because of their combination of extreme cranial and facial breadth, and facial length) show far and away the greatest amount of change (some 75% above the average), with the Eskimo and South Australians next (about 30% above the average).

Judgments as to the affinities of fossils are then obviously going to have to include a rate-correction factor, such that if, for example, some fossil were to give the same distance to Ainu and to South Australia, we would conclude that the actual affinities are with the latter. This has been done in Table 27.2. There we see that, after rate-correction, only two of the Caucasian-area fossils (Afalou 10 and Candide 1) were closest to the Norse, and both associations were marginal— just 2% and 4% closer to Norse than, respectively, Santa Cruz and Tolai. One other (Candide 5) straddled Norse and Tolai, and most of the other associations were with Tolai or Santa Cruz.

Table 27.2. The Mean *z*-Scores, Calculated as Described in the Text, between Each of 33 Fossil and Five Recent Human Populations

	Distances				
	Norse	Zulu	Tolai	Anyang	Santa Cruz
Steinheim	2.12	2.22	1.89	2.10	**1.66**
Petralona	2.31	2.46	2.24	2.45	**1.81**
Kabwe	1.75	1.85	1.67	1.84	**1.29**
Ferrassie 1	1.48	1.57	1.49	1.60	**1.12**
Monte Circeo 1	1.87	2.05	1.81	2.06	**1.39**
La Chapelle	1.79	1.88	1.74	1.99	**1.39**
Saccopastore 1	1.81	2.06	1.98	2.13	**1.48**
Shanidar 1	**1.22**	1.26	1.32	1.45	**1.19**
Gibraltar	1.63	1.80	1.56	1.87	**1.26**
Amud	1.83	1.92	1.80	2.00	**1.43**
Irhoud 1	1.37	1.44	1.24	1.48	**0.86**
Skhūl 5	1.39	1.48	1.23	1.38	**1.00**
Qafzeh 6	1.07	1.14	1.09	1.03	**0.72**
Qafzeh 9	0.87	**0.71**	**0.70**	0.93	0.82
Cro-Magnon 1	0.86	0.99	0.90	1.11	**0.79**
Mladeč 1	**0.92**	**0.89**	1.02	1.03	1.14
Predmost 3	1.01	1.08	**0.89**	1.19	**0.85**
Kostenki 14	1.03	0.79	**0.67**	1.01	1.03
Pataud	0.71	0.80	0.86	0.75	**0.56**
Chancelade	0.97	0.89	1.00	**0.73**	1.22
Oberkassel male	0.99	1.06	0.87	0.89	**0.67**
Oberkassel female	0.88	**0.82**	**0.77**	0.90	1.07
Candide	1.22	1.31	1.31	1.28	**1.01**
Candide 1	**0.81**	**0.89**	**0.85**	**0.87**	**0.86**
Candide 5	**0.87**	0.93	**0.85**	0.92	1.03
Afalou 9	1.05	1.21	1.07	**0.76**	**0.77**
Afalou 10	**0.63**	0.82	0.92	0.72	**0.64**
Afalou 29	**0.76**	**0.79**	**0.77**	**0.81**	**0.75**
Afalou 32	0.87	0.85	0.82	0.80	**0.68**
Taforalt 11	0.98	0.97	**0.69**	0.84	1.04
Taforalt 17	0.95	0.83	**0.72**	0.87	0.91
Cohuna	1.74	1.72	**1.35**	1.70	1.41
Upper Cave 101	0.97	1.06	0.88	1.15	**0.74**

Amounts-of-change adjustments present: Norse, +0.08; Zulu, 0.03; Tolai, 0; Anyang, −0.02; Santa Cruz, −0.07. The smallest distance or distances within 0.05 units of one another are in bold.

Another way to make the point is with the mean distances between the seventeen Caucasian-area fossils (Cro-Magnon to Taforalt) and five target populations: Norse, 0.914; Zulu, 0.937; Tolai, 0.88; Anyang, 0.91; Santa Cruz, 0.882. The conclusion, then, is simple: Upper Paleolithic European-area fossils do not show any phylogenetic (that is, derived character) tendency to "look European." This finding is reinforced by asking what affinities are shown by the fossils to one another. The answer is that they are very similar in degree to those shown in the original sample of fifty modern individuals when comparing individuals in one population with those in the other four (e.g., any of the Zulu to the forty non-Zulu). The only real, and not unexpected, exception to this pattern is that the three Candide individuals tended to associate with each other more closely than to most of the rest. Thus the "Caucasian-area" answer to the question, How far back in time can we trace any of the lineages at issue here? is "Not very far at all—certainly not even 20,000 years."

I note here much of this is simply putting a quantitative gloss on judgments made long ago on a qualitative, subjective basis. Various of the European-area fossils have long been seen as eskimoid, or negroid, or australoid—particularly when race was a far more important variable than it is today, and when the notion of "pure races" was still more or less viable. But that scenario would have proto-Africans, proto-Asians, proto-Europeans, and proto-Melanesians all coexisting as distinct populations some 15,000 to 35,000 years ago in one tiny corner of the world during the height of the last glacial. No modern scholar could seriously entertain such a view, but its rejection explains neither the degree of cranial and facial variation present nor the apparent affinities with diverse modern populations.

IMPLICATIONS

It appears from all this that existing racial lineages are much too young to provide support for either the Garden of Eden or regional continuity models. Our test sample seems racially to be a curiously mixed "population." Why not take that result straight? Why not accept that it is, by modern standards, overly mixed, and then ask how it might have come to be that way? To ask that question is to answer it. Mixing is the result of moving, and the most obvious motive force is something we all know about, and yet have been curiously unwilling to bring into the equation—the repeated glacial cycles of the past few million years. If these are brought into the discussion at all, it is to provide a method of isolating some population—western European Neandertals being the classic example—so that they could in effect speciate out of our consciousness and consideration. But surely the major effect of glacial movements would have been to quite literally bring people together, not to keep them apart. Glaciers advance and retreat, people retreat and advance. Sea levels fall and rise, land is exposed and covered, people come and go. Climatic zones shift; people follow their traditional ecologies. All this made human movement and, concomitantly, extensive gene flow a periodic *worldwide* phenomenon. Each major glacial movement would thus have produced a significant approach towards panmixia within *Homo*. As the climate settled down, so did human populations—and regional (racial) differentiation would have commenced anew. In this scenario, then, the situation with the

European-area fossils just discussed is simply an example of one of those obliterations of regionality. We, on the other hand, would appear to be living in one of those episodes of regional differentiation resulting from a more stable climatic regime and reduced impetus for movement. The major features of this scenario relevant to the issues at hand are, first, that most of the regional (racial) differentiation which has resulted in the varieties of humans populating the Earth today has developed since the beginning of the last glacial retreat some 15,000–20,000 years ago, and second, a decoupling of speciation and raciation within our genus. The racial lineages of *Homo sapiens* then became much younger than the species itself, even if we add the qualifier "anatomically modern" to the species name.

One obvious objection presents itself here. The above scenario would require the human face and cranium to be remarkably plastic with respect, one supposes, to adaptation to local conditions—and so one has to wonder if in fact so much change can occur so rapidly, and what direct evidence could be put forward in support of the position.

Howells again points the way. He noted, in his various interpopulational comparisons, some perhaps unexpected similarities; in particular, that

> Half the Peruvians join Europe, and half the Moriori go with American Indians, transgressions which also appear in later analyses herein. (Howells 1989:33)
>
> American Indians and Polynesians equivocate, however; Arikara and Moriori are more similar to each other than each is to its own coregionals. (Howells 1989:37)

I obtained similar results. The Moriori (from the Chatham Islands east of New Zealand) were almost certainly of Maori derivation well within the last millennium, and they are, in my analyses, substantially more similar to the Maori than to any other of the 28 populations sampled.[1] But they are also as similar to the Arikara as the latter are to the Santa Cruz population, and about as similar to Santa Cruz and the Tierra del Fuegans as the latter two are to one another.

The most straightforward interpretation of these findings is that a substantial amount of change among the Moriori lineage has been convergent on that occurring among some Amerind lineages, especially that leading to the Arikara. That last amount can be estimated by noting that Maori distances to Arikara and Santa Cruz are, respectively, 0.55 and 0.48, while the corresponding Moriori distances are 0.42 and 0.49 (all of these figures are, again, mean size-corrected z-score differences). Thus 0.14 unit of the change along the Moriori lineage (which is most of it) is convergent on the Arikara. But the Arikara lineage itself has only accumulated about 0.25 unit of change since it separated from Santa Cruz. Thus we can see that the Moriori lineage, in less than a millenioum, accumulated a bit more than half the amount of change that the Arikara did over the at least seven millennia since their separation from the Santa Cruz line. We can then extend this argument one step further by noting that the differences among the three

Amerind groups (0.42, 0.48, 0.5) are about 60% of the rate-adjusted mean among the major human populational lineages. So here no more than 11,000 years has been enough time to accumulate more than one-half the average amount of change characterizing the oldest human lineages.

The Moriori example also reminds us that we would expect the rate of adaptive change to be negatively correlated with the amount of time elapsed since the separation of the two lineages and the beginnings of independent adaptations along them. All this is consistent with the scenario just sketched, with the bottom line being that there seems to be no good reason to argue that the degree of morphological variation measured among recent populations must have taken tens or hundreds of thousands of years more to produce. True, it might have, but that was the point of beginning the argument with the Upper Paleolithic fossil material. It hints strongly at what did happen; the recent material, in the context of known dates, confirms that it could have happened that way.

THE "NEANDERTAL PROBLEM"

Finally, in the fossil realm, it might be interesting, though hardly necessary for the scenario being developed here, to consider what light these exercises shed on the "place of the Neandertals." In the current consensus, as already noted, the neandertals are seen as part of a lineage separate from that to which all anatomically modern *Homo sapiens,* from Qafzeh onward, belong. The basic reason for this is that they are judged as "too different," a judgment only rarely supported by quantitative data. Now it is true that, using the metric described above, one does find the morphological distances between the neandertals and ourselves to be substantially greater (actually, they are, on the average, about twice as large) than those separating most pairs of extant human populations (Table 27.2). But while true, it may not be especially germane in light of the fact that it tends to ignore both the amounts of time available to produce the observed differentiation and the range of morphological distances separating extant human populations. First, with the exception of Shanidar, all of the individuals listed above Qafzeh in Table 27.2 (the accepted neandertals, Skhūl 5 and Irhoud 1, Steinheim, Petralona, Kabwe) are much closer to Santa Cruz than to any of the other four (mean = 0.4 s.d.).

When I extended the survey to look for those individual modern humans most similar to a sample of three neandertals (Ferrassie, Monte Circeo, La Chapelle), the top 25 consisted of seven Moriori (out of 57), six Santa Cruz (of 51), four Tierra del Fuegans (of 24), four South Australians (of 52), three Arikara (of 42), and one Norse (of 55). That the Moriori contribute so strongly here serves to reinforce the point made above about how rapidly morphological change can occur in a human population, especially given that none of the Maori comes close to making the cut of 25.

Finally, it is quite possible to exceed neandertal-modern distances within the modern sample, without appealing to exceptional individuals. That this point has not really been made previously is likely due to the fact that the largest modern differences do not involve "more primitive" populations (i.e., ones more like the neandertals) but in fact are between the Buriats and Teita, where the mean pairwise non-rate-corrected distance between 54 Buriats and 32 Teita is 1.50 s.d. Compare this with means of 1.26, 1.31, and 1.35 between Ferrassie and, respectively, 24 Tierra del Fuegans, 57 Moriori, and 51 Santa Cruz. These figures make it difficult to see why the neandertals cannot be regarded as just another regional variant (race) of *Homo sapiens* and, perhaps, also raise the question of whether the term "anatomically modern *Homo sapiens*" should be retained at all. Is there, one has to wonder, any better reason for excluding the neandertals from anatomically modern *Homo sapiens* than there would be for doing the same to the Buriats?

OTHER LINES OF EVIDENCE

The one saving grace for all efforts at reconstructing history is that there is in fact only a single, real history to reconstruct, and that, ultimately, all the comparative data will be found to be consistent with that history. The scenario proposed here then becomes a hypothesis to be tested in terms of its ability to explain various lines of evidence. The discussion to this point has focused on the morphological data, and the fit between data and hypothesis seems excellent. For the scenario to survive, however, the fit must continue to remain excellent as it is tested in terms of other lines of evidence, and it therefore might be worthwhile to consider in some depth one area where most current practitioners would see a distinct lack of fit—historical linguistics.

PROBLEMS IN CONTEMPORARY HISTORICAL LINGUISTICS

It is my conviction and contention that this perceived lack of fit derives from the linguists and their perspectives, not from the actual data. This conclusion is illustrated by the reactions of many linguists to the proposal by Greenberg (1987) of an Amerind language phylum including all Native American languages with the exception of those belonging to Na-Dene and Eskimo-Aleut. First, there is an extremely conservative consensus among most linguists that relationships among languages that diverged more than perhaps 7,000–8,000 years ago are, using currently available data and methodology, unknowable, and, by extension, that

Greenberg's Amerind hypothesis is inherently nonsensical. It is not difficult to see that they must be wrong. All one need do is sit down with, for example, Buck's *A Dictionary of Selected Synonyms in the Principal Indo-European Languages* (1949), a basic word list, and some independent knowledge of two or more languages representing distinct Indo-European groups. I used English and Croatian, representing, respectively, its Germanic and Slavic branches. Asking then what proportion of the words in modern Croatian appear, simply by inspection (but allowing for some phonetic and semantic drift), to be cognate with the reconstructed Proto-Indo-European (PIE) form (or, where that is unavailable, with the English word), one gets a minimum figure of about 60%. For example: snow, snjeg, *sneig^{w}h; many, mnogo, *monogho; blood, krv, *kru; tree/wood, drvo, *dru; earth, zemlja, *ghem. Similar results were obtained using native speakers of Spanish and Bengali, and for Armenian and Albanian using Décsy's *The Indo-European Protolanguage: A Computational Reconstruction* (1991).

One might, therefore, then take 50% as a conservative, reasonably representative estimate for the probability of a given PIE root surviving in recognizable phonetic and semantic form in a given, extant Indo-European language. It would then be a *mighty poor linguist indeed* who could not see the similarities resulting from this retention rate in the various Indo-European languages, and infer from these the past existence of PIE. *A little more sophistication,* and our linguist ought to be able to cope with substantially lower retention rates and, thus, recognize substantially older relationships. This simple exercise suggests that there can be no doubt that the current general consensus among most linguists that relationships among languages older than about 7,000 years are unknowable is unrealistically and unreasonably pessimistic and conservative—and that this judgment is not even remotely a close call.

These data would imply a cognacy loss (again—allowing for some semantic, as well as phonetic, change) of about 9% per millennium along a lineage, implying that even at a time depth of 11,000–12,000 years, which likely gets us to proto-Amerind (Haynes 1992), one might retain perhaps 30–35% phonetic/semantic cognacy between it and an actual Amerind language. The problem, of course, is that we do not have a reconstructed proto-Amerind, and the actual cognacy rate between two randomly chosen Amerind languages whose last common ancestor was proto-Amerind would be, on the average, $(0.30–0.35)^2$, or about 10%, well within the area of chance similarity under these relatively loose constraints. On the other hand, if, instead of doing binary comparisons, you looked at ten such languages at a time, three, on the average, would have retained a particular cognate, and your chances of recognizing it, and, therefore, relationships among them, would be correspondingly increased. But the enormous number and diversity of known Amerind languages means that we will be looking for about thirty in a sample of one hundred rather than three in ten, and that makes finding them, if they are actually there, virtually a certainty. This is the strength of Greenberg's approach, and if proto-Amerind is actually something like

11,000–12,000 years old, then there could be no question that he could have *easily* seen it in his data.

At this point, however, frustration sets in. The problem is that linguists have in fact long recognized strong lexical and grammatical similarities among the languages in Greenberg's proposed Amerind. The data are not the issue. What is at issue is the consensus insistence that this may not be due to common ancestry, and the defense of this (to me, inherently bizarre) position has resulted in some of the strangest arguments present in the relevant literature.

First, Greenberg's critics claim that he has gone about it backwards. That is, instead of proceeding from lower to higher level groupings, using the comparative method as his guide, he has simply followed the approach briefly described above, increasing the probability of finding cognate retentions, and, therefore, isolating long-range relationships, by sampling as many languages as possible. The response here has to be that Greenberg has been doing precisely what was done to achieve the recognition of every other major language family, including Indo-European itself. There is no family that was recognized through the application of the comparative method as such, although the recognition of different reflexes as deriving from some ancestral form implies an ability to make choices as to which sound changes are likely, and which are not; in other words, some portions of the comparative method are implicit in approaches such as Greenberg's. Thus any suggestion that Greenberg, or anyone else, should, or even could, go about it in any other way strikes me as truly bizarre, and I use the term advisedly. What else is one to call a policy of rejecting the only approach that has ever worked?

I would argue that the situation is exactly the same in the world of organisms. It has always been higher-level groupings that were first recognized. There has been no instance where, for example, an order of mammals was recognized by starting with species, grouping them into genera, genera into subfamilies, etc., up the Linnaean hierarchy, until we finally arrive at the order. It has never happened in that direction, and there is every reason to believe that it could not. Indeed, the very idea of even trying to go that way would strike biologists as—well— bizarre. They would see a large number of bodies covered with "feathers." Granted, these feathers wouldn't all have exactly the same structure, but they would seem pretty similar in the context of the large variety of other body coverings in sight. Replace "feathers" with "hair," and we bring another large number of bodies into a second grouping. We would then note that those with feathers all have bills, lay hard-shelled eggs, have forelimbs that take the form of wings which enable most of them to fly, and so on, thus rapidly arriving at the category "birds." But now someone comes along and insists that I am going about this the "wrong way." I, in effect, must ignore these more general features in favor of ones which define much smaller numbers of bodies, and then proceed stepwise until I find them coalescing into "birds." But, I say, I have already discovered that category, and it sure was easier than it's going to be doing it your way. He responds that ease is not a relevant criterion, asserting that there is a wrong way and a right way, and that I am going about it the wrong way, and the fact that it works is irrelevant. At which point I decide that this is a badly confused fellow who really doesn't have anything to tell me, and I go on my own way.

But that is the minor problem. Far more important is the already-mentioned fact that the marked similarities among Amerindian languages that led to Greenberg's proposal of the existence of proto-Amerind have been recognized for a long time. It is thus interesting and instructive to see what some of Greenberg's critics have made of these similarities (e.g., first person singular /n/, second person singular /m/). Two of them, Lyle Campbell and Terence Kaufman, have called these similarities "pan-Americanisms" (1980). Now one would think that these generally accepted similarities were prima facie evidence for a proto-Amerind; that is, pan-American = proto-Amerind, and it becomes the null hypothesis to be tested. In that view, then, what could pan-Americanisms be other than cognates by any other name? In other words, common ancestry is always the simplest explanation (in Occam's sense, requiring fewer events) and is to be rejected only when the evidence requires it. But most American linguists simply do not look at it that way. Bright (1984:25) is representative:

> I would not be opposed to a hypothesis that the majority of recognized genetic families of American Indian languages must have had relationships of multilingualism and intense linguistic diffusion during a remote period of time, perhaps in the age when they were crossing the Bering Strait from Siberia to Alaska. We can imagine that the so-called pan-Americanisms in American Indian languages, which have attracted so much attention from "super-groupers" like Greenberg, may have originated in that period.

A similar point was made by Levine (1977:11) in his doctoral dissertation on the position of Haida:

> There are thus signs of a developing consensus within Na-Dene studies that a proto-Athapaskan-Eyak-Tlingit did indeed exist, and that extremely prolonged contact between this language or its daughter languages and the ancestors of modern Haida is entirely adequate as an explanation of resemblances between Haida and the revised Na-Dene group.

Levine's removal of Haida from Na-Dene is then quoted approvingly by Greenberg's critics as a specific example of the failure of his approach.

Note, however, that both Bright and Levine go out of their way to highlight the inherent flaw in their arguments. Bright writes of "multilingualism and *intense* linguistic diffusion," and Levine of "*extremely prolonged* contact" (emphasis added). Why the use of these adjectives? Because, presumably, it isn't a small number of similarities that link Haida with the other Na-Dene languages, and the many Amerindian languages with one another; those similarities must be many and obvious, as Greenberg (1987) and Ruhlen (1987), and their few supporters, keep emphasizing, seemingly to no avail.

Consider the situation with Haida and the other Na-Dene units (Tlingit, Eyak, and the many Athapascan languages). Levine acknowledges what was obvious to Boas in 1894, and Sapir in 1915—that there are strong similarities between the two. That leaves us with two possible explanations: common ancestry or diffusion. Levine chooses the latter for Haida and Na-Dene, as does Bright for Amerind. Why? After all, for Haida and Na-Dene the diffusion scenario requires three events/processes to explain the similarities: (1) differentiation into Haida (presumably not from a proto-Na-Dene), (2) differentiation into proto-Tlingit-Eyak-Athapascan, and (3) subsequent diffusion between the two. The common ancestry explanation requires only one: the development of proto-Na-Dene. Bright's explanation for the apparent Amerind would require many more events, against only a single one for a proto-Amerind. So I look at this and again wonder what is going on. Why deny the relevance of Occam's razor? One would think that question had been settled centuries ago. So, trying to act in good faith, I wonder if I am missing something. Perhaps there is evidence that diffusion is as likely an explanation for observed similarities as is common ancestry? But, no, that couldn't be; if it were, then there could be no historical linguistics as we have come to know it.

It thus seems to me that an untenable position is being defended here, a conclusion perhaps best illustrated by reference to an exchange in the *American Anthropologist* between Witkowski and Brown, on the one hand, and two of Greenberg's severest critics, Campbell and Kaufman, on the other, concerning the relationship, or lack thereof, between two language groups of southern Mexico and Central America: Mayan and Mixe-Zoquean. Witkowski and Brown (1978) originally presented 62 putative cognate sets linking the two groups. Campbell and Kaufman (1980), in their first rejoinder, rejected all of them for various reasons, including the assertion that fourteen of them were pan-Americanisms. Witkowski and Brown responded (1981), leading Campbell and Kaufman to the following, which has to be one of the strangest statements in the recent anthropology/linguistics literature (1983:365–366):

> We do not take at all kindly to WB's [Witkowski and Brown's] (1981:908) caricature of our reservations concerning widespread forms, called Pan-Americanisms by some, for such reservation is a standard criterion of distant genetic research in the Americas (Campbell 1973). We in no way appealed to or necessarily believe in the hypothesis attributed to us of "a gigantic Proto-Amerind phylum" (WB 1981:908), rather we made reference to the legitimate practice in the investigation of remote relationships in the Americas of avoiding widespread forms. It is generally recognized that certain forms recur with similar sound and meaning in very many American Indian languages (cf. Swadesh 1954). Acknowledgement of the widespread forms presupposes no particular explanation; while some may feel

that these support some far-flung genetic connection (cf. Swadesh 1954, 1967; Greenberg 1960; etc.), it is possible that some widely shared similarities may be due to onomatopoeia, sound symbolism, perhaps diffusion, accident, or other undetermined factors.

Campbell repeated the last part of this message in a letter published in *Scientific American* (1993:12):

> Greenberg's methods have been disproved. Similarities between languages can be the result of chance, borrowing, onomatopoeia, sound symbolism and other causes. For a proposal of remote family relationship to be plausible, one must eliminate the other possible explanations.

But, as already noted, it is not the data that are the issue here. Even Campbell and Kaufman, in the quote just given, tell us that: "It is generally recognized that certain forms recur with similar sound and meaning in very many American Indian languages." Again we see the going out of one's way to emphasize the strength of the similarities—"very many." It is thus a question of how best to interpret the data in default of an explicit conceptual framework, for data have no intrinsic meaning (Clark 1993b). So what is going to be the theory—the proposed answers to the "why" questions—here? Presumably one that we know works in other evolutionary realms; in other words, I am put in the position of assessing theory in terms of practice, and there the "theory" of Campbell and Kaufman fails miserably.

One has to be struck by how much illogic Campbell and Kaufman manage in so few words. They write, apparently oblivious to the import of their words, of "the legitimate practice in the investigation of remote relationships in the Americas of avoiding widespread forms." How remarkably convenient—if you don't like the conclusion, then just rid yourself of the only data which could possibly lead to it. What, one wonders, would they say about a zoologist who wrote of "the legitimate practice in the investigation of remote relationships among organisms of avoiding certain widespread forms such as the presence of feathers, hair, tetrapod limbs, or amniotic eggs"? Is there any conceivable way of recognizing "remote relationships" other than by documenting "widespread forms"? Could there be? I think not. Then they tell us that these "widely shared similarities may be due to onomatopoeia, sound symbolism, perhaps diffusion, accident, or other undetermined factors." Well, yes, so they might—indeed, we can be quite certain that all of these factors, including the "undetermined" ones, will have been involved, to some extent, in producing linguistic "similarities."

But the critical point here is that, in the absence of written records, there is no possible way of isolating and identifying those similarities resulting from "onomatopoeia, etc.," *until* one has developed a phylogenetic tree linking the languages under study. What that tree cannot explain, and there will always be a good deal that it cannot, is then to be looked at for evidence of onomatopoeia.

But it is obviously and inherently true that *any* similarity could be "explained" by appealing to these other factors. It is just as obviously true that this is not the case for phylogenetic explanations. The latter are falsifiable; the former are not—or, more fairly, they are not until we have a tree of relationships. That this remains an issue for most linguists is perhaps the most frustrating aspect of the controversy surrounding Amerind, for there can be no resolution of the issues involved until there is agreement on the rules of the game—and I have to say that the prospect of having to begin, in 1995, with an argument about the relevance of Occam's razor does not augur well for the success of the enterprise.

Now one might fairly ask—so what? What difference does the reality or lack thereof of a proto-Amerind 11,000–12,000 years ago make to the reality or lack thereof of the scenario being proposed in this chapter? Or, at a more personal level, why does it seem to mean so much to me? In mulling that one over, I have come to see that the answer probably has more to say about the likelihood of the correctness of the scenario than any other single accessible fact. The reason is straightforward.

The anti-Amerind sentiment is, after all, at least partly fueled by the bias or preconception of many linguists that 11,000 years is simply not enough time to have produced, starting from a single language, the known variety of Amerind languages. And if that is in fact the case, then something less than twice that time span certainly would not have been nearly sufficient to produce the known human linguistic diversity starting from one or a small number of weakly differentiated languages; our scenario would be seriously, probably fatally, wounded. The point here is that while the time scale of our scenario might at first glance appear to be too short to produce the observed morphological variation between populations in our species, it was easy, once the possibility had been raised, to show that demonstrable rates of change were in fact sufficient to have done so. But there we had the advantage of natural selection driving adaptive change. No one is contemplating any such involvement in linguistic differentiation. It would not be reasonable to assume that if a given amount of language change had taken 10,000 years to produce, we could get that much more again in, say, 2,000 years. This is why the Amerind case is critical.

But what if there actually had been a proto-Amerind around 11,000 years ago? Then we could reason as follows. Within the Americas, and excluding Na-Dene and Eskimo-Aleut, there are probably at least one hundred language groups (families) with an internal diversity comparable to other units such as Indo-European, Austronesian, and Uralic-Yukaghir (Greenberg 1987; Ruhlen 1987— other workers such as Campbell and Mithun [1979] would recognize many more).

If it takes something on the order of 6,000–7,000 years for family-level differences to develop, then in the Americas we can go from the direct ancestors of those families to proto-Amerind in perhaps 5,000 years. This might then call into question a priori judgments as to how long it takes to produce differences

beyond the level of family relationships—even among those linguists sympathetic to the possibility of discovering such higher-level relationships. Thus when I
take note of the fact that Greenberg recognizes only a further fifteen or so
suprafamilial groupings roughly equal in internal diversity (and, by implication,
age) to Amerind, I start thinking that we might not have to go much further back
in time to bring most or all of those fifteen to the level of interpopulational
diversity implied by our scenario.

I do, however, append one caveat here. While the search for proto-Indo-
European, or even a proto-Amerind "homeland," is a reasonable one (i.e., there
probably was a "homeland" in the sense of PIE being spoken by a real population
narrowly circumscribed in space and time), any such notion for Nostratic, Eurasian, Na-Dene/Sino-Tibetan/Caucasian, or any other proposed high-level grouping may be much less realistic. In other words, the scenario proposed here is
predicated on a notion like that proposed by Trubetskoy (1939) for Indo-
European:

> There is, then, no powerful ground for the assumption of an unitary Indogerman
> protolanguage, from which the individual Indogerman language groups would de-
> rive. It is just as plausible that the ancestors of Indogerman language groups were
> originally quite dissimilar, and that through continuing contact, mutual influence,
> and word borrowing became significantly closer to each other, without however
> going so far as to become identical. (Trubetskoy, quoted in Renfrew 1987:108)

This idea would in fact apply to some or all of the various recently proposed
higher-level groupings, though not, as just noted, to PIE itself. In other words,
what is being envisioned here are the consequences of the major movements of
human populations that glacial cycles would have necessitated—and some of
these must have been, in Trubetskoy's words, "continuing contact, mutual influence, and word borrowing" so that they "become significantly closer to one
another, without however going so far as to become identical."

If, thus, we were to inquire into the antiquity of, say, the Mandarin, Xhosa,
and English lineages, the answer would be that they have been distinct from one
another for >10,000 years but almost surely nothing like 25,000 years. The
reason is that the most recent major glacial retreat began some 18,000 years ago,
and this would have caused the last obliteration of linguistic regionality, and
then, soon after, the last episode of regional differentiation would have begun—
and it is that differentiation that has given us the three lineages culminating in
Mandarin, Xhosa, English, and, of course, scores of others.

What of the common ancestor of the three? It seems to me that the scenario
envisioned here makes this an example of a genre I have termed "wrong
questions"—wrong because they imply answers of the wrong form. "Common
ancestor" implies an actual language sharply delimited in space and time, and the
glacial-cycle-induced linguistic mixing can hardly qualify as such. Nor, even
more clearly, is there any point in asking about the linguistic state of affairs prior

to this most recent mish-mash. We can be quite certain that there was an enormous amount of linguistic diversity, almost certainly exceeding that of the recent past, but it is equally certain that evidence as to the nature of that diversity could not have survived. In essence, the linguistic "clock" is reset following each glacial-interglacial oscillation. This inability to monitor linguistic change in "deep time" is, however, more than balanced by the greatly increased amount of information on relationships among known languages implied by the very short time scale over which those known languages have evolved, and the greatly reduced diversity with which that evolution began.

CONCLUDING REMARKS

I have attempted to show that the morphometric data are, and the linguistic data may well be, consistent with the scenario proposed here of glacial-cycle-driven waxings and wanings of interpopulational diversity within our genus. Although there is not enough space in a single, short essay to discuss it adequately, it would appear that the mitochondrial DNA data are also interpretable within this conceptual framework (Clark 1993c; Wills 1993:53–54). But all this is no more than a bare beginning. There are many more specimens of fossil *Homo sapiens* to be investigated within a rate-controlled morphological perspective. Reasonable methods for testing the reality of proposed suprafamilial language groupings need to be developed and linguists must arrive at a consensus as to how to go about doing this. It would then be very useful to generate some approximate lexical distances among these suprafamilial units, thus providing estimates of the time depths which produced them. Finally, the mitochondrial DNA picture needs to be further clarified. I outline here a hypothesis testable through several actual and potential data sets. The potential data sets require considerable development, but mostly we need some good-faith testing. I look forward to such efforts.

ACKNOWLEDGMENTS

I thank W. W. Howells and Chris Stringer for making available the measurements for the modern and fossil specimens, respectively, and the latter and Herbert Kritscher of the Naturhistorisches Museum, Vienna, for allowing me to study the Tierra del Fuegan material in their collections. William S.-Y. Wang wrote a distance program, and Tom Schoenemann one to allow direct access to the data of Howells—both in True Basic. All statistical analyses were carried out on a MacIntosh IIsi and, later, on a LCIII, using Statview. Susan Anton asked a critical question in a Berkeley seminar which identified for me the fundamental issue involved in assigning individuals to their correct populations. Stringer and Tim White provided particularly useful discussion and commentary on the morphology section. I thank Dr. Wang for sparking, encouraging, and greatly informing my interest in the origin and evolution of languages (though he should not be blamed for the various faults of this effort and others of mine in that area). Finally, Cathy and Geoff struggled mightily with various revisions of the text, transmitted by e-mail. I appreciate their efforts.

NOTE

1.   Although Howells includes measurements on twenty Maori male crania on disk, he does not present or discuss these data in his book.

VII

PERSPECTIVES FROM EVOLUTIONARY
EPISTEMOLOGY

Having sampled the range of opinion on modern human origins research from the standpoints of the disciplines most heavily engaged in it, and mindful of need to take a broader intellectual perspective, we also include chapters on modern human origins research from philosopher/zoologist Jane Maienschein and science historian Michael Ruse.

Maienschein points out that, unlike those who explore evolutionary histories of other life forms, where relatively little is at stake, scientists and laymen alike are persistently keen on knowing about our evolutionary history. Where did we come from and how did we get here? Presumably there is only a single, "correct" answer to these questions, but modern human origins researchers are naive to expect definitive resolution in the face of competing claims about evidence and competing views about the credibility of preferred approaches. She goes on to identify some of the more prominent bones of contention in the various modern human origins research domains, in order to make the important point that within these competing epistemological traditions lie various theories—proposed answers to the "why" questions—that are communicated to the cognoscenti and to the lay public by means of different epistemic styles. Given the lack of evidence accepted as decisive by all, different epistemologies result in conclusions that are fully warranted only within the boundaries of a particular conceptual framework. All conclusions are, therefore, conditional on the unstated assumptions that underlie a particular research protocol.

A leading evolutionary epistemologist, Ruse points out that the scientific status of Darwinism has always been a subject for debate, originally as to how it measured up against the powerful, hypothetico-deductive law networks of physics, and later as to how coherent its theories and methods were, and how adequately they could explain particular instances of evolutionary process. Modern Darwinism is conceptualized as fanlike, with a central core of population genetics, and with blade-like extensions radiating out from the core representing the constituent fields of ecology, molecular biology, paleontology, archaeology, systematics, sociobiology, etc., connected to one another in varying degrees as a consequence of overlapping intellectual traditions. Paleoanthropology is an eclectic mix of several of these traditions, more like geography than physics in its logic of inference, and less like Landau's heroic narrative than an immature science still in the process of formation. Evolutionary epistemology can help paleoanthropology make explicit the inferential basis for its knowledge claims and thus provide an antidote to the strict empiricism that plagues human origins research.

28

The One and the Many

Epistemological Reflections on the Modern Human Origins Debates

JANE MAIENSCHEIN

"There is grandeur in this view of life," Darwin (1859) wrote persuasively in the first edition of his *On the Origin of Species,* "with its several powers, having been originally breathed into a few forms or into one; and that, whilst this planet has gone cycling on according to the fixed law of gravity, from so simple a beginning endless forms most beautiful and most wonderful have been, and are being, evolved." Theologians immediately took up the issue of who or what may have done that original breathing that started the whole evolutionary process. Scientists have taken up the other issues about how the process works, and to what effect. The question whether life was initially breathed into one form or into several has not seemed terribly important scientifically. Nor is it. Yet, if we substitute evolutionary processes for the breathing of an implied creator, the issue of whether humans began as one form or several takes on absolutely central importance to many. Scientists and laymen alike are persistently keen on knowing about the historical origins of modern humans. They want to know whether there was one ancestor (or one type, specifically Adam and Eve) or several (perhaps corresponding to different current races).

While we have come much further than Darwin in gathering piles of data and framing a plethora of hypotheses, we have not achieved anything approaching consensus around any one answer. Did modern humans arise once, in one place, and radiate outward from there, replacing other competing forms as they moved onward? Or did modern humans originate in more than one place and later join genetically so that their evolution was more continuous than discontinuous? How many different lines point to modern humans? And how can we know?

Science writer James Shreeve (1995:252) has taken up these questions in his fascinatingly rich exploration *The Neanderthal Enigma.* After interviewing more than 150 scientists and traveling to dozens of distant field sites, collections, and labs, he admits that he seems to have "come away with one hundred and fifty different points of view. Early modern humans appear first in Africa. No, they don't. Or it depends on what you call early. Or how you define modern. Or what you really mean by human. I indexed my notes, and indexed the indices. A city of Post-it notes grew on my office wall, each with a revelation scribbled on it. Arrows of blue chalk sprang up to link brainstorm to brainstorm. But the arrows sprouted question marks. The Post-it notes lost their sticking power and fell to the floor."

This confusion remains one of the most intriguing aspects of scientific study of this issue. Identifying the origin of modern humans is one of the most compelling questions for the public and they want to know *the* answer. For as Shreeve notes, there is a fact of the matter: "The past did, in fact, *happen,* and in only one way." Yet scientists disagree about what that one way was. They disagree not just about the conclusions or about which theory or answer they like best. They also disagree about how to go about knowing the answer. They adopt different paths to knowledge, count different claims as established and legitimate, and point to quite different objectives—or different views of what, precisely, they are trying to know. This is a case where scientists are working in the largest sense on the same broad set of questions. Yet their work differs widely because of their fundamentally divergent epistemologies, or what may fruitfully be called different epistemic styles.

Traditional paleoanthropologists concentrate on bones and teeth. Collect all the relevant fossil skulls, teeth, and bones. Observe, measure, compare, and construct a lineage or phylogenetic tree of presumed morphological characteristics to show their evolutionary relationships (which assumes, of course, that there *are* evolutionary relationships). Deciding what counts as "relevant" requires some judgments, obviously, but researchers in each tradition will agree on the basic approach. The ideal result is, rather like Othniel Marsh's famous lineage of increasingly larger horses for New York's American Museum of Natural History, a compelling sequence showing the clear trend toward the modern.

Unfortunately for this approach, researchers often have only a few samples (few relative to all the probably different samples that ever existed) and must interpret quite liberally. As a recent news story in *Science* puts it, "When paleontologists disagree, they rarely have the luxury of doing another experiment to see who's right. Instead, they must resort to a more chancy and time-consuming enterprise: returning to the field and unearthing more fossils to prove their point" (Culotta 1995:1851). Computer modeling can help with recording increasing amounts of data in the form of measurements, for example, but gaining anything that can be counted as "knowledge" from the sequence requires expert interpretation, strong persuasive powers, and probably significant scientific status if one wants to challenge existing views. Some specimens are decayed, poorly preserved, sloppily collected, and otherwise not "pure" or very useful. Individual differences also complicate the story, since some of what may look like distinctive and significant differences may simply result from odd characteristics that are essentially outliers from the norm. So resolving disputes about early lineages takes researchers back to the field for more digging, more fossils, and hence more of what they consider useful evidence.

Others look in other directions for evidence. The second traditional area focuses on behavioral or cultural phenomena. What evidence can be gotten from bits of shards, flint, and other artifacts found along with the fossil bones and teeth? What do they suggest about the life of the populations involved, and about the connections or lack of communication between diverse populations? Again, a good deal of interpretation is required to move from fragments to stories and constructed lineages. Does the apparent purposeful placement of parts of skeletons indicate a burial site? Would such a site indicate ritual and conscious respectful treatment of the dead? Does the appearance of apparently ornamental spear tips reveal sophisticated social interactions and perhaps language? Do cave paintings claim territory, represent spiritual interests, or serve as representations of experiences? Each datum offers much room for possible alternative interpretations, but within this tradition there is no doubt that it is this kind of data to which we should ultimately turn to establish relationships among different populations.

It might appear that, as scientists, all researchers would want to accept whatever evidence they can get about these past times. And some do, of course. But most favor one type or another. They have to think that the data offered is sufficiently grounded and sufficiently warranted to count as evidence of something before they consider it worthy of discussion. To the first group, the jaw bone in hand represents solid, reliable, undeniable "fact" while claims about burial sites or artifact production or hunting behavior seem highly interpreted and based on grounds too weak to be accepted at the same level of certainty. Thus, they insist, hard facts—by which they mean fossils—must prevail.

The second group disagrees. To them, the bones are not so solid. Indeed, most are fragments and all are subject to disturbance and distortion. We have only to recall the Piltdown adventure to see how easily susceptible researchers can be manipulated into believing that they have actually found just what it was they had been looking for. For these researchers, it is the cultural evidence that is clearly more reliable, because they believe that we can trace clear patterns of behavior. The same patterns of any particular behavior represented over wide areas must reveal continuous culture, and therefore continuous biology. For them, such evidence about human relations is necessarily far stronger than some easily scattered or altered physical tidbits.

So some see the primacy of fossils as the source of useful knowledge. Some look to a combination of fossils and behavior. And others insist on the primacy of behavior to establish claims about human relationships. These traditions have settled into their separate niches and thus coexist more or less peacefully within anthropology departments, and they vie with each other for support and prestige and priority with respect to particular issues. These traditions all share in such discussions as which techniques work better and are more reliable for dating, for example, recognizing that the answer may differ for dating such different phenomena as geological strata or bones or seeds. As most of the chapters in this volume show clearly, which tradition one works within correlates with a very high degree of probability to one's background and training, reinforced by continuing networks of professional contacts with researchers and ideas that share the same assumptions.

Then along came mitochondria. As Lynn Margulis has convincingly established, this little powerhouse within the cell's cytoplasm almost surely evolved symbiotically with the cell. Presumably a small organism (in the form of the mitochondrial ancestor) found it advantageous, in the sense that natural selection preserved the trait, to team up with a larger cell. The mitochondrial body came with a strand of its own DNA, which has slightly different traits than the more familiar DNA associated with nuclear chromosomes. Since the cytoplasmic material is inherited only from the mother, while the father contributes only nuclear material, all mitochondrial DNA (mtDNA) is passed on exclusively through the female line. This makes it simpler but also less complete as a record of past changes. Other characteristics of this mtDNA that make it particularly appealing for research into ancestral lineages include its relative linear stability (it does not recombine to the same extent that nuclear DNA does) and its apparently higher rate of mutation. The stability means that the DNA chain reflects more neatly the actual evolutionary past accumulation of adaptations, while the increased mutations mean that there is more to look for to establish differences in different organisms. Study of mtDNA has therefore produced great excitement in the community of scholars studying modern human origins.

Yet once again, what might look like a lovely new approach to an old problem has received mixed reviews. This new approach is valuable only for those who value it. Not everyone does, and critics point to a variety of problems. They all amount to concerns that the mtDNA is too far removed from actual physical attributes and from the phenotypic and cultural characters they are associated with. Too much interpretation is thought to be involved as researchers take the mtDNA, remove it, track it, plot it, compare it, and otherwise engage in manipulations and laboratory techniques far removed from actual, living, breathing organisms. These critics are not neat hereditary determinists, obviously, who feel that phenotype is just a small and relatively insignificant step from genotype.

How one feels about mtDNA evidence, then, all comes down to whether one regards it as evidence—and of what. Is all this information and fact really evidence of anything that matters for current considerations about modern human origins? Perhaps if this approach had not from its earliest uses been linked to one particular theoretical interpretation of modern human origins, which was itself associated with one research tradition, the dispute about its usefulness and validity might have developed differently. But it did not. Mitochondria entered the anthropological world most noisily in the company of Eve—mitochondrial Eve, presumably the mother of us all. It is this claim about a single origin and replacement that has made the approach so controversial, especially to those who hold another hypothesis associated with a different tradition.

Within these competing epistemological traditions lie various theories. With respect to modern human origins, some stress one line leading to humans while others stress multiple lines. Since there is undeniable empirical evidence that more than one line of beings that looked remarkably similar to humans existed in the past, the key question has become whether one line was sufficiently dominant that it, in effect, replaced all the others. Thus emerges the Out of Africa hypothesis, which emphasizes origins of a line in Africa that radiated outward and replaced all other pre-human lineages. While there were once arguments for a single origin in Asia, among those favoring one origin Africa has gained priority as the probable center. Those who see origins in Asia, therefore, point to multiple lines to modern humans.

Interestingly, physical anthropologists differ about which theory they prefer. So do cultural anthropologists. Thus, the competing theoretical frameworks cross epistemological lines. Other factors also enter discussions of theory. For example, there is a history of insisting that different human races have biological significance. Obviously, this view appeals to those with political and social reasons for wishing to establish that different humans are essentially and biologically different. Yet while by definition those holding this view see significant racial differences—either physically or culturally—not all are racist in the political sense which insists that particular races are, as a result of the biological differences, inferior in some important and relevant way.

William Howells (1944:220) acknowledged the tendency among anthropologists as well as the public to see races, and in his popular *Mankind So Far* he noted that

> The answer to this, and to race, would seem to lie in the "origin of species," which I have already described. Evolution is a matter of constant change. . . . Perhaps the essential thing is whether animals or men breed as a single group. If they do, they will share their bodily features, and all their changes, and remain the same race. If they do not, they will not undergo the same changes, and they will become different. This is how two or more new species are formed out of one old one. At an advanced stage in the process there will be different adaptations, but the earliest differences to appear are mostly random and of little biological significance, and these are the kind of differences which distinguish the races of man.

For Howells, the evolutionary account seemed obvious, and the breeding criterion for races (or species) equally certain. Yet, how can we establish that different lines of past beings could interbreed? What will count as evidence for the claim that they could? Will we rely on morphological comparisons and assume that similar-looking folks will like each other enough to breed? Or will we look to cultural differences and similarities to establish patterns of social and physical intercourse? Or, perhaps, we will let DNA decide the issue.

This is precisely the set of issues surrounding the fate of the neandertals. Where do the neandertals fit: on the yellow brick road to us, or off on a side track which dead-ended when another line prevailed? Did the neandertals interbreed with the line which seems most likely to have given rise to us—or not? And how can we know? For a long time the former view that neandertals were our ancestors dominated, but the view was not unanimous. As Howells put it in 1944, "The niceties of neanderthal anatomy, with which we have been grappling, are hardly a subject for bathtub reading, though indeed I have barely scratched the surface. The neanderthals are of incomparable importance to the question of our own past, even though anthropologists are to a great extent still up in the air over their significance. The problem of their origin remains obscure." Acknowledging what looked like a family resemblance, Howells asked whether the neandertals "resemble us for the reason that the two of us are closely connected—that he represents a sort of last step in evolution before our own type grew out of his? If that is so, why is our head, which is no bigger than his, so different in form? Or does he resemble us because both species have responded to the same evolutionary tendencies to have large brains and shortened jaws, even though we and he have led separate developments for ages? In sum, I am not much inclined to see neanderthal blood in these perfectly good specimens of *Homo sapiens*, and when I see how other people arrive at that idea by totally different routes I am still less inclined than before" (Howells 1944:173, 188). He preferred the view that races are insufficiently established and thus that there is continuity among all the existing human forms. Yet they derive from different lines than the neandertals.

But the view that the neandertals are *not* in line on the way to us, which suggests a different lineage, raises other questions. If the neandertals were around in what looks like the same place at what looks like the same time, why did they *not* breed with the line that evidently *did* give rise to modern humans? Was there insufficient similarity or too little attraction to interbreed? This seems implausible. Yet no other clear answer has emerged. So, once again, the debates fall back into traditional assumptions about what we should count as evidence and how to go about drawing conclusions when we are faced with uncertain situations.

We come back to the difference of epistemological norms, and what may be usefully called different epistemic styles. They exist. We see how they play out in this central anthropological question about modern human origins. It is not the case that different theories simply drive researchers to different methodologies or approaches, though that may happen of course. Rather, there are basic and deep underlying differences in epistemological assumptions.

But why? Is not science supposed to discover the one best answer? Is not the scientific method supposed to allow us to get beyond such differences and resolve disputes by testing and falsifying and otherwise determining which hypothesis is best? Or maybe even which one is true and tells us how nature really works, according to some views of the nature of science.

Perhaps the persistence of competing views for over a century suggests a failure in the robustness and innovation of the scientific approach to the study of modern human origins. But, no, not necessarily. As Willard van Orman Quine and Pierre Duhem have explained, any view of science is naive which expects definitive resolution in the face of all competing claims about evidence and competing views about preferred approaches (Quine 1981:70–72). There is no reason to think that the research is not scientific or that it is bad science simply because it has failed to reach a resolution. In other fields where there is little public interest in the answers, researchers would likely have turned to other questions more amenable to solution with the available methods and waited for new approaches to settle the debates. But this issue matters too much to too many people, and as Shreeve points out, people want to know the fact of the matter. Given the lack of evidence accepted as decisive to all sides, this is a special challenge indeed—and it causes special problems.

One problem is that there are different virtues and different values in science, and that the different traditions are weighing these differently. For example, empirical adequacy, simplicity, coherence, consistency, robustness (or the ability to explain more different phenomena), or aesthetic criteria (creativity, unique-ness, or beauty) may seem more important in a theory. Getting more data, getting it faster, having more different kinds of data, or getting "better" (cleaner, more reliable) data may serve as different criteria for "goodness" of methodology, as may a host of other considerations. Being creative and expansive beyond the basic data may be preferred in one camp, and may provide exciting stimulation for new research. Being careful and failing to rush ahead beyond the securely grounded data-in-hand may seem preferable to others. Even though everyone may agree that these are virtues, the virtues are weighted and combine in differ-ent ways.

The bones-and-teeth fossil followers want data in the form of morphological lineages which show clear progressions and which reveal predictable evolution-ary trends. They want relatively solid foundations for their interpretations. Cul-tural anthropologists look for signs of cultural exchange, which will require different types of interpretation. Molecular biologists will take the mtDNA plot-tings as strongly signifying relationships, and they will consider this solid data. Complex theories based on cultural artifacts and conjectures will seem unaccept-ably fuzzy, humanistic, and perhaps vague to them.

So, different epistemologies exist, along with different theories and different available methodologies. They coexist peacefully until they ask the same ques-tions, or questions close enough that they intersect. Then they vie for priority. Is there any way to step outside the competing views to establish what *really* should count as knowledge? Or, how can we establish claims not only that we know best about the empirical data out there, but also that we know that we know because we have somehow established that our view is best? Is there any way to produce knowledge that is objective and warranted independently of any of these epis-temic styles?

No, there is not. Not really. Asking which view is right would involve deter-mining which epistemic style is the right one. But that begs the question how we *know* it is the right one. We would have to presuppose some set of epistemic values in order to determine which epistemic style is right, and we have no way to do that. Does this, then, mean that all is lost, that science is a hopeless muddle of babbling voices, and that all scientific contenders are equally good?

No, decisively not. For as logical positivists and social relativists alike would have it, scientific conventions (concerning epistemic style) may be unavoidable and fundamental, but some are better than others. Rudolf Carnap and Thomas Kuhn alike show us that adopting a particular framework or paradigm—that is, a set of epistemological assumptions—can often work better to accomplish the goals of the community (Creath 1995–1996; Quine 1981:70–72; Reich 1991). Appealing to community goals will not settle the issues for all times and all places since the community goals can change and different conventions might work equally well. And under the most favorable circumstances such a strategy can serve only to determine which of several competing conventions is best for achieving some objective that all share.

This is a pragmatic matter. It is not "truth" or logic that will determine which view is best, but practical considerations: which best solves the problems or provides answers to the questions that the community as a whole cares most about? Which offers the most robust set of predictions or addresses the widest range of diverse issues? Such practical considerations will, in general, prevail. Yet, obviously there may be more local communities with competing views—as we have in the case of modern human origins. The larger community can tolerate such competition and can wait for further lines of research, more of what will count as evidence, or shifting epistemologies.

Given this view, can scientific debates ever end? Yes. One way is that the proponents of all but one side die (either literally or intellectually). This could occur because everyone left declares that one side wins, or because the others give up—not necessarily because they are "right" or have the "truth" but rather because they have prevailed and are declared as the winners. This, however, seems unlikely in the present case since we have the leading anthropology departments continuing to train healthy new generations of Ph.D.s to do things the same ways, to ask the same questions, and to value the same kinds of answers. With a productive and full pipeline feeding each tradition, the situation does not seem likely to change dramatically very soon. Thus it does not seem likely that any one group will clearly "win" or that the others will quit.

Another alternative is that a new player changes the rules of the game, acquires an improved technique, or simply outplays the others in some way that the larger community accepts—perhaps simply because it is compellingly different. Briefly, it looked as though mtDNA might have provided such a technique as the hypothetical Eve looked out at us from every grocery store check-out rack from the cover of popular magazines. It looked as though this interpretation might take over. But no, her limitations were quickly revealed and the competitors returned to their own traditions. We cannot, of course, predict the next likely attempt at resolution, but a compelling account of the intricacies of biochemical evolution, coupled with accurate dating techniques, could go far toward sorting out lineages and relationships. If genetic and biochemical evidence were to agree sufficiently with one or the other competing physically or culturally based interpretation, the alliance would make it much more difficult for the competition to sustain an alternate view. It would be pragmatically wise to go with what might then emerge as a clearly dominant alternative.

One final way to settle the debate is just to quit arguing and go work on some other research problems. Why do we care so much about modern human origins anyway? Why—scientifically—does it matter whether the cradle of humans was in Africa or Asia or Antarctica? Though these answers have different implications, why should we care—scientifically—which one? We could end the debate by abandoning this question altogether. Yet the stakes are probably too high. People care too much about the answer, often for social or political reasons. For example: some want to establish Africa as the center and to prove that all humans derive from black ancestors; others strive to establish the importance of either the similarities or the differences among populations of people who look different. But let us not confuse the inevitable social discourse and its insistence that the question of modern human origins should remain primary with "science"—for science, whether regarded as epistemologically valuable or rejected by its critics, is supposed to conform to a different and more rigorous set of standards than non-science.

Even those who argue about details of their preferred epistemology agree that science should be consistent with empirical study of nature, should be coherent and internally consistent, and should provide some predictive value. Eventually such a body of observations and such a coherent view emerges that the helio-centric universe pushes man and his earth out of the center, for example. Perhaps such a body of research will push early man out of the anthropological center and make us care less scientifically about the interpretations of modern human origins. Instead, we might focus on the larger intellectual issues of evolutionary development and the resulting complex patterns of morphological, genetic, and behavioral changes. Perhaps it matters little—scientifically—whether we believe that humankind began as one or many. Perhaps the anthropological community's epistemic values go far enough to show that it might be productive to postpone much interpretation and theory surrounding modern human origins until there is something more compelling to add. We could temporarily end the debates by putting them on hold until we get our various Post-it notes in order.

Darwin, after all, was content to leave us out of the *Origin* entirely, never mentioning our species even once. He knew the question of our origin would weigh foremost on his readers' imaginations, but that was a social issue and he was not ready to address it scientifically in the first volume of his famous trilogy. He did not know, nor was it a compelling scientific problem to determine the nature of the breathing or the breather at the first moment of creation of life. Nor was he scientifically prepared to address whether life was originally breathed into one form or into many. Perhaps we are not ready to make scientific progress with these questions either, and it would be pragmatically wise to make progress pursuing various productive lines of research before investing more energy insisting on problematic larger interpretations.

For those who do insist on persisting with the debates, let them recognize that different epistemic styles exist, and that just as others are making contested epistemic assumptions, so they are themselves, and that all conclusions are conditional on those various assumptions. And let them recognize as well that resolution among the alternative styles will remain difficult, and that science will nonetheless persist in finding some approaches more useful in the long run—not necessarily because they are true but because they work better.

29

Philosophy and Paleoanthropology
Some Shared Interests?

MICHAEL RUSE

What has the philosopher to say to the student of human origins, the paleo-anthropologist? There are two points of interest. On the one hand, we philosophers should look at the work of paleoanthropologists, ultimately evaluating its worth as good science. We need to know the quality of the science with which we are faced. On the other hand, we philosophers need to take the science and to see where it leads us as philosophers. What are the implications for knowledge (epistemology) and morality (ethics)? Although I write as a philosopher, I hope you will see that I speak to paleoanthropologists, for I want to persuade them that we have shared problems, interesting and fruitful to both sides.

EVOLUTIONARY THEORY UNDER THE PHILOSOPHICAL MICROSCOPE

In recent years, the big question for philosophers has been that of the status of evolutionary theory—particularly Darwinian evolutionary theory and its central concept of natural selection—as a science (Brandon 1990; Callebaut 1994; Hull 1989; Ruse 1977, 1986a, 1988c, 1989, 1994, 1996; Sober 1984). When the debate started, some thirty or more years ago, the question was framed as one of the extent to which Darwinism measured up to the standards of the best science (physics) which was then seen as manifesting hypothetico-deductive systems, that is, law networks of an axiomatic kind (Hempel 1965, 1966; Nagel 1961). And the answer was that Darwinism measures up very badly—a comforting conclusion for it meant that the neglect of biology was now philosophically justified and not simply a matter of prejudice or laziness (Goudge 1961; Smart 1963)!

I think it is fair to say that, as the debate has gone, it is now accepted that many bad reasons were offered to dismiss evolutionary theory as inadequate or second-rate. However, I think it is also fair to say that as the debate has gone, the focus of concern has shifted somewhat. Now we are no longer as convinced that scientific theories, even in physics, are sweepingly hypothetico-deductive (Giere 1988). In real life, they are much more likely to be sets of theoretical models, connected by shared principles and methods, which are applied (with luck) to specific instances. Not so much one sweeping law that applies to all planets throughout the universe, but a model that applies (say) to Mars, yet one which shares much with models intended to apply to Venus or, more distantly, to a possible planet in the Andromeda system.

Thinking of theories this way lends itself readily to evolutionary biology, for it surely does truly describe much that goes on in that science (Beatty 1978, 1981; Lewontin 1974; Thompson 1989). Somebody working on, for example, Darwin's finches in the Galapagos is not aiming for sweeping laws that necessarily apply equally to *Drosophila* in the Arizona desert. Rather some model is proposed, taking account not only of the biology of birds (as opposed to fruitflies) but also of the hazards and opportunities of island life (as opposed to landlocked desert), and then it is determined whether or not it applies to *Geospiza* (Grant 1986; Grant and Grant 1989; Mayr 1954, 1963). If it does, all well and good; if it does not, then fiddling is required or possibly outright rejection.

One thing does need emphasizing, however, which perhaps this view of theories does not stress quite as much as it should (this certainly seems true if one looks at the work of its devotees). However one may interpret them, one must appreciate the extent to which a theory—evolutionary theory in particular—is an integrated phenomenon. Without wanting to claim an exclusive lien on the case at issue—for I think interpretation is possible here—I have suggested that we should think of Darwinism as being fanlike, with a central connecting core of population genetics (molecularized now, and perhaps extended to ecological theory) which can then be applied to the various subdisciplines in the evolutionary family—palaeontology, archaeology, systematics, sociobiology, etc. (Ruse 1973, 1975, 1979b).

Stressing this integration brings out the flip side to the points made above with the Galapagos finch example. The models are going to be particular to the situation—birds, islands, tropical, absence of certain kinds of predators, and so forth. But at the same time the models are going to draw on universal theory. The Hardy-Weinberg law is as crucial on the Galapagos as it is in the Arizona desert as it is in Guelph, Ontario. It may be that some particular part of the theory, say to do with drift, is not particularly relevant; but, then, some other part of the theory, say to do with parental care (r and K selection), could be just what is needed. The point is—and I suspect that this will often prove important in the human case— one is not working in isolation, however new or strange a situation may be, and one is not working in an unscientific manner when one imports material from elsewhere, taking it on trust and applying it to one's own situation.

What I have said so far would probably not be denied by anyone working in the field. I want now to touch on something much more contentious but which is surely relevant to the student of human origins. I have been talking of *Geospiza* and *Drosophila,* not of particular birds or particular insects. Biology is not unique in having to work with kinds rather than individuals—in a sense, no science at all would be possible if one did not work like this (Hempel 1952). But for the evolutionary biologist, there is something distinctive and troubling about his/her kinds, particularly about the basic unit of the species, the interbreeding, reproductively isolated groups of nature—*Homo sapiens, Drosophila melanogaster* (Mayr 1942). Unfortunately, although it is the aim of the evolutionist to explain the emergence and persistence of species, the very essence of species is that they have no essence! The whole point about species is that they are fluid, they evolve. How then can one possibly have a science like evolutionary biology which uses them as the basic analytical unit?

Let me dig further into this problem by taking the question of law. However one conceives theories, one has to allow that there are basic regularities which govern the natural world, and that science tries to capture them—regularities which are basic in the sense that they seem to be necessary (Braithwaite 1953; Nagel 1961). Bodies have to attract each other according to the inverse square of the distance separating them, populations have to come to genetic equilibrium if there are no external disturbing forces. But are laws even possible for species? Consider one of the traditional marks of lawlikeness, the ability to bear counterfactuals.

> If this were gold, then it would have such and such a reaction in this acid, but since it doesn't, it can't be gold.
> If this population were in Hardy-Weinberg equilibrium, then the ratios would be this and that, but they aren't so there must be some impinging forces.

The objection is that such an ability to bear counter-factuals is missing in the case of law-candidates mentioning particular species.

> If this were a human then it would have forty-six chromosomes, but it doesn't— so what? It might be a person with Downs syndrome, or one of many other genetic conditions.

Critics conclude that this all goes to show that biological species are not natural kinds as one meets elsewhere in nature (Ghiselin 1974; Hull 1976, 1978; cf. Ruse 1969, 1987). They are something else, and the popular substitute is that they are "individuals," much in the same way that particular organisms are individuals. Michael Ruse and *Homo sapiens* are therefore entities of the same logical type, namely proper names for things. Thus while it is certainly true that one can have a science which applies to species—say about the way in which speciation occurs—just as one can have a science which applies to individual organisms—Mendel's laws apply to me—it is really a mistake to think that one can have a science restricted to one particular species (Rosenberg 1980).

Of course, this still leaves possible a stripped-down evolutionary theory, which applies only to the general. But what of its branches which apply to the particular? To take the immediate example of paleoanthropology, while it is certainly the case that one might separate off paleoanthropology as a discipline, in principle it could never be a science like physics or even a branch of physics, like solid-state physics. Rather, it would be the application of another science— or rather, bits and pieces of other sciences—to one subject area. Logically, paleoanthropology would be more like geography than like physics or even evolutionary biology considered as a general area of inquiry.

Here then are two issues which have concerned the philosopher. What difference would it make to one's science if one took these disputes seriously? Would it matter if one endorsed one side rather than the other? The first problem surely does matter. One's stance on the nature of theories could well influence the direction of one's work as a practicing scientist—"Should I strive to axiomatize or should I spend my time fleshing out my models empirically?" Perhaps less in paleoanthropology but certainly in the social sciences, a lot of time and effort has been wasted by people who have prematurely gone the first route rather than more pragmatically taken the second. I confess, however, to being less sure what difference it would make where one stood on the species status question. Perhaps if one thinks that species are not the sorts of things which can bear their own laws, one would spend more time on the general and less on the particular subject. So it does make a difference.

Quite apart from the merits or demerits of arguments about the status of species, however, I am inclined to take a somewhat pragmatic stance on the matter of laws—and this surely does have implications for practice. Take one of Galileo's laws, say about bodies falling to the ground with an acceleration of 981 cm per second squared. I think this is absolutely true and necessary, and yet boomerangs do not bother me. If anything they make me think more highly of the law. Likewise, if someone showed me a skeleton with a Hapsburg chin, I would be inclined to say that it could not be a neanderthal because neanderthals are chinless wonders. But if a dig in Guelph started to reveal chinned neanderthals, I would not throw up my hands in despair. Rather, I would start looking for a special adaptive explanation, and were one found I would feel as triumphant as the physicist with the boomerang. Although one hears a lot of loose Popperian talk about the essence of science being the readiness to falsify, in fact this is nonsense and a good thing too (Ruse 1988a). Scientists protect their hypotheses or laws against attack until they can no longer do so. This should be as true in paleoanthropology as it is in physics.

THE MATTER OF VALUES

Let me turn now more specifically to the study of human origins. Unfortunately, there is simply nothing in the literature by philosophers on human origins. There is, however, a growing body of work by historians of science on the subject, that is to say on the subject of scientists and others who have worked and hypothesized on the subject of human origins (e.g., Bowler 1983, 1986; Landau 1991; Rainger 1991). So let us pick up from that; although, of course, we do at once run right into the problem that historians are themselves hardly disinterested voices. They are fascinated by certain issues and this fascination influences their work, from the initial choice of subject to the final interpretations which are offered.

It is particularly important to keep this point in mind here, for it is fair to say that the most significant debate among today's historians of science is the extent to which science is a reflection of culture, perhaps being no more than a "social construction" telling of the norms and hopes of the group from within which the science was fashioned, as opposed to an objective report of external reality (Collins 1985; Desmond 1989; Shapin 1982; Shapin and Schaffer 1985; but see Davies 1986). Moreover, with very few exceptions, those historians who have turned to the science of human origins have sought and found that the enterprise is strong evidence for the side of culture, and not a few would take it as heavy evidence in favor of constructivism (but see Clark 1993a). This is not to say that the overall debate (about values in science) is trivial—it is not—nor is it to say that it is wrong to find values in the work done on human origins—they are there—but it is to advise caution before wholehearted acceptance of results.

A prime example of the kind of work to which I am referring, and for which I am advising caution, is exemplified by the paleoanthropologist-turned-critic Misia Landau, in her recent *Narratives of Human Evolution* (1991). She claims rather bluntly of her parent discipline that the stuff is made up rather than found, and that at best paleoanthropology has only coincidental connection with truth and justification:

> I suggest that all . . . paleoanthropological narratives approximate the structure of a hero tale, along the lines proposed by Vladimir Propp in his classic *Morphology of the Folktale* (1928). They feature a humble hero who departs on a journey, receives essential equipment from a helper or donor figure, goes through tests and transformations, and finally arrives at a higher state. But it is part of my argument that, as in Propp's tales, this narrative schema can accommodate widely varying sequences of events, heroes and donors corresponding to the underlying evolutionary beliefs of their authors. A main goal of this book is to show how widely the followers of Charles Darwin depart from Darwinian natural selection as the guiding force that helps the hero forward, and how their interpretations of the fossil record vary according to their convictions about this primary causal agent. (Landau 1991:x)

Supporting her case, Landau divides the evolution of humans into four key periods or events: (1) terrestriality, when we came down from the trees; (2) bipedalism, when we got up and walked; (3) encephalization, when our brains got big; and (4) civilization, the "emergence of technology, morals, and society."

She argues that all stories of origins play on these four themes, and inasmuch as they do, they show more kinship with the telling of fairy tales than with the logic of inference in science.

One immediate objection that one might have to Landau's work is that this may all be very well for those people in the period after Darwin (1860–1930), when natural selection was relegated to a subsidiary, usually very minor role, but is hardly fair today. It is one thing to see this tale-telling in the work of that out-and-out orthogeneticist, Henry Fairfield Osborn (1894, 1910, 1916, 1917, 1927). It is quite another to see it in a Darwinian selectionist like Donald Johanson, the discoverer of Lucy *(Australopithecus afarensis)* (Johanson and Edey 1981). But Landau makes clear that her argument is radical in that it is intended to apply across the rediscovery of Darwinism, and she thinks that today's workers are no less infected than were people in earlier times.

My own feeling is that while Landau clearly makes some telling points about the work of individual paleoanthropologists—one has only to glance at Osborn to see the degree to which he was a myth maker—her claims are so sweeping and powerful that they had to be loose so that they could not possibly fail. On the one hand, while it is certainly true that some folk tales fit the proposed pattern fairly readily—*Cinderella,* for instance—others do so only in part. Where is the donor in *Hansel and Gretel,* for example, other than Gretel herself, and the witch, who surely does not count? In what sense is *Little Red Riding Hood* in a better state at the end of the story?

On the other hand, given that stories of human evolution had to incorporate the four elements—one would not think much of an account that left us with monkey brains and still living in the trees—what right has one to identify these elements with the elements of folk tales? Except in a highly metaphorical manner, can one truly say that the growth of the brain was a "donor" event? Certainly to the Darwinian, brain growth is probably going to be (in part) a bloody and painful process, where success comes through cunning, stealth, strength, and much more. Not much "donorship" around at this point.

Clarifying, Landau speaks of the donor as "the motivating force or prime mover" (1991:14), but this rather trivializes the whole business. We know that there had to be a force which made the brain grow and which fueled human evolution—the question is whether it is appropriate to speak even metaphorically of this force as a donor. Perhaps one could in a non-Darwinian context, especially if one thought of the motivating force as non-material (like Bergson's élan vital). But, in a materialistic Darwinian sense, one would be stretching things.

Remember that I am taking Landau as an example. I am less interested in her arguments per se and more as they typify the kind of charges leveled against paleoanthropology. I am warning that one might justifiably approach these criticisms with caution. Moreover, I would strongly agree with Landau that much that has passed for science bears little detailed scrutiny (also see Marks, Clark, this volume). One has only to open the writings of the past, even those of people for whom one has admiration (like Thomas Henry Huxley 1863), to realize that one is in the world of (shall we say) "creative thought." And even today, values—personal and cultural—are close to, if not frequently right on, the surface. We did not need Don Johanson or Roger Lewin to tell us that paleoanthropologists often have large egos. Did we ever know of a paleoanthropologist's find that was not claimed to be *the* missing link between us and the apes? Who last proudly announced that they had found something on a now-extinct side-branch?

Or take the Norman Rockwell-like musings of Owen Lovejoy (1981), in *Science* no less, about home life down among the australopithecines: "Can the nuclear family not be viewed as a prodigious adaptation central to the success of early hominids?" Threatened alas by "continuously sexually receptive" hussies, who wait on the sidelines, ever-ready to seduce the providing male from his true biological role. Is this picture of the conventional family, with males having a bit on the side, really "fact" or simply the wishful thinking of the North American paleoanthropological male?

Even when paleoanthropologists are self-consciously aware of the matter of values, they have a nasty way of sliding in by the back door. Having read Landau's work, the late Glynn Isaac (1983) worried about the charges leveled by her. In his own writings one found no glorification of evolving man, the hero of a tale, folk or otherwise. Indeed, for him, humans are the jackals of the primate world, scavenging the kill of larger, stronger mammals. But is it completely unfair to see here a reflection of today's intellectual's pessimism about progress, a feeling that society is in a state of decline and that we humans are not the noble triumph of progress but animals like any others, perhaps worse than many others? Certainly, Isaac was producing a paleoanthropology for our time.

What do we conclude? If you ask the philosophers, you find that they are somewhat split. Some would argue that all of these values are a sign of weakness, of intellectual immaturity (e.g., McMullin 1983). They would argue that it is true that values do get into science, but this is a mark of young science, of not-yet-processed science. As it develops, the "values" become less important and are finally expelled entirely. Thus paleoanthropology, value-laden as it is and as it seems likely to remain, is ever destined to be a quasi-science, perhaps a good myth but not much more. (Although she is no philosopher, this seems to be Landau's position.) Others are much less certain that, even in the best science, values ever go out entirely (Hesse and Arbib 1986; Longino 1990). This is not to say that the facts are unimportant or that there are no checks, but rather that through its models and metaphors and choice of topics and much much more, science necessarily shows values—and no bad thing too, many would add. This being so, the judgment for the sciencelike aspirations of paleoanthropology is much more hopeful.

This is not an excuse for weakness or an invitation to sloth. There is no absolution for sloppy or inadequate work; detailed empirical investigation is not merely appropriate but needed; one has to coordinate with other sciences (like molecular biology); but, when all is said and done, in this context the scope for the production of real science surely exists. To return to Isaac's work, value-laden though it may be, it is based on the very detailed and careful study of pertinent sites and materials, using sophisticated methods of investigation, and as such should be cherished as a positive contribution to knowledge. Perhaps we were scavengers, perhaps we were not, or perhaps the categories do not truly apply, but in light of the tentative nature of all scientific inference one should not deny standing to Isaac's work simply because he is reflecting the spirit of his age.

Concluding this section, let me say that I offer these comments as reflections. Yet, I would add that for myself I simply cannot see how any science can ever be judged totally value free (Ruse 1988b, 1993, 1994, 1996). Human thought—and whatever science may be it is a human endeavor—simply does not work that way. Hence, although I suspect that my position is very much a minority view among philosophers of science (probably mostly because my interests in the history of science are much deeper than those of the average philosopher of science), in this respect at least I would not judge paleoanthropology to be seriously defective conceptually or methodologically. This is said without prejudice and in no way denies that much that is produced, even today, may well leave much to be desired.

EPISTEMOLOGY IN THE LIGHT OF EVOLUTION

Let me go the other way now—turning from philosophy looking at biology to biology looking at philosophy—and start with the problem of knowledge, epistemology. Although it is far from fashionable, there has been some work here, so-called evolutionary epistemology. Let me begin by reviewing the subject and then see where paleoanthropology might make a pertinent contribution. There are two basic approaches, and I will take them in turn (Bradie 1986; Ruse 1986a; see also Kitcher 1992).

First we have the *analogical* approach to the subject. One starts with evolution, generally Darwinism. This is a long process of continuous development, driven by a struggle for existence which leads to natural selection: some organisms make it and go on to parent the next generation, and some do not. One then argues that something very similar happens in the world of knowledge, particularly scientific knowledge. One has theories, hypotheses, models, or whatever the unit may be. They compete for scientists' attention and approval. There is a struggle for existence between ideas. Some make it and are the starting points for the next generation of scientists. Some do not. Hence, science itself evolves. It is never static. It is always in a state of becoming (Popper 1972; Toulmin 1967, 1972).

There are variations on this general theme. For instance, historical Darwinism is a gradual process. It is the essence of selection that one can move forward only by small amounts, given the randomness of the units of change (mutations). But some evolutionary epistemologists have preferred a model which postulates change by jumps, "saltationism." One who has subscribed to a view somewhat like this is Thomas Kuhn, in his celebrated *The Structure of Scientific Revolutions* (1962a). He sees the basic units of science, "paradigms," switching suddenly from one to another, without essential direction, much as he envisions the nature and course of evolution.

I will not dwell long on this position. Let me say simply that it is very illuminating, particularly inasmuch as it emphasizes the fluid nature of scientific knowledge. It has been the philosophy behind some of the most stimulating recent work in the history of science (Hull 1988b; Richards 1987, 1992). However, in the eyes of many (including myself) there are grave disanalogies. Most particularly, science (and knowledge generally), contra Kuhn, does seem to have direction, to be progressive in some sense, unlike the course of organic evolution. The basic building blocks of biology, mutations, are undirected. This is manifestly not the case with the basic building blocks of science. People go out and work and discover new ideas—not by chance but by design. Organic change and scientific change are essentially not the same, or same kind of, process (Ruse 1986a; but see also Clark 1993a; Ruse 1994).

Second we have the *literal* approach to evolutionary epistemology. Here we start from the fact that knowledge, whatever it is, is something of biological adaptive purpose. The hominid who knows that there is a lion up ahead is better off than the hominid who is ignorant of this fact, and the hominid who knows that lions have a regrettable tendency to eat humans is better off than a hominid who thinks that lions are big pussy cats. The English philosopher John Locke (1975) provided the definitive case against innate knowledge, so there can be no question of us coming into the world naked except for our fully formed nuggets of knowledge, which give us lion protection. Rather the belief of this kind of epistemologist is that there may be innate dispositions to believe in certain predefined ways, and that this structures our knowing in an adaptive fashion (Lumsden and Wilson 1981, 1983). In lion-infested territory one grows up with the mind predisposed to slot in beliefs about the perils of proximity to lions.

What would be the nature of these dispositions, and how would they manifest themselves in the developing human? Most obviously one thinks of the basic rules of reasoning—the law of non-contradiction and like principles, the basic claims of mathematics, the elementary but crucial beliefs about nature (for instance, some principle of induction), and so forth (Ruse 1986a). Although there is no "official" consensus, the belief of the evolutionary epistemological literalist is that these expressed dispositions are transcultural, part of the evolutionary legacy of all of us. Then within a particular culture they are picked up and developed into something more—such that they acquire much greater power, much greater complexity, and more precision. Lion awareness in one instance, nuclear physics in another. No one, of course, thinks that nuclear physics is directly and tightly tied to adaptive advantage, but it starts with things which "are"—which exist independent of our perceptions of them. I should say that precisely how this development (of predispositions into cultural practice) occurs is a matter of some controversy (cf. Boyd and Richerson 1985; Durham 1991).

The literalist's epistemology is intended to be an empirical position; it is supposed to be something true of the world and not an a priori deduction or inference from intuition. Hence one can legitimately ask: What about the evidence for all of this? Candidly, it has to be admitted that the evidence is not great. But it is growing. Particularly exciting today is the work of evolutionary psychologists who are making great strides in showing how much of what is widely believed to be strictly a function of our social nature is in fact an accidental consequence of brain evolution. Nevertheless, we are more skilled at making inferences where interactions with other humans are involved than in cases that are purely abstract (Cosmides 1989; Cronin 1991). Evidentially speaking, and psychological and other studies on today's humans apart, there are possibilities from comparative studies with other animals but, unfortunately, they can take us only so far (de Waal 1982). This is particularly true because our closest relatives—chimpanzees and gorillas—have taken different adaptive routes from us. Their sociality is quite unlike ours (in some respects, wolves are better models). Hence, in looking for support for the claims of the epistemologist, it is obvious that here is a place for interaction with the paleoanthropologist, with (one would hope) feedback to his/her work.

Working on the evidences of human evolution—the development of the brain, the use of tools and so forth—must surely be relevant. In what sense, for instance, would something like *Australopithecus afarensis* be a reasoning being? What of its special needs in the context of the African savannas? Would one expect to see legacies of those needs today? One would presume that an organism which evolved rapidly in a savanna region might have very different cognitive abilities than one which evolved in jungle or desert. The same is true of social questions, particularly if they are connected with the innate predisposition toward reasoning.

The point about the evolutionary approach to reasoning—and I will make only this point, before I move on to ethics—is that, as always in evolution, one should not look for the good design of an all-powerful god. Physically speaking, evolution produces a bricolage, with roots deep in the past. Do males really need nipples? The same must be true of our cognitive abilities. We may think that we have insight into absolute reality, and that our powers of reasoning must be the uniquely appropriate and all-adequate set—but that is because there are strong evolutionary reasons why we should think this. Those of our would-be ancestors who had self-confidence in their reasoning abilities did better than those who did not.

The aim of the evolutionary epistemologist is to take these beliefs apart and look at them from behind, even though one is using these very beliefs to do this. What one cannot do is deconstruct through a priori reasoning. There must be an empirical grounding. In all of this, there is surely a major role for the paleoanthropologist—just as it might repay the paleoanthropologist to see what philosophers think are the key components to understanding.

EVOLUTIONARY ETHICS

As with epistemology there are two approaches (Ruse 1986a, 1988c, 1990; see also Kitcher 1993). First, there is the evolutionary ethics which tries to deduce the foundations of ethics from the evolutionary process itself, having used the process to infer the courses of right action. This was the position of Herbert Spencer (1851, 1892) in the past century and of Julian Huxley (Huxley and Huxley 1947) in this one. Most recently, it has been endorsed by the Harvard entomologist and sociobiologist, Edward O. Wilson (1975, 1978, 1994). The position is commonly known as Social Darwinism and is generally thought to be—following on Darwinism's struggle for existence—an extension to the cultural realm of an extreme laissez-faire philosophy. In fact however there are many variations, and by no means all are philosophies of the extreme right wing (Russett 1976). Huxley, for instance, always thought in group terms, reasoning that one ought to strive for the good of the whole of society, and this is very much the position of Wilson, except he would add that since we evolved in symbiotic relationship with the rest of the organic world, we ought to strive actively to preserve nature. He himself is much involved in the campaign to save the Amazonian rain forests (Wilson and Peter 1988).

Briefly, my objections to this position stem from the attempt to deduce proper action from the course and method of evolution. In the language of the philosophers, one is trying (illicitly I believe) to go from "the way things are" or "were" ("is" statements) to the way "things ought to be" ("ought" statements) (Hudson 1983). There is a version of what G. E. Moore (1903) called the "naturalistic fallacy." The only way one can move to this position is to assume that evolution is itself value-conferring, or progressive—the assumption is that evolution produces good and that later means better. This is manifestly false.

The alternative position, again as in epistemology depending on a *literal* reading of the consequences of evolution, suggests that there are innate dispositions for sociality, and that when made manifest in the course of development some of these have a moral force. Ethics—the sense that we have of right and wrong—is therefore something which is purely an adaptation, like eyes and teeth and genitalia (Ruse and Wilson 1985, 1986). As such, it has no foundation or meaning beyond the feelings that it conveys.[1]

This position, which owes much to modern human sociobiology (the study of human social behavior from an evolutionary perspective), is rooted in modern-day Darwinism. It is structured by the conviction that, despite or rather because of the struggle for existence, much animal behavior is social and cooperative (Wilson 1975). This is because one can better serve one's own biological ends by such behavior than by competing flat out and in a violent fashion. We may be at the mercy of "selfish genes," to use Richard Dawkins's (1976) powerful metaphor, but we are not necessarily selfish humans, certainly not all of the time (also Dawkins 1982, 1986).

A number of models have been devised to explain how genetic selfishness can translate into social altruism. Well known among them is "kin selection," where help is given to relatives because they (as sharers of one's genetic heritage) promote one's own reproduction inasmuch as they reproduce themselves (Hamilton 1964a, 1964b). Also significant is "reciprocal altruism," where help is given in expectation (not necessarily conscious expectation) of help received (Trivers 1971). "You scratch my back and I'll scratch yours."

Again the key question is that of evidence. How might these dispositions to helpfulness (altruism) become visible? In what way might they vary and change from society to society, and how might this be a function of culture and not biology? How does one get these dispositions in the first place, and what might be their nature, their basic range and scope and potentiality? Philosophers and others (psychologists, anthropologists, biologists) are working on these problems, particularly those concerned primarily with the first and second questions (Betzig, Borgerhoff Mulder, and Turke 1987). The third question is more difficult to answer but no less important. Again comparative studies loom large and again there are some advances and some serious problems, particularly about the extent to which we can properly appeal to nonhuman animals, especially our closest simian relatives (Goodall 1986).

As before it would seem that there is place for the paleoanthropologist to make a contribution, conversely receiving insight on his/her peculiar problems. We want to know what the archeopaleontological record—humans, animals, artifacts—tells us about the evolution of sociality, and we want to know what else can be learned from geography and geology, from molecular biology, and more. Wilson (1978) claims we are "gerry-built on the Pleistocene" and he is surely right. For instance—to take but one instance—how significant was intra-hominid competition in our evolution? Were the biggest threats our conspecifics, as is the case today, or did the threats come from outside? How you answer this will surely have implications for how you view moral feelings today. Is xenophobia (fear of strangers) a deep part of our biological heritage, perhaps backed by moral sentiments (you ought to be wary of folk outside your group), or is it a cultural artifact, with no true moral standing, and especially not one grounded in biology?

Unfashionable though this form of evolutionary ethics may be, I regard it as one of the most exciting advances in recent philosophical thought (Ruse 1986b,

1994). I am not sure that it will solve all of the challenges of modern life, but without it I am certain that we will have no solution. Today, it is more a promise than a program. The first step is to find out how "things" are, and the paleo-anthropologist surely has a role to play here.

NOTE

1. Ethical feelings, however, are feelings of a special kind. They are feelings with moral force. There is a difference between "I find the thought of killing someone truly repugnant" and "My sense is that killing is wrong" or, more simply, "Killing is wrong."

References

Accorsi C. A., E. Aiello, C. Bartolini, L. Castelletti, G. Rodolfi, and A. Ronchitelli. 1979. Il giacimento paleolitico di Serino, Avellino: stratigrafia e paletnologia. *Atti Sociedade Toscana di Scienze Naturali* A 86:17–31.

Adams, W. Y., and E. W. Adams. 1991. *Archaeological Typology and Practical Reality.* Cambridge: Cambridge University Press.

Agogino, G. A. 1964. Comment on "The fate of the 'classic' Neanderthals: A consideration of hominid catastrophism" by C. L. Brace. *Current Anthropology* 5:19–20.

Aiello, L. 1993. The fossil evidence for modern human origins in Africa: A revised view. *American Anthropologist* 95:73–96.

Aigner, J. S. 1981. *Archaeological Remains in Pleistocene China.* Muncheu: C. H. Beck.

———. 1983. Comments. *Current Anthropology* 24:190–191.

Aitken, M. J., C. B. Stringer, and P. Mellars, eds. 1993. *The Origins of Modern Humans and the Impact of Chronometric Dating.* Princeton: Princeton University Press.

Akazawa, T. 1982a. Cultural change in prehistoric Japan: Receptivity to rice agriculture in the Japanese archipelago. In *Advances in World Archaeology,* F. Wendorf and E. Close, ed. Pp. 151–211. Orlando: Academic.

———. 1982b. Jomon people subsistence and settlements: Discriminant analysis of the later Jomon settlements. *Journal of the Anthropological Society of Nippon* 90 (Supplement):55–76.

Akazawa, Takeru, Kenichi Aoki, and Tasuku Kimura, eds. 1992. *The Evolution and Dispersal of Modern Humans in Asia.* Tokyo: Hokusen-sha.

Alimen, H. 1955. *Préhistoire de l'Afrique.* Paris: Boubée.

Allchin, B. 1963. The Indian Stone Age sequence. *Journal of the Royal Anthropological Institute* 93:210–234.

Allsworth-Jones, P. 1986. *The Szeletian and the Transition from Middle to Upper Palaeolithic in Central Europe.* Oxford: Oxford University Press.

———. 1990. The Szeletian and the stratigraphic succession in central Europe and adjacent areas: Main trends, recent results, and problems for resolution. In *The Emergence of Modern Humans,* P. Mellars, ed. Pp. 160–242. Edinburgh: Edinburgh University Press.

———. 1993. The archaeology of archaic and early modern *Homo sapiens:* An African perspective. *Cambridge Archaeological Journal* 3:21–39.

Almagro, M., R. Fryxell, H. T. Irwin, and M. Serna. 1970. Avances a la investigación arqueológica, geocronológica y ecológica de la Cueva de La Carihuela (Píñar, Granada). *Trabajos de Prehistoria* 27:45–60.

Alsoszatai-Petheo, J. 1986. An alternative paradigm for the study of early man in the New World. In *New Evidence for the Pleistocene Peopling of the Americas,* A. Bryan, ed. Pp. 15–23. Orono: Center for the Study of Early Man.

Altuna, J. 1972. Fauna de mamíferos de los yacimientos prehistóricos de Guipúzcoa. *Munibe* 24:1–464.

Altuna, J., and K. Mariezkurrena. 1988. Les macromammifères du paléolithique moyen et supérieur ancien dans la région cantabrique. *Archaeozoologia* 1:179–196.

Altuna, J., and C. de la Rua. 1989. Dataciones absolutas de los cráneos del yacimiento prehistórico de Urtiaga. *Munibe* 41:23–28.

Ammerman, A. J., and L. L. Cavalli-Sforza. 1984. *The Neolithic Transition and the Genetics of Populations in Europe.* Princeton: Princeton University Press.

Anderson, J. E. 1968. Late Paleolithic skeletal remains from Nubia. In *The Prehistory of Nubia,* Vol. II, F. Wendorf, ed. Pp. 996–1040. Dallas: Southern Methodist University Press.

Andrews, P., and L. Martin. 1987. Cladistic relationships of extant and fossil hominoids. *Journal of Human Evolution* 26:245–269.

Antunes, M. 1990. O Homen da Gruta da Figueira Brava. *Memorias da Academia das Ciências de Lisboa* 31:487–536.

Antunes, M., J. Peixoto, J. Cardoso, and A. Monge. 1989. Paleolitico medio e superior em Portugal: datas ¹⁴C, estado actual dos conhecimentos, sintese e discussão. *Ciencias da Terra* 10:127–138.

Arensburg, B. 1991. From *sapiens* to Neandertals: Rethinking the Middle East. *American Journal of Physical Anthropology* (supplement) 12:44 (abstract).

Aristotle. 1965. *Historia Animalium.* Cambridge, Massachusetts: Harvard University Press.

Armelagos, G. J., D. S. Carlson, and D. P. Van Gerven. 1982. The theoretical foundations and development of skeletal biology. In *A History of American Physical Anthropology, 1930–1980,* Frank Spencer, ed. Pp. 305–328. New York: Academic Press.

Arsuaga, J. L., A. Gracia, I. Martínez, M. Bermudez de Castro, A. Rosas, V. Villaverde, and P. Fumanal. 1989. The human remains from Cova Negra (Valencia, Spain) and their place in European Pleistocene human evolution. *Journal of Human Evolution* 18:55–92.

Avise, J. C. 1989. Gene trees and organismal histories: a phylogenetic approach to population biology. *Evolution* 43:1192–1208.

———. 1994. *Molecular Markers, Natural History and Evolution.* New York: Chapman and Hall.

Ayala, F. J. 1985. Reduction in biology: A recent challenge. In *Evolution at a Crossroads: The New Biology and the New Philosophy of Science,* D. J. Depew and B. H. Weber, eds. Pp. 65–79. Cambridge, Massachusetts: MIT Press.

———. 1995. The myth of Eve: Molecular biology and human origins. *Science* 270:1930–1936.

Bahn, P. 1983. Late Pleistocene economies of the French Pyrennees. In *Hunter-Gatherer Economy in Prehistory,* G. Bailey, ed. Pp. 168–186. Cambridge: Cambridge University Press.

Bailey, G. 1983. Economic change in Late Pleistocene Cantabria. In *Hunter-Gatherer Economy in Prehistory,* G. Bailey, ed. Pp. 149–165. Cambridge: Cambridge University Press.

Barandiarán, I. 1980. Industria ósea. In *El Yacimiento de la Cueva de El Pendo,* J. González Echegaray, ed. Pp. 151–191. Madrid: Bibliotheca Praehistorica Hispana 17.

Bard, E., B. Hamelin, R. G. Fairbanks, and A. Zindler. 1990. Calibration of the ¹⁴C timescale over the past 30,000 years using mass spectrometric U-Th ages from Barbados corals. *Nature* 345:405–410.

Barkow, J. H. 1991. *Darwin, Sex and Status.* Toronto: University of Toronto Press.

Barnicott, N. 1969. Some biochemical and serological aspects of primate evolution. *Science Progress* (Oxford) 57:459–493.

Barroso, C., M. García Sanchez, A. Ruíz Bustos, P. Medina, and J. L. Sanchidrian. 1983. Avance al estudio cultural, antropológico y paleontológico de la cueva del "Boquete de Zafarraya" (Alcaucín, Málaga). *Antropología y Paleoecología Humana* 3:3–9.

Barroso, C., J. J. Hublin, and F. Medina, F. 1993. Proyecto: Zafarraya y el reemplazamiento de los neandertales por el hombre moderno. In *Investigaciones Arqueológicas en Andalucía (1985–1992), Proyectos*. Pp. 229–238. Huelva: Servicio de Investigaciones Arqueológicas.

Barroso, C., P. Medina, J. L. Sanchidrian, A. Ruíz Bustos, and M. García Sanchez. 1984. Le gisement Mousterien de la Grotte du Boquete de Zafarraya (Alcaucín, Andalousie). *L'Anthropologie* 88:133–134.

Bartolomei G., A. Broglio, L. Cattani, M. Cremaschi, A. Guerreschi, E. Mantovani, C. Peretto, and B. Sala. 1982. I depositi würmiani del Riparo Tagliente. *Annali dell'Università di Ferrara* n.s. 3:142–161.

Bartolomei, G., A. Broglio, P. F. Cassoli, L. Castelletti, L. Cattani, M. Cremaschi, G. Giacobini, G. Malerba, A. Maspero, M. Peresani, A. Sartorelli, and A. Tagliacozzo. 1992. La Grotte de Fumane. Un site aurignacien au pied des Alpes. *Preistoria Alpina* 28:131–179.

Barton, C. H. 1895. *Outlines of Australian Physiography*. Maryborough, Queensland, Australia: Alston and Company.

Barton, C. M., N. Coinman, and D. Olszewski. 1996. Beyond the graver: Reconsidering burin function. *Journal of Field Archaeology* 23:111–125.

Barton, N. H. 1979. Gene flow past a cline. *Heredity* 43:333–339.

Barton, N. H., and A. Clark. 1990. Population structure and processes in evolution. In *Population Biology: Ecological and Evolutionary Viewpoints*, K. Wohrmann and S. K. Jain, eds. Pp. 115–173. Berlin: Springer-Verlag.

Barton, N. H., and S. Rouhani. 1993. Adaptation and the "shifting balance." *Genetic Research* 61:57–74.

Bartrolí, R., A. Cebria, I. Muro, E. Riu-Barrera, and M. Vaquero. 1995. *A Frec de Ciència: l'Atles d'Amador Romaní i Guerra*. Capellades: Ajuntament de Capellades.

Bar-Yosef, O. 1989. Geochronology of the Levantine Middle Paleolithic. In *The Human Revolution*, P. Mellars and C. Stringer, eds. Pp. 589–610. Edinburgh: Edinburgh University Press.

———. 1992. Middle Paleolithic human adaptations in the Mediterranean Levant. In *The Evolution and Dispersal of Modern Humans in Asia*, T. Akazawa, K. Aoki, and T. Kimura, eds. Pp. 189–216. Tokyo: Hokusen-sha.

Bar-Yosef, O., M. Arnold, N. Mercier, A. Belfer-Cohen, P. Goldberg, R. Housley, H. Laville, L. Meignen, N. Mercier, J. Vogel, and B. Vandermeersch. 1996. The dating of the Upper Paleolithic layers in Kebara Cave, Mt. Carmel. *Journal of Archaeological Science* 23:297–306.

Bar-Yosef, O., B. Vandermeersch, B. Arensburg, P. Goldberg, H. Laville, I. Meignen, Y. Rak, E. Tchernov, and A.-M. Tillier. 1986. New data on the origin of modern man in the Levant. *Current Anthropology* 27:63–64.

Bar-Yosef, O., D. Lieberman, and J. Shea. 1990. Comments on "Symbolism and modern human origins," by Lindly and Clark. *Current Anthropology* 31:233–261.

Basabe, J. M. 1970. Dientes humanos del paleolítico de Lezextiki (Mondragón). *Munibe* 22:113–124.

———. 1973. Dientes humanos del Musteriense de Axlor (Dima, Vizcaya). *Trabajos de Antropología* 16:187–202.

———. 1974. Restos humanos de la región Vasco-Cantábrica. *Eusko-Ikaskuntza* (Cuadernos de la Sección Antropología, Etnografía, Prehistoria, Arqueología) 1:67–83.

Bateman, R. M., I. Goddard, R. O'Grady, V. A. Funk, R. Mooi, W. J. Kress, and P. F. Cannell. 1990. Speaking of forked tongues. *Current Anthropology* 31:1–24.

Bauer, K. 1973. Age determination by immunological techniques of the last common ancestor of man and chimpanzee. *Humangenetik* 17:253–265.

Beaumont, P. B., H. de Villiers, and J. C. Vogel. 1978. Modern man in subSaharan Africa prior to 49,000 years ago B.P.: A review and evaluation with particular reference to Border Cave. *South African Journal of Science* 74:409–419.

Beatty, J. 1978. *Evolution and the Semantic View of Theories*. Ph.D. dissertation, Indiana University.

————. 1981. What's wrong with the received view of evolutionary theory? In *PSA 1980*, P. Asquith and R. M. Giere, eds. Pp. 397–426. East Lansing: Philosophy of Science Association.

————. 1987. Dobzhansky and drift: facts, values and chance in evolutionary biology. In *The Probabilistic Revolution*, Vol. 2, Lorenz Krüger, Gerg Gigerenzer, and Mary S. Morgan, eds. Pp. 271–311. Cambridge, Massachusetts: MIT Press.

Bednarik, R. 1992. Palaeolithic art found in China. *Nature* 356:116.

————. 1995. Concept-mediated marking in the Lower Paleolithic. *Current Anthropology* 36:605–616, 626–630.

Behrensmeyer, A. K. 1986. Comments. *Current Anthropology* 27:469.

Ben-Itzhak, S., P. Smith, and R. A. Bloom. 1988. Radiographic study of the humerus in Neandertals and *Homo sapiens sapiens. American Journal of Physical Anthropology* 77:231–242.

Benson, W. H. 1987. Looking at multivariate data through fuzzy sets. In *Analysis of Fuzzy Information, Vol. III: Applications in Engineering and Science*, J. Bezdek, ed. Pp. 183–189. Boca Raton: CRC Press.

Berger, R., and W. F. Libby. 1966. UCLA Radiocarbon dates V. *Radiocarbon* 8:480.

Bergson, H. 1907. *L'évolution créatrice*. Paris: Alcan.

Berlin, B., D. E. Breedlove, and P. H. Raven. 1966. Folk taxonomies and biological classification. *Science* 154:273–275.

Bernaldo de Quirós, F. 1981. Análisis matemático del paleolítico superior inicial. *Zephyrus* 32–33:41–56.

————. 1982a. L'Aurignacien en Espagne. In *Aurignacien et Gravettian en Europe*, pp. 50–63. Liège: ERAUL No. 13.

————. 1982b. The early Upper Paleolithic in Cantabrian Spain (Asturias, Santander). In *Aurignacien et Gravettien en Europe*, pp. 65–78. Liège: ERAUL No. 13.

————. 1982c. *Los Inicios del Paleolítico Superior Cantábrico*. Madrid: Memorias del CIMA No. 8.

Bernstein, R. 1983. *Beyond Objectivism and Relativism: Science, Hermeneutics and Praxis*. Philadelphia: University of Pennsylvania Press.

Bettinger, R. L., D. B. Madsen, and R. G. Elston. 1994. Prehistoric settlement categories and settlement systems in the Alashan Desert of Inner Mongolia, PRC. *Journal of Anthropological Archaeology* 13:74–101.

Betzig, C. C., M. Borgerhoff Mulder, and Turke P. W. 1987. *Human Reproductive Behavior*. Cambridge: Cambridge University Press.

Bielicki, T. 1975. Natural selection and human morphology. In *The Role of Natural Selection in Human Evolution*, F. M. Salzano, ed. Pp. 203–216. Amsterdam: North-Holland Publishing.

Bietti, A. 1991. Normal science and paradigmatic biases in Italian hunter-gatherer prehistory. In *Perspectives on the Past*, G. A. Clark, ed. Pp. 258–281. Philadelphia: University of Pennsylvania Press.

Bietti, A., S. Grimaldi, and S. Vitagliano. n.d. Small flint pebbles and Mousterian reduction sequencies: The case of southern Latium (Italy). *Proceedings of the International Round Table: Reduction Processes for the European Mousterian.* Rome, May 26–28, 1995. *Quaternaria Nova,* in press.

Bietti, A., S. Kuhn, A. G. Segre, and M. C. Stiner. 1990–1991. Grotta Breuil: a general introduction and stratigraphy. *Quaternaria Nova* 1:305–323.

Bietti, A., G. Manzi, P. Passarello, A. G. Segre, and M. C. Stiner. 1988. The 1986 excavation campaign at Grotta Breuil (Monte Circeo, Latium). *Archeologiche Laziale: Quaderni per l'Archeologiche Etrusco-Italica* 9:372–388.

Billy, G. 1970. Définition du type de Cro-Magnon "sensu stricto." In *L'Homme de Cro-Magnon, 1868–1968,* G. Camps and G. Olivier, eds. Pp. 23–32. Paris: Arts et Métiers Graphiques.

———. 1972. L'évolution humaine au paléolithique supérieur. *Homo* 72:2–12.

Binford, L. R. 1972. Models and paradigms in paleolithic archaeology. In *Models in Archaeology,* D. Clarke, ed. Pp. 109–166. London: Methuen.

———. 1973. Interassemblage variability—the Mousterian and the "functional" argument. In *The Explanation of Culture Change,* C. Renfrew, ed. Pp. 227–254. London: Duckworth.

———. 1981. *Bones: Ancient Men and Modern Myths.* New York: Academic.

———. 1983. *Working at Archaeology.* New York: Academic.

———. 1984. *Faunal Remains from Klasies River Mouth.* New York: Academic.

———. 1985. Human ancestors: Changing views of their behavior. *Journal of Anthropological Archeology* 4:292–327.

———. 1987. The hunting hypothesis, archaeological methods, and the past. *Yearbook of Physical Anthropology* 30:1–9.

———. 1989. Isolating the transition to cultural adaptation: An organizational approach. In *The Emergence of Modern Humans,* Erik Trinkaus, ed. Pp. 18–41. Cambridge: Cambridge University Press.

Binford, L. R., and C. K. Ho. 1985. Taphonomy at a distance: Zhoukoudian, the cave home of Beijing man? *Current Anthropology* 26:413–429, 436–442.

Binford, L. R., and J. Sabloff. 1982. Paradigms, systematics and archaeology. *Journal of Anthropological Research* 38:137–153.

Binford, L. R., and N. M. Stone. 1986. Zhoukoudian: A closer look. *Current Anthropology* 27:453–468, 471–475.

Bischoff, J., J. García, and L. Straus. 1992. Uranium-series isochron dating at El Castillo (Cantabria, Spain): the "Acheulean/Mousterian" question. *Journal of Archaeological Science* 19:49–62.

Bischoff, J., R. Julia, and R. Mora. 1988. Uranium-series dating of the Mousterian occupation at Abric Romaní, Spain. *Science* 332:68–70.

Bischoff, J., K. Ludwig, J. Garcia, E. Carbonell, M. Vaquero, T. Stafford, and A. Jull. 1994. Dating the basal Aurignacian sandwich at Abric Romaní (Catalunya, Spain) by radiocarbon and uranium-series. *Journal of Archaeological Science* 21:541–551.

Bischoff, J. L., N. Soler, J. Maroto and R. Julia. 1989. Abrupt Mousterian/Aurignacian boundary at c. 40 ka BP: accelerator C[14] dates from l'Arbreda Cave (Catalunya, Spain). *Journal of Archaeological Science* 16:553–576.

Black D. 1927. *On a lower molar hominid tooth from the Chou Kou Tien deposit.* Palaeontologica Sinica, Series D, Vol. 7, V. K. Ting and Y.C. Sun, eds. Beijing.

Black, D., P. Teilhard de Chardin, C. C. Young, and W. C. Pei. 1933. Fossil man in China: The Chou Kou Tien cave deposits with a synopsis of our present knowledge of the late Cenozoic in China. *Memoirs of the Geological Survey of China* A 11:1–166.

Blackwell, B., and H. P. Schwarcz. 1993. Archaeochronology and scale. In *Effects of Scale in Archaeological and Geological Perspectives,* J. Stein and A. R. Linse, ed. *Geological Society of America,* Special Paper 283:39–58.

Blanc, A. C. 1938. Una serie di nuovi giacimenti pleistocenici e paleolitici in grotte litoranee del Monte Circeo. *Rendiconti dell' Accad. Naz. dei Lincei* 28:201–209.

———. 1953. Il Riparo Mochi ai Balzi Rossi di Grimaldi. Le Industrie. *Paleontographia Italica* 50(3). Pisa.

———. 1954. Reperti fossili neandertaliani nella Grotta del Fossellone al Monte Circeo: Circeo IV. *Quaternaria* 1:171–175.

Blanc, A. C., A. G. Segre. 1953. La Grotta del Fossellone. In *Excursion au Mont Circé, Livret-Guide au 4ème Congrès International du INQUA,* pp. 37–85. Paris: UISPP.

Blumenbach, J. F. 1775 [1865]. *De Generis Humani Varietate Nativa,* Thomas Bendyshe, trans. London: Anthropological Society.

Boas, F. 1894. Classification of the languages of the North Pacific Coast. In *Memoirs of the International Congress of Anthropology,* pp. 339–346. Chicago: Schulte. Reprinted in 1974 in *The Shaping of American Anthropology: A Franz Boas Reader,* G. W. Stocking, Jr., ed. New York: Basic Books.

———. 1924. The question of racial purity. *American Mercury* 3:163–169.

Bordes, F. 1950. Review of Teilhard de Chardin, "Early Man in China." *L'Anthropologie* 54(12):82–91.

———. 1953. Essai de classification des industries moustériennes. *Bulletin de la Société Préhistorique Française* 50:457–466.

———. 1955. Le paléolithique inférieur et moyen à Yabrud (Syrie) et la question du Pré-Aurignacien. *L'Anthropologie* 59:486–511.

———. 1958. Le passage du paléolithique moyen au paléolithique supérieur. In *Hundert Jahre Neanderthaler, 1856–1956,* G. H. R. von Koenigswald, ed. Pp. 175–181. Köln-Graz: Böhlau Verlag.

———. 1960. Le Pré-Aurignacien de Yabrud (Syrie), et son incidence sur la chronologie du Quaternaire en Moyen Orient. *Bulletin of the Research Council of Israel* 96:91–103.

———. 1961a. Mousterian cultures in France. *Science* 134:803–810.

———. 1961b. *Typologie du Paléolithique Ancien et Moyen.* Bordeaux: Imprimerie Delmas.

———. 1968a. *The Old Stone Age.* New York: McGraw-Hill.

———. 1968b. *L'Ère Paléolithique.* Paris: Masson et Cie.

———. 1968c. La question Périgordienne. In *La Préhistoire: Problèmes et Tendances,* D. de Sonneville-Bordes, ed. Pp. 59–70. Paris: CNRS.

———. 1972. Du paléolithique moyen au paléolithique supérieur-continuité ou discontinuité? In *The Origin of Homo sapiens,* F. Bordes, ed. Pp. 211–218. Paris: UNESCO.

———, ed. 1972. *The Origin of Homo sapiens.* Paris: UNESCO.

———. 1973. On the chronology and contemporaneity of different Paleolithic cultures in France. In *The Explanation of Culture Change,* C. Renfrew, ed. Pp. 217–226. London: Duckworth.

———. 1992. *Leçons sur le Paleolithique,* 2 vols. Paris: Presses du CNRS.

Bordes, F., and F. Labrot. 1967. Stratigraphie de la grotte du Roc de Combe (Lôt) et ses implications. *Bulletin de la Société Préhistorique Française* 64:15–27.

Borzatti von Löwenstern, E. 1964 La Grotta di Uluzzo (campagna di scavi 1964). *Rivista Scienze Preistoriche* 19:41–51.

———. 1965. La grotta-riparo di Uluzzo C (campagna di scavi 1964). *Rivista Scienze Preistoriche* 20:1–31.

————. 1970. Prima campagna di scavi nella grotta "Mario Bernardini." *Rivista Scienze Preistoriche* 25:89–125.

Bouchud, J. 1966. Remarques sur les fouilles de L. Lartet à l'abri de Cro-Magnon (Dordogne). *Bulletin de la Société d'Étude et de Recherches Préhistoriques* 15:2–9.

Boule, M. 1911–1913. L'homme fossile de La Chapelle-aux-Saints. *Annales de Paléontologie* 6:111–172; 7:21–56, 85–192; 8:1–70.

————. 1921. *Les Hommes Fossiles: Éléments de Paléontologie Humaine.* Paris: Masson et Cie.

————. 1927. Introduction. In *Le Palaeolithique de la Chine*, M. Boule, H. Breuil, E. Licent, and P. Teilhard de Chardin, eds. Pp. i–viii. Archives de l'Institut de Paleontologie Humaine 4. Masson: Paris.

Boule, M., and H. V. Vallois. 1957. *Fossil Men.* New York: Dryden.

Bowcock, A. M., A. Ruíz-Linares, J. Jomforde, E. Minch, J. R. Kidd, and L. L. Cavalli-Sforza. 1994. High resolution of human evolutionary trees with polymorphic microsatellites. *Nature* 368:455–457.

Bowdler, S. 1976. Hook, line and dillybag: An interpretation of an Australian coastal shell midden. *Mankind* 10:248–258.

Bowler, P. J. 1983. *The Eclipse of Darwinism: Anti-Darwinian Evolution Theories in the Decades around 1900.* Baltimore: Johns Hopkins University Press.

————. 1986. *Theories of Human Evolution: A Century of Debate, 1844–1944.* Baltimore: Johns Hopkins University Press.

————. 1989. Holding your head up high: Degeneration and orthogenesis in theories of human evolution. In *History, Humanity and Evolution: Essays for John C. Greene*, J. R. Moore, ed. Pp. 329–353. Cambridge: Cambridge University Press.

————. 1991. Comment on "New models and metaphors for the Neanderthal debate" by Paul Graves. *Current Anthropology* 32:525–526.

Boyd, R., and P. J. Richerson. 1985. *Culture and the Evolutionary Process.* Chicago: University of Chicago Press.

Boyd, W. 1950. *Genetics and the Races of Man.* Boston: Little, Brown.

Brace, C. L. 1962a. Refocusing on the Neanderthal problem. *American Anthropologist* 64:729–741.

————. 1962b. Cultural factors in the evolution of the human dentition. In *Culture and the Evolution of Man*, A. Montagu, ed. Pp. 343–354. New York: Oxford University Press.

————. 1963. Structural reduction in evolution. *American Naturalist* 97:39–49.

————. 1964a. The fate of the "classic" neanderthals: A consideration of hominid catastrophism. *Current Anthropology* 5:3–43.

————. 1964b. The probable mutation effect. *American Naturalist* 98:453–455.

————. 1967. *The Stages of Human Evolution: Human and Cultural Origins.* Englewood Cliffs: Prentice-Hall.

————. 1968. Ridiculed, rejected, but still our ancestor. *Natural History* 77:38–44.

————. 1979. Krapina, "classic" Neanderthals, and the evolution of the European face. *Journal of Human Evolution* 8:527–550.

————. 1980. Australian tooth-size clines and the death of a stereotype. *Current Anthropology* 21:141–164.

————. 1985. Punctuationalism, cladistics and the legacy of medieval Neoplatonism. *American Journal of Physical Anthropology* 66:148.

————. 1988a. Punctuationism, cladistics and the legacy of medieval Neoplatonism. *Human Evolution* 3:121–138.

————. 1988b. *The Stages of Human Evolution: Human Biological and Cultural Origins*, 3rd ed. Englewood Cliffs: Prentice-Hall.

————. 1989. Medieval thinking and the paradigms of paleoanthropology. *American Anthropologist* 91:442–446.

————. 1990. The creation of specific names: *Gloria in excelsis Deo?* or praxis? or ego? *American Journal of Physical Anthropology* 81:197.

————. 1991a. Cultural innovation and the mechanism for the emergence of modern morphology. *American Journal of Physical Anthropology* (supplement) 12:52.

————. 1991b. Comment on "New models and metaphors for the Neanderthal debate" by P. Graves. *Current Anthropology* 32:526–528.

————. 1991c. *The Stages of Human Evolution,* 4th ed. Englewood Cliffs: Prentice Hall.

————. 1992. Modern Human Origins: Narrow Focus or Broad Spectrum? The David Skomp Lecture, April 16, 1992. Indiana University, Bloomington.

————. 1994. Review of *Species, Species Concepts, and Primate Evolution,* edited by W. H. Kimbel and L. B. Martin. *American Scientist* 8:484–486.

————. 1995a. Bio-cultural interaction and the mechanism of mosaic evolution in the emergence of "modern" morphology. *American Anthropologist* 97:4–11.

————. 1995b. *The Stages of Human Evolution,* 5th ed. Englewood Cliffs: Prentice Hall.

————. 1995c. The creation of specific hominid names: *Gloria in excelsis Deo?* or ego? or praxis? *Human Evolution* 8:151–166.

————. 1995d. Trends in the evolution of human tooth size. In *Aspects of Dental Biology: Palaeontology, Anthropology and Evolution,* J. Moggi-Cecchi, ed. Pp. 437–446. Florence: Istituto de Antropologia, Università di Firenze.

Brace, C. L., and K. D. Hunt. 1990. A non-racial craniofacial perspective on human variation: A(ustralia) to Z(uni). *American Journal of Physical Anthropology* 88:341–360.

Brace, C. L., and M. Nagai. 1982. Japanese tooth size, past and present. *American Journal of Physical Anthropology* 59:399–411.

Brace, C. L., M. L. Brace, and W. R. Leonard. 1989. Reflections on the face of Japan: A multivariate craniofacial and odontometric perspective. *American Journal of Physical Anthropology* 78:93–113.

Brace, C. L., K. Rosenberg, and K. D. Hunt. 1987. Gradual change in human tooth size in the Late Pleistocene and post-Pleistocene. *Evolution* 41:705–720.

Brace, C. L., S. L. Smith, and K. D. Hunt. 1991. "What big teeth you had Grandma!" Human tooth size past and present. In *Advances in Dental Anthropology,* M. A. Kelley and C. S. Larsen, eds. Pp. 33–57. New York: Wiley-Liss.

Brace, C. L., D. P. Tracer, and K. D. Hunt. 1991. Human craniofacial form as evidence for the peopling of the Pacific. *Bulletin of the Indo-Pacific Prehistory Association* 11:247–269.

Bradie, M. 1986. Assessing evolutionary epistemology. *Biology and Philosophy* 1:401–460.

————. 1994. *The Secret Chain: Evolution and Ethics.* Albany: SUNY Press.

Braithwaite, R. 1953. *Scientific Explanation.* Cambridge: Cambridge University Press.

Brandon, R. 1990. *Adaptation and Environment.* Princeton: Princeton University Press.

Bräuer, G. 1984a. A craniological approach to the origin of anatomically modern *Homo sapiens* in Africa and implications for the appearance of modern Europeans. In *The Origins of Modern Humans: A World Survey of the Fossil Evidence,* F. H. Smith and F. Spencer, eds. Pp. 327–410. New York: Alan R. Liss.

————. 1984b. The Afro-European *sapiens* hypothesis and hominid evolution in Asia during the Middle and Upper Pleistocene. In *The Early Evolution of Man,* P. Andrews and G. Frazen, eds. *Courier Forschungsinstitut Senckenberg* 69:145–165.

————. 1989. The evolution of modern humans: A comparison of the African and non-

African Evidence. In *The Human Revolution,* P. Mellars and C. Stringer, eds. Pp. 123–155. Edinburgh: Edinburgh University Press.

———. 1992a. Africa's place in the evolution of *Homo sapiens.* In *Continuity or Replacement: Controversies in Homo sapiens Evolution,* G. Bräuer and F. Smith, eds. Pp. 83–98. Rotterdam: A. A. Balkema.

———. 1992b. The origin of modern Asians: By regional evolution or by replacement? In *The Evolution and Dispersal of Modern Humans in Asia,* T. Akazawa, K. Aoki, and T. Kimura, eds. Pp. 401–413. Tokyo: Hokusen-sha.

Bräuer, G., and F. Smith, eds. 1992. *Continuity or Replacement: Controversies in Homo sapiens Evolution.* Rotterdam: A. A. Balkema.

Bräuer, K. 1973. Age determination by immunological techniques of the last common ancestor of man and chimpanzee. *Humangenetik* 17:253–265.

Breuil, H. 1912. Les subdivisions du paléolithique supérieur et leur signification. *Actes du Congrès International d'Anthropologie et d'Archéologie Préhistorique* (Geneva 1912):165–238.

———. 1927. Archéologie. In *Le Paléolithique de la Chine,* M. Boule, H. Breuil, E. Licent, and P. Teilhard de Chardin, eds. *Archives de l'Institut de Paléontologie Humaine Mémoir* 4:103–136.

———. 1931. Le feu et l'industrire lithique et osseuse á Choukoutien. *Bulletin of the Geological Society of China* 1(2):147–154.

———. 1939. The bone and antler industry of the Choukoutien *Sinanthropus* site. *Paleontologica Sinica* Series D 6:7–41.

Bright, William. 1984. *American Indian Linguistics and Literature.* Berlin and New York: Mouton.

Broca, P. 1860. Des phénomènes d'hybridité dans le genre humain. *Journal de la Physiologie de l'Homme et des Animaux* 3:392–439.

———. 1862. La linguistique et l'anthropologie. *Bulletin de la Société d'Anthropologie de Paris* 3:264–319.

———. 1868. Sur les crânes et les ossements des Eyzies. *Bulletin de la Société d'Anthropologie de Paris* 3:350–392.

Broglio, A. 1994. Il Paleolitico superiore del Friuli-Venezia Giulia. *Atti della XXIX Riunione Scientifica dell'Istituto Italiano di Preistoria e Protostoria,* pp. 37–56.

Brooks, A., D. Helgren, J. Cramer, A. Franklin, W. Hornyak, J. Keating, R. Klein, W. Rink, H. Schwarcz, J. Smith, K. Stewart, N. Todd, J. Verniers, and J. Yellen. 1995. Dating and context of three Middle Stone Age sites with bone points in the Upper Semliki Valley, Zaire. *Science* 268:548–553.

Brose, David S., and M. H. Wolpoff. 1971. Early Upper Paleolithic man and late Middle Paleolithic tools. *American Anthropologist* 73:1156–1194.

Brothwell, D. R. 1960. Upper Pleistocene human skull from Niah Cave, Sarawak. *Sarawak Museum Journal* 9:323–349.

———. 1964. Comment on "The fate of the 'classic' Neanderthals: A consideration of hominid catastrophism" by C. L. Brace. *Current Anthropology* 5:20–21.

Brown, B., and A. Walker. 1993. The dentition. In *The Nariokotome Homo Erectus Skeleton,* A. Walker and R. Leakey, eds. Pp. 161–192. Berlin: Springer-Verlag.

Brown, P. 1990. Osteological definitions of "anatomically modern" *Homo sapiens:* A test using modern and terminal Pleistocene *Homo sapiens.* In *Is Our Future Limited by Our Past?* L. Freedman, ed. Pp. 51–74. Nedlands: Centre for Human Biology, University of Western Australia.

Brown, W. M., J. M. George, and A. C. Wilson. 1979. Rapid evolution of animal mitochondrial DNA. *Proceedings of the National Academy of Sciences* 76:1967–1971.

Brues, A. M. 1960. The spearman and the archer—an essay on selection in body build. *American Anthropologist* 61:457–469.

———. 1966. "Probable mutation effect" and the evolution of hominid teeth and jaws. *American Journal of Physical Anthropology* 25:169–170.

Buck, C. D. 1949. *A Dictionary of Selected Synonyms in the Principal Indo-European Languages: A Contribution to the History of Ideas.* Chicago: University of Chicago Press.

Buckley, J. J., and W. Siler. 1988. Fuzzy numbers for expert systems. In *Fuzzy Logic in Knowledge-Based Systems, Decision and Control,* M. M. Gupta and T. Yamakawa, eds. Pp. 153–172. Amsterdam: North-Holland.

Buffon, C. de. 1749. Variétés dans l'espèce humain. In *Histoire Naturelle, Gènèrelle et Particulière,* Tome III. Paris: Imprimerie Royale.

Burjachs, F., and R. Juliá. 1994. Abrupt climatic changes during the last glaciation based on pollen analysis of the Abric Romaní, Catalonia, Spain. *Quaternary Research* 42:308–315.

Butzer, K. 1981. Cave sediments, Upper Pleistocene stratigraphy and Mousterian facies in Cantabrian Spain. *Journal of Archaeological Science* 8:133–183.

Cabrera Valdés, V. 1978. *La Cueva del Castillo (Puente Viesgo, Santander). Estudio y Revisión de los Materiales y Documentación de este Yacimiento.* Ph.D. dissertation, Universidad Complutense. Madrid.

———. 1983. Notas sobre el Musteriense Cantábrico: el "Vasconiense." In *Homenaje al Prof. Martín Almagro Basch,* Vol. I, Alberto Balil Illana, José María Blazquez, Antonio Beltran, Manuel Berges Soriano, Luis Caballero Zoreda, Emeterio Cuadrado, Manuel Fernandez-Miranda, Pedro Palol I. Salles, and Cristóbal Veny Melia, eds. Pp. 131–142. Madrid: Ministerio de Cultura.

———. 1984. *El Yacimiento de la Cueva de "El Castillo" (Puente Viesgo, Santander).* Madrid: Bibliotheca Praehistorica Hispana No. 22.

———. 1988. Aspects of the Middle Paleolithic in Cantabrian Spain. In *l'Homme de Néanderthal,* Vol. 4: La Technique, M. Otte, ed. Pp. 27–38. Liège: ERAUL No. 31.

———, ed. 1993. *El Origen del Hombre Moderno en el Suroeste de Europa.* Madrid: UNED.

Cabrera Valdés, V., and F. Bernaldo de Quirós. 1984. Die wohnstrukturen von Cueva de Chufín und Cueva del Castillo (Kantabrisches Spanien). In *Jungpaläolithische Siedlungsstrukturen in Europa,* pp. 51–58. Urgeschichtliche Materialhefte No. 6. Stuttgart: Pieter Schrecklichkeit Verlag.

———. 1985. Evolution technique et culturelle de la Cueva del Castillo. In *La Significa-tion Culturelle des Industries Lithiques,* pp. 206–220. Oxford: BAR International Series No. 239.

———. 1990. Données sur la transition entre le paléolithique moyen et le supérieur dans la region Cantabrique: revision critique. *Memoires de la Musée de Préhistoire d'Ile-de-France* 3:185–188.

———. 1992. Approaches to the Middle Paleolithic in northern Spain. In *The Middle Paleolithic: Adaptation, Behavior and Variability,* H. Dibble and P. Mellars, eds. Pp. 97–111. Philadelphia: University of Pennsylvania Museum.

Cabrera, V., and J. Bischoff. 1989. Accelerator ^{14}C dates for early Upper Paleolithic (basal Aurignacian) at El Castillo cave (Spain). *Journal of Archaeological Science* 16:577–584.

Cabrera, V., F. Bernaldo de Quirós, and M. Hoyos. 1996. Hugo Obermaier y la Cueva del Castillo. In *El Hombre Fósil, 80 Años Después,* A. Moure, ed. Pp. 177–193. Santander: Universidad de Cantabria.

Cabrera, V., M. Hoyos, and F. Bernaldo de Quirós. 1993. La transición del Paleolítico

Medio-Paleolítico Superior en la Cueva de El Castillo: características paleoclimáticas y situación cronológica. In El *Origen del Hombre Moderno en el Suroeste de Europa,* V. Cabrera, ed. Pp. 81–104. Madrid: UNED.

Callebaut, W. 1994. *Taking the Naturalistic Turn.* Chicago: University of Chicago Press.

Campbell, B. 1962. The systematics of man. *Nature* 194:225–232.

Campbell, L. 1973. Distant genetic relationship and the Maya-Chipaya hypothesis. *Anthropological Linguistics* 14:113–135.

———. 1993. Letter to the editor. *Scientific American* 255:12.

Campbell, L., and T. Kaufman. 1980. On Mesoamerican linguistics. *American Anthropologist* 82:850–857.

———. 1983. Mesoamerican historical linguistics and distant genetic relationship: Getting it straight. *American Anthropologist* 85:362–372.

Campbell, L., and M. Mithun, eds. 1979. *The Language of Native America: Historical and Comparative Assessment.* Austin: University of Texas Press.

Cann, R. L. 1988. DNA and human origins. *Annual Review of Anthropology* 17:127–143.

———. 1992. A mitochondrial perspective on replacement or continuity in human evolution. In *Continuity or Replacement: Controversies in Homo sapiens Evolution,* G. Bräuer and F. H. Smith, eds. Pp. 65–73. Rotterdam: A. A. Balkema.

———. 1993. Mitochondrial DNA and human evolution: New directions, future challenges. *Rivista di Anthropologia* 71(2):304–314.

Cann, R. L., M. Stoneking, and A. C. Wilson. 1987. Mitochondrial DNA and human evolution. *Nature* 325:31–36.

Carson, H. L. 1961. Heterosis and fitness in experimental populations of *Drosophila melanogaster. Evolution* 15:496–509.

———. 1982. Speciation as a major reorganization of polygenic balances. In *Mechanisms of Speciation,* C. Barigozzi, ed. Pp. 411–433. New York: Alan R. Liss.

Cartmill, M. 1981. Hypothesis testing and phylogenetic reconstruction. *Zeitschrift für Zoologisches Systematik und Evolutionsforschung* 19:73–96.

———. 1982. Basic primatology and prosimian evolution. In *A History of American Physical Anthropology, 1930–1980,* F. Spencer, ed. Pp. 147–186. New York: Academic.

———. 1990. Human uniqueness and theoretical content in paleoanthropology. *International Journal of Primatology* 11:173–192.

———. 1994. *A View to a Death in the Morning.* Cambridge, Massachusetts: Harvard University Press.

Cartmill, M., D. Pilbeam, and G. Isaac. 1986. One hundred years of paleoanthropology. *American Scientist* 74:410–420.

Caspari, R., and M. H. Wolpoff. 1990. The morphological affinities of the Klasies River Mouth skeletal remains (abstract). *American Journal of Physical Anthropology* 81:203.

———. 1995. The pattern of human evolution. In *Man and the Environment in the Paleolithic,* H. Ullrich, ed. Pp. 19–27. Liège: ERAUL No. 62.

———. 1996. Variation in the Late Pleistocene human remains from Klasies River Mouth Cave and the problem of defining modern humans. Manuscript in possession of authors.

Castelloe, J., and A. R. Templeton. 1994. Root probabilities for intraspecific gene trees under neutral coalescent theory. *Molecular Phylogenetics and Evolution* 3:102–113.

Casti, J. 1989. *Paradigms Lost.* New York: William Morrow.

Cavalieri, P., and P. Singer, eds. 1994. *The Great Ape Project.* New York: St. Martin's Press.

Cavalli-Sforza, L. 1991. Genes, peoples and languages. *Scientific American* 265:104–110.

Cavalli-Sforza, L. L., P. Menozzi, and A. Piazza. 1993. Demic expansions and human evolution. *Science* 259:639–646.

————. 1994. *The History and Geography of Human Genes.* Princeton: Princeton University Press.

Cavalli-Sforza, L. L., A. Piazza, P. Menozzi, and J. Mountain. 1988. Reconstruction of human evolution: Bringing together genetic, archaeological, and linguistic data. *Proceedings of the National Academy of Science* 85:6002–6006.

Chamberlain, A. T., and B. A. Wood. 1987. Early hominid phylogeny. *Journal of Human Evolution* 16:119–133.

Chase, P. G. 1986. *The Hunters of Combe Grenal: Approaches to Middle Paleolithic Subsistence in Europe.* Oxford: BAR International Series No. 286.

————. 1988. Scavenging and hunting in the Middle Paleolithic: The evidence from Europe. In *The Upper Pleistocene Prehistory of Western Eurasia,* H. Dibble and A. Montet-White, eds. Pp. 225–232. Philadelphia: University of Pennsylvania Museum.

————. 1989. How different was Middle Palaeolithic subsistence? A zooarchaeological perspective on the Middle to Upper Palaeolithic transition. In *The Human Revolution,* P. Mellars and C. Stringer, eds. Pp. 321–337. Edinburgh: Edinburgh University Press.

Chase, P. G., and H. L. Dibble. 1987. Middle Paleolithic symbolism: a review of current evidence and interpretations. *Journal of Anthropological Archaeology* 6:263–296.

————. 1990. On the emergence of modern humans. *Current Anthropology* 31:58–66.

Chen, C. 1984. The microlith in China. *Journal of Anthropological Archaeology* 3:79–115.

Chen, T., Q. Yang, and E. Wu. 1994. Antiquity of *Homo sapiens* in China. *Nature* 368:55–56.

Cheynier, A. 1967. *Comment vivait l'Homme des Cavernes à l'Age du Renne.* Paris: Editions Arnoux.

Chia, L. P., P. Kai, and Y. You. 1972. Report of excavation in Shi Yu, Shanxi—a Palaeolithic site. *Acta Archaeologica Sinica* 1:39–60.

Chia, L. P., Q. Wei, and C. R. Li. 1979. Report on the excavation of Hsuchiayao man site in 1976. *Vertebrata PalAsiatica* 17:277–293.

Churchill, S. E. 1994. *Human Upper Body Evolution in the Eurasian Later Pleistocene.* Ph.D. dissertation, University of New Mexico, Albuquerque.

Churchill, S. E., and E. Trinkaus. 1990. Neandertal scapular glenoid morphology. *American Journal of Physical Anthropology* 83:147–160.

Churchill, S. E., O. M. Pearson, F. E. Grine, E. Trinkaus, and T. W. Holliday. 1996. Morphological affinities of the proximal ulna from Klasies River main site: Archaic or modern? *Journal of Human Evolution* 31:213–237.

Cioni, O., P. Gambassini, and D. Torre. 1979. Grotta di Castelcivita: risultati delle ricerche negli anni 1975–1977. *Atti Sociedade Toscana di Scienze Naturali* A86:275–296.

Clark, G. A. 1987. From the Mousterian to the Metal Ages: Long-term change in the human diet of northern Spain. In *The Pleistocene Old World World: Regional Perspectives,* O. Soffer, ed. Pp. 293–316. New York: Plenum.

————. 1988. Some thoughts on the Black Skull: an archaeologist's assessment of WT-17000 *(A. boisei)* and systematics in human paleontology. *American Anthropologist* 90:357–373.

————. 1989a. Alternative models of Pleistocene biocultural evolution: A response to Foley. *Antiquity* 63:153–161.

————. 1989b. Paradigms and paradoxes in paleoanthropology: A response to C. Loring Brace. *American Anthropologist* 91:446–450.

————. 1989c. Romancing the stones: Biases, style and lithics at La Riera. In *Alternative*

Approaches to Lithic Analysis, D. Henry and G. Odell, ed. Pp. 27–50. Washington: AP3A No. 1.

———. 1991a. Epilogue: Paradigms, realism, adaptation and evolution. In *Perspectives on the Past,* G. Clark, ed. Pp. 411–439. Philadelphia: University of Pennsylvania Press.

———. 1991b. A paradigm is like an onion: reflections on my biases. In *Perspectives on the Past,* G. Clark, ed. Pp. 79–108. Philadelphia: University of Pennsylvania Press.

———, ed. 1991. *Perspectives on the Past: Theoretical Biases in Mediterranean Hunter-Gatherer Research.* Philadelphia: University of Pennsylvania Press.

———. 1992a. Continuity or replacement? Putting modern human origins in an evolutionary context. In *The Middle Paleolithic: Adaptation, Behavior and Variability,* H. Dibble and P. Mellars, eds. Pp. 183–205. Philadelphia: University of Pennsylvania Museum.

———. 1992b. Paleoanthropological contexts. *Science* 257:597.

———. 1993a. Paradigms in science and archaeology. *Journal of Archaeological Research* 1:203–234.

———. 1993b. Review of *History and Evolution,* edited by Nitecki and Nitecki. *American Journal of Physical Anthropology* 90:129–130.

———. 1993c. Regional roots. *American Scientist* 81:4–5.

———. 1994a. Origine de l'homme: le dialogue du sourds. *La Recherche* 25:316–321.

———. 1994b. Migration as an explanatory concept in paleolithic archaeology. *Journal of Archaeological Method and Theory* 1:305–344.

———. 1994c. Aspectos epistemológicos de la interpretación del registro arqueológico: el papel del paradigma metafísico. In *Homenaje al Doctor Joaquín González Echegaray,* J. Lasheras, ed. Pp. 1–12. Madrid: Ministerio de Cultura y CIMA.

Clark, G., and J. M. Lindly. 1988. The biocultural transition and the origins of modern humans. *Paléorient* 14:159–167.

———. 1989a. The case for continuity: observations on the biocultural transition in Europe and western Asia. In *The Human Revolution,* P. Mellars and C. Stringer, eds. Pp. 626–676. Edinburgh: Edinburgh University Press.

———. 1989b. Modern human origins in the Levant and western Asia. *American Anthropologist* 91:962–985.

———. 1991. Paradigmatic biases and paleolithic research traditions. *Current Anthropology* 32:577–587.

Clark, G. A., and L. Straus. 1986. Synthesis and conclusions, part I: Upper Paleolithic and Mesolithic hunter-gatherer subsistence in northern Spain. In *La Riera Cave: Stone Age Hunter-Gatherer Adaptations in Northern Spain,* L. Straus and G. Clark, eds. Pp. 351–366. Tempe: Arizona State University Anthropological Research Papers No. 36.

Clark, G. A., and C. M. Willermet. 1995. In search of the Neanderthals: some conceptual issues with special reference to the Levant. *Cambridge Archaeological Journal* 5:153–156.

Clark, G. A., and S. Yi. 1983. Niche width variation in Cantabrian archaeofaunas: a diachronic study. In *Animals and Archaeology I: Hunters and Their Prey,* J. Clutton-Brock and C. Grigson, eds. Pp. 183–208. Oxford: BAR International Series S-163.

Clark, J. D. 1992. African and Asian perspectives on the origins of modern humans. *Philosophical Transactions of the Royal Society* B 337:201–215.

———. 1993. African and Asian perspectives on the origins of modern humans. In *The Origins of Modern Humans and the Impact of Chronometric Dating,* M. J. Aitken, C. B. Stringer, and P. A. Mellars, eds. Pp. 148–178. Princeton: Princeton University Press.

Clark, J. D., and K. D. Schick. 1988. Context and content: Impressions of Palaeolithic sites and assemblages in the People's Republic of China. *Journal of Human Evolution* 17:439–448.

Clarke, D. 1973. Archaeology: The loss of innocence. *Antiquity* 47:6–18.

———. 1978. *Analytical Archaeology,* 2nd ed. London: Methuen.

Clarke, R. J. 1990. The Ndutu cranium and the origin of *Homo sapiens. Journal of Human Evolution* 19:699–736.

Clottes, J., J-M. Chauvet, E. Brunel-Deschamps, D. Hillaire, J-P. Daugas, M. Arnold, H. Cachier, J. Evin, P. Fortin, C. Oberlin, N. Tisnerat, and H. Valladas. 1995. Les peintures paléolithiques de la Grotte Chauvet-Pont d'Arc, à Vallon-Pont-d'Arc (Ardèche, France): Datations directes et indirectes par la méthodes du radiocarbone. *Comptes-Rendus de l'Académie des Sciences de Paris* 320, IIa:1133–1140.

Clutton-Brock, T. H., and P. H. Harvey. 1979. Comparison and adaptation. *Proceedings of the Royal Society of London* B 205:547–565.

Coinman, N. 1990. *Refiguring the Levantine Upper Paleolithic.* Ph.D. dissertation, Arizona State University, Tempe.

Collins, H. 1985. *Changing Order.* London: Sage.

Combier J. 1990. De la fin du mousterien au paléolithique superieur: les données de la region rhodanienne. In *Paléolithique moyen recent et paléolithique superieur ancien en Europe,* C. Farizy, ed. Pp. 267–277. Actes de Colloque International de Nemours, 9–11 Mai 1988, Memoire du Musée de Préhistoire d'Ile de France, B203.

Conard, N. 1990. Laminar lithic assemblages from the last interglacial complex in northwestern Europe. *Journal of Anthropological Research* 46:243–262.

Conkey, M. 1978. Style and information in cultural evolution: Towards a cultural model for the Paleolithic. In *Social Archeology: Beyond Subsistence and Dating,* C. Redman, P. J. Watson, and S. LeBlanc, eds. Pp. 61–85. New York: Academic.

———. 1980. The identification of prehistoric hunter-gatherer aggregation sites. *Current Anthropology* 21:609–630.

Coon, C. S. 1954 *The Story of Man.* New York: Knopf. [2d ed., 1962].

———. 1962. *The Origin of Races.* New York: Knopf.

———. 1968. Comment on "bogus science." *Journal of Heredity* 59:275.

Copeland, L. 1972. Discussion: The stratigraphical and cultural problems of the passage from Middle to Upper Paleolithic in Palestine caves (Bar-Yosef and Vandermeersch). In *The Origin of Homo sapiens,* F. Bordes, ed. Pp. 221–225. Paris: UNESCO.

Corruccini, R. S. 1992. Metrical reconsideration of the Skhūl IV and IX and Border Cave 1 crania in the context of modern human origins. *American Journal of Physical Anthropology* 87:433–445.

———. 1994. Reaganomics and the fate of the progressive Neandertals. In *Integrative Paths to the Past: Paleoanthropological Advances in Honor of F. Clark Howell,* R. S. Corruccini and R. L. Ciochon, eds. Pp. 697–708. Englewood Cliffs: Prentice Hall.

Cose, E. 1994. One drop of bloody history. *Newsweek* (13 February):70–72.

Cosmides, L. 1989. The logic of social exchange: Has natural selection shaped how humans reason? Studies with the Wason selection task. *Cognition* 31:187–276.

Crandall, K. A. 1994. Intraspecific cladogram estimation: Accuracy at higher levels of divergence. *Systematic Biology* 43:222–235.

Crandall, K. A., and A. R. Templeton. 1993. Empirical tests of some predictions from coalescent theory with applications to intraspecific phylogeny reconstruction. *Genetics* 134:959–969.

Crandall, K. A., A. R. Templeton, and C. F. Sing. 1994. Intraspecific phylogenetics: Problems and solutions. In *Models in Phylogenic Reconstruction,* R. W. Scotland, D. J. Siebert, and D. M. Williams, eds. Pp. 135–156. Oxford: Clarendon Press.

Creath, R. 1995–1996. Are dinosaurs extinct? *Foundations of Science* 1:285–297.

Cronin, J. E. 1983. Apes, humans and molecular clocks. In *Interpretations of Ape and Human Ancestry* R. L. Ciochon and R. S. Corruccini, eds. Pp. 3–20. New York: Plenum.

Cronin, H. 1991. *The Ant and the Peacock.* Cambridge: Cambridge University Press.

Cronk, L. 1993. Parental favoritism towards daughters. *American Scientist* 81:272–279.

Crow, J. F., W. R. Engels, and C. Denniston. 1990. Phase 3 of Wright's shifting-balance theory. *Evolution* 44:233–247.

Culotta, E. 1995. New finds rekindle debate over anthropoid origins. *Science* 268:1851.

Curio, E. 1973. Towards a methodology of teleonomy. *Experientia* 29:1045–1058.

Cuvier, G. 1821. *Discours sur les revolutions de la surface du globe: recherches sur les ossements fossiles.* Paris: G. Dufour and E. d'Ocagne.

Darwin, C. 1859. *On the Origin of Species by Natural Selection.* London: John Murray.

———. 1871. *The Descent of Man, and Selection in Relation to Sex.* London: John Murray.

———. 1884. *The Origin of Species,* 6th ed. London: John Murray.

———. 1899. *The Variation of Animals and Plants under Domestication.* New York: Appleton.

———. 1985. *On The Origin of Species.* London: Penguin Books. (originally published in 1859)

David, N. 1966. Perigordian V regional facies: An attempt to define Upper Palaeolithic ethnic groups. *VII Congrès International des Sciences Préhistorique et Protohistorique.* Prague: UISPP.

Davidson, D. S. 1933. Australian netting and basketry techniques. *Journal of the Polynesian Society* 42:257–299.

Davidson, D. S., and F. D. McCarthy. 1957. The distribution and chronology of some important types of stone implements in Western Australia. *Anthropos* 52:390–458.

Davidson, I., and W. Noble. 1989. The archaeology of perception: Traces of depiction and language. *Current Anthropology* 30:125–137.

———. 1992. Why the first colonization of the Australian region is the earliest evidence of modern human behavior. *Archaeology in Oceania* 27:113–119.

Davies, B. 1986. *Storm Over Biology: Essays on Science, Sentiment, and Public Policy.* Buffalo: Prometheus.

Dawkins, R. 1976. *The Selfish Gene.* Oxford: Oxford University Press.

———. 1982. *The Extended Phenotype: The Gene as the Unit of Selection.* Oxford: W. H. Freeman.

———. 1986. *The Blind Watchmaker.* London: Longman.

———. 1994. Gaps in the mind. In *The Great Ape Project,* P. Cavalieri and P. Singer, eds. Pp. 80–87. New York: St. Martin's Press.

Day, M. H., and C. B. Stringer. 1982. A reconsideration of the Omo Kibish remains and the *erectus-sapiens* transition. In *l'Homo erectus et la Place de l'Homme de Tautavel parmi les Hominidés Fossiles,* Vol. 2, H. de Lumley, ed. Pp. 814–846. Nice: Louis-Jean Scientific and Literary Publications.

———. 1991. Les restes crâniens d'Omo-Kibish et leur classification à l'intérieur de Genre *Homo. L'Anthropologie* 95:573–594.

de Waal, F. 1982. *Chimpanzee Politics: Power and Sex among Apes.* London: Cape.

Deacon, H. J. 1992. Southern Africa and modern human origins. *Philosophical Transactions of the Royal Society* B 337:177–183.

Décsy, G. 1991. *The Indo-European Language: a Computational Reconstruction.* Bloomington, Indiana: Eurolingua.

Delporte, H. 1970. Le passage du moustérien au paléolithique supérieur. In *L'homme de Cro-Magnon: 1868–1968*, pp. 129–139. Paris: Conseil de Recherche Scientifique en Algérie.

de Lumley, H. 1969. Étude de l'outillage Moustérien de la Grotte de Carigüela (Píñar, Grenade). *L'Anthropologie* 78:165–206, 325–364.

de Lumley, M. A. 1973. *Anténeandertaliens et néandertaliens du bassin Méditérranéen occidental Européen*. Provence: Études Quaternaires de l'Université de Provence No. 2.

de Lumley, M. A., and M. García Sanchez. 1971. L'enfant néandertalien de Carigüela à Piñar (Andalousie). *L'Anthropologie* 75:29–55.

DeSalle, R., A. Templeton, I. Mori, S. Pletscher, and J. S. Johnston. 1987. Temporal and spatial heterogeneity of mtDNA polymorphisms in natural populations of *Drosophila mercatorum*. *Genetics* 116:215–223.

Desmond, A. 1989. *The Politics of Evolution: Morphology, Medicine and Reform in Radical London*. Chicago: University of Chicago Press.

De Vos, J., S. Sartono, S. Hardja-Sasmita, and P. Y. Sondaar. 1982. The fauna from Trinil, type locality of *Homo erectus*: A reinterpretation. *Geol. Mijnbouw* 61:207–211.

De Vos, J., P. Y. Sondaar, and C. C. Swisher III. 1994. Dating hominid sites in Indonesia. *Science* 266:1726–1727.

Diamond, J. 1966. Zoological classification system of a primitive people. *Science* 151:1102–1104.

———. 1992. *The Third Chimpanzee*. New York: Harper Collins.

———. 1994. The third chimpanzee. In *The Great Ape Project*, P. Cavalieri and P. Singer, eds. Pp. 88–101. New York: St. Martin's Press.

Dibble, H. 1987. The interpretation of Middle Paleolithic scraper morphology. *American Antiquity* 52:109–117.

Dibble, H. L., and P. G. Chase. 1990. Comment on "Symbolism and modern human origins" by J. M. Lindly and G. A. Clark. *Current Anthropology* 31:241–243.

Djindjian, F. 1993. L'Aurignacien du Périgord: une révision. *Préhistoire Européenne* 3:29–54.

Dobyns, H. F. 1966. Estimating aboriginal American population. 1: An appraisal of techniques with a new hemispheric estimate. *Current Anthropology* 7:395–449.

———. 1983. *Their Number Become Thinned: Native American Population Dynamics in Eastern North America*. Knoxville: University of Tennessee Press.

Dobzhansky, T. 1944. On species and races of living and fossil man. *American Journal of Physical Anthropology* 2:251–265.

———. 1950. The problem of the earliest claimed representatives of *Homo sapiens*, comment by T. D. Stewart. *Cold Spring Harbor Symposia on Quantitative Biology* 15:106–107.

———. 1968. More bogus "science" of race prejudice. *Journal of Heredity* 59:102–104.

Dong, Z. 1989. Microliths from Dabusu, Western Jiling Province. *Acta Anthropologica Sinica* 8:49–58.

Dorit, R. L., H. Akashi, and W. Gilbert. 1995. Absence of polymorphism at the ZFY locus on the human Y chromosome. *Science* 268:1183–1185.

Douglas, M. 1966. *Purity and Danger*. New York: Routledge.

Duff, A., G. Clark, and T. Chadderdon. 1992. Symbolism in the Early Paleolithic: A conceptual odyssey. *Cambridge Archaeological Journal* 2:211–229.

Dunnell, R. 1971. *Systematics in Prehistory*. Glencoe: Free Press.

Dupré, M. 1988. *Palinología y Paleoambiente*. Valencia: Servicio de Investigación Prehistórica.

Durham, W. H. 1991. *Coevolution: Genes, Culture and Human Diversity.* Stanford: Stanford University Press.

Edgington, E. S. 1986. *Randomization Tests,* 2nd ed. New York and Basel: Marcel Dekker.

Edwards, L. R., J. H. Chen, T.-L. Ku, and G. J. Wasserburg. 1987. Precise timing of the last interglacial period from mass spectrometric determination of thorium-230 in corals. *Science* 236:1547–1553.

Eldredge, N. 1993. What, if anything, is a species? In *Species, Species Concepts, and Primate Evolution,* W. H. Kimbel and L. B. Martin, eds. Pp. 3–20. New York: Plenum Press.

Eldredge, N., and S. J. Gould. 1972. Punctuated equilibria: An alternative to phyletic gradualism. In *Models in Paleobiology,* R. J. M. Schopf, ed. Pp. 82–115. San Francisco: Freeman Cooper.

Eldredge, N., and I. Tattersall. 1982. *The Myths of Human Evolution.* New York: Columbia University Press.

Elliot, D. G. 1913. *A Review of the Primates.* New York: Monographs of the American Museum of Natural History.

Estevez, J. 1987. Dynamique des faunes préhistoriques au N-E de la péninsule Ibérique. *Archaeozoología* 1:197–218.

Evans, F. G. 1945. The names of fossil men. *Science* 102:16–17.

Excoffier, L. 1988. *Polymorphisme de l'ADN mitochondrial et histoire du peuplement humain.* Ph.D. dissertation, University of Geneva.

————. 1990. Evolution of human mitochondrial DNA: Evidence for a departure from a pure neutral model of populations at equilibrium. *Journal of Molecular Biology* 30:125–139.

Excoffier, L., and A. Langaney. 1989. Origin and differentiation of human mitochondrial DNA. *American Journal of Human Genetics* 44:73–85.

Excoffier, L., and P. E. Smouse. 1994. Using allele frequencies and geographic subdivision to reconstruct gene trees within a species—molecular variance parsimony. *Genetics* 136:343–359.

Farrand, W. 1972. Geological correlation of prehistoric sites in the Levant. In *The Origin of Homo sapiens,* F. Bordes, ed. Pp. 227–235. Paris: UNESCO.

Felsenstein, J. 1983. Parsimony in systematics: biological and statistical issues. *Annual Review of Ecology and Systematics* 14:313–333.

Ferembach, D. 1964–1965. La molaire humaine inférieure moustérienne de Bombarral (Portugal). *Comunicaçoes dos Serviços Geologicos* 46:177–188.

————. 1974. *Le Gisement Mésolithique de Moita do Sebastião, Muge, Portugal II: Anthropologie.* Lisboa: Direcçao-Geral dos Assuntos Culturais.

Fleagle, J. G. 1988. *Primate Adaptations and Evolution.* New York: Academic Press.

Fleagle, J. G., and W. L. Jungers. 1982. Fifty years of higher primate phylogeny. In *A History of American Physical Anthropology, 1930–1980,* F. Spencer, ed. Pp. 187–230. New York: Academic Press.

Foley, R. 1987a. *Another Unique Species: Patterns in Human Evolutionary Ecology.* Essex: Longman Scientific and Technical.

————. 1987b. Hominid species and stone-tool assemblages: How are they related? *Antiquity* 61:380–392.

————. 1989. Reply to "Alternative models of Pleistocene biocultural evolution: a response to Foley" by G. A. Clark. *Antiquity* 63:159–162.

————. 1991. Comment on "New models and metaphors for the Neanderthal debate" by Paul Graves. *Current Anthropology* 32:529–530.

————. n.d. An evolutionary and chronological framework for human social behaviour. *Proceedings of the British Academy*, in press.

Forey, P. 1991. Blood lines of the coelacanth. *Nature* 351:347–348.

Frayer, D. W. 1978. *Evolution of the Dentition in Upper Paleolithic and Mesolithic Europe*. Lawrence: University of Kansas Publications in Anthropology No. 10.

————. 1981. Body size, weapon use, and natural selection in the European Upper Paleolithic and Mesolithic. *American Anthropologist* 83:57–73.

————. 1984. Biological and cultural change in the European Late Pleistocene and Early Holocene. In *The Origins of Modern Humans*, F. H. Smith and F. Spencer, eds. Pp. 211–250. New York: Alan R. Liss.

————. 1986. Cranial variation at Mladeč and the relationship between Mousterian and Upper Paleolithic hominids. *Anthropos* 23:243–256.

————. 1988. Biological Evidence for Differences in Social Patterning in the European Upper Paleolithic and Mesolithic. *Supplemento del Rivista di Antropologia* 66:127–140.

————. 1989a. Postcranial metric changes in the European Upper Paleolithic and Mesolithic. *American Journal of Physical Anthropology* 78:223–224 (abstract).

————. 1989b. Oral pathologies in the European Upper Paleolithic and Mesolithic. In *People and Culture in Change*, I. Herskovitz, ed., Pp. 255–282. Oxford: BAR International Series No. 508(i).

————. 1992a. Evolution at the European edge: Neanderthal and Upper Paleolithic relationships. *Préhistoire Européene—European Prehistory* 2:9–69.

————. 1992b. The persistence of Neanderthal features in post-Neanderthal Europeans. In *Continuity or Replacement: Controversies in Homo sapiens evolution*, G. Bräuer and F. H. Smith, eds. Pp. 179–188. Rotterdam: A. A. Balkema.

————. n.d. The dental remains from Krskany and Vedrovice. In *Vedroviče and Nitra–Krskany: Anthropology of the Two Early Neolithic Central European Populations*, J. Jelínek, ed. Brno: Moravské Museum, in press.

Frayer, D. W., M. H. Wolpoff, A. G. Thorne, F. H. Smith, and G. G. Pope. 1993. Theories of modern human origins: The paleontological test. *American Anthropologist* 95:14–50.

————. 1994a. Getting it straight. *American Anthropologist* 96:424–438.

————. 1994b. Reply to "Resolving the archaic-to-modern transition" by Grover Krantz. *American Anthropologist* 96:152–155.

Freeman, L. 1973. The significance of mammalian faunas from Paleolithic occupations in Cantabrian Spain. *American Antiquity* 38:3–44.

————. 1977. Paleolithic archeology and paleoanthropology in China. In *Paleoanthropology in the People's Republic of China*, W. W. Howells and P. Tsuchitani, eds. Pp. 79–113. Washington: National Academy of Sciences Press.

————. 1981. The fat of the land: notes on Paleolithic diet in Iberia. In *Omnivorous Primates*, R. Harding and G. Teleki, eds. Pp. 104–165. New York: Columbia University.

————. 1993. La "transición" en Cantabria y la importancia de Cueva Morín y sus vecinos en el debate actual. In *El Origen del Hombre Moderno en el Suroeste de Europa*, V. Cabrera, ed., Pp. 171–194. Madrid: UNED.

Freeman, L. G., and J. Gonzalez Echegaray. 1973. *Los Enterramientos Paleolíticos de Cueva Morín (Santander)*. Santander: Patronato de las Cuevas Prehistóricas.

Friedenthal, H. 1900. Uber einen experimentellen Nachweiss von Blutsverwantschaft. *Archiv fur Anatomie und Physiologie Leipzig, Physiologische Abteilung*, pp. 494–508.

Fusté, M. 1957. Molde intracraneal de un nuevo resto del hombre de neandertal en España. *Cursos y Conferencias del Instituto Lucas Mallada* 4:41–43.

Gábori, M. 1988. Nouvelles découvertes dans le paléolithique d'Asie centrale soviétique. In *Upper Pleistocene Prehistory of Western Eurasia*, H. Dibble and A. Montet-White, eds. Pp. 287–296. Philadelphia: University of Pennsylvania Museum.

Gábori-Szánk, V. 1970. C-14 dates of the Hungarian paleolithic. *Acta Archaeologica Academiae Scientarum Hungaricae* 22:3–11.

Gai, P. 1985. Microlithic industries in China. In *Palaeoanthropology and Palaeolithic Archaeology in the People's Republic of China*, R. Wu and J. Olsen, eds. Pp. 225–241. London: Academic Press.

Gai, P., and Q. Wei. 1989. Discovery of the Late Palaeolithic site at Hutouliang. In *Selected Treatises on Nihewan*, Q. Wei and P. Gai, eds. Pp. 48–63. Beijing: Cultural Relics Publishing House.

Gallay, A. 1989. Logicism: A French view of archaeological theory founded in computational perspective. *Antiquity* 63:27–40.

Gambassini, P. 1982. Le Paléolithique supérieur ancien en Campanie. *Actes des Réunions de la Xᵉ Commission "Aurignacien et Gravettien" UISPP (1976–1981)* 13:139–151. Liège: ERAUL.

Gambier, D. 1989. Fossil hominids from the Upper Paleolithic (Aurignacian) of France. In *The Human Revolution*, P. Mellars and C. Stringer, eds. Pp. 194–211. Edinburgh: Edinburgh University Press.

———. 1992. Origine de l'homme moderne en Europe: Comparaison des données en Europe centrale et occidentale. In *Actes du Colloque International "Cinquante Mille Siècles d'Aventure Humaine,"* M. Toussaint, ed. Pp. 269–284. Liège: ERAUL No. 56.

———. 1993. Les hommes modernes du début du Paléolithique supérieur en France: Bilan des données anthropologiques et perspectives. In *El Origen del Hombre Moderno en el Suroeste de Europa*, V. Cabrera, ed. Pp. 409–430. Madrid: UNED.

———. n.d. Inventaire des vestiges humains découverts dans un contexte aurignacien ou gravettien. Manuscript in possession of author.

Gambier, D., and D. Sacchi, 1991. Sur quelques restes humains léptolithiques de la Grotte de la Crouzade, Aude. *L'Anthropologie* 95:155–180.

Gambier, D., F. Houet, and A. M. Tillier. 1990. Dents de Font de Gaume (Châtelperronien et Aurignacien) et de La Ferrassie (Aurignacien ancien) en Dordogne. *Revue d'Archéologie Préhistorique* 2:143–151.

Gamble, C. 1982. Interaction and alliance in Palaeolithic society. *Man* 17:92–107.

———. 1986. *The Palaeolithic Settlement of Europe*. Cambridge: Cambridge University Press.

———. 1993. *Timewalkers: The Prehistory of Global Colonization*. Far Thrupp (Stroud): Alan Sutton.

Gamble, C., and O. Soffer, eds. 1990. *The World at 18,000 B.P., Vol. 2: Low Latitudes*. London: Unwin Hyman.

Gao, J. 1975. Australopithecine teeth associated with *Gigantopithecus*. *Vertebrata Pal-Asiatica* 13:81–88.

García Sanchez, M. 1960. Restos humanos del Paleolítico Medio y Superior y del Neo-Eneolítico de Píñar (Granada). *Trabajos del Instituto. Bernardino de Sahagún* 15:17–78.

———. 1984. Estudio préliminar de los restos neandertalenses del Boquete de Zafarraya (Alcaucín, Málaga). In *Homenaje a Luís Siret*, Pp. 17–78. Málaga: Consejo Cultural de la Junta de Andalucía.

García Sanchez, M., and J. Carrasco. 1981. Cráneo-copa eneolítico de la Carigüela de Píñar, (Granada). *Zephyrus* 32–33:131–131.

García Sanchez, M., A. M. Tillier, M. D. Garralda, and G. Vega. 1994. Les dents d'enfant des niveaux Moustériens de la Grotte de Carihuela (Grenade, Espagne). *Paléo* 6: 79–88.

Gardin, J.-Cl., M.-S. Lagrange, J.-M. Martin, J. Molino, and J. Natali. 1981. *La Logique du Plausible: Essai d'Épistémologie Pratique*. Paris: Maison des Sciences de l'Homme.

Gardiner, B. G., P. Janvier, C. Patterson, P. L. Forey, P. H. Greenwood, R. S. Miles, and R. P. S. Jefferies. 1979. The salmon, the lungfish, and the cow: A reply. *Nature* 277:175–176.

Gargett, R. H. 1989. Grave shortcomings: The evidence for Neandertal burial. *Current Anthropology* 30:157–190.

Garralda, Mª. D. 1974. *La Población del Neolítico y Bronce I en la Península Ibérica.* Tesis Doctoral, Departamento de Biología, Universidad Complutense de Madrid.

———. 1978. Datación absoluta y restos humanos en la Península Ibérica. In *C-14 y la Prehistoria de la Península Ibérica.* Madrid: Fundación Juán March, Serie Universitaria 77:7–15.

———. 1986. The Azilian man from Los Azules I Cave (Cangas de Onís, Oviedo). *Human Evolution* 1:431–448.

———. 1992. Evolution of human height. In *Human Growth: Basic and Clinical Aspects,* M. Hernández and J. Argente, eds., pp. 135–142. Amsterdam: Excerpta Medica, International Congress Series No. 973.

———. 1993. La transición del Paleolítico Medio al Superior en la Península Ibérica: perspectivas antropológicas. In *El Orígen del Hombre Moderno en el Suroeste de Europa,* V. Cabrera, ed. Pp. 373–389, Madrid: UNED.

Garralda, Mª. D., and B. Vandermeersch. 1993. L'évolution de la stature. *Bulletins et Mémoires de la Société d'Anthropologie de Paris* 5:269–281.

Garralda, Mª. D., A. M. Tillier, B. Vandermeersch, V. Cabrera, and D. Gambier. 1992. Restes humains de l'Aurignacien archaïque de la Cueva de El Castillo (Santander, Espagne). *Anthropologie* (Brno) 30:159–164.

Garrod, D. A. E. 1956. Acheulo-Jabroudien et "Pré-Aurignacien" de la Grotte du Taboun (Mont Carmel): étude stratigrafique et chronologique. *Quaternaria* 3:39–59.

Garrod, D. A. E., and G. Henri-Martin. 1961. Rapport préliminaire sur les fouilles en Ras el-Kelb, Liban 1959. *Bulletin du Musée de Beyrouth* 16:61–66.

Garrod, D. A. E., and D. Kirkbride. 1961. Excavation of the Abri Zumoffen, a paleolithic rock-shelter near Adlun, south Lebanon. *Bulletin du Musée de Beyrouth* 16:7–45.

Garrod, D. A. E., L. H. D. Buxton, G. Elliot Smith, and D. M. A. Bate. 1928. Excavation of a Mousterian rock-shelter at Devil's Tower, Gibraltar. *Journal of the Royal Anthropological Institute* 58:37–48.

Gascoyne, M., H. P. Schwarcz, and D. C. Ford. 1983. Uranium series ages of speleothem from northwest England: Correlation with Quaternary climate. *Philosophical Transactions of the Royal Society of London* B 301:143–164.

Gates, R. R. 1944. Phylogeny and classification of hominids and anthropoids. *American Journal of Physical Anthropology* 2:279–292.

———. 1948. *Human Ancestry from a Genetical Point of View.* Cambridge, Massachusetts: Harvard University Press.

Genovés, S. 1964. Comment on "The fate of the 'classic' Neanderthals: A consideration of hominid catastrophism" by C. Loring Brace. *Current Anthropology* 5:22–25.

Gerber, A. 1994. *The semiotics of subdivision: an empirical study of the population structure of* Trimerotropis saxatilis *(Acrididae).* Ph.D. dissertation, Washington University, St. Louis.

Ghiselin, M. 1974. A radical solution to the species problem. *Systematic Zoology* 23:536–544.

Gibbons, A. 1991. Looking for the father of us all. *Science* 251:378–380.

———. 1995a. The mystery of humanity's missing mutations. *Science* 267:35–36.

———. 1995b. Out of Africa—at last? *Science* 267:1272–1273.

———. 1995c. Old dates for modern behavior. *Science* 268:495–496.

Giere, R. 1988. *Explaining Science: A Cognitive Approach.* Chicago: University of Chicago Press.

Gilead, I., and O. Bar-Yosef. 1993. Early Upper Paleolithic sites in the Qadesh Barnea area, northeast Sinai. *Journal of Field Archaeology* 20:265–280.

Gingerich, P. D. 1983. Rates of evolution: effects of time and temporal scaling. *Science* 222:159–161.

———. 1985. Species in the fossil record: Concepts, trends, and transitions. *Paleobiology* 11:27–41.

Gioia, P. 1988. An aspect of the transition between Middle and Upper Paleolithic in Italy: the Uluzzian. *Mémoires de la Musée de Préhistoire d'Ile-de-France* 3:241–250.

Glen, E., and K. Kaczanowski. 1982. Human remains. In *The Excavation of Bacho Kiro Cave: Final Report,* J. Kozlowski, ed. Pp. 75–79. Warsaw: University of Warsaw Press.

Godfrey, L., and J. Marks. 1991. The nature and origins of primate species. *Yearbook of Physical Anthropology* 34:39–68.

Golson, J. 1974. Land connections, sea barriers and the relationship of Australian and New Guinea prehistory. In *Bridge and Barrier,* 2nd ed., D. Walker, ed. Pp. 375–397. Canberra: Australian National University.

González Echegaray, J. 1969. El paso del Paleolítico Medio al Superior en la Costa Cantábrica. *Anuario de Estudios Atlánticos* 15:273–279.

———. 1978. Notes toward a systematization of the Upper Paleolithic in Palestine. In *Views of the Past,* L. Freeman, ed. Pp. 177–191. Mouton: The Hague.

———. 1993. La evolución histórica del concepto de la "transición a los cazadores" del Paleolítico Superior. In *El Origen del Hombre Moderno en el Suroeste de Europa,* V. Cabrera, ed. Pp. 105–116. Madrid: Universidad Nacional de Educación a Distancia.

González Echegaray, J., and L. G. Freeman. 1971. *Cueva Morín: Excavaciones 1966–1968.* Santander: Patronato de Cuevas Prehistoricas.

———. 1973. *Cueva Morín: Excavaciones 1969.* Santander: Patronato de Cuevas Prehistoricas.

———. 1978a. *Vida y Muerte en Cueva Morín.* Santander: Institución Cultural de Cantabria.

———. 1978b. Los restos humanos Auriñacienses de Cueva Morín. In *I° Simposio de Antropología Biológica de España,* M. D. Garralda and R. M. Grande, eds. Pp. 145–148. Madrid:UISPP.

González Echegaray, J., L. G. Freeman, I. Barandiarán, J. M. Apellániz, K. W. Butzer, C. Fuentes Vidarte, B. Madariaga, J. A. González Morales, and A. Leroi-Gourhan.1980. *El Yacimiento de la Cueva de "El Pendo" (Excavaciones 1953–1957).* Madrid: CSIC.

Goodall, J. 1986. *The Chimpanzees of Gombe: Patterns of Behavior.* Cambridge, Massachusetts: Belknap.

Goodman, M. 1962. Immunochemistry of the primates and primate evolution. *Annals of the New York Academy of Sciences* 102:219–234.

———. 1976. Towards a geneological description of the Primates. In *Molecular Anthropology,* M. Goodman and R. E. Tashian, eds. Pp. 321–353. New York: Plenum Press.

Goodman, M., A. E. Romero-Herrera, H. Dene, J. Czelusniak, and R. Tashian. 1983.

Amino acid sequence evidence on the phylogeny of primates and other eutherians. In *Macromolecular Sequences in Systematic and Evolutionary Biology,* M. Goodman, ed. Pp. 115–192. New York: Plenum.

Goodship, A. E., L. E. Lanyon, and H. Mcfie. 1979. Functional adaptation of bone to increased stress. *Journal of Bone and Joint Surgery* 61:539–546.

Gorjanović-Kramberger, D. 1904. Potijče li moderni čovjek ravnó od dilúvijalonga *Homo* primigeniusa? *I Kongress Srpskih Lekara i Prirodnjaka,* pp. 1–8. Zagreb.

———. 1906. *Der Diluviale Mensch von Krapina in Kroatien: Ein Beitrag zur Palóoanthropologie.* Wiesbaden: Kreidel.

Gottwald, S. 1993. *Fuzzy Sets and Fuzzy Logic: The Foundations of Application—from a Mathematical Point of View.* Weisbaden: Vieweg.

Goudge, T. 1961. *The Ascent of Life.* Toronto: University of Toronto Press.

Gould, S. J. 1995. Age-old fallacies of thinking and stinking. *Natural History* 104:6–13.

Gould, S. J., and R. C. Lewontin. 1979. The spandrels of San Marco and the Panglossian paradigm: A critique of the adaptationist programme. *Proceedings of the Royal Society of London* B 205:581–598.

Gowlett, J. A. J. 1987. The coming of modern man. *Antiquity* 61:210–219.

Grant, M. 1916. *The Passing of the Great Race.* New York: Scribner's.

Grant, P. R. 1986. *Ecology and Evolution of Darwin's Finches.* Princeton: Princeton University Press.

Grant, R. B., and P. R. Grant. 1989. *Evolutionary Dynamics of a Natural Population: The Large Cactus Finch of the Galapagos.* Chicago: University of Chicago Press.

Graves, P. 1991. New models and metaphors for the Neanderthal debate. *Current Anthropology* 32:513–536.

Greenberg, J. H. 1960. The general classification of Central and South American languages. In *Men and Cultures: Selected Papers of the 5th International Congress of Anthropological and Ethnological Sciences,* F. C. Wallace, ed. Pp. 791–794. Philadelphia: University of Pennsylvania Press.

———. 1987. *Language in the Americas.* Stanford: Stanford University Press.

Greene, D. L., G. H. Ewing, and G. J. Armelagos. 1967. Dentition of a Mesolithic population from Wadi Halfa, Sudan. *American Journal of Physical Anthropology* 27:41–56.

Greene, J. C. 1959. *The Death of Adam: Evolution and its Impact on Western Thought.* Ames: Iowa State University Press.

Grine, F. E. 1993. Australopithecine taxonomy and phylogeny: historical background and recent interpretation. In *The Human Evolution Source Book,* R. L. Ciochon and J. G. Fleagle, ed. Pp. 189–210. Englewood Cliffs: Prentice Hall.

Groves, C. P. 1989a. A regional approach to the problem of the origin of modern humans in Australasia. In *The Human Revolution,* P. Mellars and C. B. Stringer, eds. Pp. 274–285. Princeton: Princeton University Press.

———. 1989b. *A Theory of Human and Primate Evolution.* Oxford: Clarendon Press.

Groves, C. P., and M. M. Lahr. 1994. A bush not a ladder: Speciation and replacement in human evolution. In *Perspectives in Human Biology,* L. Freedman, N. Jablonski, and N. W. Bruce, eds. Pp. 1–11. New York: Wiley-Liss.

Grün, R., and C. B. Stringer. 1991. Electron spin resonance dating and the evolution of modern humans. *Archaeometry* 33:153–199.

Grün, R., P. B. Beaumont, and C. B. Stringer. 1990. ESR dating evidence for early modern humans at Border Cave in South Africa. *Nature* 344:537–539.

Grün, R., H. P. Schwarcz, and S. Zymela. 1987. ESR dating of tooth enamel. *Canadian Journal of Earth Sciences* 24:1022–1037.

Guba, E. 1990. The alternative paradigm dialog. In *The Paradigm Dialog,* E. Guba, ed. Pp. 17–30. Newbury Park: Sage Publications.

Guba, E., ed. 1990. *The Paradigm Dialog.* Newbury Park: Sage Publications.

Habgood, P. J. 1985. The origin of the Australian Aborigines: an alternative approach and view. In *Hominid Evolution: Past, Present and Future,* P. V. Tobias, ed. Pp. 367–380. New York: Alan R. Liss.

———. 1989. The origin of anatomically modern humans in Australasia. In *The Human Revolution,* P. Mellars and C. B. Stringer, eds. Pp. 245–273. Edinburgh: Edinburgh University Press.

———. 1992. The origin of anatomically modern humans in East Asia. In *Continuity or Replacement: Controversies in* Homo sapiens *evolution,* G. Bräuer and F. H. Smith, ed. Pp. 273–288. Rotterdam: A. A. Balkema.

Haeckel, E. 1905. *The Wonders of Life.* New York: Harper.

Hahn, J. 1995. Neue Beschleuniger: ^{14}C-daten zum Jungpaläolithikum in Südwestdeutschland. *Eiszeitalter und Gegenwart* 45:86–92.

Haldane, J. B. S. 1949. Suggestions as to the quantitative measurements of the rates of evolution. *Evolution* 3:51–56.

Hamilton, W. D. 1964a. The genetical evolution of social behaviour I. *Journal of Theoretical Biology* 7:1–16.

———. 1964b. The genetical evolution of social behaviour II. *Journal of Theoretical Biology* 7:17–32.

Hammer, M. F. 1995. A recent common ancestry for human Y chromosomes. *Nature* 378:376–378.

Haraway, D. J. 1988. Remodelling the human way of life: Sherwood Washburn and the new physical anthropology. In *Bones, Bodies, Behavior: Essays on Biological Anthropology,* G. W. Stocking Jr., ed. Pp. 206–259. Madison: University of Wisconsin Press.

Harpending, H. 1994. Signature of ancient population growth in a low resolution mitochodrial DNA mismatch distribution. *Human Biology* 66:591–600.

Harpending, H., S. T. Sherry, A. R. Rogers, and M. Stoneking. 1993. The genetic structure of ancient human populations. *Current Anthropology* 34:483–496.

Harrold, F. 1981. New perspectives on the Châtelperronian. *Ampurias* 43:1–51.

———. 1983. The Châtelperronian and the Middle-Upper Paleolithic transition. In *The Mousterian Legacy: Human Biocultural Change in the Upper Pleistocene,* E. Trinkaus, ed. Pp. 123–140. Oxford: BAR International Series No. 164.

———. 1989. Mousterian, Châtelperronian and early Aurignacian in western Europe: continuity or discontinuity? In *The Human Revolution,* P. Mellars and C. Stringer, eds. Pp. 677–713. Edinburgh: Edinburgh University.

———. 1991a. Comment on "New models and metaphors for the Neanderthal debate" by Paul Graves. *Current Anthropology* 32:530.

———. 1991b. The elephant and the blind men: Paradigms, data gaps, and the Middle-Upper Paleolithic transition in southwestern France. In *Perspectives on the Past,* G. Clark, ed. Pp. 164–182. Philadelphia: University of Pennsylvania Press.

———. 1992. Paleolithic archeology, ancient behavior, and the transition to modern *Homo.* In *Continuity or Replacement: Controversies in Homo Sapiens Evolution,* G. Bräuer and F. Smith, eds. Pp. 219–230. Rotterdam: A. A. Balkema.

Hasegawa, M., T. Yano, and H. Kishino. 1984. A new molecular clock of mitochondrial DNA and the evolution of the hominoids. *Proceedings of the Japanese Academy of Science* 60:95–105.

Hausman, A. J. 1982. The biocultural evolution of Khoisan populations of southern Africa. *American Journal of Physical Anthropology* 58:315–330.

Hayden, B. 1981. Research and development in the stone age: Technological transitions among hunter-gatherers. *Current Anthropology* 22:519–548.

———. 1982. Interaction parameters and the demise of Paleo-Indian craftmanship. *Plains Anthropologist* 27:109–123.

Haynes, C. 1992. Contributions of radiocarbon dating to the geochronology of the peopling of the New World. In *Radiocarbon after Four Decades,* A. Long and R. S. Kra, eds. Pp. 73–84. New York: Springer-Verlag.

Hedges, R. E. M., R. A. Housley, C. R. Bronk, and G. J. Van Klinken. 1992. Radiocarbon dates from the Oxford AMS system: Archaeometry datelist 14. *Archaeometry* 34:141–159.

Hedges, R., R. Housley, C. B. Ramsey, and G. Van Klinken. 1994. Radiocarbon dates from the Oxford AMS system: Archaeometry datelist 18. *Archaeometry* 36:337–374.

Hedges, S. B., S. Kumar, K. Tamurs, and M. Stoneking. 1992. Human origins and analysis of mitochondrial DNA sequences. *Science* 255:737–739.

Hedin, S. 1925. *My Life as an Explorer.* Hong Kong: Oxford University Press.

Hempel, C. 1952. *Fundamentals of Concept Formation in Empirical Science.* Chicago: University of Chicago Press.

———. 1965. *Aspects of Scientific Explanation.* New York: Free Press.

———. 1966. *Philosophy of Natural Science.* Englewood Cliffs: Prentice-Hall.

Hennig, W. 1965. Phylogenetic systematics. *Annual Review of Entomology* 10:97–116.

———. 1966. *Phylogenetic Systematics.* Urbana: University of Illinois Press.

Herrnstein, R., and C. Murray. 1994. *The Bell Curve.* New York: The Free Press.

Herskovits, M. 1924. What is a race? *American Mercury* 2:207–210.

Hesse, M., and M. Arbib. 1986. *The Construction of Reality.* Cambridge: Cambridge University Press.

Hierneaux, J. 1964. The concept of race and the taxonomy of mankind. In *The Concept of Race,* A. Montagu, ed. Pp. 29–45. New York: The Free Press.

Hodson, F. 1982. Some aspects of archaeological classification. In *Essays on Archaeological Typology,* R. Whallon and J. Brown, eds. Pp. 21–29. Evanston: Center for American Archeology.

Hofstadter, R. 1944. *Social Darwinism in American Life.* Philadelphia: University of Pennsylvania Press.

Holloway, R. 1966. Structural reduction through the "probable mutation effect": A critique with questions regarding human evolution. *American Journal of Physical Anthropology* 25:7–11.

Hooton, E. A. 1936. Plain statements about race. *Science* 83:511–513.

Horai, S., and K. Hayasaka. 1990. Intraspecific nucleotide sequence differences in the major noncoding region of human mitochondrial DNA. *American Journal of Human Genetics* 46:828–842.

Horai, S., K. Hayasaka, R. Kondo, K. Tsugane, and N. Takahata. 1995. Recent African origin of modern humans revealed by complete sequences of hominoid mitochondrial DNAs. *Proceedings of the National Academy of Science* 92:532–536.

Howell, F. C. 1951. The place of neanderthal man in human evolution. *American Journal of Physical Anthropology* 9:379–416.

———. 1960. European and northwest African Pleistocene hominids. *Current Anthropology* 1:195–232.

———. 1994. A chronostratigraphic and taxonomic framework of the origins of modern humans. In *Origins of Anatomically Modern Humans,* M. Nitecki and D. Nitecki, eds. Pp. 253–319. New York: Plenum.

Howells, W. W. 1942. Fossil man and the origin of races. *American Anthropologist* 44:182–193.

———. 1944. *Mankind So Far.* Garden City: Doubleday.

————. 1967. *Mankind in the Making: The Story of Human Evolution.* Garden City: Doubleday.

————. 1973. *Cranial Variation in Man.* Cambridge, Massachusetts: Papers of the Peabody Museum of Archaeology and Ethnology, Vol. 67.

————. 1974. Neanderthals: Names, hypotheses, and scientific method. *American Anthropologist* 76:24–38.

————. 1975. Neanderthal man: Facts and figures. In *Paleoanthropology: Morphology and Paleoecology,* R. H. Tuttle, ed. Pp. 389–407. The Hague: Mouton.

————. 1976. Explaining modern man: Evolutionists versus migrationists. *Journal of Human Evolution* 5:477–496.

————. 1989. *Skull Shapes and the Map.* Cambridge, Massachusetts: Papers of the Peabody Museum of Archaeology and Ethnology, Vol. 79.

————. 1991. Current theories on the origin of *Homo sapiens sapiens.* In *Les Processus de l'Hominisation,* D. Ferembach, ed. Pp. 73–78. Paris: CNRS.

Hoyos Sainz, L. 1947. Antropología prehistórica e histórica. In *Historia de España* I, pp. 97–241. Madrid: Espasa Calpe.

Hrdlička, A. 1911. Human dentition from the evolutionary standpoint. *Dominion Dental Journal* 23:403–422.

————. 1913. A search in eastern Asia for the race that peopled America. *Smithsonian Miscellaneous Collections* 60:10–13.

————. 1914. The most ancient skeletal remains of man. *Smithsonian Institution Annual Report 1913,* pp. 491–552. Washington: Government Printing Office.

————. 1915. The peopling of America. *Journal of Heredity* 6:79–91.

————. 1918. Recent discoveries attributed to early man in America. *Bulletin of the Bureau of American Ethnology* 66:1–67.

————. 1925. The origin and antiquity of the American Indian. *Annual Report of the Board of Regents of the Smithsonian Institution for 1923,* pp. 481–494. Washington: Government Printing Office.

————. 1927. The Neanderthal phase of man. *Journal of the Royal Anthropological Institute* 57:249–274.

————. 1929. The Neanderthal phase of man. *Annual Report of the Board of Regents of the Smithsonian Institution for 1928,* pp. 593–621. Washington: Government Printing Office.

————. 1930. The skeletal remains of early man. *Smithsonian Miscellaneous Collections* 83:1–379.

Huang, W. W., J. W. Olsen, R. W. Reeves, S. Miller-Antonio, and J. Q. Lei. 1988. New discoveries of stone artifacts on the southern edge of the Tarim Basin, Xinjiang. *Acta Anthropologica Sinica* 7:294–301.

Hublin, J. J. 1978. *Le Torus Transverse et les Structures Associées: Evolution dans le Genre Homo.* Ph.D. dissertation, Université Pierre et Marie Curie, Paris.

————. 1990. Les peuplements Paléolithiques de l'Europe: un point de vue paléobiogéographiques. In *Paléolithique Moyen Récent et Paléolithique Supérieur Ancien en Europe,* pp. 29–37. Paris: Memoires de la Musée de Préhistoire d'Ile de France No. 3.

————. 1994. The Zafarraya Mousterian site: New evidence on the contemporaneity of modern humans and Neanderthals in southwestern Europe. *Palaeoanthropology Society Abstracts* 3:7–8.

Hublin, J. J., C. Barroso, P. Medina, M. Fontugne, and J-L. Reyss. 1995. The Mousterian site of Zafarraya (Andalucia, Spain): dating and implications on the Paleolithic peopling processes of western Europe. *Comptes-Rendus de l'Académie des Sciences de Paris* 321:931–937.

Hudson, W. D. 1983. *Modern Moral Philosophy,* 2nd ed. London: Macmillan.

Hué, E. 1937. Crânes paléolithiques. *Actes du Congrès Préhistorique de France.* Toulouse-Foix: CPF.

Hull, D. 1970. Contemporary systematic philosophies. *Annual Review of Ecological Systems* 1:19–53.

———. 1976. Are species really individuals? *Systematic Zoology* 25:174–191.

———. 1978. A matter of individuality. *Philosophy of Science* 45:335–360.

———. 1988a. A mechanism and its metaphysics: An evolutionary account of the social and conceptual development of science. *Biology and Philosophy* 3:123–155.

———. 1988b. *Science as a Process.* Chicago: University of Chicago Press.

———. 1989. *The Metaphysics of Evolution.* Albany: SUNY Press.

Hulse, F. 1955. Technological advance and major racial stocks. *Human Biology* 27:184–192.

Huxley, T. H. 1863 (1959). *Man's Place in Nature.* Ann Arbor: University of Michigan Press.

———. 1872. *A Manual of the Anatomy of Vertebrated Animals.* New York: Appleton.

Huxley, T. H., and J. S. Huxley. 1947. *Evolution and Ethics, 1893–1943.* London: Pilot.

Hyodo, M., N. Watanabe, W. Sunata, E. E. Susanto, and H. Wahyono. 1993. Magnetostratigraphy of hominid fossil bearing formations in Sangiran and Mojokerto, Java. *Anthropological Science* 101:157–196.

Institute of Vertebrate Paleontology and Paleoanthropology (IVPP). 1980. *Atlas of Primitive Man in China.* Beijing: Science Press.

Isaac, G. L. 1977. *Olorgesailie.* Chicago: University of Chicago Press.

———. 1983. Aspects of human evolution. In *Evolution from Molecules to Men,* D. S. Bendall, ed. Pp. 509–543. Cambridge: Cambridge University Press.

Iturbe, G., M. Fumanal, J. Carrion, E. Cortell, R. Martinez, P. Guilem, M. Garralda, and B. Vandermeersch. 1993. Cova Beneito (Muro, Alicante): una perspectiva interdisciplinar. *Recerques del Museu d'Alcoi* 2:23–88.

Jacob, T. 1981. Solo man and Peking man. In *Homo erectus: Papers in Honour of Davidson Black,* B. Sigmon and J. Cybulski, eds. Pp. 87–104. Toronto: University of Toronto Press.

Jacob, T., and G. Curtis. 1971. Preliminary potassium-argon dating of early man in Java. *Contributions of the University of California Archaeological Research Facility* 12:23–32.

Jelinek, A. J. 1982. The Tabūn Cave and Paleolithic man in the Levant. *Science* 216:1369–1375.

———. 1994. Hominids, energy, environment, and behavior in the Late Pleistocene. In *Origins of Anatomically Modern Humans,* M. H. Nitecki and D. V. Nitecki, eds. Pp. 67–92. New York: Plenum Press.

Jelinek, A. J., W. R. Farrand, G. Haas, A. Horowitz, and P. Goldberg. 1973. New excavations at the Tabun cave, Mount Carmel, Israel, 1967–1972: A preliminary report. *Paléorient* 1:151–183.

Jelínek, J. 1969. Neanderthal man and *Homo sapiens* in eastern and central Europe. *Current Anthropology* 10:475–503.

———. 1978. *Homo erectus* or *Homo sapiens? Recent Advances in Primatology* 3:419–429.

———. 1983. The Mladeč finds and their evolutionary importance. *Anthropologie* 21:57–64.

Jia, L. P. 1980. *Early Man in China.* Beijing: Foreign Languages Press.

———. 1985. China's earliest Palaeolithic assemblages. In *Paleoanthropology and Palaeolithic Archaeology in the People's Republic of China,* R. Wu and J. Olsen, eds. Pp. 135–145. London: Academic Press.

Jia, L. P., and W. W. Huang. 1985a. The Late Palaeolithic of China. In *Palaeoanthropology and Palaeolithic Archaeology in the People's Republic of China,* R. Wu and J. Olsen, eds. Pp. 211–223. London: Academic Press.

———. 1985b. On the recognition of China's Palaeolithic cultural traditions. In *Palaeoanthropology and Palaeolithic Archaeology in the People's Republic of China,* R. Wu and J. Olsen, eds. Pp. 259–265. London: Academic Press.

———. 1990. *The Story of Peking Man.* Beijing: Foreign Languages Press and Hong Kong: Oxford University Press.

Jia, L., and Q. Wei. 1976. A Palaeolithic site at Hsue-chia-yao in Yangkao County, Shansi Province. *K'ao Ku Hsueh Pao* 2:97–114.

Jia, L., Q. Wei, and C. Li. 1979. Report on the excavation of Hsuchiayao man site in 1976. *Vertebrata PalAsiatica* 17(4):277–293.

Jobling, M. A., and C. Tyler-Smith. 1995. Fathers and sons: the Y chromosome and human evolution. *Trends in Genetics* 11:449–456.

Johanson, D., and M. Edey. 1981. *Lucy: The Beginnings of Humankind.* New York: Simon and Schuster.

Johnston, J. S., and A. R. Templeton. 1982. Dispersal and clines in *Opuntia*-breeding *Drosophila mercatorum* and *D. hydei* at Kamuela, Hawaii. In *Ecological Genetics and Evolution: The Cactus-Yeast-Drosophila Model System,* J. S. F. Barker and W. T. Starmer, eds. Pp. 241–256. Sydney: Academic.

Jolly, C. J. 1993. Species, subspecies, and baboon systematics. In *Species, Species Concepts, and Primate Evolution,* W. H. Kimbel and L. B. Martin, eds. Pp. 67–107. New York: Plenum.

Jolly, C. J., T. Disotell, T. Barker, S. Beyene, and J. E. Phillips-Conroy. 1995. Intergeneric hybrid baboons. *American Journal of Physical Anthropology* (supplement) 20:120.

Jordá Cerdá, F. 1952. El problema del Chatelperroniense (Auriñaciense Inferior) en España. *Actas del VI° Congreso Arqueológico del Sudeste.* Alcoy-Cartagena.

———. 1955. *El Solutrense y sus Problemas.* Oviedo: Diputación Provincial de Asturias.

———. 1963. El Paleolítico Superior cantábrico y sus industrias. *Saitabi* 13:3–22.

———. 1969. Los comienzos del Paleolítico Superior en Asturias. *Anuario de Estudios Atlánticos* 15:281–321.

Jorde, L. B., M. J. Bamshad, W. S. Watkins, R. Zenger, A. E. Fraley, P. A. Krakowiak, K. D. Carpenter, H. Soodyall, T. Jenkins, and A. R. Rogers. 1995. Origins and affinities of modern humans: a comparison of mitochondrial and nuclear genetic data. *American Journal of Human Genetics* 57:523–538.

Juliá, R., and J. Bischoff. 1991. Radiometric dating of Quaternary deposits and the hominid mandible of Lake Banyolas, Spain. *Journal of Archaeological Science* 18:707–722.

Kandel, A., and B. Heshmathy. 1988. Using fuzzy linear regression as a forecasting tool in intelligent systems. In *Fuzzy Logic in Knowledge-Based Systems, Decision and Control,* M. M. Gupta and T. Yamakawa, eds. Pp. 361–366. Amsterdam: North-Holland.

Karavanic, I. 1995. Upper Paleolithic occupation levels and late-occurring Neandertal at Vindija Cave (Croatia) in the context of Central Europe and the Balkans. *Journal of Anthropological Research* 51:9–35.

Keates, S. G. 1991. Aspects of hominid behaviour in Pleistocene China. Paper presented at the 4th International Senckenberg Conference, "100 Years of *Pithecanthropus*—The *Homo erectus* Problem," Frankfurt, Germany, 2–6 December.

———. 1994a. The current state of Palaeolithic research in the Far East. Paper presented at the 15th Indo-Pacific Prehistory Association Congress, Chiang Mai, Thailand, 5–12 January.

————. 1994b. Archaeological evidence of hominid behaviour in Pleistocene China and Southeast Asia. In *100 Years of Pithecanthropus—The Homo erectus Problem*, J. L. Franzen, ed. *Courier Forschungsinstitut Senckenberg* 171:141–150.

————. 1995. *The Significance of the Older Palaeolithic Occurrences in the Nihewan Basin, Northern China, in the Context of Important Early and Middle Pleistocene Northern Chinese Localities*. Ph.D. dissertation, Oxford University.

————. n.d.a. Nihewan fieldwork 1990. Manuscript in possession of author.

————. n.d.b. Dali. In *Enciclopedia Archaeologica*. Roma: Istituto della Enciclopedia Italiana, in press.

————. n.d.c. Pleistocene archaeological localities in China: A critique of their chronological and cultural significance. Manuscript in possession of author.

————. n.d.d. A compilation of archaeological localities and their materials in the People's Republic of China. Unpublished ms. in the author's possession.

Keates, S. G., and G.-J. Bartstra. 1994. Island migration of early modern *Homo sapiens* in Southeast Asia: The artifacts from the Walanae depression, Sulawesi, Indonesia. *Palaeohistoria* 33/34:19–30.

Keith, A. 1936. *History from Caves. A New Theory of the Origin of Modern Races of Mankind*. London: British Speleological Association.

Kevles, D. 1985. *In the Name of Eugenics*. Berkeley: University of California Press.

Kimbel, W. H., and Y. Rak. 1993. The importance of species taxa in paleoanthropology and an argument for the phylogenetic concept of the species category. In *Species, Species Concepts, and Primate Evolution*, W. H. Kimbel and L. Martin, eds. Pp. 461–484. New York: Plenum.

King, W. 1864. The reputed fossil man of the Neanderthal. *Quarterly Journal of Science* 1:88–97.

Kitcher, P. 1992. The naturalists return. *Philosophical Review* 101:53–114.

————. 1993. Four ways to "biologicize" ethics. In *Evolution und Ethik*, K. Bayertz, ed. Pp. 114–123. Köln–Graz: Pieter Springer Verlag.

Kitching, I. 1992. The determination of character polarity. In *Cladistics: A Practical Course in Systematics*, P. Forey, C. J. Humphries, I. J. Kitching, R. W. Scotland, D. J. Siebert, and D. M. Williams, eds. Pp. 22–43. Oxford: Clarendon Press.

Klein, R. G. 1973. *Ice-Age Hunters of the Ukraine*. Chicago: University of Chicago Press.

————. 1983. The Stone Age prehistory of southern Africa. *Annual Review of Anthropology* 12:25–48.

————. 1987. Reconstructing how early people exploited animals: problems and prospects. In *The Evolution of Human Hunting*, M. H. Nitecki and D. V. Nitecki, eds. Pp. 11–45. New York: Plenum.

————. 1989a. Biological and behavioural perspectives on modern human origins in southern Africa. In *The Human Revolution*, P. Mellars and C. B. Stringer, eds. Pp. 529–546. Edinburgh: Edinburgh University Press.

————. 1989b. *The Human Career*. Chicago: University of Chicago Press.

————. 1992. The archaeology of modern human origins. *Evolutionary Anthropology* 1:5–14.

————. 1995. Anatomy, behaviour, and modern human origins. *Journal of World Prehistory* 9:167–197.

Klein, R., and K. Cruz-Uribe. 1994. The Paleolithic mammalian fauna from the 1910–14 excavations at El Castillo Cave. In *Homenaje al Dr. Joaquín González Echegaray*, J. Lasheras, ed. Pp. 141–158. Madrid: Museo y Centro de Investigación de Altamira Monografías 17.

Kluge, A. 1983. Cladistics and the classification of the great apes. In *New Interpretations of Ape and Human Ancestry*, R. Ciochon and R. Corruccini, eds. Pp. 151–177. New York: Plenum Press.

Knecht, H. 1993. Early Upper Paleolithic approaches to bone and antler projectile technology. In *Hunting and Animal Exploitation in the Later Palaeolithic and Mesolithic of Eurasia,* G. Peterkin, H. Bricker, and P. Mellars, eds. Pp. 33–48. Washington: Archeological Papers of the American Anthropological Association No.4.

Knüsel, C. 1992. Variable views of Middle Paleolithic adaptation, behaviour and variability. *Antiquity* 66:981–986.

Kortlandt, A. 1972. *New Perspectives on Ape and Human Evolution.* Amsterdam: Stichting voor Psychobiologie.

Kosko, B. 1993. *Fuzzy Thinking: The New Science of Fuzzy Logic.* New York: Hyperion.

Kozłowski, J. 1982. *Excavation in the Bacho Kiro Cave (Bulgaria).* Warsaw: Panstwowe Wydawnictwo Naukowe.

———. 1988a. The transition from the Middle to the Early Upper Paleolithic in central Europe and the Balkans. In *The Early Upper Palaeolithic: Evidence from Europe and the Near East,* J. Hoffecker and C. Wolf, eds. Pp. 193–235. Oxford: BAR International Series No. 437.

———. 1988b. Problems of continuity and discontinuity between the Middle and Upper Paleolithic of central Europe. In *Upper Pleistocene Prehistory of Western Eurasia,* H. Dibble and A. Montet-White, eds. Pp. 349–360. Philadelphia: University of Pennsylvania Museum.

Kozłowski, J., and S. Kozłowski, eds. 1979. *Upper Paleolithic and Mesolithic in Europe: Taxonomy and Paleohistory.* Wroclaw: Polska Akademia Nauk.

Kozłowski, J., H. Laville, and B. Ginter. 1992. *Temnata Cave,* vol. 1, part 1. Kraków: Jagellonian University.

Kramer, A. 1991. Modern human origins in Australasia: Replacement or evolution? *American Journal of Physical Anthropology* 86:455–473.

Kramer, A., T. Crummett, and M. Wolpoff. n.d. Morphological diversity in the Upper Pleistocene hominids of the Levant: two species? *L'Anthropologie* (in press).

Krantz, G. 1973. Cranial hair and brow ridges. *Mankind* 9:109–111.

Kuhn, S. 1990. *Diversity within Uniformity: Tool Manufacture and Use in the "Pontinian" Mousterian of Latium (Italy).* Ph.D. dissertation, University of New Mexico, Albuquerque.

———. 1991. "Unpacking" reduction: Lithic raw material economy in the Mousterian of west-central Italy. *Journal of Anthropological Archaeology* 10:76–106.

———. 1995. *Mousterian Lithic Technology.* Princeton: Princeton University Press.

Kuhn, S., and M. Stiner. 1992. New research on Riparo Mochi, Balzi Rossi (Liguria): preliminary results. *Quaternaria Nova* 2:77–90.

Kuhn, T. 1962a. *The Structure of Scientific Revolutions.* Chicago: University of Chicago Press.

———. 1962b. The historical structure of scientific discovery. *Science* 136:760–764.

———. 1970. *The Structure of Scientific Revolutions,* 2nd ed. Chicago: University of Chicago Press.

———. 1974. Second thoughts on paradigms. In *The Structure of Scientific Theories,* F. Suppé, ed. Pp. 459–482. Urbana: University of Illinois Press.

———. 1977. *The Essential Tension.* Chicago: University of Chicago Press.

Kukla, G., Z. An, J. Melice, J. Gavin, and J. Xiao. 1990. Magnetic subsceptibility record of Chinese Loess. *Transactions of the Royal Society* 81:263–288. Edinburgh.

Lahr, M. M. 1992. *The Origins of Modern Humans: A Test of the Multiregional Hypothesis.* Ph.D. dissertation, University of Cambridge.

———. 1994. The multiregional model of modern human origins: a reassessment of its morphological basis. *Journal of Human Evolution* 26:23–56.

———. 1996. *The Evolution of Modern Human Diversity.* Cambridge: Cambridge University Press.

Lahr, M., and R. Foley. 1994. Multiple dispersals and modern human origins. *Evolutionary Anthropology* 3:48–60.

Lahr, M. M. and R. V. S. Wright. 1996. The question of robusticity and the relationship between cranial size and shape. *Journal of Human Evolution* 31:157–191.

Lamdan, M., and A. Ronen. 1989. Middle and Upper Paleolithic blades in the Levant. In *People and Culture in Change,* part 1, I. Hershkovitz, ed. Pp. 29–36. Oxford: BAR International Series 508.

Landau, M. 1991. *Narratives of Human Evolution.* New Haven: Yale University Press.

Lande, R. 1976. Natural selection and random genetic drift in phenotypic evolution. *Evolution* 30:314–334.

Lanyon, L. E., and C. T. Rubin. 1984. Regulation of bone formation by applied dynamic loads. *Journal of Bone and Joint Surgery* 66A:397–402.

Lanyon, L. E., P. T. Magee, and D. G. Baggott. 1979. The relationship of functional stress and strain to the processes of bone remodelling. An experimental study on the sheep radius. *Journal of Biomechanics* 12:593–600.

Lanzinger, M. 1984. Risultati preliminari delle ricerche nel sito preistorico di Campo di Monte Avena (Alpi Feltrine). *Rivista Scienze Preistoriche* 39:287–299.

Laplace, G. 1966. *Récherches sur l'origine et l'évolution des complexes leptolithiques.* École Française de Rome. Mélanges d'Archéologie et d'Histoire No. 4. Paris: De Boccard.

———. 1970. Les niveaux aurignaciens et l'hipothése du synthetotype. In *L'Homme de Cro-Magnon: 1868–1968,* pp. 141–163. Paris: Conseil de Recherche Scientifique en Algérie.

———. 1977. Il Riparo Mochi ai Bakzi Rossi di Grimaldi (Fouilles 1938–1949). Les industries leptolithiques. *Riv. Scienze Preist.* 32:3–131.

Lartet, L. 1868a. Mémoire sur une sépulture des anciens troglodytes du Périgord. *Annales des Sciences Naturelles (Zoologie)* 10:133–145.

———. 1868b. A burial by prehistoric cave dwellers in Périgord. *Reliquiae Aquitanicae* 6:62–72.

Lartet, E., and H. Christy. 1865. *Reliquiae Aquitanicae.* London: Williams and Norgate.

LAUT (Laboratori d'Arqueologia de la Universitat de Tarragona). 1992. Abric Romaní, Nivell H: un model d'estratègia ocupacional al Pleistocè superior mediterrani. *Estrat* 5:157–308.

Laville, H., J.-P. Rigaud, and J. Sackett. 1980. *Rockshelters of the Perigord.* New York: Academic Press.

Laville, H., J. Raynal, and J. Texier. 1983. Histoire paléoclimatique de l'Aquitaine et du Golfe du Gascogne au pleistocène supérieur depuis le dernier interglaciaire. *Bulletin de l'Institut du Bassin d'Aquitaine* 34:219–241.

Leakey, R. E. F. 1969. Faunal remains. In Early *Homo sapiens* remains from the Omo River region of southwest Ethiopia, by R. E. Leakey, K. W. Butzer, and M. H. Day. *Nature* 222:1132–1133.

———. 1984. *One Life: An Autobiography.* Salem: Salem House/Merrimack Publishers Circle.

Leakey, R. E., K. W. Butzer, and M. H. Day. 1969. Early *Homo sapiens* remains from the Omo River region of southwest Ethiopia. *Nature* 222:1132–1138.

Lee, J. S. 1939. *The Geology of China.* New York: Nordeman.

Le Gros Clark, W. E. 1934. *Early Forerunners of Man: A Morphological Study of the Evolutionary Origin of the Primates.* London: Bailliere, Tindall and Cox.

Le Mort, F. 1987. Incisions volontaires sur un arrière-crâne de néandertalien de Marillac (Charente, France). In *Préhistoire de Poitou-Charentes: Problèmes Actuels,* pp. 151–156. Paris: CTHS.

Leroi-Gourhan, A. 1958. Étude des restes humains fossiles provenant des grottes d'Arcy-sur-Cure. *Annales de Paléontologie* 44:87–148.

———. 1959. Études des restes humaines fossiles provenant des Grottes d'Arcy-sur-Cure. *Annales de Paléontologie* 44:87–148.

———. 1963. Châtelperronien et Aurignacien dans le nordest de la France (d'après la stratigraphie d'Arcy-sur-Cure, Yonne). *Bulletin de la Société Méridionale de Spéléologie et de Préhistoire* 6–9:75–84.

Leroyer, C. 1987. Les gisements castelperroniens de Quinçay et de St. Césaire: quelques comparaisons préliminaires des études palynologiques. *Actes du IIIème Congrès des Sociétés Savants (Poitiers)*, pp. 125–134.

Leroyer, C., and Arl. Leroi-Gourhan. 1983. Problèmes de la chronologie: le Castelperronien et l'Aurignacien. *Bulletin de la Société Préhistorique Française* 80:41–44.

Lévêque, F., and B. Vandermeersch. 1980. Découverte de restes humains dans un niveau Castelperronien à St. Césaire (Charente-Maritime). *Comptes Rendus de l'Academie des Sciences de Paris* 270:42–45.

———. 1981. Le néandertalien de Saint-Césaire. *La Recherche* 12:242–244.

Lévêque, F., A. Backer, and M. Guilbaud. 1993. *Context of a Late Neandertal.* Madison: Prehistory Press.

Levine, R. D. 1977. *The Skidegate Dialect of Haida.* Ph.D. dissertation, Columbia University, New York.

Levins, R., and R. Lewontin. 1985. *The Dialectical Biologist.* Cambridge, Massachusetts: Harvard University Press.

Lewin, R. 1988a. Conflict over DNA clock results. *Science* 241:1598–1600.

———. 1988b. DNA clock conflict continues. *Science* 241:1756–1759.

———. 1988c. Trees from genes and tongues. *Science* 242:514.

———. 1993. *The Origin of Modern Humans.* New York: Scientific American Library.

Lewontin, R. C. 1972. The apportionment of human diversity. In *Evolution Biology,* Vol. 6, T. Dobzhansky, M. K. Hecht, and W. S. Steere, eds. Pp. 381–398. New York: Appleton-Century-Crofts.

———. 1974. *The Genetic Basis of Evolutionary Change.* New York: Columbia University Press.

Li, W., and L. Sadler. 1991. Low nucleotide diversity in man. *Genetics* 129:513–523.

Li, W., J. Lundberg, A. P. Dickin, D. C. Ford, H. P. Schwarcz, and D. Williams. 1989. High-precision mass-spectrometric dating of speleothem from a submerged Bahamian cave. *Nature* 339:534–536.

Li, Yanxian. 1993. On the division of the Upper Paleolithic industries of China. *Acta Anthropologica Sinica* 12(3): 214–223. (in Chinese with English abstract)

Lieberman, D. E. 1993. The rise and fall of seasonal mobility among hunter-gatherers: The case of the southern Levant. *Current Anthropology* 34:599–614.

———. 1995. Testing hypotheses about recent human evolution from skulls. *Current Anthropology* 36:159–197.

Lieberman, D. E., and J. J. Shea. 1994. Behavioral differences between archaic and modern humans in the Levantine Mousterian. *American Anthropologist* 96:300–332.

Lindly, J. M., and G. A. Clark. 1990a. On the emergence of modern humans. *Current Anthropology* 31:59–66.

———. 1990b. Symbolism and modern human origins. *Current Anthropology* 31:233–261.

Little, K. L. 1961. Race and society. In *The Race Question in Modern Science,* K. Little, ed. Pp. 57–105. New York: Columbia University Press.

Liu, T. 1991a. *Loess, Environment and Global Change.* Beijing: Science Press.

———. 1991b. *Quaternary Geology and Environment.* Beijing: Science Press.

Livingstone, F. B. 1962. On the non-existence of human races. *Current Anthropology* 3:279–281.

Lo Bello, P., G. Feraud, C. M. Hall, D. York, P. Lavina, and M. Bernat. 1987. [40]Ar/[39]Ar step heating and laser fusion dating of a Quaternary pumice from Neschers, Massif Central, France: The defeat of xenocrystic contamination. *Chemical Geology (Isotope Geosciences Section)* 66:61–71.

Locke, J. 1975. *An Essay Concerning Human Understanding,* P. H. Nidditch, ed. New York: Oxford University Press. (originally published in 1690)

Long, J. C. 1993. Human molecular phylogenetics. *Annual Reviews of Anthropology* 22:251–272.

Longino, H. 1990. *Science as Social Knowledge.* Princeton: Princeton University Press.

Lorenzo, J. L. 1978. Early man research in the American hemisphere: Appraisal and perspectives. In *Early Man in America from a Circum-Pacific Perspective,* A. Bryan, ed. Pp. 1–9. Edmonton: Occasional Papers of the Department of Anthropology, University of Alberta, No. 1.

Lovejoy, C. 1981. The origin of man. *Science* 211:341–350.

Lü, Z. 1985. Comments. *Current Anthropology* 26:432–433.

Lull, R. S. 1922. The antiquity of man. In *The Evolution of Man,* R. S. Lull, H. B. Ferris, G. H. Parker, J. R. Angell, A. G. Keller, and E. G. Conklin, eds. Pp. 1–38. New Haven: Yale University Press.

Lumley, H. de. 1969. Le paléolithique inférieur et moyen du Midi Méditerranéen dans son cadre géologique I: Ligurie et Provence. *Gallia Préhistoire,* V supplement. Paris: CNRS.

Lumley, H. de, and G. Isetti. 1965. Le Moustérien à denticulés tardif de la station de San Francesco (San Remo) et de la Grotte Tournal (Aude). *Cahiers Ligures de Préhistoire et d'Archéolologie* 14:29–44.

Lumsden, C., and E. Wilson. 1981. *Genes, Mind, and Culture.* Cambridge, Massachusetts: Harvard University Press.

———. 1983. *Promethean Fire: Reflections on the Origin of Mind.* Cambridge, Massachusetts: Harvard University Press.

Luz, B., A. Cormie, and H. Schwarcz. 1990. Oxygen isotope variations in phosphate of deer bones. *Geochimica et Cosmochimica Acta* 54:1723–1728.

Lynch, M. 1988. The rate of polygenic mutation. *Genetical Research* 51:137–148.

Lynch, T. 1966. The Lower Perigordian in French archaeology. *Proceedings of the Prehistoric Society* 32:156–198.

MacCurdy, G. G. 1914. La Combe, a paleolithic cave in the Dordogne. *American Anthropologist* 16:157–184.

———. 1932. *The Coming of Man.* New York: The University Society.

Macintosh, M. W. G., and S. L. Larnach. 1973. A cranial study of the aborigines of Queensland with a contrast between Australian and New Guinea crania. In *The Human Biology of Aborigines in Cape York,* R. L. Kirk, ed. Pp. 1–12. Canberra: Australian Institute of Aboriginal Studies.

———. 1976. Aboriginal affinities looked at in world context. In *The Origin of Australians,* R. L. Kirk and A. G. Thorne, eds. Pp. 113–126. Canberra: Australian Institute of Aboriginal Studies.

Madden, M. 1983. Social network systems among hunter-gatherers considered within southern Norway. In *Hunter-Gatherer Economy in Prehistory,* G. Bailey, ed. Pp. 191–200. Cambridge: Cambridge University Press.

Maddison, D. R. 1991. African origin of human mitochondrial DNA reexamined. *Systematic Zoology* 40:355–363.

Maddison, D. R., M. Ruvolo, and D. L. Swofford. 1992. Geographic origins of human mitochondrial DNA: Phylogenetic evidence from control region sequences. *Systematic Biology* 41:111–124.

Majno, G. 1975. *The Healing Hand: Man and Wound in the Ancient World.* Cambridge, Massachusetts: Harvard University Press.

Malez, M., F. Smith, J. Radovčic, and D. Rukavina. 1980. Upper Pleistocene hominids from Vindija, Croatia, Yugoslavia. *Current Anthropology* 21:365–367.

Mallegni, F., and A. Ronchitelli. 1989. Deciduous teeth of the neandertal mandible from Molare Shelter, near Scario (Salerno, Italy). *American Journal of Physical Anthropology* 79:475–482.

Mallegni, F., and E. Segre-Naldini. 1992. A human maxilla (Fossellone 1) and scapula (Fossellone 2) recovered in the Pleistocene layers of the Fossellone Cave, Mt. Circeo, Italy. *Quaternaria Nova* 2:211–255.

Manzi, G., and P. Passarello. 1990–1991. The human remains from Grotta Breuil (M. Circeo, Italy): Comparative analysis of the parietal fragment Breuil 1. *Quaternaria Nova* 1:429–439.

———. 1995. At the archaic/modern boundary of the genus *Homo:* The neandertals from Grotta Breuil. *Current Anthropology* 36:355–366.

Manzi, G., S. Grimaldi, and G. Destro-Bisol. n.d. The Puzzling Evidence: Perspectives on the Origins of *Homo sapiens.* Unpublished manuscript in possession of authors.

Marks, A. E. 1983. The Middle to Upper Paleolithic transition in the Levant. In *Advances in World Archaeology,* Vol. 2, F. Wendorf, ed. Pp. 51–98. New York: Academic Press.

———. 1988. The Middle to Upper Paleolithic transition in the southern Levant. In *L'Homme de Neandertal,* Vol. 8, M. Otte, ed. Pp. 109–124. Liège: ERAUL No. 35.

———. 1990. The Middle and Upper Paleolithic of the Near East and the Nile Valley: the problem of cultural transformations. In *The Emergence of Modern Humans,* P. Mellars, ed. Pp. 56–80. Edinburgh: University of Edinburgh.

———. 1992. Upper Pleistocene archaeology and the origin of modern humans: A view from the Levant and adjacent areas. In *The Evolution and Dispersal of Modern Humans in Asia,* T. Akazawa, K. Aoki and T. Kimura, eds. Pp. 229–252. Tokyo: Hokusen-sha.

———. 1994. A Levantine Perspective. *Cambridge Archaeological Journal* 4:104–106.

Marks, A., N. Bicho, J. Zilhão, and C. R. Ferring. 1994. Upper Pleistocene prehistory in Portuguese Estremadura: Results of preliminary research. *Journal of Field Archaeology* 21:53–68.

Marks, J. 1995. *Human Biodiversity: Genes, Race, and History.* New York: Aldine de Gruyter.

Maroto, J., ed. 1993. *La mandíbula de Banyoles en el context dels fòssils humans del Pleistocè.* Arqueològiques, Monografica No. 13. Girona: Centre d'Investigaciones.

Marshack, A. 1989. Evolution of the human capacity: The symbolic evidence. *Yearbook of Physical Anthropology* 32:1–34.

———. 1990. Early hominid symbols and the evolution of human capacity. In *The Emergence of Modern Humans,* P. Mellars, ed. Pp. 457–498. Edinburgh: Edinburgh University Press.

———. 1996. A Middle Paleolithic symbolic composition from the Golan Heights: the earliest known depictive image. *Current Anthropology* 37:357–365.

Martin, L. 1986. Relationships among extant and extinct great apes and humans. In *Major Topics in Primate and Human Evolution,* B. Wood, L. Martin, and P. Andrews, eds. Pp. 161–187. Cambridge: Cambridge University Press.

Martin, P. S. 1973. The discovery of America. *Science* 179:969–974.

Martin, R. D. 1990. *Primate Origins and Evolution.* London: Chapman and Hall.

————. 1993. Primate origins: Plugging the gaps. *Nature* 363:223–234.

Martinson, D. G., N. G. Pisias, J. D. Hays, J. I. Imbrie, T. C. Moore Jr., and N. J. Shackleton. 1987. Age dating and the orbital theory of the iceages: Development of a high-resolution 0 to 300,000-year chronostratigraphy. *Quaternary Research* 27:1–29.

Marzke, M. 1983. Joint function and grips of the *A. afarensis* hand, with special reference to the region of the capitate. *Journal of Human Evolution* 12:197–211.

Marzke, M., K. Wullstein, and F. Viegas. 1994. Variability in the carpo-metacarpal and mid-carpal joints involving the fourth metacarpal, hamate and lunate in Catarrhini. *American Journal of Physical Anthropology* 93:229–240.

Masset, C. 1982. *Estimation de l'âge au décès par les sutures crâniennes.* Ph.D. dissertation. Université de Paris, Paris.

Masterman, M. 1970. The nature of the paradigm. In *Criticism and the Growth of Knowledge,* I. Lakatos and A. Musgrave, eds. Pp. 59–90. Cambridge: Cambridge University Press.

Matsu'ura, S. 1982. A chronological framing for the Sangiran hominids: fundamental study by the fluorine dating method. *Bulletin of the National Science Museum, Tokyo* Series D (Anthropology) 8:1–53.

————. 1985. A consideration of the stratigraphic horizons of hominid finds from the Sangiran by the fluorine method. In *Quaternary Geology of the Fossil Bearing Formations in Java,* N. Watanabe and D. Kadar, eds. Pp. 359–367. Toyko: Geological Research and Development, Center for Special Publications 4.

Mayr, E. 1942. *Systematics and the Origin of Species.* New York: Columbia University Press.

————. 1954. Change of genetic environment and evolution. In *Evolution as a Process,* J. Huxley, A. Hardy, and E. Ford, eds. Pp. 157–180. London: Allen and Unwin.

————. 1963. *Animal Species and Evolution.* Cambridge, Massachusetts: Harvard University Press.

————. 1970. *Populations, Species and Evolution.* Cambridge, Massachusetts: Belknap.

————. 1974. Cladistic analysis or cladistic classification? *Zeitschrift für Zoologische Systematik und Evolutions-forschung* 12:94–128.

————. 1982. *The Growth of Biological Thought.* Cambridge, Massachusetts: Harvard University Press.

————. 1988. *Toward a New Philosophy of Biology: Observations of an Evolutionist.* Cambridge, Massachusetts: Harvard University Press.

McBurney, C. 1972. Regional differences in the dating of the earliest leptolithic traditions. In *The Origin of Homo sapiens,* F. Bordes, ed. Pp. 237–249. Paris: UNESCO.

McCown, T. D., and A. Keith. 1939. *The Stone Age of Mount Carmel: The Fossil Human Remains from the Levalloiso-Mousterian,* Vol. II. Oxford: Clarendon Press.

McMullin, E. 1983. Values in Science. In *PSA 1982,* P. D. Asquith and T. Nickles, eds. Pp. 3–28. East Lansing: Philosophy of Science Association.

McNeill, D., and P. Freiberger. 1993. *Fuzzy Logic.* New York: Simon and Schuster.

Meignen, L., and O. Bar-Yosef. 1988. Variabilité technologique au Proche Orient: l'exemple de Kebara. In *L'Homme de Néanderthal—La Technique,* M. Otte, ed. Pp. 81–95. Liège: ERAUL No. 31.

Mellars, P. 1986. A new chronology for the French Mousterian period. *Nature* 322:410–411.

————. 1989a. Technological changes across the Middle-Upper Paleolithic transition: Technological, social and cognitive perspectives. In *The Human Revolution,* P. Mellars and C. Stringer, eds. Pp. 338–365. Edinburgh: Edinburgh University Press.

————. 1989b. Major issues in the emergence of modern humans. *Current Anthropology* 30:349–385.

―――. 1990. Comment on "Symbolism and modern human origins" by J. Lindly and G. Clark. *Current Anthropology* 31:245–246.

―――, ed. 1990. *The Emergence of Modern Humans: An Archaeological Perspective.* Edinburgh: Edinburgh University Press.

―――. 1991. Cognitive changes and the emergence of modern humans in Europe. *Cambridge Archaeological Journal* 1:63–76.

―――. 1992. Archaeology and the population-dispersal hypothesis of modern human origins in Europe. *Philosophical Transactions of the Royal Society of London* 337:225–234.

―――. 1993. Archaeology and the population-dispersal hypothesis of modern human origins in Europe. In *The Origins of Modern Humans and the Impact of Chronometric Dating,* M. J. Aitken, C. B. Stringer, and P. Mellars, eds. Pp. 196–216. Princeton: Princeton University Press.

Mellars, P., and C. Stringer. 1989. Introduction. In *The Human Revolution,* P. Mellars and C. Stringer, eds. Pp. 1–14. Edinburgh: Edinburgh University Press.

―――, eds. 1989. *The Human Revolution.* Edinburgh: Edinburgh University Press.

―――, eds. 1992. *Modern Human Origins.* London: Philosophical Transactions of the Royal Society of London Series B, No. 337.

Melnick, D. J., G. A. Hoelzer, R. Absher, and M. V. Ashley. 1993. mtDNA diversity in rhesus monkeys reveals overestimates of divergence time and paraphyly with neighboring species. *Molecular Biological Evolution* 10:282–295.

Merriwether, D. A., A. G. Clark, S. W. Ballinger, T. G. Schurr, H. Soodyall, T. Jenkins, S. T. Sherry, and D. C. Wallace. 1991. The structure of human mitochondrial DNA variation. *Journal of Molecular Evolution* 33:543–555.

Miller, G. F., and J. Mangerud. 1985. Aminostratigraphy of European marine interglacial deposits. *Quaternary Scientific Review* 4:215–278.

Miller, L. H. 1994. Impact of malaria on genetic polymorphism and genetic diseases in Africans and African Americans. *Proceedings of the National Academy of Science* 91:2415–2419.

Mivart, G. 1867. Additional notes on the osteology of the Lemuridae. *Proceedings of the Zoological Society of London,* pp. 960–975.

―――. 1873. On *Lepilemur* and *Cheirogaleus,* and the zoological rank of the Lemuroidea. *Proceedings of the Zoological Society of London,* pp. 484–510.

Miyamoto, S. 1990. *Fuzzy Sets in Information Retrieval and Cluster Analysis.* Dordrecht: Kluwer Academic Publishers.

Montagu, A. 1941. The concept of race in the human species in the light of genetics. *Journal of Heredity* 32:243–247.

―――. 1964. *Man's Most Dangerous Myth: The Fallacy of Race,* 4th ed. Cleveland: World Publishing.

Monteiro, C., J. Rueff, A. B. Falcão, S. Portugal, D. J. Weatherall, and A. E. Kulozik. 1989. The frequency and origin of the sickle cell mutation in the district of Coruche/Portugal. *Human Genetics* 82:255–258.

Moore, G. E. 1903. *Principia Ethica.* Cambridge: Cambridge University Press.

Moore, J. A. 1993. *Science as a Way of Knowing: The Foundations of Modern Biology.* Cambridge, Massachusetts: Harvard University Press.

Moore, J. H. 1994a. Ethnogenetic theory. *Research and Exploration* 10:10–23.

―――. 1994b. Putting anthropology back together again: The ethnogenetic critique of cladistic theory. *American Anthropologist* 96:925–948.

―――. 1995. The end of a paradigm. *Current Anthropology* 36:530–531.

Morganthau, T. 1994. What color is black? *Newsweek* 13 February:62–65.

Mortillet, G. de. 1883. *Le Préhistorique: Antiquité de l'Homme.* Paris: Reinwald.

Mortillet, G. de, and Adrian de Mortillet. 1910. *La Préhistoire.* Paris: Schleicher Frères.

Morton, N. E., and J. Lalouel. 1973. Topology of kinship in Micronesia. *American Journal of Human Genetics* 25:422–432.

Moss, M. L., and R. W. Young. 1960. A functional approach to craniology. *American Journal of Physical Anthropology* 18:281–292.

Movius, H. L. 1944. Early man and Pleistocene stratigraphy in southern and eastern Asia. *Papers of the Peabody Museum of American Archaeology and Ethnology* 19:1–125.

———. 1969. The abri of Cro-Magnon (Les Eyzies, Dordogne) and the probable age of the contained burials on the basis of nearby Abri Pataud. *Anuario de Estudios Atlánticos* 15:323–344.

———. 1978. Southern and Eastern Asia: Conclusions. In *Early Paleolithic in South and East Asia,* F. Ikawa-Smith, ed. Pp. 351–355. The Hague: Mouton.

Mueller-Wille, C., and D. B. Dickson. 1991. An examination of some models of Late Pleistocene society in southwestern Europe. In *Perspectives on the Past,* G. Clark, ed. Pp. 25–55. Philadelphia: University of Pennsylvania Press.

Müller-Beck, H. 1964. Comment on "The fate of the 'classic' Neanderthals: A consideration of hominid catastrophism" by C. Loring Brace. *Current Anthropology* 5:28–29.

Mussi, M. 1990. Continuity and change in Italy at the last glacial maximum. In *The World at 18,000 B.P.: High Latitudes,* O. Soffer and C. S. Gamble, eds. Pp. 126–147. London: Unwin Hyman.

Nagel, E. 1961. *The Structure of Science.* New York: Harcourt, Brace and World.

Nanzetta, P., and G. E. Strecker. 1971. *Set Theory and Topology.* Tarrytown-on-Hudson: Bogden and Quigley.

National Science Foundation (NSF). 1991. Evaluation of Paleoanthropological research at Xiaochangliang. NSF Anthropology Program. Washington, D.C.

Neeley, M., and C. M. Barton. 1994. A new approach to interpreting microlith industries in southwest Asia. *Antiquity* 68:275–288.

Nei, M. 1987. *Molecular Evolutionary Genetics.* New York: Columbia University Press.

———. 1992. Age of the common ancestor of human mitochondrial DNA. *Molecular Biology and Evolution* 9:1176–1178.

Nei, M., and A. K. Roychoudhury. 1974. Genetic variation within and between the three major races of man, Caucasoids, Negroids, and Mongoloids. *American Journal of Human Genetics* 26:421–443.

———. 1993. Evolutionary relationships of human populations on a global scale. *Molecular Biology and Evolution* 10:927–943.

Nelson, N. C. 1926. Prehistoric man of central China. *Natural History* 26:570–579.

Nitecki, M. H., and D. V. Nitecki., eds. 1994. *Origins of Anatomically Modern Humans.* New York: Plenum.

Novák, V. 1989. *Fuzzy Sets and their Applications.* Bristol: Adam Hilger.

Nuttall, G. H. F. 1904. *Blood Immunity and Blood Relationships.* Cambridge: Cambridge University Press.

Oakley, K., B. Campbell, and T. Molleson. 1971. *Catalogue of Fossil Hominids: Europe II.* London: British Museum.

Obermaier, H. 1924. *Fossil Man in Spain.* New Haven: Yale University.

———. 1925. *El Hombre Fósil,* 2nd ed. Madrid: Junta para la Amplificación de Estudios e Investigaciones Científicas.

O'Connor, S., P. Veth and N. Hubbard. 1993. Changing interpretations of postglacial human subsistence and demography in Sahul. In *Sahul in Review: Pleistocene Archaeology in Australia, New Guinea and Island Melanesia,* M. A. Smith, M. Spriggs, and B. Fankhauser, eds. Pp. 95–105. Canberra: Australian National University. Occasional Papers in Prehistory No. 24.

O'Grady, R. T., I. Goddard, R. M. Bateman, W. A. Dimichelle, V. A. Funk, W. J. Kress, R. Mooi, and P. F. Cannell. 1989. Genes and tongues. *Science* 43:1651.

O'Hara, R. J. 1993. Systematic generalization, historical fate, and the species problem. *Systemic Biology* 42:231–246.

Olsen, J. W. 1986. Comments. *Current Anthropology* 27:470–471.

Olsen, J. W., and S. Miller-Antonio. 1992. The Palaeolithic in southern China. *Asian Perspectives* 31:129–160.

Osborn, H. 1894. The hereditary mechanism and the search for the unknown factors of evolution. In *Defining Biology: Lectures from the 1890s,* J. Maienschein, ed. Pp. 83–104. Cambridge, Massachusetts: Harvard University Press.

———. 1910. *The Age of Mammals in Europe, Asia and North America.* New York: Macmillan.

———. 1916. *Men of the Old Stone Age: Their Environment, Life, and Art.* London: George Bell.

———. 1917. *The Origin and Evolution of Life on the Theory of Action Reaction and Interaction of Energy.* New York: Charles Scribner's Sons.

———. 1926. The evolution of human races. *Natural History* 26:3–13.

———. 1927. *Man Rises to Parnassus: Critical Epochs in the Prehistory of Man.* Princeton: Princeton University Press.

Osman Hill, W. C. 1940. Classification of Hominidae. *Nature* 146:402–403.

Otte, M., and L. Straus. 1995. *Le Trou Magrite.* Liège: ERAUL.

Owen, R. 1866. *On the Anatomy of the Vertebrates.* London: Longman, Green.

Owen, R. C. 1965. The patrilocal band: A linguistically and culturally hybrid social unit. *American Anthropologist* 67:675–690.

Oxnard, C. E. 1978. The problem of convergence and the place of *Tarsius* in primate phylogeny. *Recent Advances in Primatology* 3:239–247.

———. 1981. The place of man among the primates: anatomical, biomolecular and morphometric evidence. *Homo* 32:149–176.

———. 1983. Anatomical, biomolecular and morphometric views of the living primates. In *Progress in Anatomy,* R. J. Harrison and V. Navaratnam, eds. Pp. 113–142. Cambridge: Cambridge University Press.

———. 1983–1984. *The Order of Man.* Hong Kong: Hong Kong University Press (1983); New Haven: Yale University Press (1984).

———. 1987. *Fossils, Teeth and Sex: New Perspectives on Human Evolution.* Seattle: University of Washington Press.

———. 1990. *Animal Lifestyles and Anatomies.* Seattle: University of Washington Press.

Päabo, S. 1995. The Y Chromosome and the Origin of All of Us (Men). *Science* 268:1141–1142.

Palma di Cesnola, A. 1965–1966. Il paleolitico superiore arcaico (faciès Uluzziana) della grotta del Cavallo, Lecce. *Rivista di Scienze Preistoriche* 20:33–62, 21:3–59.

———. 1967. Il Paleolitico della Puglia (giacimenti, periodi, problemi). *Memoria del Museo Civico di Storia Naturale Verona* XV.

———. 1987. Panorama del Musteriano Italiano. In *I Neandertaliani,* A. C. Blanc, ed. Pp. 139–174. Viareggio: Museo Preistoria e Archeologia.

———. 1991. *Gli Scavi a Grotta Paglicci durante Il 1990.* Atti X[i] Convegno della Preistoria, Protostoria e Storia della Daunia. Foggia: S. Severo.

———. 1992. *Paglicci-Rignano Garganico.* Foggia: Mostra della Regione Puglia.

———. 1993. *Il Paleolitico Superiore in Italia.* Florence: Garlatti e Razzai.

Palma di Cesnola, A., and P. Messeri. 1967. Quatre dents humaines paléolithiques trouvées dans des cavernes de l'Italie méridionale. *L' Anthropologie* 71:249–262.

Paterson, H. E. H. 1985. The recognition concept of species. In *Species and Speciation,* E. S. Vrba, ed. Transvaal Museum Monographs 4:21–29. Pretoria: Transvaal Museum.

Pei, W. C. 1932. Preliminary note on some incised, cut and broken bones found in association with *Sinanthropus* remains and lithic artifacts from Choukoutien. *Bulletin of the Geological Society of China* 12:105–108.

———. 1937. Palaeolithic industries in China. In *Early Man*, G. G. McCurdy, ed. Pp. 221–232. Philadelphia: Academy of Natural Sciences.

———. 1938. Le rôle des animaux et des causes naturelles dans la cassure des os. *Palæontologia Sinica* n.s. D 7:1–16.

———. 1939. The Upper Cave industry of Choukoutien. *Palæontologia Sinica* n.s. D 9:1–41.

Peyrony, D. 1933. Les industries "aurignaciennes" dans le bassin de la Vezere. *Bulletin de la Société Préhistorique Française* 32:418–443.

Pike-Tay, Anne. 1990. Comment on "Symbolism and modern human origins" by J. M. Lindly and G. Clark. *Current Anthropology* 31:246–247.

Pitti, C., and C. Tozzi. 1971. La Grotta del Capriolo e la Buca della Jena presso Mommio (Camaiore): sedimenti, fauna, industria litica. *Rivista Scienze Preistoriche* 26:213–258.

Pitti C., C. Sorrentino, and C. Tozzi. 1976. L'industria di tipo paleolitico superiore arcaico della Grotta la Fabbrica (Grosseto): nota preliminare. *Atti Sociedade Toscana Scienze Naturali* A83:174–203.

Poloni, E. S., L. Excoffier, J. L. Mountain, A. Langaney and L. L. Cavalli-Sforza. 1995. Nuclear DNA polymorphism in a Mandenka population from Senegal: Comparison with eight other human populations. *Annals of Human Genetics* 59:43–61.

Pope, G. G. 1983. Evidence on the age of the Asian hominidae. *Proceedings of the National Academy of Sciences,* 80:4988–4992.

———. 1984. The antiquity and paleoenvironment of the Asian hominidae. In *The Evolution of the East Asian Environment,* Vol. II, R. O. Whyte, ed. Pp. 822–847. Hong Kong: Chinese University Press.

———. 1985a. Taxonomy, dating, and paleoenvironment: the paleoecology of the early Far Eastern hominids. *Modern Quaternary Research in Southeast Asia* 9:65–80.

———. 1985b. Evidence of Early Pleistocene hominid activity from Lampang, northern Thailand. *Indo-Pacific Prehistory Association Bulletin* 6:2–9.

———. 1988. Recent advances in Far Eastern Paleoanthropology. *Annual Review of Anthropology* 17:43–77.

———. 1989a. Bamboo and human evolution. *Natural History* 10:48–57.

———. 1989b. Paleoanthropological research at Xiaochangliang. Proposal submitted to the National Science Foundation, Washington D.C.

———. 1989c. Asian *Homo erectus* and the emergence of *Homo sapiens*: an alternative to replacement models. In *Perspectives in Human Evolution*, A. Sahni and R. Gaur, eds. Pp. 23–31. Delhi: Renaissance Publishing House.

———. 1991. Evolution of the zygomaticomaxillary region in the genus *Homo* and its relevance to the origin of modern humans. *Journal of Human Evolution* 21:189–213.

———. 1992a. Craniofacial evidence for the origin of modern humans in China. *Yearbook of Physical Anthropology* 35:243–298.

———. 1992b. Replacement versus regional continuity models: The paleobehavioral and fossil evidence from East Asia. In *The Evolution and Dispersal of Modern Humans in Asia,* T. Akazawa, K. Aoki, and T. Kimura, eds. Pp. 3–14. Tokyo: Hokusen-sha.

———. 1994. The Howellian perspective: Its development and influence on the study of human evolution and behavior. In *Integrative Paths to the Past,* R. S. Corruccini and R. L. Ciochon, eds. Pp. 1–15. Englewood Cliffs: Prentice Hall.

Pope, G. G., and S. G. Keates. 1994. The evolution of human cognition and cultural capacity: A view from the Far East. in *Integrative Paths to the Past,* R. S. Corruccini and R. L. Ciochon, eds. Pp. 531–567. New Jersey: Prentice Hall.

Pope, G. G., Z. S. An, S. Keates, and D. Bakken. 1990. New discoveries in the Nihewan Basin, northern China. *The East Asian Tertiary/Quaternary Newsletter* 11:68–73.

Popper, K. 1958. *The Logic of Scientific Discovery.* London: Hutchinson.

———. 1972. *Objective Knowledge.* Oxford: Oxford University Press.

Potts, R. 1988. *Early Hominid Activities at Olduvai.* New York: Aldine de Gruyter.

Pradel, L. 1966. The transition from Mousterian to Perigordian: Skeletal and industrial evidence. *Current Anthropology* 7:33–50.

———. 1970. Le Périgordien, le Corrézien et l'Aurignacien en France. In *L'Homme de Cro-Magnon,* G. Camps and G. Olivier, eds. Pp. 165–171. Paris: Arts et Métiers Graphiques.

Proctor, R. 1988. *Racial Hygiene.* Cambridge, Massachusetts: Harvard University Press.

Propp, V. 1928. *Morphology of the Folktale.* Austin: University of Texas Press.

Prout, T. 1964. Observations on structural reduction in evolution. *The American Naturalist* 98:239–249.

Pumarejo, P., and V. Cabrera. 1992. Huellas de descarnado sobre restos de fauna procedentes del Auriñaciense de la Cueva del Castillo. *Espacio, Tiempo y Forma* 5:39–52.

Putnam, C. 1967. *Race and Reality.* Washington: Public Affairs Press.

Qian, F., G. Zhou, et al. 1991. *Quaternary Geology and Paleoanthropology of Yuanmou, Yunnan, China.* Beijing: Science Press. (in Chinese with English summary).

Queiroz, K. de, and J. Gauthier. 1992. Phylogenetic taxonomy. *Annual Review of Ecology and Systematics* 23:449–480.

Quine, W. 1961. *From a Logical Point of View.* New York: Harper and Row.

———. 1981. *Theories and Things.* Cambridge, Massachusetts: Harvard University Press.

Radovčić, J. 1985. Neanderthals and their contemporaries. In *Ancestors: The Hard Evidence,* E. Delson, ed. Pp. 310–318. New York: Alan R. Liss.

Rak, Y. 1986. The Neandertal: A new look at an old face. *Journal of Human Evolution* 15:151–164.

———. 1991. A model for morphologic and taxonomic variation in Neandertals and early *Homo sapiens. American Journal of Physical Anthropology* (supplement) 12:147–148.

———. 1993. Morphological variation in *Homo neanderthalensis* and *Homo sapiens* in the Levant: A biogeographic analysis. In *Species, Species Concepts, and Primate Evolution,* W. H. Kimbel and L. Martin, eds. Pp. 523–536. New York: Plenum.

Rainger, R. 1991. *An Agenda for Antiquity: Henry Fairfield Osborn and Vertebrate Paleontology at the American Museum of Natural History, 1890–1935.* Tuscaloosa: University of Alabama Press.

Ravosa, M. J. 1988. Browridge development on Cercopithecidae: A test of two models. *American Journal of Physical Anthropology* 76:535–555.

Read, D. W. 1975. Primate phylogeny, neutral mutations, and "molecular" clocks. *Systematic Zoology* 24:209–221.

———. 1982. Toward a theory of archaeological classification. In *Essays on Archaeological Typology,* R. Whallon and J. Brown, eds. Pp. 56–92. Evanston: Center for American Archeology.

Reader, J. 1989. *Missing Links: The Hunt for Earliest Man.* New York: Penguin Books.

Reich, G. 1991. Did Kuhn kill logical empiricism? *Philosophy of Science* 58:264–277.

Relethford, J. H., and H. C. Harpending. 1994. Craniometric variation, genetic theory, and modern human origins. *American Journal of Physical Anthropology* 95:249–270.

Renfrew, Colin. 1987. *Archaeology and Language.* London: Jonathan Cape.

Rensch, B. 1959. *Evolution above the Species Level.* New York: John Wiley and Sons.

Revillon, S. 1993. Question typologique à propos des industries laminaires du Paléolithique moyen de Seclin (Nord) et de Saint-Germain-des-Vaux/Port Racine (Manche):

lames levallois ou lames non levallois? *Bulletin de la Societé Prehistorique Fran-çaise* 77:306–316.

Revillion, S., and A. Tuffreau, eds. 1994. *Les Industries Laminaires au Paléolithique Moyen.* Paris: CNRS.

Richards, R. 1987. *Darwin and the Emergence of Evolutionary Theories of Mind and Behavior.* Chicago: University of Chicago Press.

———. 1992. *The Meaning of Evolution: The Morphological Construction and Ideological Reconstruction of Darwin's Theory.* Chicago: University of Chicago Press.

Ridley, M. 1986. *The Problems of Evolution.* Oxford: Oxford University Press.

Rigaud, J.-Ph. 1993. Passages et transitions du Paléolithique moyen au Paléolithique supérieur. In *El Origen del Hombre Moderno en el Suroeste de Europa,* V. Cabrera, ed. Pp. 117–126. Madrid: UNED.

Rightmire, G. P. 1979. Implications of the Border Cave skeletal remains for later Pleistocene human evolution. *Current Anthropology* 20:23–35.

———. 1984. *Homo sapiens* in sub-Saharan Africa. In *The Origins of Modern Humans: A World Survey of the Fossil Evidence,* F. H. Smith and F. Spencer, eds. Pp. 295–325. New York: Alan R. Liss.

———. 1990. *The Evolution of Homo Erectus: Comparative Anatomical Studies of an Extinct Human Species.* Cambridge: Cambridge University Press.

Rightmire, G. P., and H. J. Deacon. 1991. Comparative studies of late Pleistocene human remains from Klasies River Mouth, South Africa. *Journal of Human Evolution* 20:131–156.

Ripoll, E., and H. de Lumley. 1965. El Paleolítico Medio en Cataluña. *Ampurias* 26–27:1–70.

Rivière, E. 1887. *De l'Antiquité de l'Homme dans Les Alpes-Maritimes.* Paris: Flournoy.

Robinson, J. T. 1950. The evolutionary significance of the australopithecines. *Yearbook of Physical Anthropology* 00:38–41.

———. 1953. *Meganthropus,* australopithecines and hominids. *American Journal of Physical Anthropology* 11:1–38.

Rogers, A. R. n.d.a Population structure and modern human origins. Manuscript in possession of author.

———. n.d.b Genetic evidence for a Pleistocene population explosion. Manuscript in possession of author.

Rogers, A. R., and L. B. Jorde. 1995a. Genetic evidence on modern human origins. *Human Biology* 67:1–36.

———. 1995b. Ascertainment bias in estimates of heterozygosity. Manuscript in possession of authors.

Rogers, A. R., and H. Harpending. 1992. Population growth makes waves in the distribution of pairwise genetic differences. *Molecular Biology and Evolution* 9:552–569.

Ronen, A. 1992. The emergence of blade technology: cultural affinities. In *The Evolution and Dispersal of Modern Humans in Asia,* T. Akazawa, K. Aoki, and T. Kimura, eds. Pp. 217–228. Tokyo: Hokusen-sha.

Ronen, A., and B. Vandermeersch. 1972. The Upper Paleolithic sequence in the cave of Qafzeh (Israel). *Quaternaria* 16:189–192.

Ronquist, F. 1994. Ancestral areas and parsimony. *Systematic Biology* 43:267–274.

Roper, M. K. 1969. A survey of the evidence for intrahuman killing in the Pleistocene. *Current Anthropology* 10:427–458.

Rosenberg, A. 1980. *Sociobiology and the Preemption of Social Science.* Baltimore: Johns Hopkins University Press.

———. 1985. *The Structure of Biological Science.* Cambridge: Cambridge University Press.

Rossetti, P., and G. Zanzi. 1990–1991. Technological approach to reduction sequences of the lithic industry from Grotta Breuil. *Quaternaria Nova* 1:351–365.

Rouse, I. 1960. The classification of artifacts in archeology. *American Antiquity* 25:313–323.

Ruff, C. B. 1992. Biomechanical analyses of archaeological human material. In *The Skeletal Biology of Past Peoples: Advances in Research Methods,* S. R. Saunders and M. A. Katzenburg, eds. Pp. 41–61. New York: Alan R. Liss.

Ruhlen, M. 1987. *A Guide to the World's Languages,* Vol. 1: Classification. Stanford: Stanford University Press.

Ruse, M. 1969. Definitions of species in biology. *The British Journal for the Philosophy of Science* 20:97–119.

———. 1973. *The Philosophy of Biology.* London: Hutchinson.

———. 1975. Charles Darwin's theory of evolution: An analysis. *Journal of the History of Biology* 8:219–241.

———. 1977. Is biology different from physics? *Laws, Logic, and Life,* R. Colodny, ed. Pp. 89–127. Pittsburgh: University of Pittsburgh Press.

———. 1979a. *The Darwinian Revolution: Science Red in Tooth and Claw.* Chicago: University of Chicago Press.

———. 1979b. *Sociobiology: Sense or Nonsense?* Dordrecht: Reidel.

———. 1986a. *Taking Darwin Seriously.* Oxford: Blackwell.

———. 1986b. Evolutionary ethics: A phoenix arisen. *Zygon* 21:95–112.

———. 1987. Biological species: Natural kinds, individuals, or what? *The British Journal for the Philosophy of Science* 38:225–242.

———. 1988a. *But Is It Science? The Philosophical Question in the Creation/Evolution Controversy.* Buffalo: Prometheus.

———. 1988b. Molecules to men: the concept of progress in evolutionary biology. *Evolutionary Progress,* M. Nitecki, ed. Pp. 97–128. Chicago: University of Chicago Press.

———. 1988c. *Philosophy of Biology Today.* Albany: State University of New York Press.

———. 1989. *The Darwinian Paradigm: Essays on its History, Philosophy, and Religious Implications.* London: Routledge.

———. 1990. Evolutionary ethics and the search for predecessors: Kant, Hume, and all the way back to Aristotle? *Social Philosophy and Policy* 8:59–87.

———. 1993. Evolution and progress. *Trends in Ecology and Evolution* 8:55–59.

———. 1994. *Evolutionary Naturalism: Selected Essays.* London: Routledge.

———. 1996. *Monad to Man: The Concept of Progress in Evolutionary Biology.* Cambridge, Massachusetts: Harvard University Press.

Ruse, M., and E. Wilson. 1985. The evolution of morality. *New Scientist* 1478:108–128.

———. 1986. Moral philosophy as applied science. *Philosophy* 61:173–192.

Russell, E. S. 1916. *Form and Function: A Contribution to the History of Animal Morphology.* London: John Murray.

Russell, M. 1985. The supraorbital torus: "a most remarkable peculiarity." *Current Anthropology* 26:337–360.

Russett, C. 1976. *Darwin in America: The Intellectual Response, 1865–1912.* San Francisco: Freeman.

Rust, A. 1951. *Die Höhlenfunde von Jabrud (Syrien).* Neumunster: Karl Wachholtz.

Ryan, A. S. 1980. *Anterior Dental Microwear in Hominid Evolution: Comparisons with Human and Nonhuman Primates.* Ph.D. dissertation, University of Michigan, Ann Arbor.

Sackett, J. 1991. Straight archeology French style: The phylogenetic paradigm in historic perspective. In *Perspectives on the Past,* G. Clark, ed. Pp. 109–139. Philadelphia: University of Pennsylvania Press.

Sahlins, M. D., and E. R. Service. 1960. *Evolution and Culture.* Ann Arbor: University of Michigan Press.

Sakura, H. 1985. Pleistocene human fossil remains from Pinza-Abu (Goat Cave), Miyako Island, Okinawa, Japan. *Reports on Excavation of the Pinza-Abu Cave,* pp. 161–176. Naha: Department of Education, Okinawa Prefectural Government (in Japanese).

Saller, K. 1926. Die Rassen der juengeren Steinzeit in der Mittelmeerlandern. *Bulleti Associació Catalana d'Antropologia, Etnologia i Prehistoria* 4:1–36.

Sanchez, F. 1990. *La Mandibule Moustériénne Trouvée au Boquete de Zafarraya (Alcaucín, Málaga, Espagne): Étude Biométrique Comparative.* Paris: Université P. et M. Curie, Mémoires de DEA No. 6.

Santa-Luca, A. P. 1980. *The Ngandong Fossil Hominids.* New Haven: Yale University Publications in Anthropology No. 78.

Sapir, E. 1915. The Na-Dene languages: A preliminary report. *American Anthropologist* 17:534–558.

Sarich, V. M. 1971a. Human variation in an evolutionary perspective. In *Background for Man: Readings in Physical Anthropology,* P. Dolhinow and V. M. Sarich, eds. Pp. 182–191. Boston: Little, Brown.

————. 1971b. A molecular approach to the question of human origins. In *Background for Man: Readings in Physical Anthropology,* P. Dolhinow and V. M. Sarich, eds. Pp. 60–81. Boston: Little, Brown.

————. 1995. Human races are very real and very young. *American Journal of Physical Anthropology* (supplement) 20:189.

Sarich, V. M., and A. C. Wilson. 1967. Immunological time scale for hominid evolution. *Science* 158:1200–1204.

Schick, K., N. Toth, Q. Wei, J. D. Clark, and D. Etler. 1991. Archaeological perspectives in the Nihewan Basin, China. *Journal of Human Evolution* 21:13–26.

Schiliro, G., M. Spena, E. Giambelluca, and A. Maggio. 1990. Sickle haemoglobinpathies in Sicily. *American Journal of Hematology* 33:81–85.

Schlick, Moritz. 1953. The philosophy of organic life. In *Readings in the Philosophy of Science,* H. Feigl and M. Brodbeck, eds. Pp. 523–536. New York: Appleton, Century, Crofts.

Schultz, A. H. 1936. Characters common to higher primates and characters specific for man. *Quarterly Review of Biology* 11:259–283.

Schumann, B. 1995a. Population change and continuity in the European Upper Palaeolithic. *American Journal of Physical Anthropology* (supplement) 20:192.

————. 1995b. *Biological Evolution and Population Change in the European Upper Palaeolithic.* Ph.D. dissertation, University of Cambridge.

Schwalbe, G. 1906. Studien zur Vorgeschichte des Menschen. *Zeitschrift für Morphologie und Anthropologie* 1:5–228.

Schwarcz, H. P. 1993. Problems and limitations of absolute dating of the appearance of modern man in southwestern Europe. In *El Origen del Hombre Moderno en el Suroeste de Europa,* V. Cabrera, ed. Pp. 23–46. Madrid: UNED.

Schwarcz, H. P., and B. Blackwell. 1991. Archaeometry. In *Uranium Series Disequilibrium: Application to Environment Problems in the Earth Sciences,* 2nd ed., M. Ivanovich and R. S. Harmon, eds. Pp. 513–552. Oxford: Oxford University Press.

Schwarcz, H. P., A. Bietti, W. Buhay, M. Stiner, R. Grün, and A. Segre. 1991. U-series and ESR age data for the Neanderthal site of Monte Circeo, Italy. *Current Anthropology* 32:313–316.

Schwarcz, H. P., W. M. Buhay, R. Grün, M. Stiner, S. Kuhn, G. H. Miller. 1990–1991. Absolute dating of sites in coastal Lazio. *Quaternaria Nova* I:51–67.

Schwarcz, H. P., W. M. Buhay, R. Grün, H. Valladas, E. Tchernov, O. Bar-Yosef and B. Vandermeersch. 1989. ESR dating of the neanderthal site, Kebara Cave, Israel. *Journal of Archaeological Science* 16:653–659.

Schwarcz, H. P., R. Grün, B. Vandermeersch, O. Bar-Yosef, H. Vallois and E. Tchernov. 1988. ESR dates for the hominid burial site of Qafzeh in Israel. *Journal of Human Evolution* 17:733–737.

Schwartz, J. H. 1984. On the evolutionary relationships of humans and orangutans. *Nature* 308:501–505.

Scotland, R. 1992. Cladistic theory. In *Cladistics: A Practical Course in Systematics,* P. Forey, C. J. Humphries, I. J. Kitching, R. W. Scotland, D. J. Siebert, and D. M. Williams, eds. Pp. 3–13. Oxford: Clarendon Press.

Sealy, J. 1989. *Reconstruction of Later Stone Age Diets in the South-Western Cape, South Africa: Evaluation and Application of Five Isotopic and Trace Element Techniques.* Cape Town: University of Cape Town.

Shanks, M., and C. Tilley. 1987. *Social Theory and Archaeology.* Cambridge: Polity Press.

Shapere, D. 1971. The paradigm concept. *Science* 172:706–709.

Shapin, S. 1982. History of science and its social reconstructions. *History of Science* 20:157–211.

Shapin, S., and S. Schaffer. 1985. *Leviathan and the Air Pump.* Princeton: Princeton University Press.

Shapiro, H. 1961. Race mixture. In *The Race Question in Modern Science,* {need name}, ed. Pp. 343–389. New York: Columbia University Press.

Sharrock, S. R. 1974. Crees, Cree-Assiniboines, and Assiniboines: interethnic social organization on the far Northern Plains. *Ethnohistory* 21:95–122.

Shaw, A. B. 1964. *Time in Stratigraphy.* New York: McGraw-Hill.

Sherry, S. T., A. R. Rogers, H. Harpending, H. Siidyall, T. Jenkins, and M. Stoneking. 1994. Mismatch distributions of mtDNA reveal recent human population expansions. *Human Biology* 66:761–775.

Shimkin, D. 1978. The Upper Paleolithic in north-central Eurasia: evidence and problems. In *Views of the Past,* L. Freeman, ed. Pp. 193–315. Mouton: The Hague.

Shreeve, J. 1995. *The Neanderthal Enigma: Solving the Mystery of Modern Human Orgins.* New York: William Morrow.

Sibley, A. G., and J. E. Ahlquist. 1984. The phylogeny of the hominoids as indicated by DNA-DNA hybridisation. *Journal of Molecular Evolution* 20:2–22.

Simpson, G. G. 1943. Criteria for genera, species, and subspecies in zoology and paleozoology. *Annals of the New York Academy of Sciences* 44:145–178.

———. 1945. The principles of classification and a classification of mammals. *Bulletin of the American Museum of Natural History* 85:1–350.

———. 1961a. *Principles of Animal Taxonomy.* New York: Columbia University Press.

———. 1961b. Lamarck, Darwin and Butler. *American Scholar* 30:238–249.

———. 1963. The meaning of taxonomic statements. In *Classification and Human Evolution,* S. L. Washburn, ed. Pp. 1–31. Chicago: Aldine.

———. 1964. Organisms and molecules in evolution. *Science* 146:1535–1538.

Singer, R., and J. Wymer. 1982. *The Middle Stone Age at Klasies River Mouth in South Africa.* Chicago: Chicago University Press.

Skala, H. J. 1988. On fuzzy probability measures. In *Fuzzy Logic in Knowledge-Based Systems, Decision and Control,* M. M. Gupta and T. Yamakawa, eds. Pp. 123–131. Amsterdam: North-Holland.

Smart, J. 1963. *Philosophy and Scientific Realism.* London: Routledge and Kegan Paul.

Smith, F. 1978. Evolutionary significance of the mandibular foramen area in Neandertals. *American Journal of Physical Anthropology* 48:523–534.

—————. 1982. Upper Pleistocene hominid evolution in south-central Europe: A review of the evidence and analysis of trends. *Current Anthropology* 23:667–703.

—————. 1984. Fossil hominids from the Upper Pleistocene of central Europe and the origins of modern Europeans. In *The Origins of Modern Humans,* F. Smith and F. Spencer, eds. Pp. 137–209. New York: Alan R. Liss.

—————. 1985. Continuity and change in the origin of modern *Homo sapiens. Zeitschrift für Morphologie und Anthropologie* 75:197–222.

—————. 1991. The Neandertals: Evolutionary dead ends or ancestors of modern people? *Journal of Anthropological Research* 47:219–238.

—————. 1992a. The role of continuity in modern human origins. In *Continuity or Replacement: Controversies in Homo sapiens Evolution,* G. Bräuer and F. H. Smith, eds. Pp. 145–156. Rotterdam: A. A. Balkema.

—————. 1992b. Models and realities in modern human origins: the African fossil evidence. *Philosophical Transactions of the Royal Society* 337:243–250.

—————. 1994. Samples, species, and speculations in the study of modern human origins. In *Origins of Anatomically Modern Humans,* M. H. Nitecki and D. V. Nitecki, eds. Pp. 227–249. New York: Plenum Press.

Smith, F., and J. Ahern. 1994. Additional cranial remains from Vindija Cave, Croatia. *American Journal of Physical Anthropology* 93:275–280.

Smith, F., and G. Ranyard. 1980. Evolution of the supraorbital region in Upper Pleistocene fossil hominids from south-central Europe. *American Journal of Physical Anthropology* 53:589–610.

Smith, F., and E. Trinkaus. 1991. Les origines de l' homme moderne en Europe centrale: un cas de continuité. In *Aux Origines d' Homo sapiens,* J.-J. Hublin and A.-M. Tillier, eds. Pp. 251–290. Paris: Presses Universitaires de France.

Smith, F., A. Falsetti, and S. Donnelly. 1989. Modern human origins. *Yearbook of Physical Anthropology* 32:35–68.

Smith, F., J. Simek, and M. Harrill. 1989. Geographic variation in supraorbital torus reduction during the later Pleistocene (c. 80,000–15,000 BP). In *The Human Revolution,* P. Mellars and C. Stringer, eds. Pp. 172–193. Edinburgh: Edinburgh University Press.

Smith, J., and R. Savage. 1956. Some locomotory adaptations in mammals. *Journal of the Linnean Society (Zoology)* 42:603–622.

Smith, M., M. Spriggs, and B. Fankhauser, eds. 1993. *Sahul in Review: Pleistocene Archaeology in Australian, New Guinea and Island Melanesia.* Canberra: Australia National University Occasional Papers in Prehistory No. 24.

Sober, E. 1984. *The Nature of Selection.* Cambridge, Massachusetts: MIT Press.

Soffer, O. 1985. *The Upper Paleolithic of the Central Russian Plain.* Orlando: Academic Press.

—————. 1992. Social transformations at the Middle to Upper Paleolithic transition: the implications of the European record. In *Continuity or Replacement,* G. Bräuer and F. Smith, eds. Pp. 247–257. Rotterdam: Balkema.

Soffer, O., and C. Gamble, eds. 1990. *The World at 18,000 BP, Vol. 1: High Latitudes.* London: Unwin Hyman.

Sokal, R., and P. Sneath. 1973. *Principles of Numerical Taxonomy.* London: Freeman.

Sonneville-Bordes, D. de. 1960. *Le Paléolithique supérieur en Périgord.* Bordeaux: Delmas.

———. 1966. L'evolution du paléolithique superieur en Europe occidental et sa significa-tion. *Bulletin de la Société Préhistorique Française* 63:3–34.

Spaulding, A. 1977. On growth and form in archaeology: Multivariate analysis. *Journal of Anthropological Research* 33:1–15.

———. 1982. Structure in archaeological data: Nominal variables. In *Essays in Archae-ological Typology,* R. Whallon and J. Brown, eds. Pp. 1–20. Evanston: Center for American Archeology.

Spencer, F. 1984. The Neandertals and their evolutionary significance: A brief historical survey. In *The Origins of Modern Humans,* F. Smith and F. Spencer, eds. Pp. 1–50. New York: Alan R. Liss.

Spencer, F., and F. Smith. 1981. The significance of Aleš Hrdlička's "Neanderthal phase of man": A historical and current assessment. *American Journal of Physical Anthro-pology* 56:435–459.

Spencer, H. 1851. *Social Statics: Or, the Conditions Essential to Human Happiness Specified, and the First of Them Developed.* London: Chapman.

———. 1892. *The Principles of Ethics.* London: Williams and Norgate.

Spennato, A. G. 1981. I livelli Protoaurignaziani della Grotta di Serra Cicora (Nardò, Lecce). *Studi per l'Ecologia del Quaternario* 3:61–76.

Spuhler, J. N. 1993. Population genetics and evolution in the genus *Homo* in the last two million years. In *Genetics of Cellular, Individual, Family, and Population Vari-ability,* C. F. Sing and C. L. Hanis, eds. Pp. 262–297. Oxford: Oxford University Press.

Stanley, S. M. 1975. A theory of evolution above the species level. *Proceedings of the National Academy of Sciences* 72:646–650.

Stevens, W. K. 1991. New finding moves up age of Neanderthals. *New York Times,* June 27, p. A6.

Stiner, M. 1990. *The ecology of choice: Procurement and transport of animal resources by Upper Pleistocene Hominids in West-Central Italy.* Ph.D dissertation, University of New Mexico. Ann Arbor: University Microfilms.

———. 1990–1991a. The Guattari faunas then and now. *Quaternaria Nova* 1:163–192.

———. 1990–1991b. Ungulate exploitation in the terminal Mousterian of Italy: The case of Grotta Breuil. *Quaternaria Nova* 1:333–350.

———. 1994. *Honor among Thieves: A Zooarchaeological Study of Neandertal Ecology.* Princeton: Princeton University Press.

Stoneking, M. 1993. DNA and recent human evolution. *Evolutionary Anthropology* 2:60–73.

———. 1994. In defense of "Eve"—A response to Templeton's critique. *American An-thropologist* 96:131–140.

Stoneking, M., and R. Cann. 1989. African origin of human mitochondrial DNA. In *The Human Revolution,* P. Mellars and C. Stringer, eds. Pp. 17–30. Edinburgh: Edin-burgh University Press.

Stoneking, M., K. Bhatia, and A. C. Wilson. 1986. Rate of sequence divergence estimated from restricted maps of mitochondrial DNAs from Papua New Guinea. Cold Spring Harbor Symposium. *Quantitative Biology* 51:433–439.

Stoneking, M., S. T. Sherry, A. J. Redd, and L. Vigilant. 1992. New approaches to dating suggest a recent age from the human mtDNA ancestor. *Philosophical Transactions of the Royal Society of London* B 337:167–175.

Straus, L. 1976. A new interpretation of the Cantabrian solutrean. *Current Anthropology* 17:342–343.

———. 1977. Of deerslayers and mountain men: Paleolithic faunal exploitation in Cantabrian Spain. In *For Theory Building in Archaeology*, L. Binford, ed. Pp.41–76. New York: Academic.

————. 1982. Carnivores and cave sites in Cantabrian Spain. *Journal of Anthropological Research* 38:75–96.

————. 1983a. From Mousterian to Magdalenian: Cultural evolution viewed from Vasco-Cantabrian Spain and Pyrenean France. In *The Mousterian Legacy: Human Biocultural Change in the Upper Pleistocene*, E. Trinkaus, ed. Pp. 123–140. Oxford: BAR International Series No. 164.

————. 1983b. Terminal Pleistocene faunal exploitation in Cantabria and Gascony. In *Animals and Archaeology: Hunters and Their Prey*, J. Clutton-Brock and C. Grigson, eds. Pp. 209–225. Oxford: BAR International Series No. 163.

————. 1987a. Hunting in late Upper Paleolithic western Europe. In *The Evolution of Human Hunting*, M. Nitecki and D. Nitecki, eds. Pp. 147–176. New York: Plenum.

————. 1987b. Paradigm lost: A personal view of the current state of Upper Paleolithic research. Helenium 27:157–171.

————. 1989. Age of the modern Europeans. *Nature* 342:476–477.

————. 1990. The Early Upper Paleolithic of Southwest Europe: Cro-Magnon adaptations in the Iberian Peripheries, 40,000–20,000 BP. In *The Emergence of Modern Humans*, P. Mellars, ed. Pp. 276–302. Edinburgh: Edinburgh University.

————. 1991. Whence and whither paleoanthropology: By way of introduction. *Journal of Anthropological Research* 47:125–128.

————. 1992. *Iberia before the Iberians*. Albuquerque: University of New Mexico Press.

————. 1994. Upper Paleolithic origins and radiocarbon calibration: more new evidence from Spain. *Evolutionary Anthropology* 2:195–198.

Straus, L., and C. Heller. 1988. Explorations of the twilight zone: The Early Upper Palaeolithic of Vasco-Cantabrian Spain and Gascony. In *The Early Upper Palaeolithic: Evidence from Europe and the Near East*, J. Hoffecker and C. Wolf, eds. Pp. 97–133. Oxford: BAR International Series, No. 437.

Straus, L., J. Bischoff, and E. Carbonell. 1993. A review of the Middle to Upper Paleolithic transition in Iberia. *Préhistoire Européenne* 3:11–27.

Stringer, C. B. 1984a. Fate of the Neanderthal. *Natural History* 93:6–12.

————. 1984b. Human evolution and biological adaptation in the Pleistocene. In *Hominid Evolution and Community Ecology*, R. Foley, ed. Pp. 55–83. London: Academic Press.

————. 1985. Middle Pleistocene hominid variability and the origin of Late Pleistocene humans. In *Ancestors: The Hard Evidence*, E. Delson, ed. Pp. 289–295. New York: Alan Liss.

————. 1989a. Documenting the origin of modern humans. In *The Emergence of Modern Humans: Biocultural Adaptations in the Later Pleistocene*, Erik Trinkaus, ed. Pp. 67–96. Cambridge: Cambridge University Press.

————. 1989b. The origin of early modern humans: A comparison of the European and non-European evidence. In *The Human Revolution*, P. Mellars and C. Stringer, eds. Pp. 232–244. Edinburgh: Edinburgh University Press.

————. 1990a. The emergence of modern humans. *Scientific American* 263:98–104.

————. 1990b. Comment on "Symbolism and modern human origins" by John M. Lindly and Geoffrey Clark. *Current Anthropology* 31:248–249.

————. 1990c. The Asian connection. *New Scientist* 178:33–37.

————. 1991a. Comment on "New models and metaphors for the Neanderthal debate" by Paul Graves. *Current Anthropology* 32:532.

————. 1991b. The evolution of regionality in modern humans. In *The Unity of Evolu-*

tionary Biology: Proceedings ICSEB IV, E. C. Dudley, ed. Pp. 466–468. Portland: Dioscorides Press.

———. 1992a. Replacement, continuity and the origin of *Homo sapiens.* In *Continuity or Replacement: Controversies in Homo Sapiens Evolution,* G. Bräuer and F. Smith, eds. Pp. 9–24. Rotterdam: A. A. Balkema.

———. 1992b. Reconstructing recent human evolution. *Philosophical Transactions of the Royal Society* B 337:217–224.

———. 1992c. Neanderthal dates debated, a reply. *Nature* 356:201.

———. 1993a. Secrets of the pit of the bones. *Nature* 362:501–502.

———. 1993b. New views on modern human origins. In *The Origin and Evolution of Humans and Humanness,* D. T. Rasmussen, ed. Pp. 274–285. London: Jones and Bartlett.

———. 1994. Out of Africa—a personal history. In *Origins of Anatomically Modern Humans,* M. H. Nitecki and D. V. Nitecki, eds. Pp. 149–172. New York: Plenum Press.

Stringer, C. B., and P. Andrews. 1988a. Modern human origins. *Science* 241:773–774.

———. 1988b. Genetic and fossil evidence for the origin of modern humans. *Science* 239:1263–1268.

Stringer, C. B., and G. Bräuer. 1994. Methods, misreading, and bias. *American Anthropologist* 96:416–424.

Stringer, C. B., and C. Gamble. 1993. *In Search of the Neanderthals.* New York: Thames and Hudson.

———. 1994. In search of the Neanderthals: Solving the puzzle of human origins. *Cambridge Archaeological Journal* 4:95–119.

Stringer, C. B., R. Grün, H. P. Schwarcz, and P. Goldberg. 1989. ESR dates for the hominid burial site of Es Skhūl in Israel. *Nature* 338:756–758.

Stringer, C., J.-J. Hublin, and B. Vandermeersch. 1984. The origin of anatomically modern humans in western Europe. In *The Origins of Modern Humans: A World Survey of the Fossil Evidence,* F. Smith and F. Spencer, eds. Pp. 51–135. New York: Alan Liss.

Subramanian, S. 1995. The story in our genes. *Time,* 16 January, pp. 54–55.

Suzuki, H. 1982. Skulls of the Minatogawa man. In *The Minatogawa Man: The Pleistocene Man from the Island of Okinawa,* H. Suzuki and K. Hanihara, eds. Pp. 7–50. Tokyo: University of Tokyo Museum Bulletin No. 19.

Svoboda, J. 1988. Early Upper Palaeolithic industries in Moravia: A review of the recent evidence. In *L'Homme de Neanderthal,* Vol. 8: La Mutation, M. Otte, ed. Pp. 169–192. Liège: ERAUL No. 35.

Swadesh, M. 1954. Perspectives and problems of Amerindian comparative linguistics. *Word* 10:306–332.

———. 1967. Lexicostatistic classification. In *Handbook of Middle American Indians,* Vol. 5, N. A. McQuown, ed. Pp. 79–115. Austin: University of Texas Press.

Swisher III, C. C., G. H. Curtis, T. Jacob, A. G. Getty, A. Suprijo, and Widiasmoro. 1994. Age of the earliest known hominids in Java, Indonesia. *Science* 263:1118–1121.

Szalay, F. S. 1993. Species concepts: the tested, the untestable, and the redundant. In *Species, Species Concepts, and Primate Evolution,* W. H. Kimbel and L. B. Martin, eds. Pp. 21–41. New York: Plenum Press.

Taborin, Y. 1988. Les prémices de la parure. *Mémoires de la Musée de Préhistoire d'Ile-de-France* 3:259–271.

Tanaka, H., T. Shimomura, J. Watada, and K. Asai. 1987. Fuzzy linear regression analysis of the number of staff in local government. In *Analysis of Fuzzy Information,* Vol. III: Applications in Engineering and Science, J. C. Bezdek, ed. Pp. 191–203. Boca Raton: CRC Press.

Tang, Y. 1991. The early Pleistocene mammalian fauna of China. In *Contributions to the XIII INQUA Institute of Vertebrate Paleontology and Paleoanthropology, Academica Sinica*, pp. 21–31. Beijing: Beijing Scientific and Technological Publishing House.

Tao, F. H., and X. Q. Wang. 1987. Observation of bones and flakes from Dingcun site. *Shiqian Yanjiu* 1:10–13, 96.

Tattersall, I. 1986. Species recognition in human paleontology. *Journal of Human Evolution* 15:165–175.

———. 1992. Species concepts and species identification in human evolution. *Journal of Human Evolution* 22:341–349.

———. 1994. How does evolution work? *Evolutionary Anthropology* 3:2–3.

———. 1995. *The Fossil Trail: How We Know What We Think We Know about Human Evolution.* New York: Oxford University Press.

Teilhard de Chardin, P. 1937. The Pleistocene of China: stratigraphy and correlations. In *Early Man*, G. G. McCurdy, ed. Pp. 211–220. Philadelphia: J. B. Lippencott.

———. 1938. Deuxième notes sur la paléontologie humaine en Asia meridionale. *L'Anthropologie* 49:251–253.

———. 1941. *Early Man in China.* Pekin: Institute de Géo-Biologie Publication No. 7.

———. 1959. *The Phenomenon of Man.* New York: Harper and Row.

Teilhard de Chardin, P., and E. Licent. 1924. On the discovery of a Palaeolithic industry in Northern China. *Bulletin of the Geological Society of China* 3:45–50.

Teilhard de Chardin, P., and W. C. Pei. 1932. The lithic industry of the *Sinanthropus* deposits in Choukoutien. *Bulletin of the Geological Society of China* 4:315–358.

Teilhard de Chardin, P., and J. Piveteau. 1930. Les mammifères fossiles de Nihowan (Chine). *Annales de Paléontologie* 19:1–134.

Templeton, A. R. 1983. Convergent evolution and nonparametric inferences from restriction data and DNA sequences. In *Statistical Analysis of DNA Sequence Data*, B. S. Weir, ed. Pp. 151–179. New York: Marcel Dekker.

———. 1992. Human origins and the analysis of mitochondrial DNA sequences. *Science* 255:737.

———. 1993. The "Eve" hypothesis: A genetic critique and reanalysis. *American Anthropologist* 95:51–72.

———. 1994a. "Eve": hypothesis compatibility versus hypothesis testing. *American Anthropologist* 96:141–147.

———. 1994b. The role of molecular genetics in speciation studies. In *Molecular Ecology and Evolution: Approaches and Applications*, B. Schierwater, B. Streit, G. P. Wagner, and R. DeSalle, eds. Pp. 455–477. Basel: Birkhäuser-Verlag.

———. 1994c. Biodiversity at the molecular genetic level: experiences from disparate macroorganisms. *Philosophical Translations of the Royal Society of London* B 345:59–64.

Templeton, A. R., and N. J. Georgiadis. 1995. Conserving evolutionary processes in African bovids: Deducing the past from the present to plan for the future. In *Conservation Genetics: Case Histories from Nature*, J. Avise and J. Hamrick, eds. Pp. 147–160. New York: Chapman and Hall.

Templeton, A. R., and B. Read. 1994. Inbreeding: One word, several meanings, much confusion. In *Conservation Genetics*, V. Loeschcke, J. Tomiuk and S. K. Jain, eds. Pp. 91–106. Basel: Birkhäuser-Verlag.

Templeton, A. R., and C. F. Sing. 1993. A cladistic analysis of phenotypic associations with haplotypes inferred from restriction endonuclease mapping. IV. Nested analyses with cladogram uncertainty and recombination. *Genetics* 134:659–669.

Templeton, A. R., E. Boerwinkle, and C. F. Sing. 1987. A cladistic analysis of phenotypic associations with haplotypes inferred from restriction endonuclease mapping. Basic theory and an analysis of Alcohol Dehydrogenase activity in *Drosophila. Genetics* 117:343–351.

Templeton, A. R., K. A. Crandall, and C. F. Sing. 1992. A cladistic analysis of phenotypic associations with haplotypes inferred from restriction endonuclease mapping and DNA sequence data. III. Cladogram estimation. *Genetics* 132:619–633.

Templeton, A. R., H. Hollocher, S. Lawler and J. S. Johnston. 1989. Natural selection and ribosomal DNA in *Drosophila. Genome* 31:296–303.

Templeton, A. R., E. Routman, and C. Phillips. 1995. Separating population structure from population history: a cladistic analysis of the geographical distribution of mito-chondrial DNA haplotypes in the tiger salamander, *Ambystoma tigrinum. Genetics* 140:767–782.

Templeton, A. R., C. F. Sing, A. Kessling, and S. Humphries. 1988. A cladistic analysis of phenotypic associations with haplotypes inferred from restriction endonuclease mapping. II. The analysis of natural populations. *Genetics* 120:1145–1154.

Thomas, A. 1972. L'origine des Cro-Magnoïdes. In *Les Origines Humaines et les Epoques de l'Intelligence,* J. Piveteau, ed. Pp. 261–271. Paris: Masson et Cie.

Thompson, P. 1989. *The Structure of Biological Theories.* Albany: State University of New York University Press.

Thorne, A. G. 1981. The centre and the edge: the significance of Australasian hominids to African paleoanthropology. In *Proceedings of the 8th Panafrican Congress of Pre-history and Quaternary Studies,* R. E. Leakey and B. A. Ogot, eds. Pp. 180–181. Nairobi: TILLMIAP.

———. 1993. Introduction. Session: The End of Eve? Fossil Evidence from Africa. *American Association for the Advancement of Science Abstracts,* p. 173.

Thorne, A. G., and M. H. Wolpoff. 1981. Regional continuity in Australasian Pleistocene hominid evolution. *American Journal of Physical Anthropology* 55:337–350.

———. 1991. Conflict over modern human origins. *Search* 22:175–177.

———. 1992. The multiregional evolution of humans. *Scientific American* 266:76–83.

Thorne, A. G., M. H. Wolpoff, and R. G. Eckhardt. 1993. Genetic variation in Africa. *Science* 261:1507–1508; 262:973–974.

Tillier, A. M. 1979. La dentition de l'enfant moustérien Châteauneuf 2, découvert à l'abri de Hauteroche (Charente). *L'Anthropologie* 83:417–438.

———. 1982. Les enfants néanderthaliens de Devil's Tower (Gibraltar). *Zeitschrift für Morphologie und Anthropologie* 73:125–148.

———. 1983. L'enfant néandertalien du Roc de Marsal (Compagne du Bugue, Dor-dogne): le squelette facial. *Annales de Paléontologie* 69(2):137–149.

———. 1984. L'enfant *Homo* 11 de Qafzeh (Israël) et son apport à la compréhension des modalités de croissance des squelettes moustériens. *Paléorient* 10:7–47.

———. 1986. Quelques aspects de l'ontogénèse du squelette crânien des néandertaliens. *Anthropos* 23:207–216.

———. 1989. Quelques rémarques à propos de la relation biologie-comportement chez les Néandertaliens. In *Hominidae,* G. Giacobini, ed. Pp. 319–330. Turin: Jaca Book.

Tobias, P. V. 1981. The emergence of man in Africa and beyond. *Philosophical Transac-tions of the Royal Society of London* B 292:43–56.

———. 1991. *Olduvai Gorge,* Vol. 4: The Skulls, Endocasts and Teeth of *Homo habilis.* Cambridge: Cambridge University Press.

Tobias, P. V., and G. H. R. von Koenigswald. 1964. A comparison between Olduvai hominines and those of Java, and some implications for hominid phylogeny. *Nature* 204:515–518.

Toulmin, S. 1967. The evolutionary development of science. *American Scientist* 57:456–471.

————. 1972. *Human Understanding.* Oxford: Clarendon Press.

Trigger, B. 1989. *A History of Archaeological Thought.* Cambridge: Cambridge University Press.

Trinkaus, E. 1982. A history of *Homo erectus* and *Homo sapiens* paleontology in America. In *A History of American Physical Anthropology, 1930–1980,* F. Spencer, ed. Pp. 261–280. New York: Academic Press.

————. 1983. Neandertal postcrania and the adaptive shift to modern humans. In *The Mousterian Legacy,* E. Trinkaus, ed. Pp. 165–200. Oxford: BAR International Series No. 164.

————. 1984. Western Asia. In *The Origins of Modern Humans: A World Survey of the Fossil Evidence,* H. Smith and F. Spencer, eds. Pp. 251–294. New York: Alan R. Liss.

————. 1986. The Neandertals and modern human origins. *Annual Review of Anthropology* 15:193–218.

————. 1987. The neanderthal face: Evolutionary and functional perspectives on a recent hominid face. *Journal of Human Evolution* 16:429–443.

————. 1989. The Upper Pleistocene transition. In *The Emergence of Modern Humans: Biocultural Adaptations in the Later Pleistocene,* E. Trinkaus, ed. Pp. 42–66. Cambridge: Cambridge University Press.

————, ed. 1989. *The Emergence of Modern Humans.* Cambridge: Cambridge University Press.

————. 1990. Comment on "Symbolism and modern human origins" by J. M. Lindly and G. Clark. *Current Anthropology* 31:249–250.

————. 1992a. Paleontological perspectives on Neandertal behavior. In *Cinq Millions d'Années, l'Aventure Humaine,* M. Toussaint, ed. Pp. 151–176. Liège: ERAUL No. 56.

————. 1992b. Morphological contrasts between the Near Eastern Qafzeh-Skhūl and late archaic human samples: Grounds for a behavioral difference? In *The Evolution and Dispersal of Modern Humans in Asia,* T. Akazawa, K. Aoki, and T. Kimura, eds. Pp. 277–294. Tokyo: Hokusen-sha.

————. 1992c. Cladistics and later Pleistocene human evolution. In *Continuity or Replacement: Controversies in Homo sapiens Evolution,* G. Bräuer and F. H. Smith, eds. Pp. 274–285. Rotterdam: Balkema.

————. 1993a. Femoral neck-shaft angles of the Qafzeh-Skhūl early modern humans, and activity levels among immature Near Eastern Paleolithic hominids. *Journal of Human Evolution* 25:393–416.

————. 1993b. Variability in the position of the mandibular mental foramen and the identification of Neandertal apomorphies. *Rivista di Antropologia* 71:259–274.

Trinkaus, E., and W. W. Howells. 1979. The Neanderthals. *Scientific American* 241:118–133.

Trinkaus, E., and P. Shipman. 1993a. Neandertals: Images of ourselves. *Evolutionary Anthropology* 1:194–201.

————. 1993b. *The Neandertals: Changing the Image of Mankind.* New York: Alfred A. Knopf.

Trivers, R. 1971. The evolution of reciprocal altruism. *Quarterly Review of Biology* 46:35–57.

Trubetskoy, N. S. 1939. Gedanken uber das Indogermanproblem, *Acta Linguistica* 1:81–89.

Tucker, W. 1994. *The Science and Politics of Racial Research.* Urbana: University of Illinois Press.

Tuttle, R. 1988. What's new in African paleoanthropology? *Annual Review of Anthropology* 17:391–426.

Valladas, H., V. Cabrera, M. Hoyos, and F. Bernaldo de Quirós, n.d. Datation radiocarbonique de l'Aurignacien ancien de la grotte de "El Castillo" (Cantabrie, Espagne). *Comptes Rendues de l'Academie des Sciences de Paris,* in press.

Valladas, H., J. L. Joron, G. Valladas, B. Arensburg, O. Bar-Yosef, A. Belfer-Cohen, P. Goldberg, H. Laville, L. Meignen, Y. Rak, E. Tchernov, A. M. Tillier, and B. Vandermeersch. 1987. Thermoluminescence dates for the Neanderthal burial site at Kebara in Israel. *Nature* 303:159–160.

Valladas, H., J. Reyss, J. L. Joron, G. Valladas, O. Bar-Yosef, and B. Vandermeersch. 1988. Thermoluminescence dating of the Mousterian "Proto-Cro-Magnon" remains from Israel and the origin of modern man. *Nature* 313:614–616.

Vallois, H., and G. Billy. 1965. Nouvelles recherches sur les hommes fossiles de l'abri de Cro-Magnon. *L'Anthropologie* 69:47–74, 249–272.

Vallois, H., and H. Movius. 1952. Catalogue des hommes fossiles. *Comptes rendus de la XIXéme Session du Congrès Géologique International,* pp. 64–376. Paris.

Van Valen, L. M. 1988. Species, sets, and the derivative nature of philosophy. *Biology and Philosophy* 3:49–66.

Vandermeersch, B. 1981a. Les premiers *Homo sapiens* du Proche Orient. In *Les Processus de l'Hominisation,* D. Ferembach, ed. Pp. 97–100. Paris: Colloques Internationaux du CNRS No. 599.

———. 1981b. *Les Hommes Fossiles de Qafzeh (Israël).* Paris: CNRS.

———. 1988. Réflections d'un anthropologue à propos de la transition Moustérien-Paléolithique supérieur. *Mémoires de la Musée de Préhistoire d'Ile-de-France* 3:25–27.

———. 1989. The evolution of modern humans: recent evidence from Southwest Asia. In *The Human Revolution,* P. Mellars and C. Stringer, eds. Pp. 155–163. Edinburgh: Edinburgh University Press.

———. 1992. The Near Eastern hominids and the origins of modern humans in Eurasia. In *The Evolution and Dispersal of Modern Humans in Asia,* Takeru Akazawa, Ken-ichi Aoki, and Tasuku Kimura, eds. Pp. 29–38. Tokyo: Hokusen-sha.

———. 1993. Le Proche Orient et l'Europe: Continuité ou discontinuité? In *El Origen del Hombre Moderno en el Suroeste de Europa,* V. Cabrera, ed. Pp. 361–372. Madrid: UNED.

Vandermeersch, B., and Mª. D. Garralda. 1994. El origen del hombre moderno en Europa. In *Biología de Poblaciones Humanas: Problemas Metodológicos e Interpretación Ecológica,* C. Bernis, C. Varea, F. Robles, A. González, eds. Pp. 35–34. Madrid: Universidad Autónoma de Madrid.

Vandermeersch, B., and F. Lévêque. 1989. Découverte de restes humains dans un horizon castelperronien à St. Césaire (Charente-Maritime). *Comptes Rendus de l'Academie des Sciences de Paris* 291:187–189.

Vaquero, M. 1992. Abric Romaní: procesos de canvi techològical voltant del 40,000 BP. Continuitat o ruptura. *Estrat* 5:6–156.

Vega del Sella, Conde de la 1921. *El Paleolítico de Cueva Morín (Santander) y notas para la climatología cuaternaria.* Comisión de Investigaciones Paleontológicas y Prehistóricas, Mem. 29. Madrid.

Vega Toscano, L. 1990. La fin du Paléolithique moyen au sud de l'Espagne. In *Paléolithique Moyen Récent et Paléolithique Supérieur Ancien en Europe,* C. Farizy, ed. Pp. 169–176. Nemours: Mémoires du Musée de Préhistoire de l'Ile de France 3.

Vega Toscano, L., M. Hoyos, A. Ruíz-Bustos, and H. Laville. 1988. La séquence de la grotte de la Carihuela (Píñar, Granade): chronostratigraphie et paléoécologie du Pleistocène supérieur au sud de la Péninsule Ibérique. In *l'Homme de Néanderthal,* Vol. 2: l'Environnement, M. Otte, ed. Pp. 169–180. Liège: ERAUL No. 29.

Verneau, R. 1892. Nouvelle découverte de squelettes préhistoriques aux Baoussé-Roussé près de Menton. *L'Anthropologie* 3:123–141.

————. 1899. Les nouvelles trouvailles de M. Abbo dans la Barma Grande près de Menton. *L'Anthropologie* 10:16–31.

Veth, P. M. 1993. *Islands in the Interior.* Ann Arbor: International Monographs in Prehistory No. 8.

Vigilant, L., M. Stoneking, H. Harpending, K. Hawkes, and A. C. Wilson. 1991. African populations and the evolution of human mitochondrial DNA. *Science* 253:1503–1507.

Villaverde, V., and M. Fumanal. 1990. Relations entre le Paléolithique moyen et le Paléolithique supérieur dans le versant méditerranéen espagnol. In *Paléolithique Moyen Récent et Paléolithique Supérieur Ancien en Europe,* C. Farizy, ed. Pp. 177–183. Nemours: Mémoires du Musée de Préhistoire de l'Ile de France 3.

Vitagliano, S., and M. Piperno. 1990–1991. Lithic industry of Level 27β of the Fossellone Cave (S. Felice Circeo, Latina). *Quaternaria Nova* 1:289–304.

von Koenigswald, G. H. R. 1973. *Australopithecus, Meganthropus* and *Ramapithecus. Journal of Human Evolution* 2:487–491.

Voorrips, A. 1982. Mambrino's helmet: A framework for structuring archaeological data. In *Essays in Archaeological Typology,* R. Whallon and J. Brown, eds. Pp. 93–126. Evanston: Center for American Archeology.

Wade, M. J., and C. J. Goodnight. 1991. Wright's shifting balance theory: an experimental study. *Science* 253:1015–1018.

Wake, D. B. 1991. Homoplasy: The result of natural selection, or evidence of design limitations? *American Naturalist* 138:543–567.

Washburn, S. L. 1951. The new physical anthropology. *Transactions of the New York Academy of Science* 13:298–304.

————. 1960. Tools and human evolution. *Scientific American* 222:9–21.

Washburn, S. L., and V. Avis. 1958. Evolution of human behavior. In *Behavior and Evolution,* A. Roe and G. G. Simpson, eds. Pp. 421–436. New Haven: Yale University Press.

Washburn, S. L., and R. Moore. 1974. *Ape into Man: A Study of Human Evolution.* Boston: Little, Brown.

Watanabe, H. 1985. The chopper-chopping tool complex of Eastern Asia: An ethnoarchaeological-ecological reexamination. *Journal of Anthropological Archaeology* 4:1–18.

Watson, P., S. LeBlanc, and C. Redman. 1984. *Archaeological Explanation: The Scientific Method.* New York: Columbia University Press.

Wei, Q. 1981. On the features and the development of Palaeolithic culture in China, in *Cultura y Medio Ambiente del Hombre Fósil en Asia,* A. K. Ghosh, ed. Pp. 27–34. México, D.F.: UISPP.

————. 1985. Paleoliths from the lower Pleistocene of the Nihewan beds in the Donggutuo site. *Acta Anthropologica Sinica* 4:289–300.

Weidenreich, F. 1939a. On the earliest representatives of modern mankind recovered on the soil of East Asia. *Peking Natural History Bulletin* 13:161–174.

————. 1939b. Six lectures on *Sinanthropus pekinensis* and related problems. *Bulletin of the Geological Society of China* 19:1–110.

————. 1943a. The skull of *Sinanthropus pekinensis:* A comparative study of a primitive hominid skull. *Palaeontologica Sinica* 10:1–485.

————. 1943b. The "Neanderthal man" and the ancestors of *Homo sapiens. American Anthropologist* 45:39–48.

————. 1945. Giant early man from Java and South China. *Anthropological Papers of the American Museum of Natural History* 40:1–134.

————. 1947. The trend of human evolution. *Evolution* 1:221–236.

White, L. A. 1959. *The Evolution of Culture.* New York: McGraw-Hill.

White, R. 1982. Rethinking the Middle-Upper Paleolithic transition. *Current Anthropology* 23:169–192.

———. 1989a. Visual thinking in the Ice Age. *Scientific American* 261:92–99.

———. 1989b. Production complexity and standardization in Early Aurignacian bead and pendant manufacture: Evolutionary implications. In *The Human Revolution,* P. Mellars and C. Stringer, eds. Pp. 366–390. Edinburgh: Edinburgh University Press.

———. 1990. Comment on "Symbolism and modern human origins" by J. M. Lindly and G. Clark. *Current Anthropology* 31:250–251.

———. 1993a. The dawn of adornment. *Natural History* 5:60–67.

———. 1993b. Technological and social dimensions of "Aurignacian-age" body ornaments across Europe. In *Before Lascaux,* H. Knecht, A. Pike-Tay, and R. White, eds. Pp. 277–300. Boca Raton: CRC.

Wiley, E. O. 1981. *Phylogenetics: The Theory and Practice of Phylogenetic Systematics.* New York: Wiley-Interscience.

Willermet, C. M. 1993. The debate over modern human origins: A scientific tug-of-war. M.A. thesis, Arizona State University, Tempe.

———. 1994. Craniometric data and phylogenetic issues in hominid evolution. *American Journal of Physical Anthropology* (supplement) 18:207.

Willermet, C. M., and G. A. Clark. 1995. Paradigm crisis in modern human origins research. *Journal of Human Evolution* 29:487–490.

Wills, C. 1993. *The Runaway Brain: The Evolution of Human Uniqueness.* New York: Basic Books.

Wilson, A. C., and R. L. Cann. 1992. The recent African genesis of humans. *Scientific American* 266:66–73.

Wilson, E. O. 1975. *Sociobiology: The New Synthesis.* Cambridge, Massachusetts: Harvard University Press.

———. 1978. *On Human Nature.* Cambridge: Cambridge University Press.

———. 1994. *Naturalist.* Washington: Island Books/Shearwater Books.

Wilson, E. O., and F. Peter, eds. 1988. *Biodiversity.* Washington: National Academy Press.

Witkowski, S. R., and C. H. Brown. 1978. Mesoamerican: A proposed language phylum. *American Anthropologist* 80:942–944.

———. 1981. Mesoamerican historical linguistics and distant genetic relationship. *American Anthropologist* 83:905–911.

Wobst, H. M. 1976. Locational relationships in Palaeolithic society. *Journal of Human Evolution* 5:49–58.

———. 1990. Minitime and megaspace in the Paleolithic at 18K and otherwise. In *The World at 18,000 BP,* O. Soffer and C. Gamble, eds. Pp. 331–343. London: Unwin Hyman.

Woillard, G. 1978. Grande Pile peat bog: Continuous pollen record for the last 140,000 years. *Quaternary Research* 9:1–21.

Woillard, G., and W. Mook. 1982. Carbon-14 dates at Grande Pile: correlation of land and sea chronologies. *Science* 215:159–161.

Wolf, E. R. 1994. Perilous ideas: Race, culture, people. *Current Anthropology* 35:1–12.

Wolpoff, M. H. 1975. Discussion. In *Paleoanthropology, Morphology and Paleoecology,* R. Tuttle, ed. P. 15. The Hague: Mouton.

———. 1978. Analogies and interpretation in palaeoanthropology. In *Early Hominids of Africa,* C. Jolly, ed. Pp. 461–503. London: Duckworth.

———. 1979. The Krapina dental remains. *American Journal of Physical Anthropology* 50:67–113.

———. 1980. *Paleoanthropology.* New York: Alfred A. Knopf.

————. 1981. Allez Neanderthal. *Nature* 289:823.

————. 1982. *Ramapithecus* and hominid origins. *Current Anthropology* 23:501–522.

————. 1985. Human evolution at the peripheries: The pattern at the eastern edge. In *Hominid Evolution: Past, Present and Future*, P. V. Tobias, ed. Pp. 355–365. New York: Alan R. Liss.

————. 1986. Describing anatomically modern *Homo sapiens:* A distinction without a definable difference. In *Fossil Man. New Facts, New Ideas. Papers in Honor of Jan Jelínek's Life Anniversary*, V. V. Novotný and A. Mizerová. *Anthropos* 23:41–53.

————. 1989a. The place of Neandertals in human evolution. In *The Emergence of Modern Humans: Biocultural Adaptations in the Later Pleistocene*, E. Trinkaus, ed. Pp. 97–141. Cambridge: Cambridge University Press.

————. 1989b. Multiregional evolution: The fossil alternative to Eden. In *The Human Revolution*, P. Mellars and C. B. Stringer, eds. Pp. 62–108. Edinburgh: Edinburgh University Press.

————. 1989c. Early "modern" humans from the Levant? The problem of adaptation. *American Journal of Physical Anthropology* 78:326 (abstract).

————. 1992. Theories of modern human origins. In *Continuity or Replacement: Controversies in Homo sapiens Evolution*, G. Bräuer and F. H. Smith, eds. Pp. 25–63. Rotterdam: A. A. Balkema.

————. 1994a. *Paleoanthropology, Preliminary Publication*, 2nd ed. New York: McGraw-Hill.

————. 1994b. The calm before the storm. *Cambridge Archaeological Journal* 4:97–103.

————. 1994c. What does it mean to be human—and why does it matter? *Evolutionary Anthropology* 3:116–117.

Wolpoff, M. H., and R. Caspari. 1990a. On Middle Paleolithic/Middle Stone Age hominid taxonomy. *Current Anthropology* 31:394–395.

————. 1990b. Metric analysis of the skeletal material from Klasies River Mouth, Republic of South Africa. *American Journal of Physical Anthropology* 81:319.

Wolpoff, M. H., and A. G. Thorne. 1991. The case against Eve. *New Scientist* 130:37–41.

————. 1993. The end of Eve? Fossil evidence from Africa. *American Association for the Advancement of Science Abstracts*, p. 173.

Wolpoff, M. H., D. W. Frayer, and R. Caspari. 1991. Form and function: Fact or fiction? *American Journal of Physical Anthropology* (supplement) 12:186.

Wolpoff, M., F. Smith, M. Malez, J. Radovčić, and D. Rukavina. 1981. Upper Pleistocene human remains from Vindija Cave, Croatia, Yugoslavia. *American Journal of Physical Anthropology* 54:499–545.

Wolpoff, M. H., J. N. Spuhler, F. H. Smith, J. Radovčić, G. Pope, D. W. Frayer, R. Eckhardt, and G. Clark. 1988. Modern human origins. *Science* 241:772–773.

Wolpoff, M. H., A. G. Thorne, J. Jelínek, and Z. Yinyun. 1994. The case for sinking *Homo erectus:* 100 years of *Pithecanthropus* is enough! In *100 years of Pithecanthropus: The Homo erectus problem*, J. L. Franzen, ed. *Courier Forschungsinstitut Senckenberg* 171:341–361.

Wolpoff, M. H., X. Wu, and A. G. Thorne. 1984. Modern *Homo sapiens* origins: A general theory of hominid evolution involving the fossil evidence from East Asia. In *The Origins of Modern Humans: A World Survey of the Fossil Evidence*, F. H. Smith and F. Spencer, eds. Pp. 411–484. New York: Alan R. Liss.

Wolpoff, M. H., A. G. Thorne, F. H. Smith, D. W. Frayer, and G. G. Pope. 1994. Multiregional evolution: A world-wide source for modern human populations. In *Origins of Anatomically Modern Humans*, M. Nitecki and D. Nitecki, eds. Pp. 175–199. New York: Plenum.

Woo, S. L., S. C. Kuei, D. Amiel, M. A. Gómez, W. C. Hayes, F. C. White, and A. K.

Akeson. 1981. The effect of prolonged physical training on the properties of long bones: a study of Wolff's Law. *Journal of Bone and Joint Surgery* 63:781–787.

Wood, B. A. 1992. Origin and evolution of the genus *Homo. Nature* 355:783–790.

———. 1993. Early *Homo:* How many species? In *Species, Species Concepts and Primate Evolution,* W. Kimbel and L. Martin, eds. Pp. 23–38. New York: Plenum.

———. 1994. The problems of our origins. *Journal of Human Evolution* 27:519–529.

Wood Jones, F. 1929. *Man's Place among the Mammals.* London: Edward Arnold.

Woodger, J. H. 1952. *Biology and Language.* Cambridge: Cambridge University Press.

Workman, P. L., B. S. Blumberg, and A. J. Cooper. 1963. Selection, gene migration and polymorphic stability in a U.S. white and negro population. *American Journal of Human Genetics* 15:71–84.

Wright, L. 1994. One drop of blood. *The New Yorker,* July 25, pp. 46–55.

Wright, R. V. S. 1992. Correlation between cranial form and geography in *Homo sapiens:* CRANID—a computer program for forensic and other applications. *Archaeology in Oceania* 27(3):128–134.

Wright, S. 1964. Pleiotropy in the evolution of structural reduction and dominance. *American Naturalist* 98:65–69.

Wu, R., and J. W. Olsen. 1985. *Palaeoanthropology and Palaeolithic Archaeology in the People's Republic of China.* London: Academic Press.

Wu, R. K., X. Wu, and S. S. Zhang. 1989. *Early Humankind in China.* Beijing: Science Press.

Wu, X. 1988. Comparative study of early *Homo sapiens* from China and Europe. *Acta Anthropologica Sinica* 7:287–293.

———. 1989. Early *Homo sapiens* in China. In *Early Humankind in China,* R. Wu, X. Wu, and S. Zhang, eds. Pp. 24–41. Beijing: Science Press.

———. 1990. The evolution of humankind in China. *Acta Anthropologica Sinica* 9:312–321.

———. 1992a. Origin and affinities of the stone age inhabitants of Japan. In *The Japanese as a Member of the Asian and Pacific Populations,* K. Hanihara, ed. Pp. 1–8. Kyoto: International Research Centre for Japanese Studies.

———. 1992b. The origin and dispersal of anatomically modern humans in east and southeast Asia. In *The Evolution and Dispersal of Modern Humans in Asia,* Takeru Akazawa, Kenichi Aoki, and Tasuku Kimura, eds. Pp. 373–378. Tokyo: Hokusensha.

Wu, X., and G. Bräuer. 1993. Morphological comparison of archaic *Homo sapiens* crania from China and Africa. *Zeitschrift für Morphologie und Anthropologie* 79:241–259.

Wu, X., and L. H. Wang. 1985. Chronology in Chinese palaeoanthropology. In *Palaeoanthropology and Palaeolithic Archaeology in the People's Republic of China,* R. Wu and J. Olsen, eds. Pp. 29–51. London: Academic Press.

Wu, X., and M. Wu. 1985. Early *Homo sapiens* in China. In *Palaeoanthropology and Paleolithic Archaeology in the People's Republic of China,* W. Rukang and J. W. Olsen, eds. Pp. 91–106. New York: Academic Press.

Wu, X. Z., and Y. Z. You. 1980. Dali man and its culture. *Kaogu Yu Wenwu* 1:2–6.

Xie, F., and S. Q. Cheng. 1989. Report on the excavation of microliths site at Youfang, Yangyuan County, Hebei Province. *Acta Anthropologica Sinica* 8:59–68.

Xiong, W., W. Li, I. Posner, T. Yamamura, A. Yamamoto, A. M. Gotto Jr., and L. Chan. 1991. No severe bottleneck during human evolution: Evidence from two apolipoprotein C-II deficiency alleles. *American Journal of Human Genetics* 48:383–389.

Yang, X. N. 1988. *Sculpture of Prehistoric China.* Hong Kong: Tai Dao Publishers.

Yellen, J. 1977a. *Archaeological Approaches to the Present: Models for Reconstructing the Past.* New York: Academic Press.

————. 1977b. Cultural patterning in faunal remains: Evidence from the !Kung Bushmen. In *Experimental Archaeology,* D. Ingersoll, J. E. Yellen, and W. Macdonald, eds. Pp. 271–331. New York: Columbia University Press.

Yellen, J., and H. Harpending. 1972. Hunter-gatherer populations and archaeological inference. *World Archaeology* 4:244–253.

Yellen, J., A. Brooks, E. Cornelissen, M. Mehlman, and K. Stewart. 1995. A Middle Stone Age worked bone industry from Katanda, Upper Semliki Valley, Zaire. *Science* 268:553–556.

Yi, S., and G. A. Clark. 1983. Observations on the Lower Paleolithic of Northeast Asia. *Current Anthropology* 24:181–202.

————. 1985. The "Dyuktai Culture" and New World origins. *Current Anthropology* 26:1–13, 19–20.

Yokoyama, Y., G. Shen, H. V. Nguyen, and C. Falgueres. 1987. Datation du travertin de Banyoles à Girone, Espagne. In *Quadre Cronologic del Pleistocé Superior a Catalunya: Paleoambients i Cultures Prèhistoriques,* N. Soler and J. Maroto, eds. Pp. 155–159. Barcelona: Cypsela No. 6.

You, Y. Z. 1984. Preliminary study of a Palaeolithic bone engraving. *Kexue Tongbao* 29:80–82.

Yuan, X. F., S. B. Xu, and R. J. Wu. 1989. The discovery of stone artifacts from Junan, Shandong Province. *Acta Anthropologica Sinica* 8:32–38.

Zadeh, L. 1965. Fuzzy sets. *Information and Control* 8:338–353.

Zehna, P. W., and R. L. Johnson. 1962. *Elements of Set Theory.* Boston: Allyn and Bacon.

Zeuner, F. 1958. The replacement of neandertal man by *Homo sapiens.* In *Hundert Jahre Neanderthal,* G. H. R. von Koenigswald, ed. Pp. 313–315. Utrecht: Kemink en Zoon.

————. 1961. The shoreline chronology of the paleolithic of Abri Zumoffen. *Bulletin de la Musée de Beyrouth* 16:49–60.

Zhang, S. S. 1985. The Early Palaeolithic of China. In *Palaeoanthropology and Palaeolithic Archaeology in the People's Republic of China,* R. K. Wu and J. Olsen, eds. Pp. 147–186. London: Academic Press.

————. 1990. Regional industrial gradual advance and cultural exchange of Paleolithic in North China. *Acta Anthropologica Sinica* 9:322–333.

Zhang, Y. 1991. An examination of the temporal variation in the hominid dental sample from Locality One, Zhoukoudian. *Acta Anthropologica Sinica* 10:88–95.

Zilhão, J. 1993. Le passage du Paléolithique moyen au Paléolithique supérieur dans le Portugal. In *El Origen del Hombre Moderno en el Suroeste de Europa,* V. Cabrera, ed. Pp. 127–146. Madrid: UNED.

Zimmerman, H.-J. 1990. *Fuzzy Set Theory and Its Applications,* 2nd ed. Boston: Kluwer Academic.

Zollikofer, C. P. E., M. S. Ponce de León, R. D. Martin, and P. Stuckl. 1995. Neanderthal computer skulls. *Nature* 375:283–285.

Zubrow, E. 1989. The demographic modelling of Neanderthal extinction. In *The Human Revolution,* P. Mellars and C. B. Stringer, eds. Pp. 212–231. Edinburgh: Edinburgh University Press.

Zuckerkandl, E. 1963 Perspectives in molecular anthropology. In *Classification and Human Evolution,* S. L. Washburn, ed. Pp. 243–272. Chicago: Aldine.

Zuckerman, S. 1932. *Social Life of Monkeys and Apes.* London: Kegan Paul.

————. 1933. *Functional Affinities of Man, Monkeys and Apes.* London: Kegan Paul.

Index

A

Acheulean industry, 162, 177, 272, 277, 281
adaptationist programme, 212
adaptive radiation, 65, 66, 69, 70
admixture, 5, 34, 56, 65, 66, 69, 73, 115, 129, 130, 191, 192, 194, 197, 277, 312, 313, 356, 373, 382–386
Afro-European *sapiens* model, *see also* replacement, 219
age at death, determining, 122
Age of Enlightenment, 63
Ahmarian industry, 161, 166, 240
altruism, 434
Amerind people, 13, 15–17
anagenesis, 4, 5, 65, 66, 79, 81, 87, 256, 281, 306–310, 315–317, 327
Andrews, Roy Chapman, 274
apes, 48–51, 200, 320, 354, 369–372, 375, 380, 381
  human rights for, 48, 49, 51
apomorphy, *see also* derived character, 151, 315, 322–325
Aquitaine, 166, 186, 187
archaeology, 63, 64, 67–69, 89–104, 132–147, 161–188, 235–266, 294–303
Ardipithecus ramidus, 381
Aristotelian logic, 10, 72, 81, 83, 87, 88, 206
Aristotle, 206, 212, 254, 255
art, 163, 236, 238, 239, 244, 251, 267, 302, 303, 415
Aterian industry, 170, 252
attribute, archaeological definition of, 259

Aurignacian
  Cantabrian Typical, 181
  hominids, 115, 119–130, 136, 149, 155–160, 174, 235–252
  industry, 26, 68, 89, 98, 105–125, 129, 133–137, 140–147, 152, 155–188, 221, 235–252, 259, 261, 263
Aurignaco-Mousterian industry, 173
Australian aborigines, 14–17, 39, 41–43, 195, 313, 319, 323, 324
australopithecines, 277–282, 379–381, 429
Australopithecus afarensis, 381, 428, 432
Australopithecus anamensis, 381
autapomorphy, 110, 223–225, 233, 256
autonomists, in biology, 203, 215–218
Azilian industry, 154, 177, 238
Azinipodian industry, 163

B

Bachokirian industry, 165, 241
Badegoulian industry, 161
Baradostian industry, 174
Bauplan, 207
Bayes' Theorem, 337
Bell Curve, The, 55, 56
biocultural approach, 253–266
biological determinism, 47, 48, 55–57, 209, 210
biological structuralism, 212
biospecies, 78, 193
bivalent, 82, 83, 88
Black, Davidson, 274
boats, 42

Site/Specimen Index

A

Abri Pataud, 174, 264, 397
Abric Romaní (Abric Agut), 148–150,
 156, 159, 235, 242–248
Afalou, 196, 396, 397
 Afalou 9, 397
 Afalou 10, 396, 197
 Afalou 29, 397
 Afalou 32, 397
Agut, 247
Altuna, 183
Amud, 171, 172, 237, 397
Andalucia, 159, 160
Anyang, 277
Arcy-sur-Cure, 113, 114, 118, 163, 164,
 172–174, 227
Arene Candide, 264
Aurignac, 119
Axlor, 148, 149, 247

B

Bacho Kiro, 114, 129, 157, 175, 240,
 241, 252
 Bacho Kiro 559, 157
 Bacho Kiro 1124, 157
Balzi Rossi caves, *see also* Barma Gran-
 de, 134, 136, 144
 Barma Grande, 136, 264
 Basanda Turc, 136
 Caviglione, 136
Bañolas (Banyoles, Bañoles), 148–150,
 159, 235, 247

Barma Grande, *see also* Balzi Rossi
 caves, 136, 264
Bayol, 119
Bellevaud, 119
Bize, 119
Blanchard, 119
Bodo d'Ar, 287, 288
Bohunice, 241
Boker Tachtit, 112, 240
Border Cave, 15, 38, 110, 224, 237
Bouil Bleu, 119
Brassempouy, 119, 122, 123
Broken Hill, *see also* Kabwe, 287, 288
 Broken Hill 1, 287, 288
 Broken Hill 2, 287, 288
Brno, 264
Bruniquel, 264
Buca della Iena, 134, 137

C

Camargo, 149, 158, 245, 247
Candide, 396, 397, 398
 Candide 1, 396, 397
 Candide 5, 396, 397
Cap Blanc, 264
Carigüela (Carihuela), *see also* Píñar,
 148, 150–154, 159, 188, 235, 246,
 247, 249
Castanet, 119
Certova Pec, 241
Chancelade, 264, 397
Chez Leix, 119, 123
Cioclovina, 264
Cohuna, 397